METHODS IN MOLECULAR BIOLOGY™

Series Editor
John M. Walker
School of Life Sciences
University of Hertfordshire
Hatfield, Hertfordshire, AL10 9AB, UK

For other titles published in this series, go to
www.springer.com/series/7651

Flow Cytometry Protocols

3rd Edition

Edited by

Teresa S. Hawley

Flow Cytometry Core Facility, George Washington University Medical Center, Washington, DC, USA

Robert G. Hawley

Department of Anatomy and Regenerative Biology, George Washington University Medical Center, Washington, DC, USA

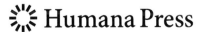 Humana Press

Editors
Teresa S. Hawley
Flow Cytometry Core Facility
George Washington University
Medical Center
Washington, DC
USA
anatsh@gwumc.edu

Robert G. Hawley
Department of Anatomy
and Regenerative Biology
George Washington University
Medical Center
Washington, DC
USA
rghawley@gwu.edu

ISSN 1064-3745   e-ISSN 1940-6029
ISBN 978-1-61737-949-9   e-ISBN 978-1-61737-950-5
DOI 10.1007/978-1-61737-950-5
Springer New York Dordrecht Heidelberg London

Humana Press is part of Springer Science+Business Media (www.springer.com)

Preface

Flow cytometry is riding the crest of unprecedented advances in innovative technologies. Improvement in instrumentation, lasers, fluorophores, and data analysis software have facilitated the development of new applications as well as the optimization of existing applications. This thoroughly revised up-to-date edition highlights the expanding contribution of flow cytometry to basic biological research and diagnostic medicine.

The introductory chapter presents a historical perspective documenting valuable contributions of pioneers in the field. An eloquent synopsis of the principles of flow cytometry provides a solid foundation for the understanding of basic applications. Modern flow cytometry has been evolving toward high-dimensional complexity. A novel concept, which underlies an accurate and efficient strategy for analyzing complex multiparametric data, is introduced. Great strides have been made toward quantitative fluorescence measurement, bead-based multiplexed analysis, semiautomated high-throughput flow cytometry, and fluorescence resonance energy transfer the analysis of protein interactions. Other applications range from polychromatic phenotypic characterizations to genomic and proteomic analyses. Technologies utilized encompass conventional flow cytometry and imaging cytometry. The prevalence of aerosol-generating cell/particle sorting warrants a detailed description of standard safety measures. The last chapter poignantly asserts that while flow and imaging cytometry evolve on a platform of costly sophisticated technologies, minimalist imaging cytometry holds promise for field-research applications in resource-challenged environments.

The utility of multiparametric flow cytometry is best demonstrated in function-based studies. Assessment of cytotoxic effector activity and regulatory T cell functions can be determined using cell tracking dyes, phenotypic markers, and viability probes. Intracellular cytokine staining is routinely used to visualize antigen-specific T cells. Combined with functional and phenotypic markers, specific cellular subsets can be further dissected. Phospho flow deciphers intracellular kinase signaling cascades by using highly specific antibodies which differentiate between the phosphorylated and nonphosphorylated states of proteins. Combined with immunophenotyping, discrete biochemical signaling events in individual cells within heterogeneous populations can be carefully examined. The complex progression of apoptotic death can be evaluated by monitoring multiple apoptotic characteristics simultaneously. Employing multiple antibodies to detect epitopes on cell cycle-regulated proteins provides more information than the measurement of DNA content alone.

Single-cell resolution and high-throughput capability make flow cytometry amenable to the identification of rare cells within a population. Technical aspects of rare event detection are discussed in the context of practical examples. Direct investigation of distinct cellular subsets in normal hematopoietic development versus hematologic diseases is critical to the understanding of disease initiation and progression. High-resolution polychromatic fractionation of hematopoietic precursors dissects developmental stages and identifies cellular subsets with defined lineage potentials. Carefully devised protocols in the study of human hematologic disorders enable the diagnosis and monitoring of patients with leukemia and lymphoma, or primary immunodeficiency diseases. Cell-derived microparticles, which are implicated in pathogenesis, can be analyzed by conventional as well as imaging flow cytometry.

The impact of fluorescent proteins (FPs) on bioscience is underscored by the award of the 2008 Nobel Prize in Chemistry to three scientists involved in the discovery and subsequent optimization of the green fluorescent protein (GFP). Unlike other bioluminescent reporters, fluorescence of GFP and GFP-related proteins does not require exogenous substrates or cofactors. GFP from jellyfish and GFP-like proteins from other marine organisms, as well as recently engineered nontoxic red-shifted variants, have played vital roles in the noninvasive detection of genes transferred into living cells. Several schemes for the simultaneous detection of multiple FPs in living cells were presented in the second edition. The applications described in this edition further illustrate the broad utility of FPs in flow cytometry. The study of plant biology has been accelerated by the use of FP expression constructs. Other downstream applications involve genome-wide expression profiling and mass spectrometry, two analysis platforms that readily interface with flow cytometry. Such integration bodes well for the study of systems biology.

We would like to thank John Walker for inviting us to participate in this exciting endeavor again and for his expert editorial guidance. We are especially grateful to all of the contributors for their enthusiasm and generosity. Their willingness to impart their knowledge exemplifies the spirit of cooperation that is pervasive in the cytometry community. As the "Notes" section is a hallmark of this series, we are proud to present chapters that contain up to 59 notes!

Washington, DC   *Teresa S. Hawley*
Washington, DC   *Robert G. Hawley*

Contents

Preface . . . . *v*

Contributors . . . . *ix*

1 Flow Cytometry: An Introduction 1
   Alice L. Givan

2 Breaking the Dimensionality Barrier 31
   C. Bruce Bagwell

3 Quantitative Fluorescence Measurements
   with Multicolor Flow Cytometry 53
   Lili Wang, Adolfas K. Gaigalas, and Ming Yan

4 Quantum Dots for Quantitative Flow Cytometry 67
   Tione Buranda, Yang Wu, and Larry A. Sklar

5 Bead-Based Multiplexed Analysis of Analytes by Flow Cytometry 85
   Henri C. van der Heyde and Irene Gramaglia

6 Flow-Based Combinatorial Antibody Profiling:
   An Integrated Approach to Cell Characterization 97
   *Shane Bruckner, Ling Wang, Ruiling Yuan, Perry Haaland,
   and Amitabh Gaur*

7 Tracking Immune Cell Proliferation and Cytotoxic
   Potential Using Flow Cytometry 119
   *Joseph D. Tario Jr., Katharine A. Muirhead, Dalin Pan,
   Mark E. Munson, and Paul K. Wallace*

8 Multiparameter Intracellular Cytokine Staining 165
   Patricia Lovelace and Holden T. Maecker

9 Phospho Flow Cytometry Methods for the Analysis of Kinase
   Signaling in Cell Lines and Primary Human Blood Samples 179
   *Peter O. Krutzik, Angelica Trejo, Kenneth R. Schulz,
   and Garry P. Nolan*

10 Multiparametric Analysis of Apoptosis by Flow Cytometry 203
   *William G. Telford, Akira Komoriya,
   Beverly Z. Packard, and C. Bruce Bagwell*

11 Multiparameter Cell Cycle Analysis 229
   James W. Jacobberger, R. Michael Sramkoski, and Tammy Stefan

12 Rare Event Detection and Analysis in Flow Cytometry:
   Bone Marrow Mesenchymal Stem Cells, Breast Cancer
   Stem/Progenitor Cells in Malignant Effusions,
   and Pericytes in Disaggregated Adipose Tissue 251
   Ludovic Zimmerlin, Vera S. Donnenberg, and Albert D. Donnenberg

13 Flow Cytometry-Based Identification of Immature
    Myeloerythroid Development.. 275
    Cornelis J.H. Pronk and David Bryder

14 Flow Cytometry Immunophenotyping
    of Hematolymphoid Neoplasia 295
    *Katherine R. Calvo, Catharine S. McCoy,
    and Maryalice Stetler-Stevenson*

15 Flow Cytometry Assays in Primary Immunodeficiency Diseases 317
    Maurice R.G. O'Gorman, Joshua Zollett, and Nicolas Bensen

16 Flow Cytometric Analysis of Microparticles 337
    *Henri C. van der Heyde, Irene Gramaglia, Valéry Combes,
    Thaddeus C. George, and Georges E. Grau*

17 Noncytotoxic DsRed Derivatives for Whole-Cell Labeling 355
    Rita L. Strack, Robert J. Keenan, and Benjamin S. Glick

18 Flow Cytometric FRET Analysis of Protein Interaction. 371
    György Vereb, Péter Nagy, and János Szöllősi

19 Fluorescent Protein-Assisted Purification for Gene Expression Profiling 393
    M. Raza Zaidi, Chi-Ping Day, and Glenn Merlino

20 Multiparametric Analysis, Sorting, and Transcriptional Profiling
    of Plant Protoplasts and Nuclei According to Cell Type 407
    David W. Galbraith, Jaroslav Janda, and Georgina M. Lambert

21 Lentiviral Fluorescent Protein Expression Vectors
    for Biotinylation Proteomics.. 431
    Irene Riz, Teresa S. Hawley, and Robert G. Hawley

22 Standard Practice for Cell Sorting in a BSL-3 Facility 449
    *Stephen P. Perfetto, David R. Ambrozak, Richard Nguyen,
    Mario Roederer, Richard A. Koup, and Kevin L. Holmes*

23 The Cytometric Future: It Ain't Necessarily Flow! 471
    Howard M. Shapiro

Index *483*

Contributors

DAVID R. AMBROZAK • *Vaccine Research Center, National Institute of Allergy and Infectious Diseases, National Institutes of Health, Bethesda, MD, USA*

C. BRUCE BAGWELL • *Verity Software House, Topsham, ME, USA*

NICOLAS BENSEN • *Department of Pathology and Laboratory Medicine, The Children's Memorial Hospital, Chicago IL, USA*

SHANE BRUCKNER • *BD Biosciences, San Diego, CA, USA*

DAVID BRYDER • *Department of Immunology, Institution for Experimental Medical Science, Lund University, Lund, Sweden*

TIONE BURANDA • *Department of Pathology and Cancer Center, University of New Mexico School of Medicine, Albuquerque, NM, USA*

KATHERINE R. CALVO • *Department of Laboratory Medicine, Hematology Section, National Institutes of Health Clinical Center, Bethesda, MD, USA*

VALÉRY COMBES • *Department of Pathology, University of Sydney, Camperdown, NSW, Australia*

CHI-PING DAY • *Laboratory of Cancer Biology and Genetics, Center for Cancer Research, National Cancer Institute, National Institutes of Health, Bethesda, MD, USA*

ALBERT D. DONNENBERG • *Division of Hematology/Oncology, Department of Medicine, University of Pittsburgh School of Medicine and University of Pittsburgh Cancer Institute, Pittsburgh, PA, USA*

VERA S. DONNENBERG • *Department of Surgery, University of Pittsburgh School of Medicine and University of Pittsburgh Cancer Institute, Pittsburgh, PA, USA*

ADOLFAS K. GAIGALAS • *National Institute of Standards and Technology, Gaithersburg, MD, USA*

DAVID W. GALBRAITH • *Department of Plant Sciences, University of Arizona, Tucson, AZ, USA*

AMITABH GAUR • *BD Biosciences, San Diego, CA, USA*

THADDEUS C. GEORGE • *Amnis Corporation, Seattle, WA, USA*

ALICE L. GIVAN • *Englert Cell Analysis Laboratory of the Norris Cotton Cancer Center and Department of Physiology, Dartmouth Medical School, Lebanon, NH, USA*

BENJAMIN S. GLICK • *Department of Molecular Genetics and Cell Biology, Cummings Life Science Center, The University of Chicago, Chicago, IL, USA*

IRENE GRAMAGLIA • *La Jolla Infectious Disease Institute, San Diego, CA, USA*

GEORGES E. GRAU • *Department of Pathology, University of Sydney, Camperdown, NSW, Australia*

PERRY HAALAND • *BD Technologies, Research Triangle Park, NC, USA*

ROBERT G. HAWLEY • *Department of Anatomy and Regenerative Biology, The George Washington University Medical Center, Washington, DC, USA*

TERESA S. HAWLEY • *Flow Cytometry Core Facility, The George Washington University Medical Center, Washington, DC, USA*

HENRI C. VAN DER HEYDE • *Cell Analysis Core Facility, Flow Cytometry, La Jolla Infectious Disease Institute, San Diego, CA, USA*

KEVIN L. HOLMES • *Flow Cytometry Section, Research Technologies Branch, National Institute of Allergy and Infectious Diseases, National Institutes of Health, Bethesda, MD, USA*

JAMES W. JACOBBERGER • *Cytometry and Imaging Microscopy Core, Case Comprehensive Cancer Center, Case Western Reserve University, Cleveland, OH, USA*

JAROSLAV JANDA • *Department of Plant Sciences, University of Arizona, Tucson, AZ, USA*

ROBERT J. KEENAN • *Department of Biochemistry and Molecular Biology, Gordon Center for Integrated Sciences, The University of Chicago, Chicago, IL, USA*

AKIRA KOMORIYA • *OncoImmunin, Inc., Gaithersburg, MD, USA*

RICHARD A. KOUP • *Vaccine Research Center, National Institute of Allergy and Infectious Diseases, National Institutes of Health, Bethesda, MD, USA*

PETER O. KRUTZIK • *Baxter Laboratory in Stem Cell Biology, Department of Microbiology and Immunology, Stanford University, Stanford, CA, USA*

GEORGINA M. LAMBERT • *Department of Plant Sciences, University of Arizona, Tucson, AZ, USA*

PATRICIA LOVELACE • *Institute for Immunity, Transplantation, and Infection, Stanford University, Stanford, CA, USA*

HOLDEN T. MAECKER • *Human Immune Monitoring Center, Institute for Immunity, Transplantation, and Infection, Stanford University, Stanford, CA, USA*

CATHARINE S. MCCOY • *Flow Cytometry Laboratory, Laboratory of Pathology, National Cancer Institute, National Institutes of Health, Bethesda, MD, USA*

GLENN MERLINO • *Laboratory of Cancer Biology and Genetics, Center for Cancer Research, National Cancer Institute, National Institutes of Health, Bethesda, MD, USA*

KATHARINE A. MUIRHEAD • *SciGro, Inc., Madison, WI, USA*

MARK E. MUNSON • *Verity Software House, Topsham, ME, USA*

PÉTER NAGY • *Department of Biophysics and Cell Biology, Research Center for Molecular Medicine, Medical and Health Science Center, University of Debrecen, Debrecen, Hungary*

RICHARD NGUYEN • *Vaccine Research Center, National Institute of Allergy and Infectious Diseases, National Institutes of Health, Bethesda, MD, USA*

GARRY P. NOLAN • *Baxter Laboratory in Stem Cell Biology, Department of Microbiology and Immunology, Stanford University, Stanford, CA, USA*

MAURICE R.G. O'GORMAN • *Departments of Pathology and Pediatrics, Feinberg School of Medicine, Northwestern University, Chicago, IL, USA; Department of Pathology and Laboratory Medicine, The Children's Memorial Hospital, Chicago, IL, USA*

BEVERLY Z. PACKARD • *OncoImmunin, Inc., Gaithersburg, MD, USA*

DALIN PAN • *Department of Flow and Image Cytometry, Roswell Park Cancer Institute, Buffalo, NY, USA*

STEPHEN P. PERFETTO • *Vaccine Research Center, National Institute of Allergy and Infectious Diseases, National Institutes of Health, Bethesda, MD, USA*

CORNELIS J.H. PRONK • *Department of Paediatrics, Section of Immunology/Stem Cell Center, Lund University Hospital, Lund, Sweden*

IRENE RIZ • *Department of Anatomy and Regenerative Biology, The George Washington University Medical Center, Washington, DC, USA*

MARIO ROEDERER • *Vaccine Research Center, National Institute of Allergy and Infectious Diseases, National Institutes of Health, Bethesda, MD, USA*

KENNETH R. SCHULZ • *Baxter Laboratory in Stem Cell Biology, Department of Microbiology and Immunology, Stanford University, Stanford, CA, USA*

HOWARD M. SHAPIRO • *The Center for Microbial Cytometry, West Newton, MA, USA*

LARRY A. SKLAR • *Department of Pathology and Cancer Center, University of New Mexico School of Medicine, Albuquerque, NM, USA*

R. MICHAEL SRAMKOSKI • *Cytometry and Imaging Microscopy Core, Case Comprehensive Cancer Center, Case Western Reserve University, Cleveland, OH, USA*

TAMMY STEFAN • *Cytometry and Imaging Microscopy Core, Case Comprehensive Cancer Center, Case Western Reserve University, Cleveland, OH, USA*

MARYALICE STETLER-STEVENSON • *Flow Cytometry Unit, Laboratory of Pathology, National Cancer Institute, National Institutes of Health, Bethesda, MD, USA*

RITA L. STRACK • *Department of Biochemistry and Molecular Biology, Gordon Center for Integrated Sciences, The University of Chicago, Chicago, IL, USA*

JÁNOS SZÖLLŐSI • *Department of Biophysics and Cell Biology, Research Center for Molecular Medicine, Medical and Health Science Center, University of Debrecen, Debrecen, Hungary*

JOSEPH D. TARIO • *Department of Flow and Image Cytometry, Roswell Park Cancer Institute, Buffalo, NY, USA*

WILLIAM G. TELFORD • *FACS Core Facilty, Experimental Transplantation and Immunology Branch, National Cancer Institute, National Institutes of Health, Bethesda, MD, USA*

ANGELICA TREJO • *Baxter Laboratory in Stem Cell Biology, Department of Microbiology and Immunology, Stanford University, Stanford, CA, USA*

GYÖRGY VEREB • *Department of Biophysics and Cell Biology, Research Center for Molecular Medicine, Medical and Health Science Center, University of Debrecen, Debrecen, Hungary*

PAUL K. WALLACE • *Department of Flow and Image Cytometry, Roswell Park Cancer Institute, Buffalo, NY, USA*

LILI WANG • *National Institute of Standards and Technology, Gaithersburg, MD, USA*

LING WANG • *BD Technologies, Research Triangle Park, NC, USA*

YANG WU • *Department of Pathology and Cancer Center, University of New Mexico School of Medicine, Albuquerque, NM, USA*

MING YAN • *BD Biosciences, San Jose, CA, USA*

RUILING YUAN • *BD Technologies, Research Triangle Park, NC, USA*

M. RAZA ZAIDI • *Laboratory of Cancer Biology and Genetics, Center for Cancer Research, National Cancer Institute, National Institutes of Health, Bethesda, MD, USA*

LUDOVIC ZIMMERLIN • *Division of Hematology/Oncology, Department of Medicine, University of Pittsburgh School of Medicine and University of Pittsburgh Cancer Institute, Pittsburgh, PA, USA; Université Paris Diderot – Paris 7, École Doctorale Biologie et Biotechnologie, Paris, France*

JOSHUA ZOLLETT • *Department of Pathology and Laboratory Medicine, The Children's Memorial Hospital, Chicago, IL, USA*

Chapter 1

Flow Cytometry: An Introduction

Alice L. Givan

Abstract

A flow cytometer is an instrument that illuminates cells (or other particles) as they flow individually in front of a light source and then detects and correlates the signals from those cells that result from the illumination. In this chapter, each of the aspects of that definition will be described: the characteristics of cells suitable for flow cytometry, methods to illuminate cells, the use of fluidics to guide the cells individually past the illuminating beam, the types of signals emitted by the cells and the detection of those signals, the conversion of light signals to digital data, and the use of computers to correlate and analyze the data after they are stored in a data file. The final section of the chapter will discuss the use of a flow cytometer to sort cells. This chapter can be read as a brief, self-contained survey. It can also be read as a gateway with signposts into the field. Other chapters in this book will provide more details, more references, and even an intriguing view of the future of cytometry.

Key words: Flow cytometry, Laser, Fluidics, Fluorescence

1. Introduction

An introductory chapter on flow cytometry must first confront the difficulty of defining a flow cytometer. The instrument described by Andrew Moldavan in 1934 (1) is generally acknowledged to be an early prototype. Although it may never have been built, in design it looked like a microscope but provided a capillary tube on the stage so that cells could be individually illuminated as they flowed in single file in front of the light emitted through the objective. The signals coming from the cells could then be analyzed by a photodetector attached in the position of the microscope eyepiece. Following work by Coulter and others in the next decades to develop instruments to count particles in suspension (2–5), a design was implemented by Kamentsky and Melamed in 1965 and 1967 (6, 7) to produce a microscope-based

Teresa S. Hawley and Robert G. Hawley (eds.), *Flow Cytometry Protocols*, Methods in Molecular Biology, vol. 699,
DOI 10.1007/978-1-61737-950-5_1, © Springer Science+Business Media, LLC 2011

flow cytometer for detecting light signals distinguishing the abnormal cells in a cervical sample. In the years after publication of the Kamentsky papers, work by Fulwyler, Dittrich and Göhde, Van Dilla, and Herzenberg (8–11) led to significant changes in over-all design, resulting in a cytometer that was largely similar to today's cytometers. Like today's cytometers, a flow cytometer in 1969 did not look at all like a microscope, but was still based on Moldavan's prototype and on the Kamentsky instrument in that it illuminated cells as they progressed in single file in front of a beam of light and it used photodetectors to detect the signals that came from the cells [see Shapiro (12) and Melamed (13, 14) for more complete discussions of this historical development]. Our definition of a flow cytometer even today involves an instrument that illuminates cells as they flow individually in front of a light source and then detects and correlates the signals from those cells that result from the illumination.

In this chapter, each of the aspects in that definition will be described: the cells; methods to illuminate the cells; the use of fluidics to make sure that the cells flow individually past the illuminating beam; the use of detectors to measure the signals coming from the cells; and the use of computers to correlate the signals after they are stored in data files. As an introduction, this chapter can be read as a brief survey; it can also be read as a gateway with signposts into the field. Other chapters in this book (and in other books (12, 15–24)) will provide more details, more references, and even an intriguing view of the future of cytometry.

2. Cells (or Particles or Events)

Before discussing "cells," we need to qualify even that basic word. "Cytometer" is derived from two Greek words, "κντος," meaning container, receptacle, or body (taken in modern formations to mean cell), and "μετρον," meaning measure. Cytometers today, however, often measure things other than cells. "Particle" can be used as a more general term for any of the objects flowing through a flow cytometer. "Event" is a term that is used to indicate anything that has been interpreted by the instrument, rightly or wrongly, to be a single particle. There are subtleties here; for example, if the cytometer is not quick enough, two particles close together may actually be detected as one event. Because most of the particles sent through cytometers and detected as events are, in fact, single cells, those words will be used here somewhat interchangeably.

Because flow cytometry is a technique for the analysis of individual particles, a flow cytometrist must begin by obtaining a suspension of particles. Historically, the particles analyzed by flow cytometry were often cells from the blood; these are ideally suited for this technique because they exist as single cells and require no manipulation before cytometric analysis. Cultured cells or cell lines have also been suitable although adherent cells need some treatment to remove them from the surface on which they are grown. More recently, bacteria (25, 26), sperm (27, 28), and plankton (29) have been analyzed. Flow techniques have also been used to analyze individual particles that are not cells at all (e.g., viruses (30), nuclei (31), chromosomes (32), DNA fragments (33), and latex beads (34)). In addition, cells that do not occur as single particles can be made suitable for flow cytometric analysis by the use of mechanical disruption or enzymatic digestion; tissues can be disaggregated into individual cells and these can be run through a flow cytometer. The advantage of a method that analyzes single cells is that cells can be scanned at a rapid rate (500 to more than 5,000/s) and the individual characteristics of a large number of cells can be enumerated, correlated, and summarized. The disadvantage of a single cell technique is that cells that do not occur as individual particles will need to be disaggregated; when tissues are disaggregated for analysis, some of the characteristics of the individual cells can be altered and all information about tissue architecture and cell distribution is lost.

In flow cytometry, because particles flow in a narrow stream in front of a narrow beam of light, there are size restrictions. In general, cells or particles must fall between approximately 1 μm and approximately 30 μm in diameter. Special cytometers may have the increased sensitivity to handle smaller particles (like DNA fragments (33) or small bacteria (35)) or may have the generous fluidics to handle larger particles (such as plant cells (36)). But ordinary cytometers will, on the one hand, not be sensitive enough to detect signals from very small particles and will, on the other hand, become obstructed with very large particles.

Particles for flow cytometry should be suspended in buffer at a concentration of about 5×10^5 to 5×10^6/mL. In this suspension, they will flow through the cytometer mostly one by one. The light emitted from each particle will be detected and stored in a data file for subsequent analysis. In terms of the emitted light, particles will scatter light and this scattered light can be detected. Some of the emitted light is not scattered light, but is fluorescence. Many particles (notably phytoplankton) have natural background (auto-) fluorescence and this can be detected by the cytometer. In most cases, particles without intrinsically interesting autofluorescence will have been stained with fluorescent dyes during preparation – in order to make non-fluorescent compounds "visible" to the cytometer. A fluorescent dye is one which absorbs light of certain specific colors and then emits light of a different color (usually of a longer wavelength). The fluorescent dyes may be conjugated to antibodies and, in this case, the fluorescence of a cell will be a read-out for the amount of protein/antigen (on the cell surface or in the cytoplasm or nucleus) to which the antibody has bound. Some fluorochrome-conjugated molecules can be used to indicate apoptosis (37). Alternatively, the dye itself may fluoresce when it is bound to a cellular component. Staining with DNA-sensitive fluorochromes can be used, for example, to look at multiploidy in mixtures of malignant and normal cells (31); in conjunction with mathematical algorithms, to study the proportion of cells in different stages of the cell cycle (38); and, in restriction-enzyme digested material, to type bacteria according to the size of their fragmented DNA (39). There are other fluorochromes that fluoresce differently in relation to the concentration of calcium ions (40) or protons (41, 42) in the cytoplasm, or to the potential gradient across a cell or organelle membrane (43). In these cases, the fluorescence of the cell may indicate the response of that cell to stimulation. Other dyes can be used to stain cells in such a way that the dye is partitioned between daughter cells upon cell division; the fluorescence intensity of the cells will reveal the number of divisions that have occurred (44). Chapters in this book will provide detailed information about fluorochromes and their use. In addition, the Molecular Probes (now part of Invitrogen, Carlsbad, CA) handbook by Richard Haugland is an excellent, if occasionally overwhelming, source of information about fluorescent molecules.

The important thing to know about the use of fluorescent dyes for staining cells is that the dyes themselves need to be appropriate to the cytometer. This requires knowledge of the wavelength of the illuminating light, knowledge of the wavelength specificities of the filters in front of the instrument's photodetectors, and knowledge of the absorption and emission characteristics of the dyes themselves. The fluorochromes used to stain cells must be able to absorb the particular wavelength of the illuminating light and the detectors must have appropriate filters to detect the fluorescence emitted. For the purposes of this introductory chapter, we will now assume that we have particles that are individually suspended in medium at a concentration of about one million/mL and that they have been stained with (or naturally contain) fluorescent molecules with appropriate wavelength characteristics.

3. Illumination

In most flow cytometers, fluorescent cells are illuminated with the light from a laser. Lasers are useful because they provide intense light in a narrow beam. Particles in a stream of fluid can move through this light beam rapidly; under ideal circumstances, only

Table 1
Common types of lasers in current use

Laser	Emission wavelength(s)
Argon ion	Usually 488, 514, UV(351/363)
Red helium neon (HeNe)	633
Green helium neon (HeNe)	543
Krypton ion	Usually 568, 647
Violet diode	408
Blue solid state	488

one particle will be illuminated at a time, and the illumination is bright enough to produce scattered light or fluorescence of detectable intensity.

Lasers in today's cytometers are either gas lasers (e.g., argon ion, krypton ion, or helium–neon lasers) or solid state lasers (e.g., blue solid state laser, red or green diode lasers, violet or near-UV diode lasers, or the relatively new orange fiber laser). In all cases, light of specific wavelengths is generated (see Table 1). The wavelengths of the light from a given laser are defined and inflexible, based on the characteristics of the lasing medium. Until recently, the most common laser found on the optical benches of flow cytometers was an argon ion laser; it was chosen for early flow cytometers because it provides turquoise light (488 nm) that is absorbed efficiently by fluorescein, a fluorochrome that had long been used for fluorescence microscopy. Argon ion lasers can also produce green light (at 514 nm), ultraviolet light (at 351 and 364 nm), and a few other colors of light at low intensity. Some cytometers will use only 488 nm light from an argon ion laser; other cytometers may permit selection of several of these argon ion wavelengths from the laser. Nowadays, the blue solid state laser is prevalent.

Whereas the early flow cytometers used a single argon ion laser at 488 nm to excite the fluorescence from fluorescein (and later to include, among many possible dyes, phycoerythrin, propidium iodide, peridinin chlorophyll protein [PerCP], and various tandem transfer dyes – all of which absorb 488-nm light), there was an increasing demand for fluorochromes with different emission spectra so that cells could be stained for many characteristics at once and the fluorescence from the different fluorochromes distinguished by color. This led to the requirement for illumination light of different wavelengths and therefore for an increasing number of lasers on the optical bench. Current research flow cytometers may include, for example, two or three lasers from those listed in Table 1.

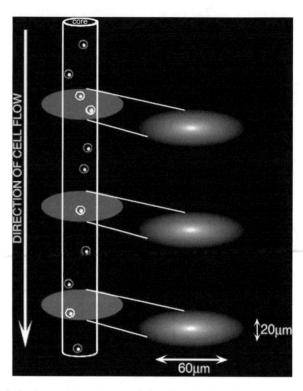

Fig. 1. Cells flowing past laser beam analysis points (in a three-laser cytometer). Beams with elliptical cross-sectional profiles allow cells to pass into and out of the beam quickly, mainly avoiding the coincidence of two cells in the laser beam at one time (but see the coincidence event in the first analysis point). In addition, an elliptical laser beam provides more uniform illumination if cells stray from the bright center of the beam.

Flow cytometers with more than one laser focus the beam from each laser at a different spot along the stream of flowing cells (Fig. 1). Each cell passes through each laser beam in turn. In this way, the scatter and fluorescence signals elicited from the cells by each of the different lasers will arrive at the photodetectors in a spatially or temporally defined sequence. Thus, the signals from the cells can be associated with a particular excitation wavelength.

All the information that a flow cytometer reveals about a cell comes from the period of time that the cell is within the laser beam. That period begins at the time that the leading edge of the cell enters the laser beam and continues until the time that the trailing edge of the cell leaves the laser beam. The place where the laser beam intersects the stream of flowing cells is called the "interrogation point," the "analysis point," or the "observa-tion point." If there is more than one laser, there will be several analysis points. In a standard flow cytometer, the laser beam(s) will have an elliptical cross-sectional area, brightest in the center and measuring about $10–20 \times 60$ μm to the edges. The height of the laser beam (10–20 μm) marks the height of the analysis point and the dimension through which each cell will pass. In commercial and research cytometers, cells will flow through each analysis point at a velocity of 5–50 m/s. They will, therefore, spend approx 0.2–4 µs in the laser beam. Because fluorochromes typically absorb light and then emit that light in a time frame of several nanoseconds, a fluorochrome on a cell will absorb and then emit light approximately a thousand times while the cell is within each analysis point.

4. Fluidics: Cells Through the Laser Beam(s)

In flow cytometry, as opposed to traditional microscopy, particles flow. In other words, the particles need to be suspended in fluid and each particle is then analyzed over the brief and defined period of time that it is being illuminated as it passes through the analysis point. This means that many cells can be analyzed and statistical information about large populations of cells can be obtained in a short period of time. The downside of this flow of single cells, as mentioned above, is that the particles need to be single and in suspension. But even nominal single-cell suspensions contain cells in clumps if the cells tend to aggregate; or there may be cells in "pseudo-clumps" if they are in a concentrated suspension and, with some probability, coincide with other cells in the analysis point of the cytometer. Even in suspensions of low cell concentration, there is always some probability that coincidence events will occur (Fig. 2). The fluidics in a cytometer are

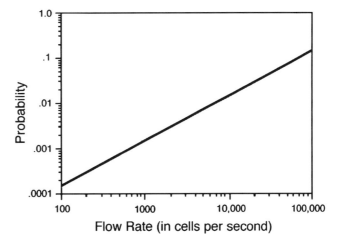

Fig. 2. The probability of a flow cytometric "event" actually resulting from more than one cell coinciding in the analysis point. For this model, the laser beam was considered to be 30 µm high and the stream flowing at 10 m/s. (Reprinted with permission of John Wiley & Sons, Inc. Copyright 2001 from Givan, A. L. [2001] *Flow Cytometry: First Principles*, 2nd edition. Wiley-Liss, New York, NY.).

designed to decrease the probability that multiple cells will coincide in the analysis point; in addition, the fluidics must facilitate similar illumination of each cell, must be constructed so as to avoid obstruction of the flow tubing, and must do all of this with cells flowing in and out of the analysis point as rapidly as possible (consistent with the production of sufficiently intense scattered and fluorescent light for sensitive detection).

One way to confine cells to a narrow path through the uniformly bright center of a laser beam would be to use an optically clear chamber with a very narrow diameter or, alternatively, to force the cells through the beam from a nozzle with a very narrow orifice. The problem with pushing cells from a narrow orifice or through a narrow chamber is that the cells, if large or aggregated, tend to clog the pathway. The hydrodynamics required to bring about focussed flow without clogging is based on principles that go back to work by Crosland-Taylor in 1953 (45). He noted that "attempts to count small particles suspended in fluid flowing through a tube have not hitherto been very successful. With particles such as red blood cells the experimenter must choose between a wide tube which allows particles to pass two or more abreast across a particular section, or a narrow tube which makes microscopical observation of the contents of the tube difficult due to the different refractive indices of the tube and the suspending fluid. In addition, narrow tubes tend to block easily." Crosland-Taylor's strategy for confining cells in a focussed, narrow flow stream, but preventing blockage through a narrow chamber or orifice involved injecting the cell suspension into the center of a wide, rapidly flowing stream (the sheath stream), where, according to hydrodynamic principles, the cells will remain confined to a narrow core at the center of the wider stream (46). This so-called hydrodynamic focussing results in coaxial flow (a narrow stream of cells flowing in a core within a wider sheath stream); it was first applied to cytometry by Crosland-Taylor, who realized that this was a way to confine cells to a precise position without requiring a narrow stream that was susceptible to obstruction.

The "flow cell" is the place in the cytometer where the sample stream is injected into the sheath stream (Figs. 3 and 4). After joining the sheath stream, the velocity of the cell suspension (in meters per second) either increases or decreases so that it becomes equal to the velocity of the sheath stream. The result is that the cross-sectional diameter of the core stream containing the cells will either increase or decrease to bring about this change in the velocity of flow while maintaining the same sample volume flow rate (in mL/s). The injection rate of the cell suspension will, therefore, directly affect the width of the core stream and the stringency by which cells are confined to the center of the illumination beam.

Having used hydrodynamic focussing to align the flow of the cells within a wide sheath stream so that blockage is infrequent,

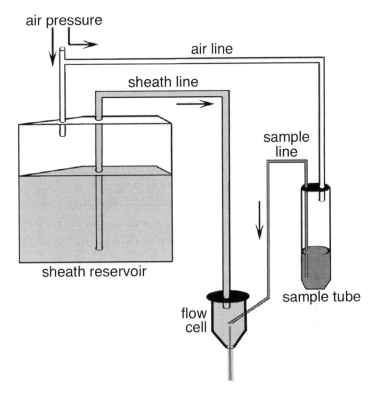

Fig. 3. The fluidics system of a flow cytometer, with air pressure pushing both the sample with suspended cells and the sheath fluid into the flow cell. (Reprinted with permission of John Wiley & Sons, Inc. Copyright 2001 from Givan, A. L. [2001] *Flow Cytometry: First Principles*, 2nd edition. Wiley-Liss, New York, NY; and also from Givan, A. L. [2001] Principles of flow cytometry: an overview, in *Cytometry*, 3rd edition [Darzynkiewicz, Z., et al., eds.], Academic Press, San Diego, CA, pp. 415–444).

there is still a requirement for rapid analysis, for better confining of the flow of cells to the very bright center of the laser beam, and for the avoidance of coincidence of multiple cells in the analysis point. These characteristics are provided by the design of the flow cell (47). Some cytometers illuminate the stream of cells within an optically clear region of the flow cell (as in a cuvette). Other systems use flow cells where the light beam intersects the fluid stream after it emerges from the flow cell through an orifice ("jet-in-air"). In all cases, the flow cell increases the velocity of the stream by having an exit orifice diameter that is narrower than the diameter at the entrance. The differences in diameter are usually between 10- and 40-fold, bringing about an increase in velocity equal to 100- to 1,600-fold (47). As the entire stream (with the cell suspension in the core of the sheath stream) progresses toward the exit of the flow cell, it narrows in diameter and increases in velocity. With this narrowing of diameter and increasing of velocity, the path of the cells becomes tightly confined to the center of the laser beam so that all cells are illuminated similarly and the cells move through the laser beam rapidly. In addition, cells are

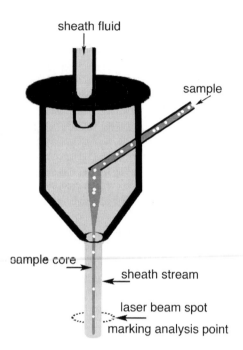

Fig. 4. A flow cell, with the sample suspension injected into the sheath fluid and forming a central core in the sheath stream. The small diameter of the flow cell at its exit orifice causes the sheath stream and sample core to narrow so that cells flow rapidly and are separated from each other and less likely to coincide in the analysis point.

spread out at greater distances from each other in the now very narrow stream and are, therefore, less likely to coincide in the analysis point.

In summary, with regard to the fluidics of the flow cytometer, the hydrodynamic focussing of a core stream of cells within a wider sheath stream facilitates the alignment of cells in the center of the laser beam without the clogging problems associated with narrow tubing and orifices. In addition, the flow cell itself increases the velocity of the stream; as well as increasing the rate at which cells are analyzed, this increased velocity also narrows the core stream to align it more precisely in the uniformly bright center of the laser beam and, at the same time, increases the distance between cells in the stream so that coincidence events in the analysis point are less frequent.

5. Signals from Cells

Lenses around the analysis point collect the light coming from cells as a result of their illumination by the laser(s). Typically there are two lenses, one in the forward direction along the path of the laser beam and one at right angles (orthogonal) to this direction

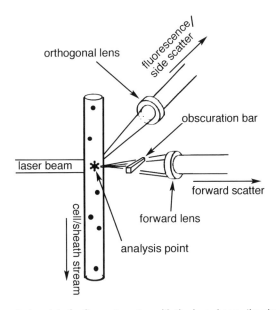

Fig. 5. The analysis point of a flow cytometer, with the laser beam, the sheath stream, and the lenses for collection of forward scatter and side scatter/fluorescence all at orthogonal angles to each other.

(Fig. 5). These lenses collect the light (the signals) given off by each cell as it passes through the analysis point. The lens in the forward direction focuses light onto a photodiode. Across the front of this forward lens is a blocker (or "obscuration") bar, approximately 1 mm wide, positioned so as to block the laser beam itself as it passes through the stream. Only light from the laser that has been refracted or scattered as it goes through a particle in the stream will be diverted enough from its original direction to avoid the obscuration bar and strike the forward-positioned lens and the photodiode behind it. Light hitting this forward scatter photodetector is therefore light that has been bent to small angles by the cell: the three-dimensional range of angles collected by this photodiode falls between those obscured by the bar and those lost at the limits of the outer diameter of the lens. The light hitting this photodetector is called forward scatter light ("fsc") or forward angle light scatter ("fals"). Although precisely defined in terms of the optics of light collection for any given cytometer, forward scatter light is not well-defined in terms of the biology or chemistry of the cell which generates this light. A cell with a large cross-sectional area will refract a large amount of light onto the photodetector. But a large cell with a refractive index quite close to that of the medium (e.g., a dead cell with a permeable outer membrane) will refract light less than a similarly large cell with a refractive index quite different from that of the medium. Because of the rough relationship between the amount of light refracted past the bar and the size of the particle, the forward scattered light signal is sometimes (misleadingly) referred to as a "volume" signal. This term belies its complexity (48).

The lens at right angles to the direction of the laser beam collects light that has been scattered to wide angles from the original direction. Light collected by this lens is defined by the diameter of the lens and its distance from the analysis point and is called side scatter light ("ssc") or 90° light scatter. Laser light is scattered to these angles primarily by irregularities or texture in the surface or cytoplasm of the cell. Granulocytes with irregular nuclei scatter more light to the side than do lymphocytes with spherical nuclei. Similarly, more side scatter light is produced by fibroblasts than by monocytes.

The signals that have been described so far (forward scatter and side scatter) are signals of the same color as the laser beam striking the cell. In a single laser system, this is usually 488 nm light from an argon ion or a blue solid state laser; in a system with two or more lasers, the scattered light is also usually 488 nm, because it is collected from the primary laser beam. This scattered light provides, as we have described, information about the physical characteristics of the cell. In addition to this scattered light, the cell may also give off fluorescent light: fluorescent light is defined as light of a relatively long wavelength that is emitted when a molecule absorbs high energy light and then emits the energy from that light as photons of somewhat lower energy. Fluorescein absorbs 488 nm light and emits light of about 530 nm. Phycoerythrin absorbs 488 nm light and emits light of around 580 nm. Therefore, if a cell has been stained with antibodies of some particular specificity conjugated to fluorescein and with antibodies of different specificity conjugated to phycoerythrin, and then the cell passes through a 488 nm laser beam, it will emit light of 530 and 580 nm. Some cells may also contain endogenous fluorescent molecules (like chlorophyll, or pyridine or flavin nucleotides). In addition, cells can be stained with other probes that fluoresce more or less depending on the DNA content of the cell or the calcium ion content of the cell, for example. In all these cases, the intensity of the fluorescent light coming from the cell is, to an arguable extent, related to the abundance of the antigen, or the DNA, or the endogenous molecule, or the calcium ion concentration of the cell. Measuring the intensity of the fluorescent light will give, then, some indication of the phenotype or function of the cells flowing through the laser beam(s).

Detecting fluorescent light is similar to detecting side-scatter light, but with the addition of wavelength-specific mirrors and filters (49). These mirrors and filters are designed so that they transmit and reflect light of well-defined wavelengths. The light emitted to the side from an analysis point is focussed by the lens onto a series of dichroic mirrors and band-pass filters that partition

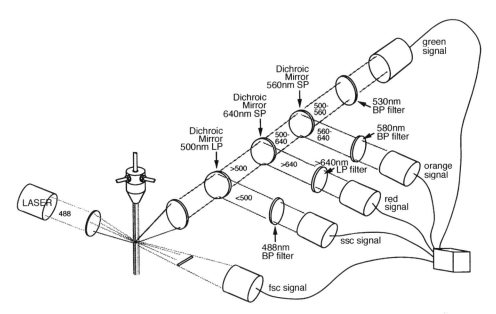

Fig. 6. The use of wavelength-specific bandpass and longpass filters, and dichroic mirrors to partition the light signal from cells to different photodetectors according to its color. In the system depicted here, there are five photodetectors; they detect forward-scatter light, side-scatter light, and fluorescence light in the green, orange, and red wavelength ranges. (Reprinted with permission of John Wiley & Sons, Inc. Copyright 2001 from Givan, A. L. [2001] *Flow Cytometry: First Principles*, 2nd edition. Wiley-Liss, New York, NY.).

this multicolor light, according to its color, onto a series of separate photomultiplier tubes (Figs. 6 and 7). In a simple example, side scatter light (488 nm) is directed toward one photomultiplier tube (PMT); light of 530 nm is directed toward another PMT; light of 580 nm is directed toward another PMT; and light >640 nm is directed toward another PMT (Fig. 6). In this example, the system will have four PMTs at the side, individually specific for turquoise, green, orange, or red light. Adding in the additional photodetector for forward scatter light, this instrument would be called a five-parameter flow cytometer. Flow cytometers today have, typically, anywhere from 3 to 15 photodetectors and thus are capable of detecting and recording the intensity of forward scatter light, side scatter light, and fluorescent light of 1–13 different colors. Because multiple excitation wavelengths are required to excite a large range of fluorochromes (distinguishable by their fluorescence emission wavelengths), high-parameter cytometers will normally have several lasers. The cells will pass, in turn, through each of the laser beams and the photodetectors will be arranged spatially so that some of the detectors will measure light excited by the first laser, some will measure light excited by the second laser, and so forth.

The signals emitted by a cell as it passes through each laser beam are light pulses that occur over time, with a beginning, an end, a height, and an integrated area. On traditional cytometers,

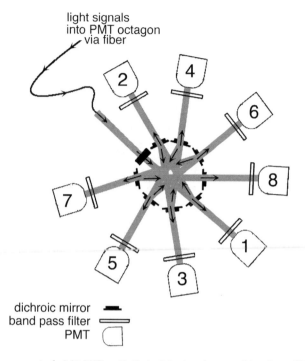

Fig. 7. Arrangement of eight PMTs with their dichroic mirrors and bandpass filters to partition light from the analysis point. Light enters the octagon via a fiber and progresses toward PMT #1; at the dichroic mirror in front of PMT #1, light of some colors is transmitted and reaches PMT #1, but light of other colors is partitioned and reflected toward PMT #2. At the dichroic mirror in front of PMT #2, light of some colors is transmitted toward PMT #2, but light of other colors is reflected toward PMT #3. In this way, light progresses through the octagon, with specific colors directed toward specific photodetectors. This octagon system has been developed by Becton Dickinson (BD Biosciences, San Jose, CA); this figure is a modification of graphics from that company.

the signal is "summarized" either by the height of the signal (related to the maximum amount of light given off by the cell at any time as it passes through the laser beam) or by the integrated area of the signal (related to the total light given off by the cell as it passes through the laser beam). Some newer cytometers analyze the signal repeatedly (ten million times per second) during the passage of the cell through the beam and those multiple numbers are then processed to give a peak height or integrated area read-out or can be used to describe the pattern of light as it is related to the structure of the cell along its longitudinal axis. With these options, there will be one or more values that are derived from each photodetector for each cell. In a simple case, for example, only the integrated (area) fluorescence detected by each photodetector will be used; the values stored for these integrated intensities from, for example, a five-photodetector system, will form the five-number flow cytometric description of each cell. A 15 parameter system will have, in this simple example, a 15-number flow description for each cell.

6. From Signals to Data

In a so-called "analog" or traditional cytometer, the current from each photodetector will be converted to a voltage, will be amplified, may be processed for ratio calculations or spectral cross-talk correction, and then, finally, will be digitized by an analog-to-digital converter (ADC) so that the final output numbers will have involved binning of the analog (continuous) amplified and processed values into (digital) channels. The amplification can be linear or logarithmic. If linear amplification is used, the intensity value is usually digitized and reported on a 10-bit or 1,024-channel scale and is displayed on axes with linear values; the numbers on the axis scale are proportional to the light intensity (Fig. 8, upper graph). By contrast, if logarithmic amplifiers are used, the

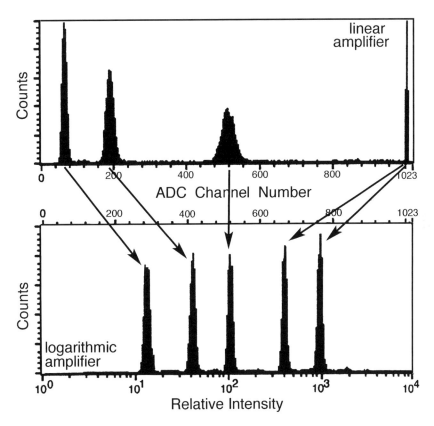

Fig. 8. The graphing of flow cytometric signals from fluorescent beads of five different intensities. The signals in the *upper graph* were acquired on an analog cytometer with linear amplification; the intensity values reported by an analog-to-digital converter with 1,024 channels have been plotted directly on the histogram axis. The two brightest beads are off scale (at the *right hand edge* of the graph). The signals in the *lower graph* were acquired on an analog cytometer with logarithmic amplification; intensity values reported by the 1,024-channel ADC can be plotted according to channel number, but are conventionally plotted according to the derived value of "relative intensity." With logarithmic amplification, beads of all five intensities are "on scale." (Reprinted from Givan, A. L. [2001] Principles of flow cytometry: an overview, in *Cytometry*, 3rd edition [Darzynkiewicz, Z., et al., eds.], Academic Press, San Diego, CA, pp. 415–444).

output voltage from the amplifier is proportional to the logarithm of the original light intensity; it will be digitized, again usually on a 1,024-channel scale, and the digitized final number will be proportional to the log of the original light intensity. In this case, the usual display involves the conversion of the values to "relative intensity units" and the display of these on axes with a logarithmic scale (Fig. 8, bottom axis of lower graph). When logarithmic amplifiers are used, they divide the full scale into a certain number of decades: that is, the full scale will encompass three or four or more tenfold increases; a four decade scale is common (that is, something at the top end of the scale will be 10^4 times brighter than something at the bottom end of the scale). It is important here to understand the reasons for choice of linear or logarithmic amplification. Linear amplification displays a limited range of intensities; logarithmic amplification, by contrast, will display a larger range of intensities (compare upper and lower graphs in Fig. 8). For this reason, linear amplifiers are conventionally used for measurements of DNA content – where the cells with the most DNA in a given analysis will generally have about twice the DNA content of the cells with the lowest DNA content. By contrast, cells that express proteins may have, after staining, 100 or 1,000 times the brightness of cells that do not express those proteins; logarithmic amplification allows the display of both the positive and negative cells on the same graph.

Although the terminology is confusing (because all cytometers report, in the end, digitized channel numbers), some newer flow systems are referred to as "digital systems" because the current from the photodetectors is converted to a voltage and then digitized immediately without prior amplification and processing. There are advantages to this early digitization that relate to time and to the elimination of some less than perfect electronic components (12, 50). By using high resolution analog-to-digital converters, the intensity values from the signal (possibly sampled at 10 MHz) can be reported on a 14-bit or 16,384-channel scale. The reported digitized numbers can then be processed to describe the integrated area, the height, or the width of the signal, where the integrated area, in particular, is usually most closely proportional to the intensity of the original light signal. Because the numbers are reported to high resolution, they can then be plotted on either linear or logarithmic axes. These high resolution ADCs obviate the need for logarithmic amplifiers, avoiding some of the problems that derive from their non-linearity (51–53).

What we now have described is a system that detects light given off by a particle in the laser beam; the light is detected according to what laser has excited the fluorescence or scatter, according to the direction the light is emitted (forward or to the side), and according to its color. The intensity of the light hitting each photodetector may be analyzed according to its peak height intensity, its integrated area (total) intensity, or according to the many numbers

CELL#	fsc	ssc	integrated area of signal		
			green fl.	orange fl.	red fl.
1	784	1233	10344	476	300
2	700	1145	11657	334	435
3	698	1289	13228	476	436
4	877	990	10453	335	478
5	789	1119	12897	501	512
6	690	998	14987	375	423
7	777	1309	14376	349	584
8	689	1401	13765	360	474
9	2089	3022	543	299	14099
10	786	1322	10367	474	499
11	688	1034	11438	356	375
12	1998	3400	464	487	15833
13	2134	3289	502	503	14998
14	745	1008	13245	499	416
15	300	432	321	321	431
16	876	1204	11498	509	485
17	775	1023	11749	464	458
18	2109	3356	387	375	15684
19	799	1039	12149	399	396

et cetera

Fig. 9. A flow cytometric data file is a list of cells in the order in which they passed through the analysis point. In this five-parameter file, each cell is described by five numbers for the signals from the forward scatter, side scatter, green fluorescence, orange fluorescence, and red fluorescence photodetectors.

that describe the signal over time. The numbers derived from each photodetector can be recorded digitally or can be amplified either linearly or logarithmically before digitization. Each cell will then have a series of numbers describing it. And the data file contains the collection of numbers describing each cell that has been run through the cytometer during the acquisition of a particular sample (Fig. 9). If 10,000 cells have been run through the cytometer and it is a five-detector cytometer, there will be five numbers describing each cell (or ten, if, e.g., signal width AND signal height are recorded). If five parameters have been recorded for each cell, the data file will consist of a list of 50,000 numbers, describing in turn all the cells in the order in which they have passed through the laser beam. The data file structure will conform more-or-less to a published flow cytometry standard (FCS) format (54).

7. From Data to Information

After the storage of the data about a group of cells into a data file, all the rest of flow cytometry is computing. Instrument vendors write software for the analysis of data acquired on their instruments;

independent programmers write software for the analysis of data acquired on any instrument. Software packages vary in price, in elegance, and in sophistication, but they all do some of the same things: they all allow a display of the distribution of any one parameter value for the cells in the data file (in a histogram); they all allow a display of correlated data between any two parameters' values (in a dot plot, contour plot, or density plot); and they all allow the restriction of the display of information to certain cells in the data file ("gating").

A histogram (Fig. 10) is used to display the distribution of one parameter over the cells in the data file. With the data from a

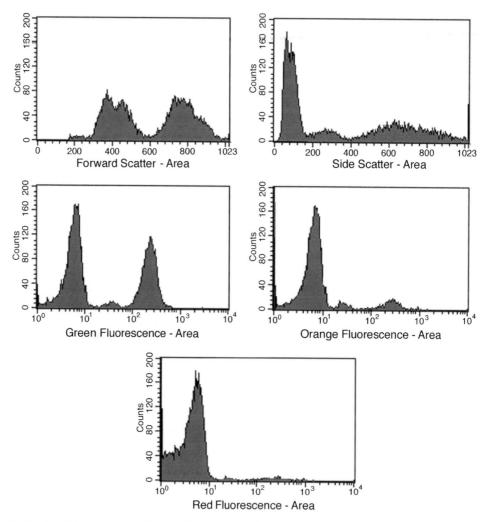

Fig. 10. The five histograms derived from a five-parameter flow cytometric data file. The scatter and fluorescence distributions are plotted individually for all the cells in the file. Forward and side scatter were acquired with linear amplification and the integrated intensity values ("area") are plotted according to the ADC channel number. The fluorescence signals were acquired with logarithmic amplification and their integrated intensity values are plotted according to the derived value of relative intensity.

five-parameter flow cytometer, there will be five numbers describing each cell (e.g., the intensities of forward scatter, side scatter, green fluorescence, orange fluorescence, and red fluorescence). A histogram (really a bar graph, with fine resolution between categories so that the bars are not visible) can display the distribution of each of those five parameters (in five separate histograms), so that we can see whether the distribution for each parameter is unimodal or bimodal; what the average relative intensities are of the cells in the unimodal cluster or in the two bimodal clusters; what proportion of the cells have intensities brighter or duller than a certain value. Numbers can be derived from these histogram distributions; by using "markers" or "cursors" to delineate ranges of intensity, software can report the proportion of cells with intensities in each of the ranges.

Because a flow data file provides several numbers (in the case of a five-parameter flow cytometer, five numbers) describing each cell, plotting all the data on five separate histograms does not take advantage of the ability we have to reveal information about the correlation between parameters on a single cell basis. For example, do the cells that fluoresce bright green also fluoresce orange or do they not fluoresce orange? For the display of correlated data, flow software provides the ability to plot dot plots, contour plots, or density plots. While these alternative two-dimensional plots have different advantages in terms of visual impact and graphical authenticity to the hard data, they all display two parameters at once and report the same quantitative analysis (Fig. 11). By using a 2-dimensional plot, a scientist can see whether the cells that fluoresce green also fluoresce orange. Further, this information can be reported quantitatively, using markers to delineate intensity regions in two dimensions. These two-dimensional markers are called quadrants if they have been used to delineate so-called double negative, single positive (for one color), single positive (for the second color), and double positive cells (Fig. 11).

One of the unique aspects of flow cytometry is the possibility of "gating." Gating is the term used for the designation of cells of interest within a data file for further analysis. It permits the analysis of sub-sets of cells from within a mixed population. It also provides a way to analyze cells in high levels of multi-parameter space (55, 56). Figure 12 shows an example of a mixed population of white blood cells that have been stained with fluorescent antibodies. Because the white blood cells of different types can be distinguished from each other by the separate clusters they form in a plot of forward scatter versus side scatter light, the fluorescence of the monocytes can be analyzed without interference from the fluorescence of the lymphocytes or neutrophils in the data file. Similarly, the neutrophils and lymphocytes can also be analyzed independently of the other populations of cells. The procedure for doing this involves drawing a "region" around the cluster of, for example, monocytes in the fsc versus ssc plot.

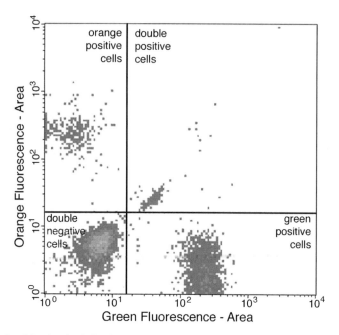

Fig. 11. A two-dimensional density plot, indicating the correlation of green and orange fluorescence for the cells in a data file. Quadrants divide the intensity distributions into regions for unstained ("double negative") cells, cells that are stained both green and orange ("double positive cells"), and cells stained for each of the colors singly. In this example, it can be seen that most of the green-positive cells are not also orange-positive (that is, they are not double-positive).

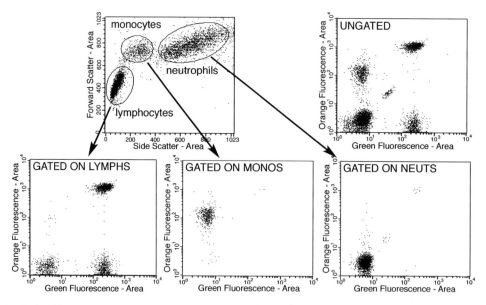

Fig. 12. A plot of forward versus side scatter for leukocytes from the peripheral blood indicates that regions can be drawn around cells with different scatter characteristics, marking lymphocytes, monocytes, and neutrophils. These regions can be used to define gates. The four plots of green versus orange fluorescence are either ungated displaying the fluorescence intensities of all the cells (*upper right*); or are gated on each of the three leukocyte populations, revealing that the three types of leukocytes stain differently with the orange and green antibodies.

That region defines a group of cells with particular characteristics in the way they scatter light. The region can then be used as a gate for subsequent analysis of the fluorescence of cells. A gated dot plot of, for example, green fluorescence versus orange fluorescence can display the fluorescence data from only the cells that fall into the "monocyte" region. Gates can be simple in this way. Or they can be more complex, facilitating the analysis of cells that have been stained with many reagents in different colors: for example a gate could be a combination of many regions, defining cells with certain forward and side scatter characteristics, certain green fluorescence intensities, and certain orange fluorescence intensities. The final step in analysis could use a gate that combines all these regions and asks how the cells with bright forward scatter, medium side scatter, bright green intensity, and little or no orange intensity are distributed with regard to red fluorescence.

8. Sorting

It might seem that flow cytometers would have developed first with the ability to detect many colors of fluorescence from particles or cells and that it then might have occurred to someone that it would be useful to separate cells with different fluorescent or scatter properties into separate test tubes for further culture, for RNA or DNA isolation, or for physiological analysis. History, however, went in the opposite direction. Early flow cytometers were developed to separate cells from each other (7, 8, 11). A cytometer was developed in 1965 by Mack Fulwyler at the Los Alamos National Laboratory to separate red blood cells with different scatter signals from each other to see if there were actually two separate classes of erythrocytes or, alternatively, if the scatter differences were artifactual based on the flattened disc shape of the cells. The latter turned out to be true – and the same instrument was then used to separate mouse from human erythrocytes and a large component from a population of mouse lymphoma cells (8). For several years, flow cytometers were thought of as instruments for separating cells (the acronym FACS stands for "fluorescence-activated cell *sorter*"). It was only slowly that these instruments began to be used primarily for the assaying of cells without separation. Modern cytometers most often do not even possess the capability of separating cells.

Although there are many methods available for separating/isolating sub-populations of cells from a mixed population (e.g., adherence to plastic, centrifugation, magnetic bead binding, complement depletion) and these methods are usually significantly more rapid than flow sorting, flow sorting may be the best separation technique available when cells differ from each other by the way they scatter light, by slight differences in antigen intensity, or by multiparameter criteria. Cells sorted by flow cytometry are routinely used for functional assays, for PCR replication of cell-type specific DNA sequences, for artificial insemination by sperm bearing X or Y chromosomes, and for cloning of high-expressing transfected cells. In addition, sorted chromosomes are used for the generation of DNA libraries.

The strategy for electronic flow sorting involves the modification of a non-sorting cytometer in three ways: the sheath stream is vibrated so that it breaks up into drops; the stream (now in drops) flows past two charged (high voltage) plates; and the electronics of the instrument are modified so that the drops can be charged or not according to the characteristics of any contained cell, as detected at the analysis point (Fig. 13). In a sorting cytometer, cells flow through the analysis point where they are illuminated and their scatter and fluorescence signals detected as in a non-sorting instrument. They then continue to flow downstream where, as the stream breaks up into drops, they become enclosed in individual drops. If the cells are far enough apart from each

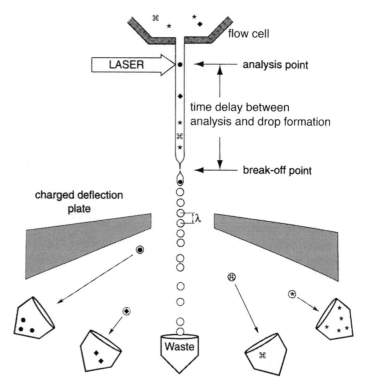

Fig. 13. Droplet formation for sorting. A vibrating flow cell causes the sheath stream to break up into drops at the break-off point. Cells flowing in the stream are enclosed in drops and, in the example shown here, drops can be charged strongly or less strongly positive or negative, so that four different types of cells (as detected at the analysis point) can be sorted into four receiving containers.

other in the stream, there will be very few cases in which there is more than one cell in each drop; there will be, by operator choice, cells in, on average, only every third or fifth or tenth drop – and there will be empty drops between the drops containing cells. The number of cells per drop (and the number of empty drops) will be determined by the number of cells flowing per second, by the vibration frequency that is creating the drops, and by the impact of the mathematics of a Poisson distribution, whereby cells are never perfectly distributed along the stream, but can cluster with some probability (the Fifth Avenue bus phenomenon).

A vibrating stream breaks up into drops according to the following equation

$$v = f\lambda$$

which defines the fixed relationship between the velocity of the stream (v), the frequency of the drop generation vibration (f), and the distance between the drops (λ). Drop formation is stable when the distance between drops equals 4.5 times the diameter of the stream, so a given stream diameter (flow cell exit orifice) will determine the λ value. Rapid drop formation (high f) is desirable because rapid drop formation allows rapid cell flow rate (without multiple cells in a drop). Therefore this condition is facilitated by using a high stream velocity and also a narrow stream diameter (but being aware that a narrow flow cell orifice gets clogged easily). Common conditions for sorting, with a 70-μm flow cell orifice, involve stream velocities of about 10 m/s and drop drive frequencies of about 32 kHz (meaning that cells flowing at 10,000 cells per second will be, on average, in every third drop). Conditions for so-called high speed sorting involve stream velocities of about 30 m/s and drop drive frequencies of 95 kHz (cells flowing at 30,000 cells per second will be, on average, in every third drop).

Because drops break off from the vibrating stream at a distance from the fixed point of vibration, the stream of cells can be illuminated and the signals from the cells collected with little perturbation as long as the analysis point is relatively close to the point of vibration and far away from the point of drop formation. In a sorting cytometer, cells are illuminated close to the flow cell (or within an optically clear flow cell); their signals are collected, amplified, and digitized in ways similar to those in nonsorting cytometers. A sorting cytometer differs from a nonsorting cytometer because cells become enclosed in drops after they move down the stream. At points below the drop break off point, the stream will consist of a series of drops, all separate from each other, with some drops containing cells. The flow operator will have drawn sort regions around cells "of interest" according to their flow parameters. If a cell in the analysis point has been determined to be a cell of interest according to the sort regions, the drop containing that cell will be charged positively or negatively so that it will be deflected either to the left or right as it passes the positive and negative deflection plates. Modern cytometers have the ability to charge drops in four ways (strongly or weakly positive and strongly or weakly negative), so that four sort regions can be designated and four sub-populations of cells can be isolated from the original population. Collecting tubes are placed in position, one or two more or less at the left and one or two more or less at the right, and the deflected drops, containing the cells of interest, will be collected in the tubes. Uninteresting cells will be in uncharged drops; they will not be deflected out of the main stream and they will pass down the center and into the waste container.

Sorted cells will be pure when only those drops containing the cells of interest are charged. This happens because the sort operator determines the length of time required for a cell to move from the analysis point to the position of its enclosure in a drop (at the drop "breakoff point" downstream); the stream is charged for a short period of time at exactly this time delay after a cell of interest has been detected. This drop delay time can be determined empirically, by testing different drop delay times on a test sort with beads. Alternatively, it can be measured using drop separation units (knowing the drop generation frequency, the reciprocal is equal to the number of seconds per drop; therefore, the distance between drops (which can be measured) has an equivalence in time units). Given the time that it takes between analysis of a cell in the laser beam and the enclosure of that cell into its own self-contained drop, the flow cytometer can be programmed to apply a charge to the stream for a short interval, starting at the time just before the cell of interest is about to detach from the main stream into the drop. If the charge is applied for a short interval, only one drop will be charged. If it is charged for a longer interval, then drops on either side of the selected drop can be charged (for security, in case the cell moves faster or slower than predicted). The charge on the stream can be positive or negative (or weakly or strongly positive or negative) and therefore the drops containing two or four mutually exclusive subsets of cells can be deflected into separate collection vessels.

Cell sorting is validated by three characteristics: the efficiency of the sorting of the cells of interest from the original mixed population; the purity of those cells according to the selection criteria; and the time it has taken to obtain a given number of sorted cells (57). In most cytometers, purity is protected because drop charging is aborted when cells of the wrong phenotype are enclosed with cells of interest in a single drop. In this way, high cell flow rates will compromise the sorting efficiency, but will not compromise purity until the cell flow rate is so fast that there is

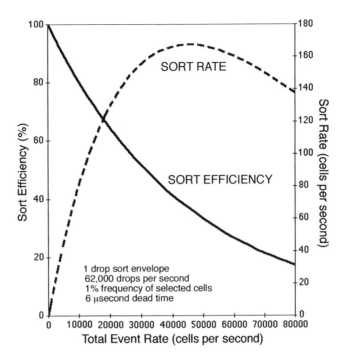

Fig. 14. The efficiency and speed of sorting are affected by the flow rate of cells through the analysis point. High speed sorting decreases the efficiency, but increases the speed at which sorted cells are collected (until so many sorts are aborted due to multiple cells in drops that the sort rate begins to decline). The model from which these graphs were derived is from Robert Hoffman.

significant coincidence of multiple cells in the laser beam. As the cell flow rate increases (using cells at higher concentrations), the speed of sorting will increase until the number of aborted sorts gets so high that the actual speed of sorting of the desired cells starts to drop off (Fig. 14). Some sorting protocols will be designed to obtain rare cells: these sorts will be done at relatively low speed, but with resulting high efficiency. Other protocols will start with buckets of cells and will be concerned most with getting cells in a short period of time, without regard to the efficiency of the sort; these sorts can be done at high speed where the efficiency is low, but the speed of sorting is high. The bottom line is that speed and efficiency interact and cannot, both, be optimized at the same time.

9. Conclusions

Flow cytometry has, arguably, remained a relatively constant technology since its entry in 1969 (10, 11) into the modern era. The main technological changes that have occurred over the past 34 years have been quantitative rather than qualitative.

More parameters can now be analyzed simultaneously (58), more cells can be analyzed or sorted per second, and more sensitivity is available to detect fewer fluorescent molecules; in addition, flow cytometers are both smaller and less expensive than they were. However, a flow cytometer still involves particles flowing one by one past a laser beam, with photodetectors nestled around the analysis point to detect fluorescent or scattered light coming from the particles.

By contrast, with the less-than-radical changes in flow instrument technology, the increasing diversity of applications has been striking. For a start, the use of flow cytometry has increased remarkably; Fig. 15 indicates that references to "flow cytometry" in the Medline data base were zero in 1970, 113 in 1980, 2,286 in 1990, and 4,893 in the year 2000. In 2009, that number in the PubMed database went up to 8,811. But more important than the number of references is the range of applications that are now being addressed by flow cytometry. In the 1970s, leukocytes and cultured cells were the main particles analyzed by flow cytometry; now plankton, bacteria, disaggregated tissues (see Chapter 12, *this volume*), plant cells (see Chapter 20, *this volume*), viruses, chromosomes, latex beads, quantum dots (see Chapter 4, *this volume*), and DNA fragments are all, more-or-less, taken for granted. In addition, early flow cytometry looked at fluorescence emanating primarily from the stained proteins on the surface of cells or from stained DNA in their nuclei. Now plankton are

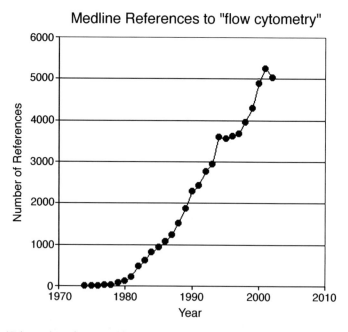

Fig. 15. Increasing reference to "flow cytometry" in the Medline-indexed literature over the past three decades. The actual use of flow cytometers predates the use of the term itself.

examined for their autofluorescence, animal and plant cells are assayed for fluorescence that reflects their proliferative or metabolic function, sperm are sorted based on their X or Y genotype, many aspects of protein or DNA synthesis are assayed to indicate stage of cell cycle or of apoptosis, and bacteria are typed according to the size of their DNA fragments after restriction enzyme digestion. Indeed, flow cytometry has also been applied to beads, which are used to capture soluble analytes from the blood or culture medium; the beads, not the cells, are analyzed by flow cytometry to see how much of the analyte has been bound to their surface and, by comparison with standard curves, to indicate the concentration of analyte in the medium (see (34) and Chapter 5, *this volume*).

Through its history, flow cytometry has sparked the collaboration of mathematicians, engineers, chemists, biologists, and physicians, working together to provide instrumentation that now probes not just our bodies and our culture flasks, but also the depths of the ocean (59) and, potentially, life in space (or, at least, life in space ships(60, 61)). If nothing else, the example of flow cytometry should inspire us toward collaboration and to an open mind.

Acknowledgments

Howard Shapiro and Ben Verwer have provided me with advice on some of the electronic and illumination aspects of flow cytometry. I thank them for their help; all mistakes are very much my own.

References

1. Moldavan, A. (1934) Photo-electric technique for the counting of microscopical cells. *Science* **80**, 188–189.

2. Gucker, F. T., Jr., O'Konski, C. T., Pickard, H. B., and Pitts J. N., Jr. (1947) A photoelectronic counter for colloidal particles. *J. Am. Chem. Soc.* **69**, 2422–2431.

3. Cornwall, J. B. and Davison, R. M. (1950) Rapid counter for small particles in suspension. *J. Sci. Instrum.* **37**, 414–417.

4. Coulter, W. H. (1956) High speed automatic blood cell counter and analyzer. *Proc. Natl. Electron. Conf.* **12**, 1034–1040.

5. Bierne, T. and Hutcheon, J. M. (1957) A photoelectric particle counter for use in the sieve range. *J. Sci. Instrum.* **34**, 196–200.

6. Kamentsky, L. A. and Melamed, M. R. (1965) Spectrophotometer: New instrument for ultrarapid cell analysis. *Science* **150**, 630–631.

7. Kamentsky, L. A. and Melamed, M. R. (1967) Spectrophotometric cell sorter. *Science* **156**, 1364–1365.

8. Fulwyler, M. J. (1965) Electronic separation of biological cells by volume. *Science* **150**, 910–911.

9. Dittrich, W. and Göhde, W. (1969) Impulsfluorometrie dei einzelzellen in suspensionen. *Z. Naturforsch.* **24b**, 360–361.

10. Van Dilla, M. A., Trujillo, T. T., Mullaney, P. F., and Coulter, J. R. (1969) Cell microfluorimetry: A method for rapid fluorescence measurement. *Science* **163**, 1213–1214.

11. Hulett, H. R., Bonner, W. A., Barrett, J., and Herzenberg, L. A. (1969) Cell sorting: Automated separation of mammalian cells as a function of intracellular fluorescence. *Science* **166**, 747–749.

12. Shapiro, H. M. (2003) *Practical Flow Cytometry*, 4th editon. Wiley-Liss, New York.

13. Melamed, M. R., Mullaney, P. F., and Shapiro, H. M. (1990) An historical review of the development of flow cytometers and sorters, in *Flow Cytometry and Sorting*, 2nd edition. (Melamed, M. R., Lindmo, T., and Mendelsohn, M. L., eds.), Wiley-Liss, New York, pp. 1–8.

14. Melamed, M. R. (2001) A brief history of flow cytometry and sorting. *Methods in Cell Biology* **63**(part A), pp. 3–17.

15. Van Dilla, M. A., Dean, P. N., Laerum, O. D., and Melamed, M. R. (eds.) (1985) *Flow Cytometry: Instrumentation and Data Analysis*. Academic Press, London.

16. Melamed, M. R., Lindmo, T., and Mendelsohn, M. L. (eds.) (1990) *Flow Cytometry and Sorting*, 2nd edition. Wiley-Liss, New York.

17. Watson, J. V. (2004) *Introduction to Flow Cytometry*. Cambridge University Press, Cambridge.

18. Watson, J. V. (2005) *Flow Cytometry Data Analysis: Basic Concepts and Statistics*. Cambridge University Press, Cambridge.

19. Robinson, J. P., Darzynkiewicz, Z., Dean, P. N., Dressler, L. G., Rabinovitch, P. S., Stewart, C. C., Tanke, H. J., and Wheeless, L. L. (eds.) (1997) *Current Protocols in Cytometry*. John Wiley & Sons, New York.

20. Diamond, R. A. and DeMaggio, S. (eds.) (2000) *In Living Color: Protocols in Flow Cytometry and Cell Sorting*. Springer-Verlag, Berlin.

21. Durack, G. and Robinson, J. P. (eds.) (2000) *Emerging Tools for Single-Cell Analysis*, Wiley-Liss, New York.

22. Ormerod, M. G. (ed.) (2000) *Flow Cytometry: A Practical Approach*, 3rd edition. Oxford University Press, Oxford.

23. Darzynkiewicz, Z., Roederer, M., and Tanke, H. J. (eds.) (2004) *Cytometry: Methods in Cell Biology*, 4th edition, Vol. 75. Academic Press, San Diego.

24. Givan, A. L. (2001) *Flow Cytometry: First Principles*, 2nd edn. Wiley-Liss, New York.

25. Lloyd, D. (1993) *Flow Cytometry in Microbiology*. Springer-Verlag, London.

26. Alberghina, L., Porro, D., Shapiro, H., Srienc, F., and Steen, H. (eds.) (2000) *Analysis of Microbial Cells at the Single Cell Level: J. Microbiol. Methods*, Vol. 42.

27. Fugger, E. F., Black, S. H., Keyvanfar, K., and Schulman, J. D. (1998) Births of normal daughters after MicroSort sperm separation and intrauterine insemination, in-vitro fertilization, or intracytoplasmic sperm injection. *Hum. Reprod.* **13**, 2367–2370.

28. Gledhill, B. L., Evenson, D. P., and Pinkel, D. (1990) Flow cytometry and sorting of sperm and male germ cells, in *Flow Cytometry and Sorting*, 2nd edition (Melamed, M. R., Lindmo, T., and Mendelsohn, M. L., eds.), Wiley-Liss, New York, pp. 531–551.

29. Reckerman, M. and Collin, F. (eds.) (2000) *Aquatic Flow Cytometry: Achievements and Prospects, Scientia Marina*, Vol. 64.

30. Marie, D., Brussard, C. P. D., Thyrhaug, R., Bratbak, G., and Vaulot, D. (1999) Enumeration of marine viruses in culture and natural samples by flow cytometry. *Appl. Environ. Microbiol.* **59**, 905–911.

31. Hedley, D. W. (1989) Flow cytometry using paraffin-embedded tissue: Five years on. *Cytometry* **10**, 229–241.

32. Gray, J. W. and Cram, L. S. (1990) Flow karyotyping and chromosome sorting, in *Flow Cytometry and Sorting*, 2nd edition (Melamed, M. R., Lindmo, T., and Mendelsohn, M. L., eds.), Wiley-Liss, New York.

33. Habbersett, R. C., Jett, J. H., and Keller, R. A. (2000) Single DNA fragment detection by flow cytometry, in *Emerging Tools for Single-Cell Analysis* (Durack, G. and Robinson, J. P., eds.), Wiley-Liss, New York, pp. 115–138.

34. Carson, R. T. and Vignali, D. A. A. (1999) Simultaneous quantitation of 15 cytokines using a multiplexed flow cytometric assay. *J. Immunol. Methods* **227**, 41–52.

35. Steen, H. B. (2000) Flow cytometry of bacteria: Glimpses from the past with a view to the future. *J. Microbiol. Methods* **42**, 65–74.

36. Harkins, K. R. and Galbraith, D. W. (1987) Factors governing the flow cytometric analysis and sorting of large biological particles. *Cytometry* **8**, 60–70.

37. Darzynkiewicz, Z., Juan, G., Li, X., Gorczyca, W., Murakami, T., and Traganos, F. (1997) Cytometry in cell necrobiology: Analysis of apoptosis and accidental cell death (necrosis). *Cytometry* **27**, 1–20.

38. Gray, J. W., Dolbeare, F., and Pallavicini, M. G. (1990) Quantitative cell-cycle analysis, in *Flow Cytometry and Sorting*, 2nd edition (Melamed, M. R., Lindmo, T., and Mendelsohn, M. L., eds.), Wiley-Liss, New York, pp. 445–467.

39. Kim, Y., Jett, J. H., Larson, E. J., Penttila, J. R., Marrone, B. L., and Keller, R. A. (1999) Bacterial fingerprinting by flow cytometry:

Bacterial species discrimination. *Cytometry* **36**, 324–332.

40. June, C. H., Abe, R., and Rabinovitch, P. S. (1997) Measurement of intracellular calcium ions by flow cytometry, in *Current Protocols in Cytometry* (Robinson, J. P., Darzynkiewicz, Z., Dean, P. N., Dressler, L. G., Rabinovitch, P. S., Stewart, C. C., and Tanke, H. J., eds.), Wiley, New York, pp. 9.8.1–9.8.19.

41. Li, J. and Eastman, A. (1995) Apoptosis in an interleukin-2-dependent cytotoxic T lymphocyte cell line is associated with intracellular acidification. *J. Biol. Chem.* **270**, 3203–3211.

42. Chow, S. and Hedley, D. (1997) Flow cytometric measurement of intracellular pH, in *Current Protocols in Cytometry* (Robinson, J. P., Darzynkiewicz, Z., Dean, P. N., Dressler, L. G., Rabinovitch, P. S., Stewart, C. C., and Tanke, H. J., eds.), Wiley, New York, pp. 9.3.1–9.3.10.

43. Shapiro, H. M. (1997) Estimation of membrane potential by flow cytometry, in *Current Protocols in Cytometry* (Robinson, J. P., Darzynkiewicz, Z., Dean, P. N., Dressler, L. G., Rabinovitch, P. S., Stewart, C. C., and Tanke, H. J., eds.), Wiley, New York, pp. 9.6.1–9.6.10.

44. Lyons, A. B. (1999) Divided we stand: Tracking cell proliferation with carboxyfluorescein diacetate succinimidyl ester. *Immunol. Cell Biol.* **77**, 509–515.

45. Crosland-Taylor, P. J. (1953) A device for counting small particles suspended in a fluid through a tube. *Nature* **171**, 37–38.

46. Kachel, V., Fellner-Feldegg, H., and Menke, E. (1990) Hydrodynamic properties of flow cytometry instruments, in *Flow Cytometry and Sorting*, 2nd edition (Melamed, M. R., Lindmo, T., and Mendelsohn, M. L., eds.), Wiley-Liss, New York, pp. 27–44.

47. Pinkel, D. and Stovel, R. (1985) Flow chambers and sample handling, in *Flow Cytometry: Instrumentation and Data Analysis.* (Van Dilla, M. A., Dean, P. N., Laerum, O. D., and Melamed, M. R., eds.), Academic Press, London, pp. 77–128.

48. Salzman, G. C., Singham, S. B., Johnston, R. G., and Bohren, C. F. (1990) Light scattering and cytometry, in *Flow Cytometry and Sorting*, 2nd edition (Melamed, M. R., Lindmo, T., and Mendelsohn, M. L., eds.), Wiley-Liss, New York, pp. 81–107.

49. Waggoner, A. (1997) Optical filter sets for multiparameter flow cytometry, in *Current Protocols in Cytometry* (Robinson, J. P., Darzynkiewicz, Z., Dean, P. N., Dressler, L. G., Rabinovitch, P. S., Stewart, C. C., and Tanke, H. J., eds.), Wiley, New York, pp. 151–158.

50. Verwer, B. (2002) BD FACSDiVa Option, BD Biosciences.

51. Bagwell, C. B., Baker, D., Whetstone, S., Munson, M., Hitchox, S., Ault, K. A., and Lovett, E. J. (1989) A simple and rapid method of determining the linearity of a flow cytometer amplification system. *Cytometry* **10**, 689–694.

52. Muirhead, K. A., Schmitt, T. C., and Muirhead, A. R. (1983) Determination of linear fluorescence intensities from flow cytometric data accumulated with logarithmic amplifiers. *Cytometry* **3**, 251–256.

53. Wood, J. C. S. (1997) Establishing and maintaining system linearity, in *Current Protocols in Cytometry* (Robinson, J. P., Darzynkiewicz, Z., Dean, P. N., Dressler, L. G., Rabinovitch, P. S., Stewart, C. C., and Tanke, H. J., eds.), Wiley, New York, pp. 1.4.1–1.4.12.

54. Seamer, L. (2000) Flow cytometry standard (FCS) data file format, in *In Living Color: Protocols in Flow Cytometry and Cell Sorting* (Diamond, R. A. and DeMaggio, S., eds.), Springer-Verlag, Berlin.

55. Loken, M. R. (1997) Multidimensional data analysis in immunophenotyping, in *Current Protocols in Cytometry* (Robinson, J. P., Darzynkiewicz, Z., Dean, P. N., Dressler, L. G., Rabinovitch, P. S., Stewart, C. C., and Tanke, H. J., eds.), Wiley, New York, pp. 10.4.1–10.4.7.

56. Roederer, M., De Rosa, S., Gerstein, R., Anderson, M., Bigos, M., Stovel, R., Nozaki, T., Parks, D., and Herzenberg, L. (1997) 8 color, 10-parameter flow cytometry to elucidate complex leukocyte heterogeneity. *Cytometry* **29**, 328–39.

57. Hoffman, R. A. and Houck, D. W. (1998) High speed sorting efficiency and recovery: Theory and experiment. *Cytometry Suppl.* **9**, 142.

58. Perfetto, S. P., Chattopadhyay, P., and Roederer, M. (2003) Seventeen-colour flow cytometry: Unravelling the immune system. *Nat. Rev. Immunol.* **4**, 648–655.

59. Dubelaar, G. B., Gerritzen, P. L., Beeker, A. E., Jonker, R. R., and Tangen, K. (1999) Design and first results of CytoBuoy: A wireless flow cytometer for in situ analysis of marine and fresh waters. *Cytometry* **37**, 247–254.

60. Sams, C. F., Crucian, B. E., Clift, V. L., and Meinelt, E. M. (1999) Development of a whole blood staining device for use during space shuttle flights. *Cytometry* **37**, 74–80.

61. Crucian, B. E. and Sams, C. F. (1999) The use of a spaceflight-compatible device to perform WBC surface marker staining and whole-blood mitogenic activation for cytokine detection by flow cytometry. *J. Gravit. Physiol.* **6**, 33–34.

Chapter 2

Breaking the Dimensionality Barrier

C. Bruce Bagwell

Abstract

Recent advances in biotechnology have resulted in cytometers capable of performing numerous correlated measurements of cells, often exceeding ten. In the near future, it is likely that this number will increase by fivefold and perhaps even higher. Traditional analysis strategies based on examining one measurement versus another are not suitable for high-dimensional data analysis because the number of measurement combinations expands geometrically with dimension, forming a kind of complexity barrier. This dimensionality barrier limits cytometry and other technologies from reaching their maximum potential in visualizing and analyzing important information embedded in high-dimensional data.

This chapter describes efforts to break through this barrier and allow the visualization and analysis of any number of measurements with a new paradigm called Probability State Modeling (PSM). This new system creates a virtual progression variable based on probability that relates all measurements. PSM can produce a single graph that conveys more information about a sample than hundreds of traditional histograms. These PSM overlays reveal the rich interplay of phenotypic changes in cells as they differentiate. The end result is a deeper appreciation of the molecular genetic underpinnings of ontological processes in complex populations such as found in bone marrow and peripheral blood.

Eventually these models will help investigators better understand normal and abnormal cellular progressions and will be a valuable general tool for the analysis and visualization of high-dimensional data.

Key words: Flow cytometry, Multidimensional analysis, Dimensionality barrier, Probability state model

1. Introduction

1.1. Measurements, Parameters, Markers, Colors, Particles, and Events

Cytometers perform measurements on cells. Whether a cytometer is an image or flow system, these measurement signals are normally digitized, processed in some manner, and stored. Unfortunately, these particle measurements have historically been referred to as "parameters." Since the term "parameter" has a very different statistical meaning, especially in the context of modeling, it will only be used in a modeling context for this chapter.

Teresa S. Hawley and Robert G. Hawley (eds.), *Flow Cytometry Protocols*, Methods in Molecular Biology, vol. 699,
DOI 10.1007/978-1-61737-950-5_2, © Springer Science+Business Media, LLC 2011

The word "measurement" or "variable" will be used to mean a specific digitized attribute of a particle. Measurements derived from fluorescence photons are commonly referred to as "colors." For example, six-color data means that the data contains six correlated measurements from different fluorescence colors. Usually other measurements such as time, forward scatter, and side scatter accompany these colors. For both flow and image cytometry, most color detectors can generate numerous types of measurements by means of pulse or statistical processing.

The objects that are measured by a cytometer are referred to as "particles." These particles are often cells, but they don't need to be. Once the cytometer's pulse processing algorithms decide that a modulation in a signal is probably a real particle and not noise, they digitize all the measurements and the particle becomes an electronically identified "event." All measurements derived from events are generally stored on computer storage media as a "listmode" file.

1.2. Difficulties with Visualization and Analysis of High-Dimensional Data

Cells are largely composed of interdependent chemical machines moving from state to state along thermodynamic gradients. Technologies that perform numerous measurements on single cells have the potential of uncovering the specifics of how these machines perform their various functions. Recently the technology of cytometry has evolved to making 17 correlated measurements from cells (1, 2) and is on the precipice of dramatically increasing this number to over 50 (3–5).

The rate-limiting issue that has impeded this evolution toward high-dimensional displays and analyses is what has commonly been termed the dimensionality barrier or curse (6, 7). This barrier is one of overwhelming complexity in both viewing and analyzing large numbers of correlated measurements. Visualization of multiple-dimensional data is traditionally done by viewing numerous bivariate dot or contour plots. As the number of correlated measurements, n, increases, the number of bivariate plots expands geometrically by the relation, $n \times (n-1)/2$. For seventeen-dimensional data, a total of 136 bivariates are necessary to view all correlations in data, which severely limits visual interpretation of the embedded information.

Analysis methods that involve measurement partitioning schemes (8, 9), also run into this barrier because the number of partitions can also increase geometrically with measurement number, ultimately increasing the required number of events to unattainable values or requiring unreasonably large partition sizes. Nevertheless, these methods do provide valuable information on quantitative differences between high-dimensional data files.

1.3. Other Approaches to the Dimensionality Barrier Problem

There have been numerous schemes that have been proposed to ameliorate this dimensionality problem (10–16). Many of these methods work quite well, but most will eventually fail when the number of measurements becomes too high. A review of current approaches to high-dimensional cytometry data visualization has recently been published (17). In general, any approach to high-dimensional data visualization that attempts to examine one measurement or combination of measurements versus another eventually runs into this barrier.

Numerous clustering methods have been used to successfully classify cytometric data into relevant populations (14, 18–23). Since clustering methods generally use distance or proximity functions to find neighboring events, they obviate many of the complexity issues associated with the dimensionality barrier. Unfortunately, most clustering routines have two weaknesses. The first is that they usually are not good at finding small continuums between clusters. These continuums are many times far more important than the clusters themselves. The second problem is that after the clustering routine is finished, visual inspection of clusters is generally done by examining bivariates with color-coded dots representing found clusters. As already discussed, as soon as bivariates involving measurements are used to inspect data, the dimensionality barrier interferes with the complete interpretation of the data's embedded information.

1.4. Why Write this Chapter

This chapter provides a brief historical account of how this dimensionality curse, as it relates to complex processes, was solved by a radically different approach to both visualization and analysis of high-dimensional data. The technique is termed Probability State Modeling and its specifics have been described elsewhere (24–26).

The rest of this chapter will be devoted to an exploration of some PSM applications. The main reason for writing this chapter is to introduce the reader to this new and powerful visualization and analysis tool and show how it leads to a more thorough understanding of complex cellular systems.

2. Probability State Modeling: A Radically Different Approach

2.1. The Ginger Root View of Bone Marrow

Work on this project began in earnest early in 2001. The general idea was to develop new visualization methods that could reveal the underlying complex measurement relations found in tissues like bone marrow. At the time, the commonly held image of populations in bone marrow, proposed by Howard Shapiro in the early 1980s (27), was that they were similar to the variegated bulges of a ginger root. Populations that were close to each other from an ontological point of view would end up being close to each other in the ginger root three-dimensional solid structure. This concept suggested that the proper approach to the problem of high-dimensional display and analysis would involve some kind of nearest-neighbor clustering routine that could produce this kind of solid structure from the analysis of correlated measurements.

2.2. Creating a Sophisticated Model Representing Complex Cytometry Data

Investigation of new clustering and visualization algorithms is facilitated by software that can simulate realistic data. It's always important to have data sets where important truths about the data are known in order to understand how well computer algorithms approximate these truths. For this project, it was necessary to design a computer simulator that could produce high-dimensional data that was similar if not identical to that produced by a cytometer.

At the core of the simulator is a model. A model is a mathematical construct that normally simulates some real-world process or processes. The model needs to be complex enough to represent all lineages in bone marrow and general enough to handle other types of cytometry-derived populations. It needs to be able to define measurement uncertainty as well as population heterogeneity. It also has to represent complex continuums between developmental stages of multiple lineages in bone marrow.

Building a complex model is much like writing a book. In order to deal with the complexity, a book is normally divided into logical chapters. Within each chapter, there may be a number of subheadings. These subheadings may be further nested. This nested approach takes a complex system and breaks it into simpler and more manageable parts. A model that can represent something like bone marrow has the following nested format.

Model {bone marrow}

  B-Cell Lineage {Type}

    CD19 Parameter Profile {Constant}

      Control Definition Points

      Means (r states)

      SD's (r states)

    CD34 Parameter Profile {One Step-Down}

    CD20 Parameter Profile {One Step-Up}

    CD10 Parameter Profile {Three Step-Down's}

    CD45 Parameter Profile {Three Step-Up's}

    …

    Common Progression Scale {0–100}

    Frequency {r states}

  T-Cell Lineages {Type}

  Monocytic Lineage {Type}

  Myeloid Lineages {Type}

  Erythroid Lineage {Type}

  …

Unfortunately, it is necessary to develop some syntactical terms to describe the various parts of this model design. The model chapters are the major lineages in bone marrow and are called "Types." For simplicity, only the B-Cell Lineage section is expanded above. Each lineage or Type is further divided into subsection's or "Parameter Profiles" that describe how particular measurements change as a function of lineage progression. These changes are well-known for normal bone marrow development (28–30).

The CD19 Parameter Profile, for example, is relatively constant in B-cell maturation; whereas, CD34 has an early step-down in intensity. A critical element in the model is the common progression axis or scale that relates all the Parameter Profiles to each other. In this particular model, progression starts at zero and ends at 100%, and is divided into r separate bins or states. The reason for this construction will become clearer when the model is used for analysis instead of simulation.

The shape of a Parameter Profile is controlled by a set of important inflection points called Control Definition Points (CDP). The CDP's are the real parameters for this model. A CDP contains a position along the progression axis, a relative measurement intensity value, and a standard deviation (SD). The SD defines the measurement uncertainty and heterogeneity at a particular progression value. Each Parameter Profile also had a set of r means and SD's for each state along the progression.

Finally, the model Type contains a frequency value for each state. We won't directly need these frequencies for simulation, but they will become important later during analysis.

2.3. Simulating Realistic Data

Now that the model is defined, simulating realistic data is straightforward. The first step is to randomly pick a Type from the model. In order to do this, the approximate percentage of each Type in the model needs to be specified. The Type needs to be randomly chosen in order that the data appear as realistic as possible. An example best describes how this stochastic selection is done. Suppose the percentages for B-cells (2.5%), T-cells (10.0%), Monocytes (25%), Myelocytes (50%), and Erythroids (12.5%) are stored in the following list,

{2.5, 10.0, 25.0, 50.0, 12.5}

A stochastic selection function (31) takes this list and randomly chooses one of the elements based on these weights. It will pick the first element approximately 2.5% of the time, the second element 10.0% of the time, the third element 25.0% of the time, and so on. A good analogy to this type of selection is "The Wheel of Fortune" game show where a wheel is spun by a contestant and then randomly stops at various points along the wheel. If the wheel segments are proportional to the above Type percentages, then a Type will be stochastically selected with each spin of the wheel.

Once a Type is stochastically selected, the system randomly picks one of the states along the selected Type's progression. This type of selection is uniform which means that each state has an equal chance of being chosen. Once the algorithm knows the state, it can examine each of the Parameter Profiles and estimate a measurement value and SD by interpolation between the appropriate CDP's. The last step is to use a normally distributed random number generator defined by the SD and add this uncertainty value to the interpolated measurement value. This final number is stored on disk in a listmode type of structure. This process is continued for the remainder of the Parameter Profiles and then repeated over and over again until the desired number of simulated events is achieved. The end result is the creation of a listmode file that looks very similar to those produced by cytometers.

2.4. Three Insights

There is a direct relationship between the simulator's model and synthesized data. The information contained in the model is reflected in the synthesized data. These seemingly obvious statements are at the heart of this new technology. A plot of all the model Type's Parameter Profile's vs. progression represents all measurement correlations that are in the synthesized data file. This fact means that if there is a way to reverse directions and find the model from the data, then we can represent all the correlations in the data without running into the dimensionality barrier.

Since a uniform random number generator picked the progression states within a Type, the progression axis represents cumulative fraction or probability. This type of axis allows subpopulation percentages to be directly read from the axis scale. This insight was very important since it precisely defined the concept of progression without the use of time. Progression in this context is defined probabilistically.

The third insight was that any combination of Parameter Profiles could be synthesized and the frequency of all the states would always be uniform. The probability of an event representing a state is equal for all states since that is how the generator was designed. In many ways, this last insight was the most important since it meant that it might be possible to simplify an analysis strategy by examining each measurement in succession rather than looking at all measurements concurrently.

2.5. Synthesizer to Classifier

What these three insights really meant was that if it were possible to create a model from observed data, then the problem of representing any number of correlated measurements in one graph was solved. The question at this point was whether it was feasible to use the model to accurately classify observed events into progression states. More specifically, was there a way of classifying all synthesized events such that all the frequencies of the chosen states were as uniform as possible? When a model is in this uniform frequency condition, it can synthesize data that is indistinguishable from the observed data and therefore can be conceived of as representing the data. This type of model is called a Probability State Model (PSM).

2.6. Choosing a State to Associate with an Event

How is an event classified into a specific state? The key to answering this question is to associate a probability for each state that represents the likelihood that the event belongs to that state. Once this list of probability weights is known for an event, a stochastic selection process picks the state to associate with the event. Once the state is known, the algorithm keeps track of how many events are in each state. If the model represents the data, then these frequencies will be uniform.

2.7. Non-uniform Frequency Gradient and the Minimization Searching Algorithm

At this point, our probability state model will have uniform state frequencies if it classifies data that it has synthesized. How does this relate to finding a new model with unknown data? Consider what happens if one or more of the CDP's are moved from the positions they were in when the data was synthesized. When the data is reclassified with this different model, the uniform nature of the state frequencies will be perturbed. The further the CDP's are moved from their original positions, the greater the perturbation.

If the degree of non-uniformity of the frequencies is quantified by something like reduced chi-square (32), standard minimization routines (33, 34) can find the optimal positions of the CDP's that minimize this non-uniformity. Thus, by finding the optimal positions of the CDP's that minimize the non-uniformity of the frequencies, a model can be determined from the data. As already stated, when the model represents the data, all the measurement correlations can be presented graphically.

2.8. The Big Surprise

The ability to show all measurement correlations is a very important step in fully understanding high-dimensional data, but there is more to the story. It turns out that if the data is partitioned by the positions of the CDP's, the resultant percentages account for population overlap due to measurement error (26). Thus, the overlay graphs that show state means versus progression also show percentages that account for population overlap.

2.9. Multiple Types

Since each Type generates a list of probability weights, the maximum weight from each vector can then be used in a stochastic selection routine to pick the Type. Once the Type is chosen, then another stochastic selection method picks a specific state. Thus, two cascading stochastic selection routines are used to pick a Type and state within that Type for every event or cell in a listmode file. This means that a single model can represent every lineage in a sample from normal bone marrow.

2.10. Overlay Graphics

As the system classifies events into specific Types and states, it keeps a running mean and standard deviation for each state. These means and standard deviations can then form bands for each measurement showing all measurement correlations and variances along the progression axis. Since the progression axis is in %events units, this type of format also shows all the percentages of contained subpopulations. As mentioned earlier, these percentages account for population overlap due to measurement error. This single easy-to-understand graph can represent the same information as hundreds of bivariates.

2.11. Modeling Strategy

In order to synthesize a listmode file representing complex populations found in bone marrow, every one of the Parameter Profiles needs to be defined. However, when reversing directions and modeling data, very little a priori knowledge is necessary for a solution. A good analogy is the crossword puzzle. When a crossword puzzle is created, the author must know the answer to everyone of the clues. When solving a crossword puzzle, the answer to the first clue needs to be known, but later in the solution, the extra information from previous answers enables intelligent guesses for the other clues. Thus, less and less information is necessary during the process of solving the puzzle. In fact, at a certain point, the final solution becomes apparent without the need of any of the remaining clues. Probability State Modeling works in the same way.

2.12. Simple to Complex, Known to Unknown

The strategy in modeling data is to define the simplest Parameter Profiles that are known and move toward more complex Parameter Profiles that are less known. In many cases, once a few of these profiles are defined, every one of the remaining will be obvious from the classified data.

A subtle, but important characteristic of this system is that each Parameter Profile will theoretically distribute the events equally across the states whether it is considered alone or in combination with others. Therefore, a solution to a very complex set of measurements can be approached one measurement at a time. It is not necessary to consider all the measurements at one time. This characteristic makes the analysis normally quite simple and straightforward.

2.13. PSM Results are Understandable to Everyone

Another important point is that the a priori knowledge necessary for the modeling process is identical to the knowledge necessary for interpretation of traditional gating analyses. The only difference between the two approaches is that information is used early when modeling and later for traditional gating interpretation. The advantage of using the knowledge early is that after the modeling is represented as an overlay graph, very little further interpretation is necessary. This attribute is important because it enables non-cytometrists to directly appreciate the results generated by cytometry.

2.14. Compensation is a Requirement for PSM

If the data are not properly compensated (35, 36) for signal cross-over, then the reasons behind coordinated changes in measurements can be ambiguous. Changes in measurement intensities due to important biological processes can be indistinguishable from simple signal cross-over. The general idea behind PSM is to better understand high-dimensional data. Lack of compensation tends to confuse rather than illuminate and therefore compensated data is a requirement for PSM.

3. PSM Applications

3.1. B-Cell Lineage

3.1.1. B-Cell Progression in Human Bone Marrow

Figure 1 shows a PSM overlay from two four-color listmode files derived from "uninvolved" human bone marrow that were provided by Michael Loken at Hematologics, Inc., Seattle, WA. The files' measurements are File 1: CD19 APC, CD34 PE, CD22 FITC, CD45 PerCP, SSC, and FSC; and File 2: CD19 APC, CD10 PE, CD20 FITC, CD45 PerCP, SSC, and FSC. The colors of the fluorescence Parameter Profiles are the emission colors of the above fluorochromes. FSC and SSC are light and dark gray respectively. The common measurements among the files are CD19, SSC, CD45, and FSC.

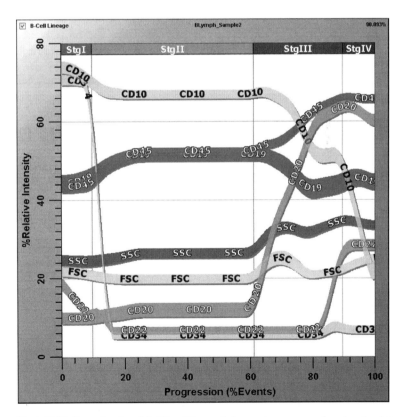

Fig. 1. Eight-dimensional model of B-cell lineage in human bone marrow from two 4-color listmode files. File 1 contained the measurements: CD19, CD34, CD22, CD45, SSC, and FSC. File 2 contained CD19, CD10, CD20, CD45, SSC, and FSC. Four stages of B-cell development are observed characterized by high coordinated changes in marker expression.

CD19 and SSC are used as simple selection measurements to classify all the B-cells in the marrow. For the first file, CD34, CD22, and CD45 stratify many of the B-cell events along the progression axis. For the second file, CD19, SSC, and CD45 select and stratify the events as in file 1, but then CD10 and CD20 further stratify the events to form four apparent stages.

The end of Stage 1 (red in electronic version) is characterized by a downregulation of CD34, CD10, and CD22 and an upregulation of CD45 and CD19. Arrows can be used to more compactly denote this kind of stage boundary definition: CD34↓↓, CD10↓, CD22↓, CD45↑, and CD19↑. A double arrow denotes a dramatic change in marker intensity. The underlined marker indicates that it was solely used for defining the boundary position. The end of Stage 2 (green in electronic version) is characterized by CD20↑↑, CD45↑, SSC↑, and FSC↑. The end of Stage III (blue in electronic version) and beginning of Stage IV (purple in electronic version) are defined by CD10↓↓, CD19↓, and CD22↑.

3.1.2. Common Measurement Scaffolds and High-dimensional Modeling

If common measurements modulate uniquely along a progression, they form a kind of scaffold that all other measurements can relate to whether they are in the same file or different files. With properly designed common measurements, multiple files can be analyzed to form a single model that is representative of the measurement correlations in all files. The CD19, SSC, and CD45 common measurements are good enough to create this scaffold and thus a composite model representing six colors is possible from the analysis of two four-color files (Fig. 1). As evidence of this integration, Fig. 2 shows a bivariate display of CD20 versus CD22; where each measurement is derived from a separate file. In this figure, the CD20 values come from raw measurement values and CD22 values are calculated from the model. The arrows in the figure show the model's predicted progression and the event dot colors match the stage colors.

The implication of this common measurement scaffold is that cytometry is no longer limited to the confines of detector or fluorochrome availability. High-dimensional models are possible from low-dimensional listmode files. The key, of course, is to choose correct common measurements to tie everything together. As discussed later (see Subheading 3.1.6), this modeling system can also help in the selection of the best common measurements.

3.1.3. High-Dimensional B-Cell Progression

Figure 3 shows an eleven-dimensional PSM analysis of "uninvolved" human bone marrow data from Brent Wood at University of Washington, Seattle, WA. The measurements and fluorochromes are kappa FITC, lambda PE, CD19 PE-Texas Red, CD34 PerCP-Cy5.5, CD20 PE-Cy7, CD45 Pacific Blue, CD38 A594, CD10 APC, CD5 APC-Cy7, SSC, and FSC. The colors of the fluorescence Parameter Profiles are the emission colors of the above fluorochromes. FSC and SSC are light and dark gray respectively.

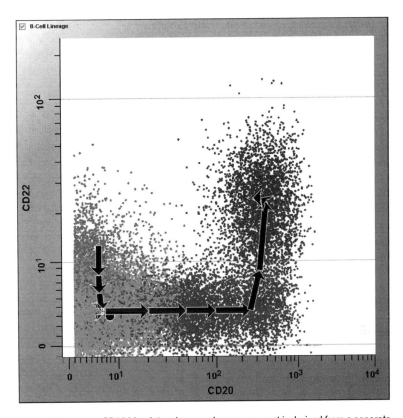

Fig. 2. CD20 versus CD22 bivariate where each measurement is derived from a separate file. CD20 values come from raw measurement values and CD22 values are calculated from the model. The arrows show the model's predicted progression and the event dot colors match the stage colors shown in Fig. 1 (in electronic version).

CD19 and SSC are again used as selection measurements. CD34 and CD38 are modeled as step-down Parameter Profiles and CD20, a step-up. Both CD45 and CD10 are three-level Parameter Profiles. The rest of the Parameter Profiles are not modeled and are distributed along the progression axis by the modeled measurements.

The variations of the additional markers in this file appear tightly coordinated with the B-cell stages defined earlier. Specifically, CD38 elevates slightly while CD34 downregulates and then decreases dramatically while CD10 downregulates. Kappa, lambda, and CD5 begin to be expressed when CD45 and CD20 are upregulated. CD5 is partially downregulated while CD10 and CD38 down regulate.

3.1.4. Importance of the Transitions

There are a number of comments to make concerning the above PSM analyses of B-cell data from bone marrow. The most important point is that when all the measurement correlations are represented in a PSM overlay, B-cell development appears to occur in distinct steps or transitions (Figs. 1 and 3). Each transition involves numerous synchronized changes in specific markers.

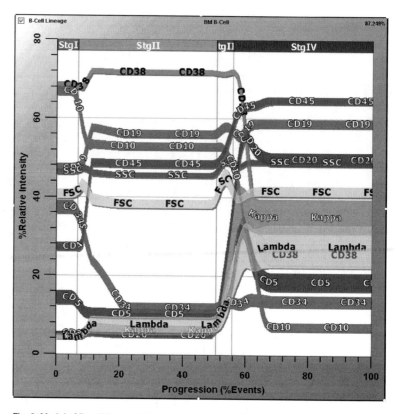

Fig. 3. Model of B-cell lineage in human bone marrow from eleven-dimensional listmode file. The file contains the measurements: kappa, lambda, CD19, CD34, CD20, CD45, CD38, CD10, CD5, SSC, and FSC. The variations of these markers appear to be tightly coordinated with the stages shown in Fig. 1. Specifically, CD38 elevates slightly while CD34 downregulates and then decreases dramatically, while CD10 downregulates. Kappa, lambda, and CD5 begin to be expressed when CD45 and CD20 are upregulated. CD5 is partially downregulated while CD10 and CD38 downregulate.

These observations are consistent with the hypothesis that B-cell ontogeny is controlled by master regulatory genes (37). The ability to clearly visualize these transitions and sort them may eventually give molecular biologists the tools they need to determine which master regulatory genes are involved in the different stages of B-cell ontogeny. Other lineages may also have similar control mechanisms (38–40).

3.1.5. The Ideal
B-Cell Panel

Identifying the important transitions in B-cell lineage progression also helps in the prediction of which markers are likely to be most important in staging B-cells. CD34 and TdT (not shown) are important in identifying the first transition (Stage I to II). CD10, CD45, and CD20 help identify all three transitions. In general, markers that have relatively small line-spreads and that dramatically change intensity during one or more of these transitions are ideal candidates for common measurements in a multi-tube B-cell panel.

3.1.6. Empirical Determination of the Best Antibody Panel

PSM provides another approach to the determination of an optimal set of common measurements. When an event is stochastically selected into a specific state along the model's progression axis, the program estimates the probability of the decision being correct. The average of this probability for all classified events can be expressed as a percent and labeled as %Fidelity Index (%FI). This percentage quantifies how well all measurements in the model work together to categorize events along the progression axis.

Since each modeled measurement can be including or excluding from this classification process, it is possible to examine all combinations of relevant measurements and rank them according to their %FI. Table 1 shows the result of such an analysis on the above eleven-dimensional bone marrow B-cell data.

Selection measurements such as CD19 and SSC are not part of this analysis since they don't stratify events along the progression axis. This analysis suggests that along with CD19, the markers CD34 and CD45 would be the best three-color choice to stage bone marrow B-cells (%FI = 4.17). The best four-color

Table 1
Ranking of Antibody Panels

Number non-constant markers	Markers	% Fidelity index
0		1.00
1	CD20	2.10
1	CD34	2.22
1	CD38	2.50
1	CD45	3.07
2	CD34 CD20	3.44
2	CD20 CD45	3.58
2	CD38 CD20	3.59
2	CD38 CD34	3.70
2	CD38 CD45	4.07
2	CD34 CD45	4.17
3	CD34 CD20 CD45	4.54
3	CD38 CD20 CD45	4.66
3	CD38 CD34 CD20	4.68
3	CD38 CD34 CD45	4.95
4	CD38 CD34 CD20 CD45	5.40

choice would be CD19, CD38, CD34, and CD45 (%FI = 4.95). These choices are only relevant to this particular sample. If a number of samples were analyzed in this manner, it would be possible to engineer an ideal panel of antibodies based on the objective %FI measurements.

3.2. Other Lineages

The advantages of PSM analyses of B-cell lineages are also applicable to other lineages or progressions in bone marrow, peripheral blood, and other tissue sites. The remaining portion of this chapter will show a sampling of these PSM analyses.

3.2.1. Myeloid Progression in Human Bone Marrow

Figure 4a, b show a seventeen-dimensional PSM analysis of "uninvolved" human bone marrow data from Brent Wood at University of Washington. The measurements and fluorochromes from the first twelve-dimensional file are CD15 FITC, CD33 PE,

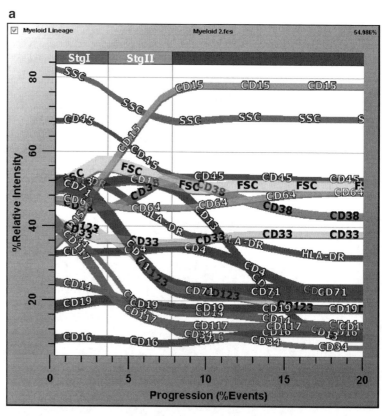

Fig. 4. Seventeen-dimensional model of human myeloid lineage in bone marrow from two twelve-dimensional listmode files. File 1 contains the measurements CD15, CD33, CD19, CD117, CD13, HLA-DR, CD38, CD34, CD71, CD45, FSC, and SSC. File 2 contains the measurements CD64, CD123, CD4, CD14, CD13, HLA-DR, CD38, CD34, CD16, CD45, FSC, and SSC. Myeloid lineage appears to also occur in distinct steps or transitions, suggesting a similar type of controlling mechanism as hypothesized for B-cells. (**a**) is zoomed in on the first 20% of the progression axis to better visualize early stages. (**b**) is the complete progression.

b

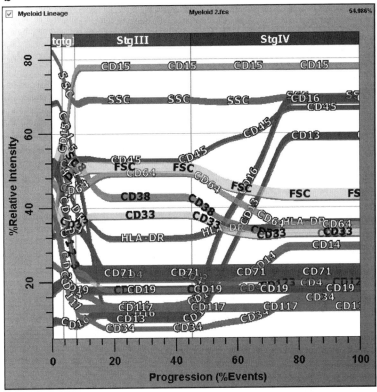

Fig. 4. (continued)

CD19 PE-TR, CD117 PE-Cy5, CD13 PE-Cy7, HLA-DR Pacific Blue, CD38 A594, CD34 APC, CD71 APC-A700, CD45 APC-Cy7, FSC, and SSC. The second twelve-dimensional file's markers are CD64 FITC, CD123 PE, CD4 PE-TR, CD14 PE-Cy5.5, CD13 PE-Cy7, HLA-DR Pacific Blue, CD38 A594, CD34 APC, CD16 APC-A700, and CD45 APC-Cy7, FSC, and SSC. The colors of the fluorescence Parameter Profiles are the emission colors of the above fluorochromes. FSC and SSC are light and dark gray respectively. The stage boundaries for this myeloid data are defined as

End of stage I (Fig. 4a): <u>HLA-DR</u>↓↓, CD34↓↓, and CD15↑↑

End of stage II (Fig. 4a): <u>CD13</u>↓↓, CD123↓↓, CD4↓, CD45↓, CD38↓, FSC↑, and SSC↑

End of stage III, start of stage IV (Fig. 4b): <u>CD16</u>↑↑, CD13↑↑, CD45↑, CD38↓, CD33↓, and CD14↑

The above patterns represent only neutrophil maturation and the relatively high intensity of HLA-DR is due to autofluorescence.

Figure 4a is zoomed in on the first 20% of the progression axis to better visualize early stages. Figure 4b is the complete progression. When all measurement correlations are viewable by PSM, the myeloid lineage also appears to occur in distinct steps or transitions, suggesting a similar type of controlling mechanism as hypothesized for B-cells.

3.2.2. CD4 T-Cell Antigen-Dependent Development in Human Peripheral Blood

Figure 5 shows a ten-dimensional PSM analysis of CD4 T-cell's antigen-dependent progression in normal human peripheral blood data from Margaret Inokuma at BD Biosciences, San Jose, CA. The measurements and fluorochromes are CD3 Pacific Blue, CD8 APC-Cy7, CD4 AmCyan, CD27 APC, CD28 PerCP-Cy5.5, CD57 FITC, CCR7 PE, CD45RA PE-Cy7, SSC, and FSC. The colors of the fluorescence Parameter Profiles are the emission colors of the above fluorochromes. FSC and SSC are light and dark gray respectively. The stage boundaries for this data are defined as

End of Naïve stage: <u>CD45RA↓↓</u>, CD28↑, CCR7↓, and CD3↓

End of T1 stage: <u>CCR7↓↓</u>, and CD45RA↓

End of T2 stage: <u>CD27↓↓</u>

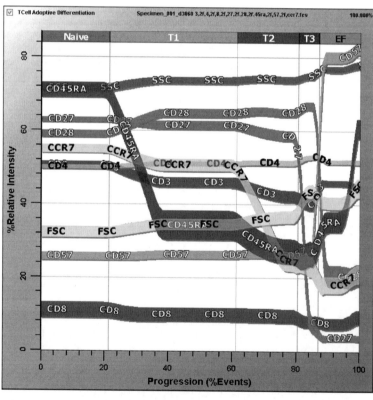

Fig. 5. Ten-dimensional model of CD4 T-cells antigen-dependent progression in human normal peripheral blood – smoothed. The listmode file contains the measurements: CD3, CD8, CD4, CD27, CD28, CD57, CCR7, CD45RA, SSC, and FSC. The stage boundaries for this data are defined as end of Naïve stage: CD45RA↓↓, CD28↑, CCR7↓, and CD3↓; end of T1 stage: CCR7↓↓, and CD45RA↓; end of T2 stage: CD27↓↓; and end of T3 stage, start of EF (effector) stage: CD28↓↓, CD57↑↑, and CD45RA↑↑.

End of T3 stage, start of EF (effector) stage: <u>CD28</u>↓↓, CD57↑↑, and CD45RA↑↑

The cascading nature of the measurement intensity changes in the antigen-dependent T-cell maturation suggests a different type of underlying mechanism of development than bone marrow B-cells and myeloid cells.

3.2.3. Erythroid Progression in Mouse Bone Marrow

Figure 6 shows a nine-dimensional PSM analysis of erythroid development in mouse bone marrow data from Kathleen McGrath at University of Rochester, Rochester, NY. The measurements and when appropriate, fluorochromes, from an Amnis Image Stream X listmode file are Area Cell, Area Nucleus, Mean Vybrant Violet (Mean VV), CD44 PE-Cy5.5, CD71 PE, c-Kit PE-Cy7, Ter119 APC, Mean Ter119 APC, and Thiazole Orange (TO).

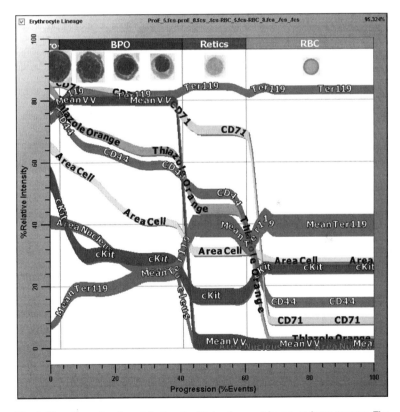

Fig. 6. Nine-dimensional model of erythroid development in mouse bone marrow. The listmode file contains the measurements: Area Cell, Area Nucleus, Mean Vybrant Violet (Mean VV), CD44, CD71, c-Kit, Ter119, Mean Ter119, and Thiazole Orange (TO). The morphology of the erythroid lineage stages are shown at the *top*. In the beginning of the erythroid progression, c-Kit is positive for the proerythroblasts. As progression continues, Ter119 increases slightly, while TO (RNA), Area Cell, Area Nucleus, and CD44 decrease. The point of nucleus extrusion is easily identified by a dramatic decrease in Mean VV (DNA) and Area Nucleus. At the end of the reticulocyte stage, TO, CD71, and CD44 decrease.

The colors of the fluorescence Parameter Profiles are the emission colors of the above fluorochromes. Area Cell and Area Nucleus are light and dark gray respectively. Mean VV detects double-stranded DNA in viable cells and TO detects RNA.

The basophilic, polychromatic, and orthochromatic stages of erythroid development were not clearly separated by these measurements and were grouped together as the BPO stage. The stage boundaries for this data are defined as

End of Proerythroblasts: <u>c-Kit</u>↓↓, TO↓, CD44↓, Area Cell↓, Area Nucleus↓, and Ter119↑

End of BPO: <u>Mean VV</u>↓↓, Area Nucleus↓↓, TO↓, CD71↓, CD44↓, and Area Cell↓

End of Reticulocytes, start of RBC: <u>TO</u>↓, CD71↓↓, and CD44↓

The morphology of the erythroid lineage stages are shown at the top of Fig. 6. In the beginning of the erythroid progression, c-Kit is positive for the proerythroblasts. As progression continues, Ter119 increases slightly, while TO (RNA), Area Cell, Area Nucleus, and CD44 decrease. The Mean Ter119 has been shown to have structural information in it that helps delineate the basophilic, polychromatic, and orthochromatic stages (41). The point of nucleus extrusion is easily identified by a dramatic decrease in Mean VV (DNA) and Area Nucleus. At the end of the reticulocyte stage TO, CD71, and CD44 decrease.

4. Summary

Cytometry has been seriously limited by the overwhelming complexity in viewing and analyzing high-dimensional data. The key to solving this problem was the design of a mathematical model system capable of representing complex measurement correlations normally found in living tissues. Later, it was discovered that stochastic selection coupled with a general minimization algorithm could be leveraged to construct and optimize these models to diverse high-dimensional listmode data. Since this method divides progressions into individual states and searches for a solution that makes these states equally probable for event selection, it was called Probability State Modeling or PSM. PSM breaks through the complexity barrier by creating a new progression probability-based variable that all measurements can relate to. Since all measurements relate to the same progression axis, a single graphical overlay with progression on the x-axis can represent all the correlations present in high-dimensional data.

When PSM was applied to complex bone marrow populations, these graphs revealed a rich tapestry of measurement changes secondary to underlying biochemical processes. In the case of bone marrow B-cells, the data shows only three transitions where measurement intensities change in a tightly correlated manner consistent with the hypothesis that B-cell ontogeny is controlled by master regulatory genes. A sampling of other lineages and progressions showed that the PSM approach is quite general and can be very revealing in deducing possible biological mechanisms underlying cellular ontogeny.

Enabling the visualization of high-dimensional data is only a small part of the advantages of PSM. Regions or zones defined along the progression axis automatically account for population overlap due to measurement error. A single representation of a sample can be constructed from numerous files if there are proper common measurements present in each file. PSM can also objectively quantify how well specific measurements might work as common measurements.

The major advantage, however, is that PSM forces scientists to create unambiguous models of biological systems. Model building and hypothesis testing are central to the scientific method. True understanding of these processes will ultimately lead to solutions for important problems that threaten our society.

References

1. Chattopadhyay, P. K., Price, D. A., Harper, T. F., Betts, M. R., Yu, J., Gostick, E., Perfetto, S. P., Goepfert, P., Koup, R. A., DeRosa, C., Bruchez, M. P., and Roederer, M. (2006) Quantum dot semiconductor nanocrystals for immunophenotyping by polychromatic flow cytometry. *Nat. Med.* **12**, 972.

2. Perfetto, S. P., Chattopadhyay, P. K., and Roederer, M. (2004) Unravelling the immune system. *Nat. Rev. Immunol.* **4**, 648–55.

3. Bandura, D. R., Baranov, V. I., Ornatsky, O. I., Antonov, A., Kinach, R., Lou, X., Pavlov, S., Voroviev, S., Dick, J. E., and Tanner, S. D. (2009) Mass cytometry: technique for real time single cell multitarget immunoassay based on inductively coupled plasma time-of-flight mass spectrometry. *Anal. Chem.* **81**, 6813–22.

4. Ornatsky, O. I., Lou, X., Nitz, M., Schafer, S., Sheldrick, W. S., Baranov, V. I., Bandura, D. R., and Tanner, S. D. (2008) Study of cell antigens and intracellular DNA by identification of element-containing labels and metallointercalators using inductively coupled plasma mass spectrometry. *Anal. Chem.* **80**, 2539–47.

5. Tanner, S. D., Bandura, D. R., Ornatsky, O., Baranov, V. I., Nitz, M., and Winnik, M. A. (2008) Flow cytometer with mass spectrometer detection for massively multiplexed single-cell biomarker assay. *Pure Appl. Chem.* **80**, No. 12, 2627–41.

6. Tan, P., Steinback, M. and Kumar, V. (2006) *Introduction to Data Mining*, Pearson Education, Boston, MA, pp. 51–63.

7. Baggerly, K. A. (2001) Probability binning and testing agreement between multivariate immunofluorescence histograms: extending the chi-squared test. *Cytometry* **45**, 141–50.

8. Roederer, M., Moore, W., Treister, A., Hardy, R. R., and Herzenberg, L. A. (2001) Probability binning comparison: a metric for quantifying multivariate distribution differences. *Cytometry* **45**, 47–55.

9. Rogers, W. T., Moser, A. R., Holyst, H. A., Bantly, A., Mohler II, E. R., Scangas, G., and Moore, J. S. (2008) Cytometric fingerprinting: quantitative characterization of multivariate distributions. *Cytometry* **73A**, 430–41.

10. Kosugi, Y., Sato, R., Genka, S., Shitara, N., and Takakura, K. (1988) An interactive multivariate analysis of FCM data. *Cytometry* **9**, 405–8.

11. Lugli, E., Pinti, M., Nasi, M., Troiano, L., Ferraresi, R., Mussi, C., Salvioloi, G., Patsekin, V., Robinson, J. P., Djurante, C., Cocchi, M., and Cossarizza, A. (2007) Subject classification obtained by cluster analysis and

principal component analysis applied to flow cytometric data. *Cytometry Part A* **71A**, 334–44.

12. Bagwell C. B., Horan P., and Lovett, E. (1985) A method for displaying multiparameter flow cytometric listmode data. International Conference Analytical Cytology XI, November, 17–22.

13. Leary, J. F., Ellis, S. P., McLaughlin, S. R., Corio, M. A., Hespelt, S., Gram, J. G., and Burde, S. (1991) High-resolution separation of rare-cell types, in *Cell Separation Science and Technology* (Kompala, P. and Todd, P. F., eds.) American Chemical Society Press, Washington, DC, Series No. **464**, pp. 26–40.

14. Murphy, R. (1985) Automated identification of subpopulations in flow cytometric list mode data using cluster analysis. *Cytometry* **6**, 302–9.

15. Wegman, E.J. and Luo, Q. (1997) High dimensional clustering using parallel coordinates and the grand tour. *Comput. Sci. Stat.* **28**, 352–60.

16. Preffer, I. F., Dombkowski, D., Sykes, M., Scadden, D., and Yang, Y.-G. (2002) Lineage-negative side-population (SP) cells with restricted hematopoietic capacity circulate in normal human adult blood: immunophenotypic and functional characterization. *Stem Cells* **20**, 417–27.

17. Preffer, F. and Dombkowski, D. Advances in complex multiparameter flow cytometry technology: applications in stem cell research (2009) *Cytometry Part B* **76B**, 295–314.

18. Dean, P. (1990) Data processing, in *Flow Cytometry and Sorting* (Melamed, M.R., Lindmo, T., and Mendelsohn, M.I., eds.), Wiley-Liss, Hoboken, NJ, pp. 438–40.

19. Crowell J. M., Hiebert, R. D., Salzman, G. B., Price, M. J., Cram, L. S, and Mullaney, P. F. (1978) A light-scattering system for high-speed cell analysis. *IEEE Trans. Biomed. Eng.* **BME-25**, 519–26.

20. Finn, W. G., Carter, K. M., Raich, R., Stoolman, L., and Hero, A. O. (2009) Analysis of clinical flow cytometric immunophenotyping data by clustering on statistical manifolds: treating flow cytometry data as high-dimensional objects. *Cytometry Part B* **76B**, 1–7.

21. Chan, C., Feng, F., Ottinger, J., Foster, D., West, M., and Kepler, T. B. (2008) Statistical mixture modeling for cell subtype identification in flow cytometry. *Cytometry Part A* **73A**, 693–701.

22. Irish, J. M., Hovland, R., Krutzik, P. O., Perez, O. D., Bruserud, O., Gjertsen, B. T., and Nolan, G. P. (2004) Single cell profiling of potentiated phosphor-protein networks in cancer cells. *Cell* **118**, 217–28.

23. Boedigheimer, M. J. and Ferbas, J. (2008) Mixture modeling approach to flow cytometry data. *Cytometry Part A* **73A**, 421–29.

24. Bagwell, C. B. (2007) Probability State Models. Utility Aplication No. US11/897,148, 19 Sep.

25. Bagwell, C. B. (2008) Breaking the Dimensionality Barrier, in *Laboratory Hematology Practice* (Kottke-Marchant, K. and Davis, B.H., eds.), Wiley-Blackwell, Hoboken, NJ, **Ch 12**.

26. Bagwell, C. B. (2009) Probability State Modeling – a new paradigm for cytometric analysis, in *Flow Cytometry In Drug Discovery and Development* (Litwin, V. and Marder, P., eds.), John Wiley and Sons, Inc., Hoboken, NJ, **Ch 15**.

27. Shapiro, H. M. (2003) *Practical Flow Cytometry*, 4th edition, Wiley-Liss, Hoboken, NJ, pp. 465–7.

28. Loken, M. R., Shah, V. O., Dattilio, K. L., and Civin, C. I. (1987) Flow cytometric analysis of human bone marrow. II. Normal B lymphocyte development. *Blood* **70**, 1316–24.

29. Loken M. R. and Wells, D. A. (2000) Normal antigen expression in hematopoiesis, in *Immenophenotyping* (Stewart, C. C. and Nicholson, J. K., eds.), Wiley-Liss, Hoboken, NJ, pp. 138–142.

30. Wood, B. (2004) Multicolor immunophenotyping: human immune system hematopoiesis. *Methods Cell Biol* **75**, 559–76.

31. Gentle, J. E. (2003) Transformations of uniform deviates: general methods, in *Random Number Generation and Monte Carlo Methods*, 2nd edition, Springer Science + Businesss Media, LLC, New York, NY, pp. 101–9.

32. Bevington, P. R. (1969) Data reduction and error analysis for the physical sciences. McGraw-Hill Book Company, New York, NY, p. 89.

33. Bevington, P. R. (1969) Data reduction and error analysis for the physical sciences. McGraw-Hill Book Company, New York, NY, p. 245.

34. Press, W. H., Vetterling, W. T., Teukolsky, S. A., and Flannery, B. P. (1992) Numerical recipes in C, 2nd edition, Cambridge University Press, New York, NY, pp. 408–12.

35. Bagwell, C. B. and Adams, E. G. (1993) Fluorescence spectral overlap compensation for any number of flow cytometry parameters. *Ann NY Acad Sci* **677**, 167–84.

36. Roederer, M. (2001) Spectral compensation for flow cytometry: visualization artifacts, limitations, and caveats. *Cytometry* **45**, 194–205.

37. Lawrence, H. J., Savageau, G., Largman, C., and Humphries, R. K. (2001) Homeobox gene networks and the regulation of hematopoiesis, in Hematopoiesis : A developmental approach (Zon, L. I., ed.), Oxford University Press, New York, NY, pp. 404–5.

38. Argiropoulos, B. and Humphries, R. K. (2007) Hox genes in hematopoiesis and leukemogenesis. *Oncogene* **26**, 6766–76.

39. Kim, S. I. and Bresnick, E. H. (2007) Transcriptional control of erythropoiesis: emerging mechanisms and principles. *Oncogene* **26**, 6777–94.

40. Pronk, C. J., Rossi, D. J., Mansson, R., Attema, J. L., Norddahl, G. L., Chan, C. K. et al. (2007) Elucidation of the phenotypic, functional, and molecular topography of a myeloerythroid progenitor cell hierarchy. *Cell Stem Cell* **1**, 428–42.

41. McGrath, K. E., Bushnell, T. P., and Palis, J. (2008) Multispectral imaging of hematopoietic cells: where flow meets morphology. *J. Immunol. Methods* **336**, 91–7.

Chapter 3

Quantitative Fluorescence Measurements with Multicolor Flow Cytometry

Lili Wang, Adolfas K. Gaigalas, and Ming Yan

Abstract

Multicolor flow cytometer assays are routinely used in clinical laboratories for immunophenotyping, monitoring disease and treatment, and determining prognostic factors. However, existing methods for quantitative measurements have not yet produced satisfactory results. This chapter details a procedure for quantifying surface and intracellular protein biomarkers by calibrating the output of a multicolor flow cytometer in units of antibodies bound per cell (ABC). The procedure includes the following critical steps (a) quality control (QC) of the multicolor flow cytometer, (b) fluorescence calibration using hard dyed microspheres assigned with fluorescence intensity values in equivalent number of reference fluorophores (ERF), (c) compensation for correction of fluorescence spillover, and (d) application of a biological reference standard for translating the ERF scale to the ABC scale. The chapter also points out current efforts for implementing quantification of biomarkers in a manner which is independent of instrument platforms and reagent differences.

Key words: Multicolor flow cytometry, Fluorescence calibration, Equivalent number of reference fluorophores, CD4+ lymphocytes, Antibody bound per cell, CS&T microspheres, Instrument quality control, Compensation

1. Introduction

1.1. Background

Multicolor flow cytometers are used to monitor the level of expression of multiple cell receptors that are significant in disease diagnostics and immunotherapies. The complexity of the immune response necessitates the monitoring of as many cell receptors as practical. However, to determine the levels of expression of cell receptors requires quantitative measurements, which at present are not very satisfactory. The purpose of this chapter is to detail procedures which can lead to quantitative multicolor flow cytometer measurements. These quantitative measurements rely heavily on the availability of fluorescence standards to calibrate the flow cytometer.

Teresa S. Hawley and Robert G. Hawley (eds.), *Flow Cytometry Protocols*, Methods in Molecular Biology, vol. 699, DOI 10.1007/978-1-61737-950-5_3, © Springer Science+Business Media, LLC 2011

In the past, quantitative measurements with single color flow cytometers were performed using microspheres with assigned units of MESF (molecules of equivalent soluble fluorophore) to calibrate the fluorescence signal. Reference standards were developed for assignment of MESF values to microspheres with surface labeled FITC. The use of these microspheres was described in the Clinical and Laboratory Standards Institute (CLSI) guideline for fluorescence calibration and quantitative measurements (1). However, the quantitation methodology developed for single color cytometers is not easily extended to multicolor flow cytometers. It is impractical to produce different standard reference fluorophore solutions, such as National Institute of Standards and Technology (NIST) fluorescein Standard Reference Material (SRM) 1932 (2), for every fluorophore label used in multicolor flow cytometry. An alternative approach to quantitative measurements with multicolor flow cytometers has been described (3). This approach involves two major steps and provides a scheme for converting the detected fluorescence signals in various fluorescence channels of a multicolor flow cytometer into numbers of antibodies bound per cell (ABC). The ABC numbers are good indicators of the actual number of different receptors on the cell surface. In the following, we describe the two major steps involved in the quantitation scheme.

1.2. Methodology for Quantitative Measurements

1. In the first step, a unit of fluorescence intensity is assigned to a given population of hard dyed microspheres. The assignment is based on the equality of the fluorescence signals from the microsphere suspension and a solution of reference fluorophores. This fluorescence unit is defined as the equivalent number of reference fluorophores (ERF) that gives the same fluorescence signal as one microsphere. The ERF unit is different from MESF in that the fluorophores embedded in the microspheres and the fluorophores in the reference solution can be very different and may have very different molar absorptivities. Consequently, the ERF unit applies only to a specific excitation-detection scheme associated with a fluorescence channel (FC) in a multicolor flow cytometer. However, the ERF unit assignments can be performed using a fluorimeter that mimics the response of the flow cytometer.

   The microspheres embedded with multiple fluorophores display a broad emission profile to cover all FCs of multicolor flow cytometers. These microspheres were traditionally used to monitor the daily performance of the flow instruments. However, the microspheres with ERF assignments can also be used to define a linear scale for each FC. The scale is implemented by analyzing 5–6 different microsphere populations, each implanted with different amount of fluorophores and each assigned an ERF value. (We emphasize again that the ERF values are assigned for a specified laser excitation and a specified range of wavelength in the fluorescence detection channel.) For a given FC, the fluorescence signals associated with the different microsphere populations can be plotted versus the ERF values assigned to the different microsphere populations leading to the calibration curves shown in Fig. 1a. Such a calibration curve is obtained for each of the FCs yielding an estimate of linearity of response, dynamic range, and detection sensitivity. The detection sensitivity, Q, can be defined as photoelectrons per ERF molecule, and is an important measure of instrument sensitivity for multicolor flow cytometers. The availability of a fluorescence unit, such as ERF, allows the measurement and use of relative Q values for multiple channels of a multicolor flow cytometer.

2. In the second step, a biological standard such as a lymphocyte with a known number of antibody binding sites (e.g. CD4 binding sites) is used to translate the linear ERF scale to an ABC scale. Figure 1b shows the steps taken in the preparation of the biological standard. The upper part of Fig. 1b shows five containers each with a population of standard cells (dotted line) at the bottom of each container. The antibody specific to the receptor is divided into several lots and each lot is labeled with one of the fluorophores that will be detected in each FC. The symbols Y in the upper part of Fig. 1b denote the antibody and the subscripts denote the label (FITC, PE, APC, etc.) corresponding to the FC. Figure 1b shows the case of a flow cytometer with five FCs; however, the number of FC can vary with the application. The antibodies with different labels are placed in different containers holding the biological standard cells. The mixture of the biological cell standard and the antibodies is incubated. After the incubation, the cells stained with labeled antibodies are washed, concentrated, and pooled. The single container at the bottom of Fig. 1b shows the final biological standard which consists of the pooled antibody stained lymphocytes. Passing the labeled biological standard through the flow cytometer leads to a response in each of the FC. The dashed arrows in Fig. 1a show how the calibration line together with the response from the biological standard in that FC leads to the establishment of a scale for ABC (right hand axis in Fig. 1a). Subsequent to the calibration, flow cytometer measurements on the analyte cells can be reported in terms of ABC values.

The fluorescence spectra of most label fluorophores cover a wide range of wavelengths. Consequently a given label fluorophore may give a large fluorescence signal in the FC assigned to that fluorophore and smaller fluorescence signals in FCs assigned to other label fluorophores. Clearly the fluorescence signal in the FC not assigned to a given label fluorophore is a bias and should

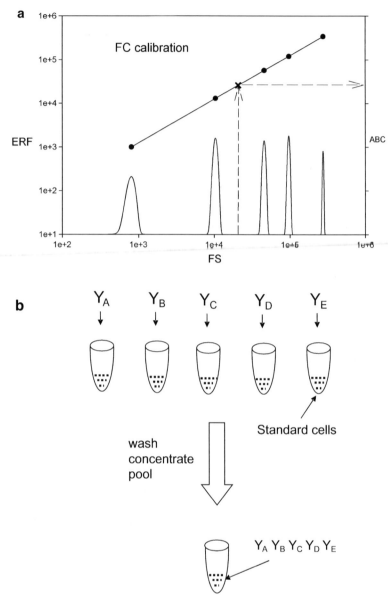

Fig. 1. (**a**) The *solid circles* represent a plot of the hypothetical numbers of equivalent reference fluorophores (ERF) assigned to the five populations of Ultra Rainbow microspheres as a function of the mean pulse height (denoted as FS, which stands for fluorescence signal) associated with the five simulated peaks in the fluorescence channel (FC). The *solid line* is a best linear fit to the five points and constitutes a calibration of the FC. (**b**) A schematic of the process used to produce a biological standard. Standard cells are incubated with the same antibody (Y) labeled with different fluorophores (A, B, C, D, and E). After incubation, the labeled cells are washed, concentrated, and pooled. The pooled cells constitute the biological sample. The *vertical dashed line* in (**a**) is drawn from the mean pulse height of the response associated with the biological standard. The point (x), where the *dashed line* crosses the *calibration line*, defines the ERF value, which corresponds to the number of labeled antibodies on the biological standard. This point (as well as the zero point defined by a negative population) sets the antibodies bound per cell (ABC) scale on the *right side* of (**a**).

be corrected. The correction procedure is called compensation and can be implemented using measurements of the individually labeled biological standard cells (4). Using the present example, five individual measurements would be carried out on each population of cells stained with an antibody against a highly expressed receptor (e.g. CD45 or CD8) which is labeled with a specific fluorophore (Fig. 1b). Appropriate mathematical computation, so-called software compensation, on the data collected for each population would provide the necessary correction factors. Note that cells labeled with antibody against CD4 are not optimal for compensation correction. Compared to CD8 or CD45, the expression levels of CD4 are relatively weak so that spillover estimates are hampered by low signal levels.

2. Materials

2.1. Staining Fresh Whole Blood

1. Freshly drawn human whole blood.
2. Phosphate buffered saline (1× PBS).
3. Lysing solution: 1× FACS™ Lysing Solution (BD Biosciences, San Jose, CA) or ACK (ammonium chloride) lysing solution.
4. Fluorescently labeled anti-CD4 antibodies covering every FC of a multicolor flow cytometer.
5. Fluorescently labeled anti-CD8 (or anti-CD45) antibodies covering every FC of a multicolor flow cytometer.

2.2. Quality Control of Flow Cytometers

1. BD™ Cytometer Setup and Tracking (CS&T) microspheres (BD Biosciences).
2. Disposable 12×75-mm BD Falcon™ capped polystyrene tubes (BD Biosciences) or equivalent.
3. BD FACSFlow solution with surfactant (BD Biosciences).
4. Flow cytometer.

2.3. Quantitative Fluorescence Measurements and Data Analysis

1. Ultra Rainbow microspheres (Spherotech Inc., Lake Forest, IL) and/or CS&T microspheres.

3. Methods

In the following we will assign ERF values to Rainbow microspheres and use the microspheres to establish a linear scale for the fluorescence response. Fresh whole blood samples will be utilized to outline the procedure for converting the ERF scale to the ABC scale, which is used in reporting quantitative flow cytometry measurements. The procedure is similar to that reported using Ultra Rainbow microspheres as described in (3). The procedures should be applicable for flow cytometers operated with 375, 405, 488, and 632 nm laser lines commonly used in most flow cytometers, and appropriate dichroic mirrors and band pass filters to define the FCs. In addition to the Rainbow microspheres, we will also use CS&T microspheres as another example for assuring instrument performance in terms of linearity, detection efficiency and optical background (as described in Subheading 1.2, step 1), and converting the linear scale to a biologically relevant scale (as described in Subheading 1.2, step 2). The procedure for the use of CS&T microspheres is different in detail, but very similar in conceptual basis to that outlined for Rainbow microspheres.

3.1. Staining Fresh Whole Blood

1. Wash heparinized normal donor blood samples (6–8 mL) twice with 1× PBS in 50-mL centrifugal tubes. After centrifugation at ~450×g for 10 min, plasma portion of the blood should be removed by aspiration. Replenish the blood volume with 1× PBS to the original blood volume.

2. Aliquot 100 µL of washed whole blood into individual tubes. Incubate the whole blood in each tube with differently labeled antibodies designated for each FC of a multicolor flow cytometer for 30 min at room temperature: one set of tubes with anti-CD4 antibody and another set of tubes with anti-CD8 (or anti-CD45) antibody. Protect from light during incubation. Users can either adopt the amount of antibody recommended by a manufacturer (e.g. 20 µL of anti-CD4 antibody from BD Biosciences per 100 µL of whole blood) or perform their own antibody titration curve for the determination of an optimal amount of each antibody used. Start titrations at 3 µg per mL of antibody, and do six 1.5-fold dilutions. Choose the lowest concentration that gives nearly maximal fluorescence.

3. Lyse the cell suspensions for 10 min with 2 mL of a lysing solution. After centrifugation at ~450×g for 10 min, remove the supernatant.

4. Wash once more with 1× PBS. After centrifugation at ~450×g for 10 min, remove the supernatant.

5. Add a small volume of 1× PBS (100 µL) in each tube and combine a half volume of differently stained whole blood cells in different tubes into a single tube to make a final sample volume of no more than 1 mL with 1× PBS.

6. Acquire samples immediately or store tubes at 4°C and acquire within 2 h.

3.2. Quality Control of Flow Cytometers

CS&T microspheres are designed for use with BD FACSDiva™ 6.0 software to provide automated cytometer characterization and performance tracking of supported BD™ digital flow cytometers. These microspheres are uniquely manufactured to be used with up to 21 different fluorescent parameters for cell analyzers as well as cell sorters. They can also be used with other (not BD) cytometers if the analysis, described below, is implemented using the software resident on the cytometer. The CS&T microspheres consist of three hard dyed fluorescence microsphere populations. The fluorescence intensity of the bright microspheres is close to the stained cells such as CD4+ stained with various fluorophores. The bright microsphere of CS&T microspheres is used to set the target median fluorescence intensity (MFI) value for cytometer tracking and quality control. The coefficient of variation (CV) of the bright microsphere population is small enough and is used for the assessment of laser alignment to the sample core stream of the flow cell in cytometer. The mid and dim microsphere intensities are designed to measure cytometer performance such as the photon detecting efficiency and the optical background. The dim microspheres mimic the unstained, negative cell intensity. These hard dyed microspheres are stable in time and with temperature. These characteristics make the microspheres ideal for cytometer performance setup and tracking.

Diluted CS&T microspheres are run on the flow cytometer. The MFI and robust CV (rCV) (5) are measured for each microsphere population in all fluorescence detectors. Algorithms within the software differentiate the fluorescence signal from each microsphere type based on the size (scattering) and fluorescence intensity in each detector. The software then uses this data to calculate and report a variety of measurement parameters. The parameters include an estimate of the linear response range, the standard deviation of the electronic noise, and cytometer settings adjusted for maximizing population resolution in each detector. These measurements provide instrument properties as described in Subheading 1.2, step 1. Each lot of bright CS&T microspheres is assigned a fluorescence intensity value in BD internal tracked fluorescence unit referred to as Assigned BD unit (ABD). The ABD value is associated with and is very roughly equivalent to freshly stained CD4+ lymphocytes, which is assumed to yield a cytometer response corresponding to a ABD value of 40,000. The essential function of the ABD unit is identical to that of ABC unit described in Subheading 1.2, step 2.

1. Prepare the CS&T microspheres suspension immediately before use. Add three drops of CS&T microspheres into a 12×75-mm tube, with BD FACSFlow solution with surfactant or $1 \times$ PBS. Vortex the microsphere suspension thoroughly.

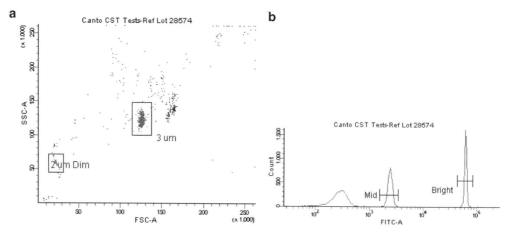

Fig. 2. (a) Gates are drawn on the population of CS&T microspheres with high scatter intensities (3 μm) and the population of CS&T microspheres with low scatter intensities (2 μm; this is the Dim microsphere population) on the forward scatter channel (FSC) vs side scatter channel (SSC) dot plot. (b) Two sub-gates of the 3 μm gate in (a) are drawn on the Bright and Mid microsphere populations for the fluorescein isothiocyanate (FITC) FC.

2. Run CS&T microspheres in suspension on a flow cytometer. Perform cytometer setup to bring the bright microsphere population close to the maximum of MFI axis and within the linear range of each FC. Record 20,000 total events.

3. Draw a gate on the population showing high scatter intensities on the forward scatter channel (FSC) versus side scatter channel (SSC) dot plot shown as the 3 μm gate in Fig. 2a. Draw two subgates on the bright and mid microsphere populations on a histogram of, as an example, the fluorescein isothiocyanate (FITC) channel shown in Fig. 2b. This procedure can be performed on any cytometer.

4. On the same FSC versus SSC dot plot, draw another gate on the population with low scatter intensities shown as the 2 μm gate in Fig. 2a. This is the dim microsphere population.

5. The detection efficiency, Q_r, is estimated by the slope of rSD_{ph}^2 vs median of mid and dim microspheres. The rSD_{ph} is the robust standard deviation from photon statistics, which is equal to the rSD corrected for the intrinsic microsphere variation. It can be calculated as $rSD_{ph} = rCV_{ph} \times (\text{median}) / 100$ for mid and dim microspheres, where $rCV_{ph} = (rCV^2 - rCV_{bright}^2)^{1/2}$.

6. Calculate the slope, k, and intercept, m, of the rSD_{ph}^2 versus median channel value plot shown in Fig. 3. Note that rSD_{ph}^2 is used to represent the intensity of the fluorescent photons. This is valid because the SD of the pulse heights from a detector is proportionate to the square root of the photon intensity impinging on the detector. The procedure described above is referred to as the SD^2 method.

Channel	FITC							
Bead type	Median	%rCV	rSD_ph^2	ABD	slope	intercept	Qr	Br
Bright	33007	2.1		132340	8.897	1467.7	0.028	661
Mid	1281	9.1	12865					
Dim	152	35	2820					

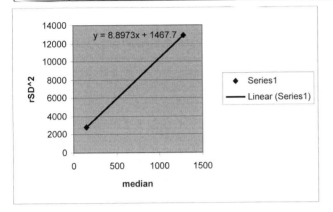

Fig. 3. The measurement parameters and derived parameters are summarized in an Excel table using CS&T microspheres for quality control of multicolor flow cytometers. The bottom plot of rSD² versus the median of Mid and Dim microspheres is used to obtain the slope and intercept of the linear fit, k and m, respectively.

7. Compute the Q_r and B_r using the following equations:

$$Q_r = 1/k \times \text{median}_{\text{bright}}/\text{ABD}, \quad B_r = m \times Q_r \times (\text{ABD}/\text{median}_{\text{bright}})^2$$

The analysis table is given as an excel table at the top of Fig. 3. The parameters, Q_r and B_r, are calculated in terms of ABD unit. It is assumed that a calibration by the bright microspheres in terms of ABD units has been performed.

The method described above provides an estimate of Q_r and B_r value. It is assumed that the rCV of the bright microsphere provides a correction for the intrinsic CV of fluorescence microsphere. The actual method involves more than the CV of the bright microsphere. In BD Biosciences' CS&T software, the actual microsphere intrinsic CV differences are taken into account (see Note 1). The intrinsic CV difference and ABD value of bright microsphere are calibrated by BD Biosciences and recorded in microsphere lot specific file.

3.3. Quantitative Fluorescence Measurements and Data Analysis

In principle, quantitative multicolor flow cytometer measurements can be carried out using BD Biosciences' approach of the ABD units described in Subheading 3.2. However, this approach contains assumptions about qualities of various antibodies and labeling fluorophores (see Note 2) and is optimally applicable to BD Biosciences' instrument platforms. The usefulness of ERF units for the calibration of fluorescence intensity measurements using multicolor flow cytometers is currently being undertaken by the fluorescence calibration task force of International Society for Advancement of Cytometry (ISAC) Standards Committee. Through this exercise, CS&T microspheres would be assigned with ERF values enhancing their utility in different, commercially available flow instrument platforms. The procedure described below follows that in (3). Pertinent steps are summarized here.

1. Add one to three drops of both blank and fluorescent calibration microspheres (Ultra Rainbow microspheres described in (3) or CS&T microspheres with assigned ERF values [in the future]) in 0.5 mL of PBS. Acquire 20,000 events within a most populated microsphere gate on the FSC versus SSC dot plot. Make sure that the brightest fluorescent microsphere population lies within the quantifiable cytometer scale. For analysis, use different gates in each FC histogram to obtain the five median (or geometric mean) channel values for the calibration microspheres.

2. Construct a calibration curve of the calibration microspheres, median channel value (x-axis) vs ERF value (y-axis). The ERF values for five microsphere populations are provided in Table 2 of (3). The linear curve fitting results in a linear fitting equation, $Y_{ERF} = a \cdot X_{median\ channel} + b$, for every FC of the multicolor flow cytometer, where parameters, a and b, refer to slope and intercept of the linear fit in a given FC, respectively.

3. FC compensation is carried out by using an unstained control and individual samples of cells stained with different fluorochrome-labeled anti-CD8 (or anti-CD45). It is assumed that appropriate software is available to perform the compensation correction.

4. Run the fluorescently anti-CD4 stained whole blood samples and obtain the respective fluorescence histogram in every FC. Apply a double gating strategy: lymphocyte gate in the FSC vs SSC dot plot, and CD4+ gate on the fluorescence histogram. Obtain the median channel value of CD4+ lymphocyte population for every fluorescence channel. The currently accepted ABC value for freshly stained CD4+ lymphocytes is about 48,000 (6, 7) (see Notes 2–4).

5. Run an unknown blood sample and obtain its median channel value. Determine the ABC value of the receptor of interest in the unknown sample by the following equation, $ABC_{unknown} = 48,000 + a \cdot (X_{unknown} - X_{CD4+})$, where a is the slope of the linear fit described in step 2, and $X_{unknown}$ and X_{CD4+} are the median channel values of the unknown and CD4+, respectively.

4. Notes

1. Q_r and B_r are calculated using actual microsphere intrinsic CV differences and extensions of the original approaches (8). The Q_r and B_r determined using the three fluorescence microsphere set can be expressed by the following equations:

$$Q_r = \frac{\text{Median}_{mid} \times \left(1 - \frac{\text{Median}_{mid}}{\text{Median}_{Bright}}\right) - \text{Median}_{Dim} \times \left(1 - \frac{\text{Median}_{Dim}}{\text{Median}_{Bright}}\right)}{\text{Median}_{mid}^2 \times \left(\text{CV}_{Mid}^2 - \text{CV}_{Bright}^2 - \Delta\text{CV}_{mid-bead}^2\right) - \text{Median}_{Dim}^2 \times \left(\text{CV}_{Dim}^2 - \text{CV}_{Bright}^2\right)} \times \frac{\text{Median}_{Bright}}{\text{ABD}_{Bright}}$$

$$B_r = \left(Q_r \times \left(\text{Median}_{Dim}^2 \times \left(\text{CV}_{Dim}^2 - \text{CV}_{Bright}^2\right) - \text{SD}_{el}^2\right) \times \frac{\text{ABD}_{bright}}{\text{Median}_{bright}} - \text{Median}_{Dim} \times \left(1 - \frac{\text{Median}_{Dim}}{\text{Median}_{Bright}}\right)\right) \times \frac{\text{ABD}_{Bright}}{\text{Median}_{Bright}}$$

where, $\Delta\text{CV}_{mid-bead}^2$ is intrinsic microsphere CV square difference and ABD_{Bright} is ABD value for the bright microsphere. These values depend on the excitation laser and emission filter. For common filters used in flow cytometer, the value of a particular CS&T microsphere lot can be found in the CS&T microsphere file provided by BD Biosciences along with the CS&T microspheres. The microsphere lot file can be opened by Microsoft Excel with comma delimited input.

2. It is recommended that a single clone of the antibody amenable to labeling with different types of fluorophores associated with various fluorescence channels is used for the scale conversion. Assuming that similarly labeled, but different antibodies against different antigens have the same average fluorescence per antibody bound, yields a direct measure of ABC. The assumption of equivalent fluorescence of bound antibodies is a significant one that needs to be verified and should include an assessment of uncertainties. A basic factor to consider is whether the effective number of fluorophores per antibody (effective F/P) is the same for the calibration antibody and test antibodies. The ideal situation would use antibody conjugates that consisted of only one fluorophore coupled to each antibody in a location that did not interfere with the ability of the antibody to bind to antigen. Frequently a fluorochrome conjugated antibody will have lower affinity than unconjugated antibody, so both the calibration antibody and any test antibodies should be purified to exclude unconjugated antibody. It is also possible that two different conjugates of the same antibody reagent conjugated with the same average number of fluorophores could give different degrees of cell staining if the distribution of F/P is different.

3. Poncelet and co-workers reported in 1991 that CD4+ T cells from HIV-infected individuals bound consistently about

46,000 CD4 Mab molecules measured according to the calibration curve generated with cell lines expressing known amounts of CD5 molecules detected via radiolabeled CD5 Mab (6). A similar CD4 expression level on fresh normal whole blood (~48,000) was measured by Davis et al. using three different methods including the most sensitive method that utilized unimolar Leu 3a-PE conjugate and Quantibrite PE microspheres (7). Nonetheless, measurements of CD4 expression using flow cytometry depend on variables such as fixation conditions, antibody clones, fluorophore and conjugation chemistries, and quantitation methods used (ref. (9) and references therein; also see Note 2). An ultimate proteomic approach is currently under investigation for quantifying the number of CD4 receptors per CD4+ T lymphocyte (10, 11) using recombinant soluble CD4 proteins as reference standards.

4. A lyophilized human blood reference cell standard is presently under development for quantification of CD4+ cell numeration and CD4+ receptor expression. This lyophilized cell standard is produced at the National Institute for Biological Standards and Control (NIBSC, UK), a WHO designated blood institute. The CD4+ count of the reference material will be measured according to the secondary reference measurement procedure of the International Council for Standardization in Hematology (ICSH). The CD4+ expression level will be measured by flow cytometry and ultimately by proteomic approaches carried out at National Institute of Standards and Technology (NIST, USA). The lyophilized reference cell standard will resolve an issue associated with the accessibility of fresh blood samples for most research laboratories.

Acknowledgments

We are indebted to Dr. Gerald Marti and Fatima Abbasi at Center for Biologics Evaluation and Research, FDA and Robert Hoffman at BD biosciences for their collaborative effort on quantitative multicolor flow cytometer measurements over the years.

Disclaimer

Certain commercial equipment, instruments, and materials are identified in this paper to specify adequately the experimental procedure. In no case does such identification imply recommendation or endorsement by the National Institute of Standards and Technology, nor does it imply that the materials or equipment are necessarily the best available for the purpose.

References

1. Fluorescence calibration and quantitative measurements of fluorescence intensity, NCCLS (I/LA24-A) (2004).

2. Certificate of analysis, standard reference material 1932, fluorescein solution. National Institute of Standards and Technology (2007) http://ts.nist.gov/measurementservices/referencematerials/index.cfm.

3. Wang, L., Gaigalas, A. K., Marti, G. E., Abbasi, F., and Hoffman, R.A. (2008) Toward quantitative fluorescence measurements with multicolor flow cytometry. *Cytometry Part A* **73A**, 279–288.

4. Roederer, M. (2002) Compensation in flow cytometry, in *Current Protocols in Cytometry.* (Robinson, J. P., ed.), John Wiley & Sons, Inc., Hoboken, NJ, pp. 1.14.1–1.14.20.

5. Huber, P. J. (1981) *Robust Statistics.* John Wiley & Sons, Inc., Hoboken, NJ. (For application of robust statistics in BD FACSDiva™ 6.0 Software, see BD's technical note in the following link, http://www.bdbiosciences.ca/cgi-bin/literature/view?part_num=23-9609-00.)

6. Poncelet, P., Poinas, G., Corbeau, P., Devaux, C., Tubiana, N., Muloko, N., Tamalet, C., Chermann, J. C., Kourilsky, F., and Sampol, J. (1991) Surface CD4 density remains constant on lymphocytes of HIV-infected patients in the progression of disease. *Research in Immunology* **142**, 291–298.

7. Davis, K. A., Abrams, B., Iyer, S. B., Hoffman, R. A., and Bishop, J. E. (1998) Determination of CD4 antigen density on cells: Role of antibody valency, avidity, clones, and conjugation. *Cytometry* **33**, 197–205.

8. Hoffman, R. A. and Wood, J. C. S. (2007) Characterization of flow cytometer instrument sensitivity, in *Current Protocols in Cytometry.* (Robinson, J. P., ed.), John Wiley & Sons, Inc., Hoboken, NJ, pp. 1.20.1–1.20.18.

9. Wang, L., Abbasi, F., Gaigalas, A. K., Hoffman, R. A., Flagler, D., and Marti, G. E. (2007) Quantifying CD4 expression on T lymphocytes using fluorescein conjugates in comparison with unimolar CD4-phycoerythrin conjugates. *Cytometry Part B, Clinical Cytometry* **72B**, 442–449.

10. Tao, W. A., Wollscheid, B., O'Brien, R., Eng, J. K., Li, X. J., Bodenmiller, B., Watts, J. D., Hood, L., and Aebersold, R. (2005) Quantitative phosphoproteome analysis using a dendrimer conjugation chemistry and tandem mass spectrometry. *Nature Methods* **2**, 591–598.

11. Brembilla, N. C., Cohen-Salmon, I., Weber, J., Rüegg, C., Quadroni, M., Harshman, K., and Doucey, M. A. (2009) Profiling of T-cell receptor signaling complex assembly in human CD4 T-lymphocytes using RP protein arrays. *Proteomics* **9**, 299–309.

Chapter 4

Quantum Dots for Quantitative Flow Cytometry

Tione Buranda, Yang Wu, and Larry A. Sklar

Abstract

In flow cytometry, the quantitation of fluorophore-tagged ligands and receptors on cells or at particulate surfaces is achieved by the use of standard beads of known calibration. To the best of our knowledge, only those calibration beads based on fluorescein, EGFP, phycoerythyrin and allophycocyanine are readily available from commercial sources. Because fluorophore-based standards are specific to the selected fluorophore tag, their applicability is limited to the spectral region of resonance. Since quantum dots can be photo-excited over a continuous and broad spectral range governed by their size, it is possible to match the spectral range and width (absorbance and emission) of a wide range of fluorophores with appropriate quantum dots. Accordingly, quantitation of site coverage of the target fluorophores can be readily achieved using quantum dots whose emission spectra overlaps with the target fluorophore.

This chapter focuses on the relevant spectroscopic concepts and molecular assembly of quantum dot fluorescence calibration beads. We first examine the measurement and applicability of spectroscopic parameters, ε, ϕ, and $\%T$ to fluorescence calibration standards, where ε is the absorption coefficient of the fluorophore, ϕ is the quantum yield of the fluorophore, and $\%T$ is the percent fraction of emitted light that is transmitted by the bandpass filter at the detector PMT. The modular construction of beads decorated with discrete quantities of quantum dots with defined spectroscopic parameters is presented in the context of a generalizable approach to calibrated measurements of fluorescence in flow cytometry.

Key words: Quantum dots, Fluorescence calibration beads, Flow cytometry, Extinction coefficient, Spectroscopy

1. Introduction

1.1. General Considerations

The use of standard calibration beads to define the quantity of fluorophore-tagged ligands or proteins on cells or beads represents an essential element of quantitative flow cytometry that has enabled the direct comparison of inter-laboratory data as well as quality control in clinical flow cytometry (1–4). Commercial standards, such as Quantum™ FITC MESF beads (http://www.bang-slabs.com), comprise a mixture of five distinct populations of

Teresa S. Hawley and Robert G. Hawley (eds.), *Flow Cytometry Protocols*, Methods in Molecular Biology, vol. 699,
DOI 10.1007/978-1-61737-950-5_4, © Springer Science+Business Media, LLC 2011

beads, where the subsets are functionalized with discrete titers of fluorescein conjugates. The equivalence of fluorescence radiance of each bead population in the mix is correlated to the fluorescence of soluble fluorescein. Thus, the surface occupancy of fluorescein on the bead standards is known as the Molecules of Equivalent Soluble Fluorophores (MESF), where the MESF value is equal to the known number of molecules in solution (3, 4). We have recently described a simple method of producing fluorescence calibration beads that bear discrete and defined quantities of quantum dots and demonstrated their applicability as standard calibration beads (5, 6) useful for multiple color calibration. The measurement model is based on a simple ratiometric formula, Eq. 1, for comparing the number of fluorophores of Samples 1 and 2. Sample 1 is an analyte capable of emitting radiation of intensity I_1 that is equivalent to a single unit of Sample 2; Sample 2 is a calibration standard emitting radiation of intensity I_2. For simplicity, both samples must be excited at the same wavelength and emit in the same spectral region as defined by the same band-pass (BP) filter. Equation 1 is derived from spectroscopic parameters: extinction coefficient measured at the excitation wavelength (ε_λ), quantum yield (ϕ), and the BP transmittance ($\%T$) of sample emission. It enables one to determine the number density (ρ_i) of sample fluorophores.

$$\frac{I_1}{I_2} = \frac{\rho_1}{\rho_2} = \frac{\varepsilon_{\lambda_{ex},1}}{\varepsilon_{\lambda_{ex},2}} \cdot \frac{\phi_1}{\phi_2} \cdot \frac{\%T_1}{\%T_2} \tag{1}$$

In Eq. 1, ε can be measured from absorption spectra for most probes. The quantum yields ϕ_i can be determined by the end-user relative to known textbook standards (7, 8).

The photophysical properties of qdots are not clearly defined like the intrinsic physical constants of molecular systems. Quantum dots are semiconductor nanoparticles with tunable optical properties that are strongly dependent on size (9). The spectroscopic properties of dots are dependent on their size dispersion, shape, and surface defects (10–23). Subtle differences in preparation can lead to batch-to-batch variation in basic spectroscopic properties, such as quantum yield (10–23). To the best of our knowledge, the commercial dots are not spectroscopically standardized nor are they well characterized in the product literature for them to be used "as made" (10). In this chapter, we refer to the previous characterization (6) of a pair of CdSe quantum dots with emission bands centered at 525 nm (QD525 purchased from Invitrogen, Carlsbad, CA; lot# 1005-0045) and 585 nm (QD585, Invitrogen lot# 0905-0031). The lot numbers are cited to emphasize the notion that the spectroscopic characteristics of these dots are strictly relevant to samples from the indicated lots (6). Our previous study of these samples showed an anomalous dependence of emission yield on excitation wavelength that was more pronounced in QD525 dots than in the QD585 quantum dots. This was attributed to size heterogeneity and surface defects in the QD525, consistent with characteristics previously described in the literature (10, 22, 23). It is therefore important to characterize new batches of dots before they can be used as calibration standards. Two lots of subsequently purchased QD585 dots (Invitrogen lot# 4526A and 49987A) were shown to have quantum yield values of 0.12 and 0.3, respectively.

1.2. Molecular Assembly of Qdot Fluorescence Calibration Beads

Qdots can be photo-excited at any wavelength that can be matched to a fluorophore of choice. Therefore, calibration standards can be assembled for any fluorophore by using qdots whose emission spectra overlap with the target fluorophore. The approach is based on the display of variable and discrete site densities of streptavidin-functionalized qdots on biotin functionalized beads. Streptavidin-functionalized quantum dots are decorated by up to eight streptavidin tetrameric units per dot. It is therefore difficult to control the valency with which each dot binds to biotin-functionalized bead. The problem of unregulated multivalency has been shown to significantly limit the effort to consistently produce discrete molecular assemblies of qdots and beads at each titer of qdots (5). We have previously shown that an intervening layer of site directed biotin-labeled antibodies attached to beads circumvents the problem of multivalency by limiting the capture of a single dot to a single antibody in a reproducible and theoretically predictable manner (see Fig. 1) (5). The molecular assembly on beads has been well characterized by us in terms of dissociation constants at each junction module (24–29). We have shown that titration of streptavidin-functionalized qdot to antibody bearing beads (Fig. 1b) can yield discrete quantities of bound qdots in a manner consistent with the weakest equilibrium binding constant between the qdots and the antibody (25, 28).

1.3. Principles and Practice: Applicability of Qdot Calibration Beads to Quantitative Flow Cytometry

Although calibration standards, such as Quantum™ FITC MESF beads, are strictly applicable to fluorescein-based probes, it is useful to present a simple correction factor that extends their applicability to other probes operating in the same spectral range. These include most fluorescein conjugates that tend to have lower quantum yields than the parent molecule where simple derivitization of fluorescein typically attenuates the yield by at least 18% (e.g., fluorescein biotin (30)) or greater (e.g., 80% for FITC antibody conjugates (f/p ratio 4:1)). The correction factor (cf) that is derived from Eq. 1 is shown in Eq. 2.

$$\text{cf} = \frac{(\varepsilon_{488}\phi\%T)_{\text{sample}}}{(\varepsilon_{488}\phi\%T)_{\text{standard}}} \tag{2}$$

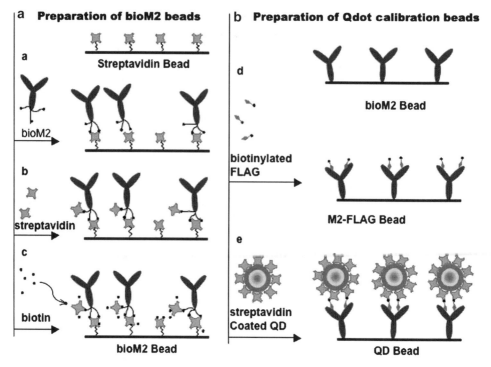

Fig. 1. Summary of the molecular assembly of quantum dot fluorescence calibration beads on streptavidin-coated beads. (a) Addition of biotinylated M2 anti-FLAG antibodies (bioM2) to 6.7 μm beads. (b) Elimination of free biotin sites with soluble streptavidin prevents direct binding of qdots. (c) Biotin is used to block streptavidin-biotinylated FLAG peptide interactions. (d) Addition and capture of biotinylated FLAG peptides by bioM2 antibodies. (e) Addition and capture of streptavidin qdots with biotinylated FLAG peptides tethered to bioM2.

Correction factors for several dyes have been calculated in Table 1. In this way, Quantum FITC MESF or QD525 Quantum dot beads are shown to be comparable calibration standards for GFP, Alexa 488, BODIPY 505FL dyes, etc. (5). By extension this approach is applicable to all other calibration schemes, as long as standards and samples use the same detectors and excitation source (5). For the readers' conveniences, correction factors for octadecylrhodamine B (R18, a commonly used lipophilic fluorescent dye) as well as rhodamine B to two different batches of QD585 are also listed in Table 1.

2. Materials

2.1. Determination of the Relative Emission Yields of Quantum Dots

1. Streptavidin coated quantum dots QD525 and QD585, fluorescein, and rhodamine B (Invitrogen).
2. Quartz cuvettes for absorption and fluorescence measurements.

Table 1
Some correction factors for "fluorescein-like" green fluorophores

Probe	$\varepsilon(\lambda_{ex, 488 nm})^a$	ϕ^b	% T^c	cf(MESF)^d	cf(QD525)^e	cf(QD585a/ QD585b)^f
Fluorescein (MESF beads)	78,000	0.93	28	1	1.37	–
Fluorescein biotin	72,600	0.74	28	0.75	1.02	–
FLAG^{FITC}	72,600	0.15	28	0.16	0.21	–
BODIPY fl	44,000	0.9	20	0.40	0.53	–
EGFP	56,000	0.6	40	0.60	0.91	–
EGF^{FITC}	72,600	0.45	28	0.46	0.61	–
QD525	130,000	0.30	38	0.73	1	–
QD585a (Lot 4526A)	530,000	0.12	65	–	–	1/0.4
QD585b (Lot 49987A)	530,000	0.3	65	–	–	2.5/1
R18	4,750	0.3	51	–	–	0.017/0.007
Rhodamine B	13,000	0.31	51	–	–	0.046/0.019

^{a}Extinction coefficient of probe at excitation wavelength
^{b}Quantum yield measured relative to fluorescein
^{c}%T is the percent fraction of emitted light that is transmitted by the bandpass (BP) filter at detector (see Fig. 5). The BP filters are those on a standard model BD FACScan flow cytometer. The maximum transmittance of 80% by the BP filter is assumed for all samples. Spectral mismatch between fluorescein and derivatives is assumed to be negligible in this table

^{d}Correction factor cf $= \dfrac{\left(\varepsilon_{488}\phi\%T\right)_{sample}}{\left(\varepsilon_{488}\phi\%T\right)_{standard}}$ needed to normalize data to spectral characteristics of standard beads

(e.g., cf×MESF value)
^{e}Correction factor needed to normalize data to spectral characteristics of QD525 (Invitrogen lot# 1005-0045)
^{f}Correction factors needed to normalize data to spectral characteristics of QD585 from two different batches (Invitrogen lot# 4526A and 49987A). These latter batches were purchased after QD585 lot# 0905-0031 was used up

3. Probes are solubilized in any of the working buffers listed below, in dilute solutions where the optical density is kept below 0.1 at the highest absorption wavelength (λ_{max}).

4. Phosphate-buffered saline (PBS) (Mediatech, Inc, Herndon, VA).

5. Tris buffer: 10 or 25 mM Tris–HCl, 150 mM NaCl, pH 7.5.

6. HHB buffer: 30 mM HEPES, 110 mM NaCl, 10 mM KCl, 1 mM MgCl$_2$, 10 mM glucose, pH 7.4.

***2.2. Preparation
of Qdot-Labeled
Microbeads***

1. Streptavidin-coated polystyrene particles (6.7 μm in diameter, 0.5% w/v) (Spherotech Inc., Libertyville, IL).

2. Streptavidin-coated quantum dots QD525 and QD585 (Invitrogen).

3. FLAG peptide (DYKDDDDK), bioM2 anti-FLAG antibody (Sigma, St. Louis, MO). *BioM2 antibody contains significant quantities of free biotin and requires purification prior to use* (see Note 1).

4. Biotinylated FLAG (FLAG^bio) and FITC-conjugated FLAG peptides (FLAG^FITC) synthesized at University of New Mexico (25).

5. PBS, Tris, and HHB buffers as described in Subheading 2.1.

6. Tris and HHB buffers supplemented with 0.1% bovine serum albumin (BSA).

7. Soluble streptavidin; soluble biotin.

2.3. Equipment

1. Spectrophotometer such as Hitachi model U-3270 (Hitachi, San Jose, CA).

2. Spectrofluorometer such as QuantaMaster™ Model QM-4/2005 (Photon Technology International, Inc., Lawrenceville, NJ).

3. Flow cytometer such as BD FACScan (BD Biosciences, San Jose, CA).

3. Methods

***3.1. Determination
of the Relative
Emission Yields
of Quantum Dots***

This protocol describes the measurement of relative emission quantum yields of quantum dots. The quantum yield of QD525 is measured relative to fluorescein, and QD585 is measured relative to rhodamine B. These dots are selected because their spectral characteristics are most suitable for analysis on a standard BD FACScan. Several factors, including size heterogeneity and potential impurities, have been shown to cause anomalous dependence of the quantum yield on the excitation wavelength when dots are excited at different wavelengths, in apparent violation of Kasha's rule (31): that the energy and yield of the emission should be independent of the excitation wavelength. It is therefore useful to collect a photoexcitation spectrum of any new batch of quantum dots to determine characteristic emission behavior over the putative spectral range of analysis. The protocol described in this chapter is applicable to measurements made using instruments comparable to the ones utilized here.

1. Measure absorption spectra of the probes following standard procedures (see Fig. 2a). The molar absorptivity of all probe

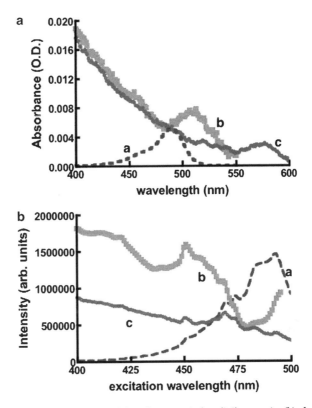

Fig. 2. Comparison of absorbance (a), and uncorrected excitation spectra (b) of quantum dots and fluorescein: (a) Fluorescein, (b) QD525 lot# 1005-0045, and (c) QD585 lot# 0905-0031. The samples' absorbances were matched at 488 nm by design (Figure 2a is reprinted from (6) with permission from Elsevier).

solutions should be at or below 0.1. In this limit, errors due to inner filter effects alone are on the order of 10% (7).

2. The slitwidths of the excitation and emission monochromators must both be set at 2 nm. Narrow slitwidths allow for better resolution of size dependent emission of poly disperse sized dots.

3. Use appropriate wavelength cutoff and BP filters for stray light rejection (5).

4. It is useful to make measurements of complete emission spectra of probes by off-resonance excitation (i.e., away from absorption maxima): at 450 nm for the green probes (fluorescein and QD525) and 540 nm for the yellow probes (Rhodamine B and QD585). Make sure that the molar absorptivity of the qdots and standard sample are matched at the excitation wavelengths.

5. Use Eq. 3 to calculate the relative quantum yields (ϕ_s) of QD525 and QD585 using fluorescein and rhodamine B as standards.

$$\phi_s = \phi_{ref} \frac{I_s}{I_{ref}} \cdot \frac{OD_{ref}}{OD_s} \cdot \frac{n_s^2}{n_{ref}^2} \tag{3}$$

Use the integrated intensity of the emission spectra of sample relative to fluorescein ($\phi_{ref} = 0.93$) or rhodamine B (0.31 in water (32)). I_s and I_{ref} are the integrated band intensities. The optical densities (OD) of the sample (s) and reference (ref) are taken at the excitation wavelength. n is the index of refraction of the solvent; 1.32 for water. For qdots, ϕ_s is accurate for the given lot number and excitation wavelength; therefore, lot number and source should always be identified (6).

6. Measure the emission intensity of all probes exciting at 488 nm (the primary excitation wavelength in a BD FACScan flow cytometer) and calculate relative quantum yields using Eq. 3.

7. To collect photoexcitation spectra, set the wavelength for measuring the emission intensity at the peak (e.g., 520 nm for fluorescein) or along the red edge of the emission spectrum (e.g., 540 nm for fluorescein). Collection at the emission peak allows for greater sensitivity while collection along the red edge allows for a broader spectral window.

8. The excitation spectra of fluorescein and rhodamine B are expected to match their corresponding absorption spectra, provided the photoexcitation spectra have been corrected for the spectral distribution (L_λ) of the lamp (7). L_λ depends on the characteristics of the lamp and monochromators. Most modern instruments automatically perform this correction provided it is enabled and properly set up. In cases where a spectrofluorometer instrument lacks the capability to automatically perform a correction of L_λ of excitation spectra, this can be done manually (see Note 2). This is an important consideration as significant distortions can be observed in the uncorrected photoexcitation spectrum (see Figs. 2b and 3).

9. Photoexcitation spectra of qdots measured relative to standards are useful for determining the quantum yield per unit wavelength of qdots over a desired spectral window (see Fig. 4).

10. Measure %T. The spectra of BP filters can be obtained from manufacturers' Web sites. Overlay the measured emission spectra with the appropriate BP filter and the calculated fraction of the emission band that overlaps with the BP filter is %T (see Fig. 5).

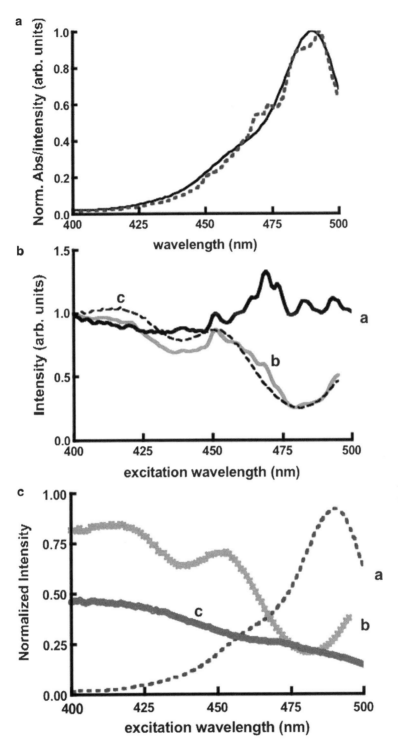

Fig. 3. Correcting excitation spectra for the lamp's spectral distribution (L_λ). (**a**) Overlay of fluorescein's uncorrected excitation spectra with the absorption spectrum. (**b**) Intensity versus wavelength plots of: (a) the spectral distribution of the Xenon arc lamp used for sample excitation. The lamp profile was derived from (**a**), by dividing the L_λ-uncorrected emission spectrum of fluorescein by its absorption spectrum; (b) L_λ-uncorrected excitation spectrum of QD525; (c) L_λ-QD525 excitation corrected spectrum. (**c**) Overlay of L_λ-corrected spectra of (a) fluorescein, (b) QD525, and (c) QD585 (Figure 3c is reprinted from (6) with permission from Elsevier).

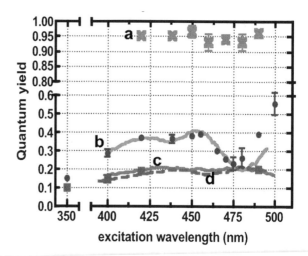

Fig. 4. Plot of quantum yields of (a) fluorescein, (b) QD525 lot# 1005-0045, (c) QD585 lot# 0905-0031, and (d) QD605 versus excitation wavelength. The quantum yields were calculated as described in Eq. 3. The lines are data derived from excitation spectra while data points represent data derived from integrated intensity (Reprinted from (6) with permission from Elsevier).

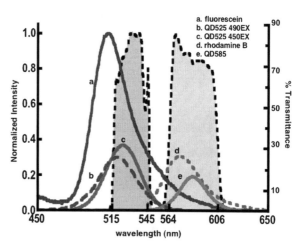

Fig. 5. Normalized emission spectra of fluorescein, rhodamine B and quantum dots. Intensity maxima of QD525 lot# 1005-0045 and QD585 lot# 0905-0031 correspond to their quantum yields relative to fluorescein and rhodamine B (32). *Dark* and *light* (*green* and *orange* in the electronic version) bars represent BP filters used in a standard flow cytometer (530/30 BP for the FL1 channel and 585/42 BP for the FL2 channel). The resonance overlap between the BP filters and emission spectra regulates the amount of light that is transmitted or rejected by the BP filter: 28% of fluorescein emission is transmitted to the FL1 channel compared to 38% of QD525, and 65% of QD585 emission is transmitted to the FL2 channel (Reprinted from (6) with permission from Elsevier).

**3.2. General
Preparation
of Qdot-Labeled
Microbeads**

*3.2.1. Preparation of M2
Beads*

The following protocol is summarized as a schematic in Fig. 1. Conceptual details can be found in the original literature (5, 25). Much of the work involves the assembly of bioM2 antibodies on beads. The M2 antibodies are critical for achieving an *effective* monovalent (one antibody:one qdot) binding interaction. This is an empirical finding from centrifugation assays in which it has been shown that the number of antibody-captured qdots is equal to the site density of antibodies on the same beads (5, 25). Based on earlier studies (5, 25, 28, 33, 34), this chapter makes the following assumptions about the stock suspensions of 6.7 μm Spherotech beads: (1) there are typically 40,000 beads/μL; (2) there are ≈4 million saturable bioM2 binding sites on these beads as determined from centrifugation binding assays of a fluorescently labeled FLAG peptide (5, 25). Nevertheless, the reader should always verify particle counts by using a hemocytometer or particle counter.

1. Use a pipette to extract the desired volume of beads from a vigorously shaken stock suspension. Centrifuge beads at $14,000 \times g$ and resuspend in desired buffer (Tris or HHB) containing 0.1% BSA in a minimum volume of 100 μL. To minimize nonspecific binding, the beads must be vortexed gently for 30 min to allow 0.1% BSA to coat beads.

2. Step a in Fig. 1a. Add bioM2 antibody at the desired stoichiometry: ≥2 bioM2:1 total streptavidin sites on all beads (see Note 1). Vortex or shake for an hour at room temperature.

3. Centrifuge and wash beads twice to remove free bioM2. The two footed binding assures that the bioM2 is tightly bound to beads.

4. Step b in Fig. 1a. Commercially available biotinylated M2 antibodies have about three to four biotin conjugates on the *Fc* domain of the antibody according to manufacturer's data sheet. It is likely that two of the biotin groups are cis-bound to the streptavidin on the beads. *To prevent the extra biotins from subsequent binding to streptavidin-functionalized quantum dots*, add a stoichiometric excess of soluble streptavidin to M2 beads, and allow to vortex mildly at room temperature for 15 min. Wash the streptavidin-saturated M2 beads once in buffer containing 0.1% BSA and centrifuge at $14,000 \times g$ for 2 min.

5. Step c in Fig. 1a. After the removal of the supernatant in step b, add 80 μL of 1 mM soluble biotin to the bead pellet, and spin down immediately after brief vortexing. Wash five times by repeated centrifugation and removal of supernatant, and finally resuspend in 100 μL of buffer. *Soluble biotin is added in order to saturate excess biotin-binding sites on the bound streptavidin* to prevent undesired binding of biotinylated FLAG peptide used to tether qdots to bioM2 (see steps d and e in Fig. 1).

6. Depending on the stock of bioM2 beads, aliquot samples into 25 μL volumes (or 1×10^6 beads) for long-term storage (see Note 3).

3.2.2. Preparation of Quantum Dot Calibration Beads

1. Step d in Fig. 1b. An aliquot of 1×10^6 bioM2 beads is first saturated with 100 nM biotinylated FLAG peptide (FLAG^{bio}) at room temperature for 1 h under mild vortexing, and washed two times in Tris–BSA buffer by repeated centrifugation and removal of supernatant. Resuspend in 100 μL of buffer. *It is important to note, that washes of the beads does not confer significant loss, by dissociation, of the peptide as one might expect.* In cells or beads, where the density of receptors is relatively high (e.g., $>1 \times 10^5$ receptors/cell), the binding and dissociation of ligands is believed to occur in the "collisional limit" regime (35). In this limit, the probability of dissociated ligand rebinding to an adjacent receptor is much higher in terms of normal bimolecular mass action kinetics, where the ligand would normally diffuse away. Therefore, the eight million peptide binding sites on these beads are expected to show anomalous low loss of ligand (5, 34).

2. Step e in Fig. 1b. Add 100 nM streptavidin-coated Qdots (final concentration) to the beads, and incubate at room temperature for an hour. The qdots bind to the biotin end of the FLAG peptide. Wash the beads, resuspend them in buffer, and analyze them at the flow cytometer. The attachment of qdots and beads is governed by the bivalent FLAG/M2 interaction $K_d = 2$ nM.

3.3. Discrete Labeling of Quantum Dots Calibration Beads from Mass Action Considerations

This protocol describes the application of mass action to the preparation of beads of known surface coverage. The quantitative display of streptavidin-coated Qdots on beads is regulated by the weakest affinity constant in the assembly step e in Fig. 1b. The quantity of bound qdots can be determined from a simple mass action model as shown in Eq. 4a where Q or $[Q]_0$, and Ab or $[Ab]_0$ represent the initial concentrations of qdots and bioM2 sites on beads, respectively. $[Q]_{free}$ is unbound qdots at equilibrium. v is the volume of buffer in which beads are suspended, n is the number of beads, and A is Avogadro's number.

$$Q + Ab \underset{k_{off}}{\overset{k_{on}}{\rightleftharpoons}} Q \cdot Ab \qquad (4a)$$

$$[Q \cdot Ab] = \frac{[Q]_{free}[Ab]_{free}}{K_D} Q \cdot Ab \qquad (4b)$$

$$qdots/bead = \frac{[Q \cdot Ab]vA}{n} \qquad (4c)$$

In keeping with standard practice of generating fluorescence calibration beads, it is desirable to decorate four distinct populations of beads with surface concentrations of quantum dots that span a wide range of surface occupancy. For convenience, here we describe the synthesis of a four-component set of fluorescence calibration bead populations decorated with discrete qdots surface densities ranging from 6×10^4 to ≈ 2 million.

Using 1×10^6 bioM2 beads in 100 µL volumes yields 66.0 nM Ab. Use Eq. 4b to establish the desired surface densities. The reader might find it useful to realize that, based on mass action considerations, the limiting reagent $[Q]_0$ is depleted by Ab, and in the concentration range of interest, $[Q \cdot Ab] \approx [Q]_0$. The reader can show that the error in this assumption is within 10% when $[Q]_0 \leq 0.7[Ab]_0$. To establish a good distribution in surface occupancy of qdots, one might consider to space $[Q]_0$ titer intervals at half log, i.e., $[Q]_0 = 1$, 3, 10, 30 nM to yield $[Q \cdot Ab] = 0.97$, 2.9, 9.6, 28.0 nM corresponding to 5.8×10^4, 1.75×10^5, 5.84×10^5, 8.76×10^5, 1.75×10^6 qdot/bead.

1. Add 25 µL of bioM2 beads (1×10^6 beads in total) to 25 µL of 20 µM FLAG[bio] and 50 µL of Tris buffer, mildly vortex the mixture at room temperature for 1 h.

2. Wash the beads twice in Tris–BSA buffer by centrifugation and resuspension (at max. speed, $14,500 \times g$).

3. Resuspend the beads in *freshly* made (see Note 3) qdot solution in Tris–BSA buffer, mildly vortex at room temperature for at least an hour (see Note 3).

4. Centrifuge the QD-M2 calibration beads at max speed for 90 s and remove 60 µL of the supernatant. Check beads on flow cytometer (see Fig. 6).

5. Discrete histograms of narrow distribution are a good indicator of a well-regulated assembly (see Note 3). Plot mean channel fluorescence (MCF) versus Qdot/Bead. A linear plot ($R^2 = 0.99$) is a good indicator of a well prepared set of calibration standards.

6. Store the beads at 4°C and use for up to 1 month (see Note 3).

3.4. Standardization of Flow Cytometry Data with Fluorescence Calibration Beads: Correction for Spectral Mismatch, Quantum Yield Between Sample and Standard Calibration Beads

It is useful to demonstrate the applicability of qdot calibration beads relative to Quantum™ FITC MESF calibration beads. This is achieved through Eq. 2, which establishes a correction factor that is used to account for the differences in the measurement parameters, ε, ϕ, and %T, between any given probe and the calibration standard. Correction factors for green wavelength fluorophores commonly used in our work are shown in Table 1.

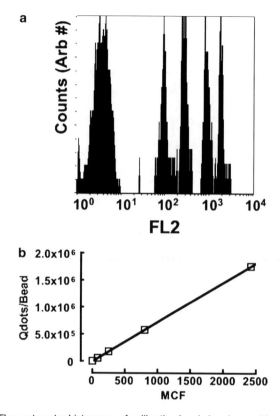

Fig. 6. (**a**) Flow cytometry histograms of calibration beads bearing a wide range of site densities of QD585 as a result of mixing one million bioM2 beads with 0, 1, 3, 10, and 30 nM QD585 dots (see Subheading 3.3). (**b**) Linear plot of Qdot sites on calibration beads versus the corresponding mean channel fluorescence (MCF) reading (Reprinted from (5) with permission from Elsevier).

3.5. Quantitative Measurement of Qdot-Labeled Virus Particles Bound to Cells Using Flow Cytometry and TEM

Virus particles are amenable for tagging with qdots. Once labeled, the degree of labeling can be resolved by transmission electron microscopy (TEM). In this way, flow cytometry can be used to quantify bound viruses. Human papillomaviruses (HPVs) are etiological agents of a number of benign and malignant tumors of the skin and mucosa (36). We have recently analyzed the binding of QD585-labeled HPV particles to A431 cells by flow cytometry. QD585 calibration beads were used to measure the total number of qdots per cell ≈12,000 QD585 per cell, see (5) using Eq. 1. In previously unpublished data from that study, TEM was used to estimate the number of qdots per virus particle, as shown in Fig. 7.

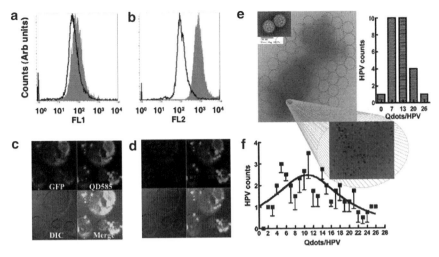

Fig. 7. Simultaneous two-color measurement of eGFP/Qdot 585 nm (QD585) double-stained Human papillomavirus (HPV) particles on A431 cells using flow cytometry and microscopy. (**a**) Flow cytometry measurement of GFP fluorescence. Solid histogram (MCF ≈ 100) corresponds to cells bearing HPV particles while open histogram (MCF ≈ 50) represents negative control cells with no virus. (**b**) Fluorescence measurement of QD585. Solid histogram (MCF ≈ 1,100) represents cells bearing HPV particles stained with QD585, and open histogram (MCF ≈ 100) represents negative control cells (no virus). (**c**) Confocal images of pseudovirion (PsV)-bearing A431 cells prepared under similar conditions to flow cytometry measurements as in (**a**). (**d**) Confocal images of the same cells after five scans show the near complete photobleaching of eGFP. (**e**) TEM micrograph of QD585 dots on transparent biotinylated PsV particles, delineated by a random honeycomb grid scaled to fit the size of 60 nm PsV. Insert shows PsV negatively stained with uranyl acetate. The bar scale is 100 nm for both micrographs. (**f**) Frequency distribution histogram of Qdots/PsV-mean: 11 ± 5 (Figure 7a–d is reprinted from: (5) with permission from Elsevier).

4. Notes

1. Commercial bioM2 is contaminated with high levels of free biotin, which is very likely to block binding of bioM2 to beads. This is removed by using a 100 kDa cutoff Microcon Centrifugal Filter Device (Millipore, Billerica, MA).

   (a) Assuming a starting material of 500 µL of anti-FLAG bioM2 antibody (in 50% glycerin, 1 mg/mL). Add to centrifugal device and centrifuge in cold room at $3,000 \times g$ for 25 min (or till there are <50 µL).

   (b) Wash the antibody in 25 mM (or 50 mM) Tris buffer 3× by repeating Note 1a. It should be noted that glycerin is very hard to filter through, therefore longer centrifugation time and/or additional washing steps might be needed.

   (c) Resuspend the purified and concentrated bioM2 in 100 µL of 25 mM Tris buffer (5 mg bioM2/mL), make 20 µL aliquots and keep frozen in –20°C freezer till needed.

(d) Assuming 4×10^{15} molecules of bioM2/mg: each aliquot of bioM2 (0.1 mg) can coat 25 million 6.7 μm streptavidin bead (1×10^{14} biotin binding sites) with 2× excess of bioM2.

2. Correcting excitation spectra for spectral distribution of the lamp. The excitation spectra are shown in Fig. 2b. Uneven features in spectra are due to the uneven spectral distribution (L_λ) of the lamp (7). To extract (L_λ) easily:

(a) Overlay the normalized absorption spectra and the excitation spectra as shown in Fig. 3a. Because emission almost always originates from the lowest vibrational level of the first excited state, the emission yield is independent of excitation wavelength for most organic fluorophores (8, 31). Therefore, one expects a resemblance between the normalized absorption and photoexcitation spectra of fluorescein. The excitation lineshape of fluorescein retains the same basic features that mirror the absorption spectrum except for structural artifacts introduced by the uneven spectral distribution of the Xenon arc lamp.

(b) Divide the normalized excitation spectrum column with the normalized absorption spectrum column. The result, L_λ, is shown in Fig. 3b.

(c) Divide all of the uncorrected excitation spectra columns with L_λ. The corrected data are shown in Fig. 3c.

3. Qdots and bioM2 antibodies have limited shelf lives. Although Invitrogen Qdots are suspended in a proprietary buffer environment that confers fluorescence, when qdots are diluted in a regular buffer, they gradually lose fluorescence intensity over time. It is therefore important to maintain qdots in this buffer as long as possible. If calibration beads are not stored in this buffer, then it is important to frequently check the fluorescence intensity of the beads. This can be done by comparing the fluorescence intensity of dots left in regular buffer with a fluorophore of known concentration over time. The reader might also consider making minimal quantities of Qdot beads (Fig. 1b). In this way, long-term storage of qdots outside their native buffer is avoided. BioM2 antibodies also lose activity – 30% over 6 months (5, 33). The activity of bioM2 beads can be assayed using FITC-labeled FLAG peptide (25) and commercial (Quantum™ FITC MESF) standard calibration beads to check on the integrity of the bioM2 beads over time. The typical assay involves periodic checking of the surface density of FITC-labeled FLAG peptide on bioM2 beads using Quantum™ FITC MESF beads or a centrifugation assay (5). The narrow distribution of histograms (see Fig. 6) associated with each titer of qdots mixed with bioM2 beads can be used as an indicator of the quality of

calibration beads. Broad histograms associated with freshly made calibration beads usually indicate: (a) Prevalence of nonspecific binding, i.e., steps a–c in Fig 1. This can be checked by running a negative control, i.e., adding qdots to bioM2 beads without the biotinylated FLAG peptide; (b) Heterogeneous binding of qdots due to loss of bioM2 activity.

Acknowledgments

This work was supported by NIH K25AI60036, U54MH074425, U54MH084690, HL081062, NSF CTS0332315, Dedicated Health Research Funds of the University of New Mexico School of Medicine (C-2294-T). Images in this paper were generated in the University of New Mexico Cancer Center Fluorescence Microscopy Facility, supported as detailed on the Web page: http://hsc.unm.edu/crtc/microscopy/Facility.html.

References

1. Sklar, L. A. (2005) *Flow Cytometry in Biotechnology*, Oxford University Press, New York, NY.

2. Nolan, J. P. and Sklar L. A. (1998) The emergence of flow cytometry for sensitive, real-time measurements of molecular interactions. *Nature Biotechnology* **16**, 633–8.

3. Wang, L., Gaigalas, A. K., Abbasi, F., Marti, G., Vogt, R., and Schwartz, A. (2002) Quantitating fluorescence intensity from fluorophores: Practical use of MESF values. *Journal of research of the National Institute of Standards and Technology* **107**, 339–53.

4. Schwartz, A., Wang, A., Early, E. et al. (2002) Quantitating fluorescence intensity from fluorophore: The definition of MESF assignment. *Journal of research of the National Institute of Standards and Technology* **107**, 83–91.

5. Wu, Y., Campos, S. K., Lopez, G. P., Ozbun, M. A., Sklar, L. A., and Buranda, T. (2007) The development of quantum dot calibration beads and quantitative multicolor bioassays in flow cytometry and microscopy. *Analytical Biochemistry* **364**, 180–92.

6. Wu, Y., Lopez, G. P., Sklar, L. A., and Buranda, T. (2007) Spectroscopic characterization of streptavidin functionalized quantum dots. *Analytical Biochemistry* **364**, 193–203.

7. Parker, C. A. (1968) *Photoluminescence of Solutions*, Elsevier, Amsterdam.

8. Lakowicz, J. R. (1999) *Principles of Fluorescence Spectroscopy*, 2nd edn., Plenum Press, New York, NY.

9. Gaponenko, S. V. (1998) *Optical Properties of Semiconductor Nanocrystals*, Cambridge University Press, Cambridge.

10. Tonti, D., van Mourik, F., and Chergui, M. (2004) On the excitation wavelength dependence of the luminescence yield of colloidal CdSe quantum dots. *Nano Letters* **4**, 2483–7.

11. Bawendi, M. G. (1994) Chemistry and Physics of Semiconductor Nanoclusters. Abstracts of Papers of *The American Chemical Society* **21(208)**, 4-INOR.

12. Brus, L. E., Bawendi, M., Wilson, W. L. et al. (1991) Primary Photophysics of Quantum Semiconductor Crystallites. Abstracts of Papers of *The American Chemical Society* **14(201)**, 409-INOR.

13. Caruge, J. M., Chan, Y. T., Sundar, V., Eisler, H. J., and Bawendi, M.G. (2004) Transient photoluminescence and simultaneous amplified spontaneous emission from multiexciton states in CdSe quantum dots. *Physical Review B* **70**, 085316.

14. Dabbousi, B. O., RodriguezViejo, J., Mikulec, F. V. et al. (1997) (CdSe)ZnS core-shell quantum dots: Synthesis and characterization of a size series of highly luminescent

nanocrystallites. *Journal of Physical Chemistry. B* **13**, 9463–75.

15. Danek, M., Jensen, K. F., Murray, C. B., and Bawendi, M. G. (1996) Synthesis of luminescent thin-film CdSe/ZnSe quantum dot composites using CdSe quantum dots passivated with an overlayer of ZnSe. *Chemistry of Materials* **8**, 173–80.

16. Efros, A. L., Rosen, M., Kuno, M., Nirmal, M., Norris, D. J., and Bawendi M. G. (1996) Band-edge exciton in quantum dots of semiconductors with a degenerate valence band: Dark and bright exciton states. *Physical Review B* **15**, 4843–56.

17. Kuno, M., Lee, J. K., Dabbousi, B. O., Mikulec, F. V., and Bawendi, M. G. (1997) The band edge luminescence of surface modified CdSe nanocrystallites: Probing the luminescing state. *Journal of Chemical Physics* **15**, 9869–82.

18. Norris, D. J., Nirmal, M., Murray, C. B., Sacra, A., and Bawendi, M. G. (1993) Size-dependent optical spectroscopy of Ii–Vi semiconductor nanocrystallites (quantum dots). *Zeitschrift Fur Physik D Atoms Molecules and Clusters* **26**, 1–4.

19. Norris, D. J., Sacra, A., Murray, C. B., and Bawendi, M. G. (1994) Measurement of the size-dependent hole spectrum in CdSe quantum dots. *Physical Review Letters* **18**, 2612–5.

20. Norris, D. J. and Bawendi, M. G. (1995) Structure in the lowest absorption feature of CdSe quantum dots. *Journal of Chemical Physics* **103**, 5260–8.

21. Norris, D. J. and Bawendi, M. G. (1996) Measurement and assignment of the size-dependent optical spectrum in CdSe quantum dots. *Physical Review B* **15**, 16338–46.

22. Rumbles, G., Selmarten, D. C., Ellingson, R. J. et al. (2001) Anomalies in the linear absorption, transient absorption, photoluminescence and photoluminescence excitation spectroscopies of colloidal InP quantum dots. *Journal of Photochemistry and Photobiology A: Chemistry* **14**, 2–3.

23. Ellingson, R. J., Blackburn, J. L., Yu, P. R., Rumbles, G., Micic, O. I., and Nozik, A. J. (2002) Excitation energy dependent efficiency of charge carrier relaxation and photoluminescence in colloidal InP quantum dots. *Journal of Physical Chemistry B* **15**, 7758–65.

24. Babbitt, S. E., Kiss, A., Deffenbaugh, A. E. et al. (2005) ATP hydrolysis-dependent disassembly of the 26 S proteasome is part of the catalytic cycle. *Cell* **20**, 553–65.

25. Buranda, T., Lopez, G. P., Simons, P., Pastuszyn, A., and Sklar, L. A. (2001) Detection of epitope-tagged proteins in flow cytometry: Fluorescence resonance energy transfer-based assays on beads with femtomole resolution. *Analytical Biochemistry* **298**, 151–62.

26. Buranda, T., Huang, J. M., Perez-Luna, V. H., Schreyer, B., Sklar, L. A, and Lopez, G. P. (2002) Biomolecular recognition on well-characterized beads packed in microfluidic channels. *Analytical Chemistry* **74**, 1149–56.

27. Piyasena, M. E., Buranda, T., Wu, Y., Huang, J. M., Sklar, L. A., and Lopez, G. P. (2004) Near-simultaneous and real-time detection of multiple analytes in affinity microcolumns. *Analytical Chemistry* **76**, 6266–73.

28. Simons, P. C., Shi, M., Foutz, T. et al. (2003) Ligand-receptor-G-protein molecular assemblies on beads for mechanistic studies and screening by flow cytometry. *Molecular Pharmacology* **64**, 1227–38.

29. Simons, P., Vines, C. M., Key, T. A. et al. (2005) Analysis of GTP-binding protein-coupled receptor assemblies by flow cytometry, in *Flow Cytometry in Biotechnology* (Sklar, L. A., ed.), Oxford University Press, New York, NY, pp. 323–46.

30. Buranda, T., Jones, G., Nolan, J., Keij, J., Lopez, G. P., and Sklar, L. A. (1999) Ligand receptor dynamics at streptavidin coated particle surfaces: A flow cytometric and spectrofluorimetric study. *The Journal of Physical Chemistry. B* **103**, 3399–410.

31. Kasha, M. (1950) Characterization of electronic transitions in complex molecules. *Discussions of the Faraday Society* **9**, 14–9.

32. Magde, D., Rojas, G. E., and Seybold, P. G. (1999) Solvent dependence of the fluorescence lifetimes of xanthene dyes. *Photochemistry and Photobiology* **70**, 737–44.

33. Buranda, T., Waller, A., Wu, Y. et al. (2007) Some mechanistic insights into GPCR activation from detergent-solubilized ternary complexes on beads. *Advances in Protein Chemistry* **74**, 95–135.

34. Babbitt, S. E., Kiss, A., Deffenbaugh, A. E. et al. (2005) ATP hydrolysis-dependent disassembly of the 26 S proteasome is part of the catalytic cycle. *Cell* **121**, 553–65.

35. Christopoulos, A. and Kenakin, T. (2002) G protein-coupled receptor allosterism and complexing. *Pharmacological Reviews* **54**, 323–74.

36. Walboomers, J. M. M., Jacobs, M. V., Manos, M. M. et al. (1999) Human papillomavirus is a necessary cause of invasive cervical cancer worldwide. *Journal of Pathology* **189**, 12–9.

Chapter 5

Bead-Based Multiplexed Analysis of Analytes by Flow Cytometry

Henri C. van der Heyde and Irene Gramaglia

Abstract

The Enzyme-Linked Immunosorbent Assay (ELISA) and Western Blot analysis have been workhorse techniques for the analysis of protein levels and state, such as phosphorylation. The ELISA is also useful for measuring the affinity of a molecule for its ligand. The disadvantage of these techniques is that only a single protein can be analyzed for ELISA and a few (up to three) proteins for Western Blotting. Exact quantification is difficult with Western Blotting and changes are often reported as fold differences. We present here protocols for using fluorescent microspheres coated with the selected capture molecule to perform in essence several hundred mini ELISAs with each microsphere representing an ELISA; this reduces the variability of the assay. In addition, it is possible to analyze up to 100 analytes simultaneously using microspheres because each fluorescent microsphere exhibits distinct fluorescence in the red and far red channels: the fluorescence intensity in the channels in the red and far red channels (up to ten different intensities for each channel leading to a 10×10 matrix of intensities) constitutes the address for each analyte.

Key words: Flow cytometry, Beads, Microspheres, Quantification, Affinity

1. Introduction

This chapter provides detailed protocols for the multiplexed analysis of user selected analytes by flow cytometry using fluorescence-labeled microspheres or beads. There are a number of commercial kits available where the assay for the analyte has been implemented, but there are gaps in the analytes assessed by these kits (1). For many applications, it is worthwhile to develop a custom kit for the analyte of interest, and protocol examples are provided here. Moreover, this microsphere-based flow cytometry assay can be used to determine other parameters, such as affinity, protease sites, SNPs amongst others; we will provide examples of how levels and affinity can be assessed by using this microsphere platform.

Teresa S. Hawley and Robert G. Hawley (eds.), *Flow Cytometry Protocols*, Methods in Molecular Biology, vol. 699,
DOI 10.1007/978-1-61737-950-5_5, © Springer Science+Business Media, LLC 2011

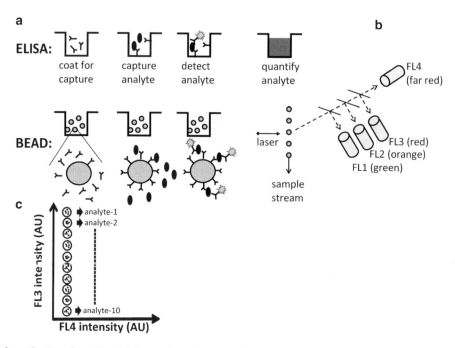

Fig. 1. Quantification of analytes. (**a**) Comparison of the steps in an ELISA and MMA. In ELISA, a capture molecule is adsorbed to the plate and used to capture the analyte. Detection antibodies usually conjugated with an enzyme are used to cleave a colorimetric substrate; the optical density (OD) measured at a characteristic wavelength is then a measure of the concentration of analyte. In MMA, each of the microsphere populations is conjugated with the selected capture molecule (up to 100) and then equal numbers of the microsphere populations are added to a well containing the solution to be analyzed. The microspheres capture the analyte, which is then detected by fluorescently labeled detection antibodies. (**b**) The fluorescence intensity measured on each bead by flow cytometry is then a measure of the concentration of the analyte. The flow cytometer interrogates each microsphere. (**c**) The intensities of fluorescence in FL3 and FL4 determine the address of microsphere: the capture molecule and hence the analyte being measured. The intensity of FL2 fluorescence for the microsphere measures the concentration of analyte. This address system allows multiple analytes to be measured in a single analysis.

In essence, any assay that is performed on a solid support with colorimetric, fluorescence, or luminescence detection can be converted to the multiplexed microsphere analysis (MMA). Figure 1 provides a comparison of the steps in an Enzyme-Linked Immunosorbent Assay (ELISA) versus MMA. In an ELISA, the capture molecule (usually antibody) is coated onto the plate by overnight incubation; alternate approaches are to use a capture biotinylated molecule, which binds tightly to streptavidin coated plates. In many cases, the capture molecule will be a monoclonal antibody (mAb) but may be any molecule (e.g. proteins or oligonucleotides) that binds the analyte with reasonable affinity. The plate is then washed to remove the unbound antibody. Dilutions of the sample are added to vary the concentration of the analyte, which binds to the capture antibody after several hours of incubation; dilutions of the sample are needed because the linear range of an ELISA is limited. The unbound analyte is washed off. The captured analyte is detected with an antibody specific for a site on the analyte distinct from the binding site of the capture molecule. This antibody can be directly conjugated with peroxidase or alkaline phosphatase or a secondary antibody coupled with an enzyme can be used. The enzyme then converts the colorimetric substrate and the reaction is terminated after a selected period of time. The absorbance is a measure of the amount of analyte captured. Because only a single color is used in this assay, only a single analyte is measured in each well.

In MMA, the beads or microspheres become the solid support in the ELISA and the detection system is fluorescence intensity assessed by flow cytometry rather than colorimetric. The first or coating step (Fig. 1a) is to conjugate the microsphere with the capture molecule. Protocols for this coating step are provided in Subheading 3.1. The analyte is then captured on the bead surface. The analyte is then detected by flow cytometric assessment of the fluorescence intensity of each bead. At least 200 individual microspheres for a particular analyte are measured. Conceptually, each microsphere analyzed represents a miniaturized ELISA. Assays assessing titer, levels, and affinity are described in Subheadings 3.2–3.4, respectively.

The analysis of multiple analytes by bead-based technologies is possible because flow cytometry measures fluorescence intensity at multiple bands of wavelengths (Fig. 1b) and the beads are impregnated with fluorescent dyes (2, 3). Each bead population exhibits one of ten different intensities in a selected channel (usually red or FL3); these different intensities can be distinguished in the red channel of the flow cytometer. The capture molecule for each of up to ten analytes is conjugated onto a population of beads with distinct fluorescence, up to a maximum of ten (Fig. 1c). The intensity of FL3 fluorescence thus provides the "address" of analyte. The orange channel (FL2) is used for detection. If more than ten analytes need to be analyzed simultaneously, beads impregnated with two fluorescent dyes can be used emitting in the red FL3 and far-red FL4 channels. Ten different intensities of fluorescence in each channel have been achieved, providing a 10×10 matrix of intensities or 100 different addresses.

The bead assay exhibits similar sensitivity as the original ELISA because similar reagents are used except possibly at the detection step. Fluorescence intensity measurement by flow cytometry using photomultiplier tubes is sensitive, providing the rationale for the sensitivity of bead assay despite the absence of an amplification step. The ELISA uses an enzyme to amplify the signal, which decreases its linear range. The bead assay also exhibits improved reproducibility: the ELISA comprises the assessment of a single well whereas the bead assay represents the median value of about 200 mini-ELISAs performed on the microspheres. In our laboratories, we therefore prefer to use bead assays in the place of conventional ELISA.

2. Materials

2.1. Microspheres, Supplies, and Equipment

1. Single color microspheres exhibiting up to ten distinct intensities. Suppliers include: Bangs Laboratories (Fisher, IN), Duke Scientific (Palo Alto, CA), and Spherotech (Lake Forest, IL).
2. Single color microspheres coated with Streptavidin (Bangs Laboratories).
3. Two color microspheres exhibiting up to 100 distinct intensities. Suppliers include: Crystalplex (Pittsburgh, PA) and Luminex (Austin, TX).
4. Quantum-PE Molecules of Equivalent Soluble Fluorochrome (MESF) calibration microspheres (Bangs Laboratories).
5. Multiscreen HTS – BV 1.2 μm non sterile filter plates (Millipore, Billerica, MA).
6. Multiscreen HTS Vacuum manifold (Millipore).
7. Siliconized microcentrifuge tubes (VWR, West Chester, PA).
8. Digital shaker; rotator.
9. Plate reading flow cytometer. Examples include: Luminex analyzer, Luminex-based analyzer (e.g. Millipore, Qiagen, Bio-Rad), FACSArray Bioanalyzer (BD Biosciences, San Jose, CA), FlowCytomix Pro (Beckman Coulter, Fullerton, CA). Luminex microspheres must be used on Luminex or Luminex-based instruments whereas other microspheres are designed for conventional flow cytometers such as BD and Beckman Coulter instruments.

2.2. Buffers and Reagents

1. Phosphate buffered saline (PBS).
2. PBS containing 0.02% Tween.20.
3. PBS containing 30 mM glycine.
4. Activation buffer: 2-(N-morpholino)ethanesulfonic acid (MES) buffer (Sigma–Aldrich, St. Louis, MO).
5. Capture buffer for proteins: PBS with 0.02% Tween 20 and 10% serum or plasma, or commercial source (BD Biosciences).
6. Capture buffer for nucleic acids (Active Motif, Carlsbad, CA).
7. Wash buffer: PBS with 0.02% Tween 20.
8. Plasma or serum (see Note 1).
9. 1-Ethyl-3-(3-dimethylaminopropyl) carbodiimide hydrochloride (EDAC; Sigma–Aldrich).

3. Methods

3.1. Coating of Beads with Molecule to Capture Analyte

The microspheres are first conjugated with the capture molecule (see Note 2). Depending on the application, the capture may be an antibody, a protein, or oligonucleotide (Fig. 2). The following protocols may be used to conjugate these molecules to microspheres.

3.1.1. Quick and Easy Conjugation of Beads with Carbodiimide Chemistry (4)

1. Resuspend 1×10^7 washed and pelleted ($14,000 \times g$) microspheres in a siliconized microcentrifuge tube with 100 µL of concentrated capture molecule (0.1 mg/mL works well) in PBS and incubate overnight at $4°C$.

2. Add 900 µL of PBS to the microspheres (1 mL total volume).

3. Weigh 10 mg of EDAC in a siliconized microcentrifuge tube and then add the solution containing microspheres. Vortex well for at least 30 s and then incubate with rotation at $4°C$ for 60 min.

4. Wash the microspheres three times with 1 mL of PBS containing 0.02% Tween 20.

5. On the final wash, count the beads by flow cytometry and resuspend at 1×10^7/mL. Store at $4°C$.

3.1.2. Controlled Conjugation of Protein to Microspheres with Carbodiimide Chemistry (5)

1. Wash 1×10^7 microspheres in a siliconized microcentrifuge tube three times with MES activation buffer and ensure the microspheres are well resuspended by vortexing vigorously.

2. Resuspend in $1,000$ µL of PBS and add 10 mg of EDAC.

3. Allow to react for 15 min at room temperature with rotation.

4. Wash twice with PBS; resuspend in $1,000$ µL of PBS and 100 µL of capture molecule solution (usually 1 mg/mL for antibodies). This step may require titration to achieve the optimal conjugation: too little – the signal is low; too much – allosteric hindrance prevents optimal binding of the analyte.

5. React microspheres with capture molecule for 2–4 h at room temperature.

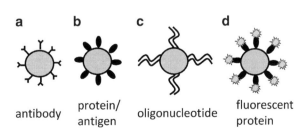

Fig. 2. Possible capture molecules to be attached to populations of microspheres.

6. Wash microspheres with PBS containing 30 mM glycine and then wash with PBS containing 0.02% Tween 20.

7. Count the microspheres and resuspend at 1×10^7/mL. Store at 4°C until needed.

3.1.3. Capture of Double Stranded Oligonucleotides to Microspheres

Proteins or oligonucleotides can be conjugated to carboxyl modified microspheres as described in Subheading 3.1.2. A simpler alternative that ensures the correct orientation of the microsphere is to use streptavidin-coated microspheres and either a biotinylated protein or an oligonucleotide synthesized with biotin at 5′ nucleotide. The biotinylated capture molecule is incubated at saturating concentrations with the microspheres. After the capture step, the microspheres are washed twice in PBS with 0.02% Tween 20. Resuspend at 1×10^7/mL. Store at 4°C.

3.2. Assessment of Antibody Levels

The assessment of analyte levels can be performed in two types of assays: titer and absolute quantification (4). The titer allows comparisons of groups without determining the actual concentration. The titer of a sample is defined as the highest dilution of sample that provides a detectable signal by the assay. Some commercial kits use the titer to assess the differences in the phosphorylation of intracellular proteins. In this case, capture antibodies specific for the selected intracellular proteins are conjugated to distinct populations of microspheres (Fig. 2a). Titer is often used to compare the antibody levels in infected versus uninfected animals, or to compare the efficacy of different vaccine protocols for eliciting antigen-specific antibody. In this example, each of the selected antigens is conjugated to a population of beads (Fig. 2b). The levels of nuclear translocation can also be assessed by titer to compare activation of transcription within cells by a selected molecule. In the nuclear transcription factor assay, nuclear extracts are made, and the transcription factors selected for analysis are captured by populations of microspheres with oligonucleotides specific for each transcription factor (Fig. 2c).

Once the selection of the capture molecule has been made and the populations of microspheres generated as described in Subheading 3.1, the titer assay can be performed as outlined in Fig. 3 and detailed below.

1. Mix the populations of microspheres for the assay with particular attention paid to blocking the microspheres well in order to minimize non specific binding to the bead (see Note 1). Microspheres (1×10^4/μL) are mixed in capture buffer with blocking proteins/nucleotides (see Note 1).

2. Wet filters by adding 200 μL of wash buffer to the well and then applying vacuum to aspirate the buffer through filter (i.e. washing).

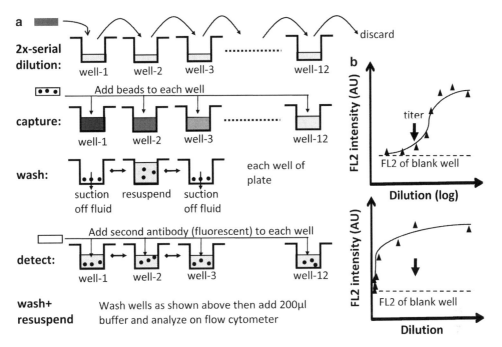

Fig. 3. (**a**) Schematic of the steps to assess the titer of the analyte by MMA. (**b**) Analyses of the fluorescence intensity to determine either the titer (*upper*) or affinity (*lower*) of the analyte.

3. On the first row of a 96-well plate, pipette 20 µL of PBS containing 0.02% Tween 20 (comprises unlabeled beads) into wells 1, 2, and 3; then pipette 20 µL of a positive control into wells 4, 5, and 6.

4. Pipette 20 µL of capture buffer into the second row of 12 wells on the 96-well plate; the number of rows equals the number of samples to be tested. Dilute the sample 1:2 with capture buffer (20 µL sample into 20 µL buffer) in the first well and then complete the serial dilution by pipetting 20 µL from each well into the next, mixing, then repeating until well 12 (Fig. 3a).

5. Add 10 µL of the microsphere mixture to each well and mix by pipetting gently.

6. Cover the plate with plastic sealer to prevent evaporation and place on a digital mixer. The incubation time varies with the molecule to be captured and this step may need to be optimized. Generally, antibodies in serum or plasma require 2 h of incubation at 4°C; intracellular proteins require overnight incubation at 4°C; and oligonucleotide capture of transcription factors requires 4 h incubation at room temperature. This step can be performed in V-bottom plates and then transferred to filter microplates by using a multichannel pipette.

7. Place the filter microplate on the suction device and apply vacuum; the supernatant should be suctioned off into the collection container below.

8. Add 200 μL of wash buffer and suction off fluid; repeat for two washes.

9. Add 30 μL of detection molecule directly conjugated with PE and incubate for 1 h at room temperature with mixing on a rotator. The concentration of the detection molecule should be optimized by titration, but 0.1 mg/mL should be saturating for detection antibodies and provides an initial starting concentration.

10. The fluid is suctioned off and the microspheres washed twice. The microspheres are then resuspended in 200 μL of wash buffer. The plate is then placed in a plate reading flow cytometer or Luminex instrument depending on the microspheres chosen.

11. Acquire the negative and positive controls to verify that the assay is working and then acquire the samples and collect at least $200 \times \#$ microsphere populations events.

12. Generate a dot plot for FL3 versus FL4. Each selected microsphere population should be present with about 200 events and exhibit distinct fluorescence from each other. Place a region around each microsphere population and label the population.

13. Determine the median intensity of PE fluorescence at each dilution of the sample for each of microsphere population. The lowest dilution should exhibit the greatest signal, which decreases as the analyte is diluted. Plot the FL2 intensity versus dilution in log scale (Fig. 3b) for each sample and each microsphere population. The lowest dilution is on the right of the x-axis; the highest dilution is on the left. The dilution value with detectable signal (usually two standard deviations above the signal of the blank microspheres (wells 1–3) then is the titer of the sample for the analyte defined by the microsphere population.

3.3. Quantification of Analyte Levels

In order to perform absolute quantification of analytes, each of the selected analytes must be available purified with the concentration of the analyte defined. An initial stock solution is made comprising a defined concentration of each analyte under investigation (=standard). Twofold dilutions of this standard are made in the first 12 wells of the 96-well filter plate (Fig. 3a). Each sample and twofold serial dilutions of the sample are pipetted into another 12 wells of the plate; replicates of the sample may be performed to measure standard deviation for each sample. The protocol described in Subheading 3.2 is followed with the standard and sample as described above. The median intensity of fluorescence is assessed for each microsphere population with the standard and sample.

The median intensity of fluorescence for each microsphere population binding a specific analyte is then plotted vs. the known concentration of analyte to generate a standard curve. In the linear portion of the curve, a regression line is calculated. The derived line is then used to determine the concentration of the analyte in samples.

3.4. Assessment of Analyte Affinity

If the capture microsphere comprises a protein or oligonucleotide (Fig. 2b, c), then it is possible to determine the affinity of the analyte for the capture molecule. This measurement is useful to assess the affinity of antigen-specific antibody elicited by infection or vaccination because Zinkernagel and colleagues (6) have reported that affinity determines whether the elicited antibody is capable of neutralizing virus. It is also useful for assessing how the affinity declines for antigenic variants of the same protein. By including a set of oligonucleotides with selected mutations, the changes in affinity of binding of transcription factors can be screened for each mutation. The procedure for affinity assessment of antibody by microsphere analysis is described below:

1. Assess the median fluorescence intensity of analyte binding to the microspheres after twofold dilutions of known concentrations of the analyte as described in the titer assay of Subheading 3.2.

2. Without changing the settings on the flow cytometer, acquire about 2,000 events using Quantum-PE MESF calibration microspheres. There are five peaks each with the number of fluorophores for each peak defined by the manufacturer. Perform linear regression on the line and then use this line to calculate the MESF for each of the median fluorescence intensities measured for each analyte concentration.

3. Determine the fluorophore/protein ratio for the detection antibody and convert the MESF to the number of molecules bound. First, assess absorbance of the protein in a 1 cm cuvette at 280 nm (A_{280}) and at excitation λ_{max} for fluorophore ($A_{\lambda_{max}}$; λ_{max} PE: 578 nm; λ_{max} FITC: 564 nm). The protein concentration is then calculated by using the following formula:

$$[\text{protein}] = \frac{A_{280} - (A_{\lambda_{max}} \times \text{correction factor}) \times \text{dilution factor}}{\varepsilon_{\text{protein}}}$$

where $\varepsilon_{\text{protein}}$ is the molar extinction coefficient in $\text{cm}^{-1}\text{M}^{-1}$ (203,000 for IgG at 280 nm); dilution is any dilution performed during the absorbance measurement. The correction factor is the fluorophore's contribution to the absorbance at

280 nm and is measured by assessing the fluorescence of an equal concentration of labeled and unlabeled protein ($A_{\lambda_{max}} = 0$); the correction factor for FITC is 0.3. The fluorophore to protein ratio (F:P) is:

$$F:P = \frac{A_{\lambda_{max}} \times dilution\,factor}{[protein] \times \varepsilon_{dye}}$$

where ε_{dye} is the extinction coefficient of fluorophore at its absorbance maximum (FITC: 68,000 cm^{-1}M^{-1}). The number of fluorescent molecules bound, which is an estimate of the number of molecules captured on the microsphere, is calculated by dividing the MESF by the F:P ratio. There is a simpler approach to assessing F:P for antibodies by using microspheres coated with precise numbers of anti-Ig antibodies (see Note 3).

4. Plot the # of molecules bound versus the concentration. This curve is then fit with a hyperbolic single site binding curve $f = a \times \dfrac{x}{b + x}$ where a is the # of molecules bound at saturation and b is the equilibrium dissociation constant.

4. Notes

1. Blocking the spaces without capture analyte is an important step to improve the signal to noise ratio of the assay. Blocking molecules bind via adsorption to the microsphere, thereby preventing other molecules (such as the fluorescence-labeled detection molecule) from binding non-specifically to the microsphere. Many companies sell blocking agents under trade names; however, the compositions of these buffers are trade secrets. For our assays analyzing the levels of Ig, we used the same species of plasma as the detecting mAb (rat) to ensure that the detecting mAb did not exhibit cross reactivity. Using 10% plasma or higher in our hands markedly decreased non-specific binding. Herring sperm (10 µg/mL) is added as a blocking agent if oligonucleotides are used.

2. There are two important parameters that need to be confirmed for the conjugated microspheres: (a) capture molecule is conjugated to the microsphere, and (b) there is low variation in the amount of molecule conjugated to the bead. The simplest approach to verifying that the microsphere is conjugated is by using a fluorescence-conjugated mAb specific for the capture molecule. Starting with a saturating concentration of mAb and then by titrating the fluorescence-labeled mAb over the microspheres as described in Subheading 3.2, it is

possible to determine the number of capture molecules that have been conjugated on the surface of the bead. Alternately, the labeling with a saturating concentration of the fluorescent mAb specific for the capture molecule is assessed by flow cytometry and compared with beads labeled with an isotype control antibody. The labeling should result in at least a two-decade shift in fluorescence intensity and the standard deviation of the labeling should be low.

Oligonucleotides usually do not have specific antibodies directed against them. To verify that the double stranded oligonucleotides are conjugated to the beads, the binding of the oligonucleotide to the microspheres is detected by incubating them with propidium iodide (100 µL of 5 µg/mL) for 0.5 h and then washing the microspheres once to remove excess propidium iodide. There should be at least a single decade shift if the oligonucleotides have bound to the microspheres and the binding should be saturated. A similar approach may be taken with fluorescent dye that labels proteins CBQCA (7).

3. Bangs laboratories sells the Quantum Simply Cellular kits, which can be used to directly quantify the number of molecules without the need for determining the protein concentration and using MESF standard beads. There are five populations of microspheres each conjugated with known amounts of anti-Ig mAbs (0; 8,000; 34,000; 96,000; and 181,000). Saturating quantities (determined by titration) of the fluorescence labeled mAb is incubated for 30 min with microspheres and the intensity of fluorescence assessed by flow cytometry. The median intensity of fluorescence for each population is then plotted versus the known number of binding sites. A linear regression line is calculated. The intensity of fluorescence determined in the experiment is then converted to # of molecules by using the linear regression line.

References

1. Nolan, J. P., Yang, L., and van der Heyde, H. C. (2006) Reagents and instruments for multiplexed analysis using microparticles, in *Current Protocols in Cytometry* (Robinson, J. P., ed.), John Wiley and Sons, Inc., Hoboken, NJ. Chapter 13:Unit13.18.

2. Nolan, J. P. and Mandy, F. F. (2006) Multiplexed and microparticle-based analyses: quantitative tools for the large-scale analysis of biological systems. *Cytometry* **69A**, 318–325.

3. Nolan, J. P., and Sklar, L. A. (2002) Suspension array technology: evolution of the flat-array paradigm. *Trends Biotechnol.* **20**, 9–12.

4. van der Heyde, H. C., Burns, J. M., Weidanz, W. P., Horn, J., Gramaglia, I., and Nolan, J. P. (2007) Analysis of antigen-specific antibodies and their isotypes in experimental malaria. *Cytometry A* **71**, 242–250.

5. Hermanson, G. T. (1995) Zero-length cross-linkers, in *Bioconjugate Techniques*, Harcourt Brace & Co., San Diego, CA, pp. 169–186.

6. Bachmann, M. F., Kalinke, U., Althage, A., Freer, G., Burkhart, C., Roost, H., Aguet, M, Heingartner, H, and Zinkernagel, R.M. (1997). The role of antibody affinity and avidity in antiviral protection. *Science* **276**, 2024–2047.

7. Graves, S. W., Woods, T. A., Kim, H., and Nolan, J. P. (2005) Direct fluorescent staining and analysis of proteins on microspheres using CBQCA. *Cytometry A* **65**, 50–58.

Chapter 6

Flow-Based Combinatorial Antibody Profiling: An Integrated Approach to Cell Characterization

Shane Bruckner, Ling Wang, Ruiling Yuan, Perry Haaland, and Amitabh Gaur

Abstract

BD FACS™ CAP (CAP = combinatorial antibody profile) is a screening tool for rapid characterization of human cell surface protein expression profiles using semi-automated high-throughput flow cytometry. The current configuration consists of 229 directly conjugated antibodies arrayed in a 96-well plate as three-color cocktails, which enables the characterization of each of the 229 individual surface markers. Each individual cell type of interest is analyzed on the 96-well screening plates and the data are acquired on a flow cytometer equipped with a high-throughput sampler. The expression level of each marker for each cell type is then calculated using semiautomated custom flow cytometry software. The process of characterizing these surface markers in a highly efficient manner using BD FACS™ CAP is enabled by automated liquid handling for staining, automated flow cytometry for data acquisition, and standardized algorithms for automated data analysis.

Key words: Cell surface protein expression profiling, High-throughput flow cytometry, Automated data analysis, Multiparametric flow cytometry

1. Introduction

Characterization of subsets of cells in a heterogeneous population has long been the forte of flow cytometry. With the identification of new markers on the cell surface, facilitated by continuing development and use of novel monoclonal antibodies, more and more unique subsets of cells are being identified. As the collection of antibodies to the various cell surface molecules grew, it became evident that simultaneous use of all or most of these antibodies could yield an information-rich cellular expression profile. BD FACS™ CAP (BD Biosciences, San Diego, CA) is one such semi-automated, high-throughput flow cytometry screening tool that

Teresa S. Hawley and Robert G. Hawley (eds.), *Flow Cytometry Protocols*, Methods in Molecular Biology, vol. 699, DOI 10.1007/978-1-61737-950-5_6, © Springer Science+Business Media, LLC 2011

enables the rapid characterization of cells through surface protein expression profiles. Similar to other high-throughput proteomics profiling methods, BD FACS™ CAP yields an abundance of information that is very useful for identifying complex protein expression profiles of cells. In this respect, BD FACS™ CAP is currently being used to characterize the surface expression of hundreds of cell surface proteins on many different cell types. This profiling technology will enable users to discover subsets of surface protein markers that can be used to identify uniquely specific cellular populations that were previously uncharacterized. Because BD FACS™ CAP is a flow cytometry-based screening technology, it has all of the advantages that flow cytometry offers, most importantly the ability to collect multiparametric protein expression data on single cells rapidly.

The current screening plate configuration consists of 229 directly conjugated antibodies, with the extension of this configuration inevitable as additional antibodies and fluorochrome conjugates become available. The antibodies in the present format are arrayed in a 96-well plate as three-color "cocktails," using the fluorochromes, FITC or Alexa488, PE, and APC or Alexa647. Several wells have been intentionally left empty, reserved for compensation and various controls. Isotype controls for selected immunoglobulin classes and subclasses have been included in the screening plates that are used for setting gating thresholds. Of the 229 surface markers, 208 are specific to a single protein, 11 bind small sets of related proteins, and 10 bind to uncharacterized proteins or carbohydrate antigens. The majority of cell surface targets (170 of 229) are surface receptors. Each individual cell type is analyzed on triplicate plates to ensure reproducibility of the data acquired on a flow cytometer equipped with a high-throughput sampler (HTS). The expression level of each marker for the stained cells is then calculated using semiautomated custom flow cytometry software. The process of characterizing these surface markers in a highly efficient manner using BD FACS™ CAP is enabled by automated liquid handling for staining, automated flow cytometry for data acquisition, and standardized algorithms for flow data analysis.

High dimensional analysis of cell surface protein profiles using BD FACS™ CAP can be applied to a wide spectrum of experimental objectives. The main objective of many of these experiments is to identify a differential set of surface markers that are present on one set of cells compared to another, whether this is due to differentiation, different treatment regimens, or other conditions such as varying culture conditions, etc. More specifically, the primary goal may be to identify a set of biomarkers that may be used to isolate different cellular subtypes from a primary cell culture or to identify a unique set of biomarkers that would aid in the clinical diagnosis or predicting prognosis of a certain disease condition.

Additionally, this technology may be used to obtain surface fingerprints for various cells or tissue types, to evaluate starting and ending cellular products for cell therapy or research purposes, or to characterize contaminating cells in a heterogeneous cell line. For stem cell therapy or cell banking, in particular, BD FACS™ CAP is a valuable tool for (1) documenting phenotypic variants or changes due to different donors, or phenotypic differences in cells derived from different sources (e.g., bone marrow vs. fat vs. umbilical cord, etc.); (2) evaluating isolation protocols, passages, and culture or storage media/processes (e.g., examining the correlation between the function and phenotype of a given primary cell type in different culture stages; therefore, identifying a set of markers that can serve as quality control of a specific culture environment); and (3) studying stem cells for differentiation, carcinogenesis, and drug targeting. In addition, this technology may be applied in tandem with the BD Discovery Platform (BD Technologies, Research Triangle Park, NC), a high-throughput screening platform used to develop and optimize cell culture environment for any given cell type. By surveying the expression of integrins and growth factor receptors on the surface of the cell type of interest, appropriate selections of candidate factors can be made to facilitate the development and optimization of the cell culture environment for any specific cell type.

In the future, the BD FACS™ CAP platform may be expanded to include more than the 229 individual antibodies currently used. With the ever-expanding availability of new antibodies and fluorochromes, it is foreseeable that other fluorescence channels could be used to add possibly hundreds of more individual antibodies, or to use "anchor" antibodies to first identify a specific cellular subset of interest, e.g., T regulatory cells, and then to analyze the surface expression of all 229 markers on the specific cellular subtype. With further development of this technology, it may also be possible to include activation state analysis of the cells by using other flow-based technologies that interrogate intracellular proteins such as cytokines or phosphorylated proteins in combination with the BD FACS™ CAP platform.

2. Materials

2.1. Supplies and Equipment

1. Lyophilized BD FACS™ CAP plates (BD Biosciences) (see Note 1).

2. Cytometer: A cytometer equipped with a 488-nm laser and a 633-nm laser, and the appropriate filter sets to detect FITC/ Alexa Fluor 488, PE, and APC/Alexa Fluor 647 are required. In addition, the cytometer must be equipped with a HTS capable of running 96-well plates. This technology has been

used successfully on the BD FACS™ Calibur equipped with CellQuest software, BD FACS™ Canto equipped with FACS-Diva™ software, and BD LSRII equipped with FACSDiva™ software (BD Biosciences).

2.2. Reagents and Cells

1. Antibodies: All monoclonal antibodies are directly conjugated to fluorescein isothiocyanate (FITC), Alexa Fluor™ 488, R-phycoerythrin (PE), allophycocyanin (APC), or Alexa Fluor™ 647. The antibodies are formulated as three-color cocktails in 79 wells of a 96-well plate at experimentally defined saturating concentrations (see Table 1).

2. Human FcR blocking reagent (Miltenyi Biotec, Auburn, CA).

3. Wash buffer and stain buffer (BD Pharmingen, San Diego, CA).

4. 1% paraformaldehyde.

5. BD CompBeads™ Plus (BD Biosciences) (see Note 2).

6. (Optional) LIVE/DEAD® fixable dead cell stain kit (Invitrogen, Carlsbad, CA) (see Note 3).

7. Any cell type that is amenable to flow cytometry may be analyzed by BD FACS™ CAP.

2.3. Analysis Software

Custom scripts written using the R software (1) were utilized for analysis, as described in Subheading 3.3.1. R is an Open Source environment for the implementation of statistical computing and graphics. It provides a large, well-integrated collection of tools for data management, data storage, data analysis, and graphical displays. There is a very large community of R users that provide custom packages to address specialized problems such as those discussed here. R includes a very flexible and effective programming language for automating many complex procedures, which provides the basis for our custom scripts. This combination of specialized analysis packages and custom scripts allows for a highly specialized, robust, efficient, semiautomated approach to flow cytometry data analysis.

3. Methods

3.1. Staining Protocol

1. Harvest the cells to be stained and determine cell number and viability. Ensure that the viability is high in each sample to be processed. If the samples have low viability, a live/dead cell discrimination dye may be added to the staining protocol to minimize the false-positive rate.

2. Resuspend cells to a concentration of 10^7 cells per 90 μL of stain buffer.

Table 1
A current list of specificities included in the BD FACS™ CAP screening plate

CD1a	CD24	CD45RB	CD66b	CD102	CD135	CD183	CD244	Lymphotoxin b receptor
CD1b	CD25	CD46	CD69	CD103	CD137	CD184	CD252	MIC A/B
CD1d	CD26	CD47	CD70	CD104	CD137 Ligand	CD185	CD256	TRA-1-60 antigen
CD2	CD27	CD48	CD71	CD105	CD138	CD191	CD267	Integrin b7
CD3	CD28	CD49a	CD72	CD106	CD140a	CD192	CD268	TNF
CD4	CD29	CD49b	CD73	CD107a	CD140b	CD193	CD271	Leukotriene B4 receptor
CD4 v4	CD30	CD49c	CD74	CD107b	CD141	CD194	CD273	SSEA3
CD5	CD31	CD49d	CD75	CD108	CD142	CD195	CD275	SSEA4
CD6	CD32	CD49e	CD77	CD109	CD146	CD196	CD278	CLA
CD7	CD33	CD49f	CD79a	CD110	CD147	CD197	CD282	NKB1
CD8	CD34	CD50	CD79b	CD112	CD150	CDw198	CD294	fMLP receptor
CD9	CD35	CD51/61	CD80	CD114	CD151	CDw199	CD305	NKp46
CD10	CD36	CD53	CD81	CD116	CD154	CD200	CD309	MET
CD11a	CD37	CD54	CD85	CD117	CD158a	CD201	CD318	TRA-1-81 antigen
CD11b	CD38	CD55	CD86	CD118	CD158b	CD205	CD321	HLA-A,B,C
CD11c	CD39	CD56	CD87	CD119	CD161	CD206	CD326	HLA-DR
CD13	CD40	CD57	CD88	CD120b	CD162	CD208	CDw329	P-glycoprotein
CD14	CD41a	CD58	CD89	CD122	CD163	CD210	CD332	PAC-1
CD15	CD41b	CD59	CD90	CD123	CD164	CD212-β1	CD333	FMC7
CD16	CD42a	CD61	CD91	CD124	CD166	CD212-β2	CD336	HER-2/neu
CD16b	CD42b	CD62E	CDw93	CD126	CD172a	CD220	CD337	Hematopoietic progenitor cell
CD18	CD43	CD62P	CD94	CD127	CD172b	CD221	CD349	EGFR
CD19	CD44	CD62L	CD95	CD130	CD178	CD226		TGFBR2
CD20	CD45	CD63	CD97	CD132	CD180	CD227		TWEAK
CD21	CD45RA	CD64	CD98	CD133	CD181	CD231		NKG2D
CD22	CD45RO	CD66	CD99	CD134	CD182	CD235a		

3. Add 10 µL of FcR blocking reagent per 10^7 cells, mix well, and incubate at 4°C for 10 min.

4. Dilute the cell suspension with BD stain buffer to a concentration of 15×10^6 cells/30 mL so that each 100 µL contains 5×10^4 cells.

5. Add 100 µL of this cell suspension to the staining wells of the 96-well plate containing the lyophilized antibody cocktails.

6. Add cells to the empty wells for assay setup and preparation of compensation controls.

7. The compensation markers should be chosen according to the specific cell type that is being analyzed.

8. Incubate the plates at 4°C for 30 min.

9. Add 100 µL of BD wash buffer to each well, centrifuge at $300 \times g$ for 5 min, and decant the supernatant.

10. Add an additional 250 µL of BD stain buffer to each well, mix, centrifuge at $300 \times g$ for 5 min, and decant the supernatant.

11. Resuspend the stained cells in 200 µL of freshly prepared 1% paraformaldehyde.

12. Store the plates at 4°C in the dark until acquisition.

13. Data should be acquired within 24 h after fixation.

3.2. Data Acquisition

Data can be acquired on a BD FACSCanto™ or BD LSRII flow cytometer equipped with a HTS. Data acquisition is performed with BD FACSDiva™ software.

1. Create a gate on a dot plot of forward scatter (FSC) versus side scatter (SSC) defining the viable cell population of interest.

2. Collect a minimum of 2,000 events per well.

3. During acquisition, closely monitor histogram plots of all the relevant fluorochromes to ensure that no anomalous events occur.

4. If a viability dye is used, first gate on the live cells to ensure that only viable cells are plotted on the FSC versus SSC plot described above.

3.3. Data Analysis

Successful data management is one of the most critical aspects of plate-based high-throughput flow cytometry. Depending on the design of the plate, there may be as many as 96 separate FCS files. For the example used in this chapter, the FCS files require about 130 Mb of storage space per plate. The authors' data management strategy requires manual copying of files from the cytometer to a central hard drive that supports our core flow laboratory. From there, the files are copied one more time to a server in our corporate data center that provides a secure, version controlled location. We use Subversion (http://www.eclipse.org/subversive/) and Eclipse (http://www.eclipse.org) along with standard backup processes implemented by our corporate data center to ensure the persistence and integrity of the original data.

3.3.1. Software

BD FACS™ CAP output from the cytometer is analyzed using semiautomated routines in R to ensure that the data are analyzed in an efficient, consistent, and objective fashion. The automated methods were developed to reflect the analysis that would be performed by an expert cytometrist using FlowJo (Tree Star Inc., Ashland, OR). The results of the R-based analysis were validated by direct comparison to FlowJo results (data not shown).

1. Use R 2.9.1 along with the R packages flowCore 1.10.0 (2), plateCore 1.2.1 (3), flowViz 1.8.0 (4), KernSmooth 2.23-2 (5), and ggplot2 0.8.3 (6). flowCore, plateCore, and flowViz are flow-specific R packages developed to make processing, analyzing, and visualizing of flow data easier. In addition, plateCore provides a variety of built-in functions for managing plate-based flow cytometry data, which greatly increases the efficiency of analysis for BD FACS™ CAP. ggplot2 is an advanced graphics plot that greatly increases our ability to manage the many complex graphical displays required for data analysis.

2. In addition to these R packages, we use a collection of R-scripts and functions (developed by BD Biosciences) to manage the work flow and to automate much of the work that would have to be done by the cytometrist using standard GUI-driven software packages. All development of R-scripts is done using the integrated development environment (IDE) Eclipse (http://www.eclipse.org) along with the Eclipse plug-in, StatET (http://www.walware.de/goto/statet/). Standardized guidelines are followed for code development, and RUnit (http://www.r-project.org) testing is performed to ensure correctness of the code. All scripts are securely maintained under version control using the technologies described earlier.

3.3.2. Compensation

Compensation can be done using either single-color staining of the same cells that are used for analysis, or BD CompBeads.

1. Conduct a preliminary study to identify the appropriate marker/dye combinations.

2. Run compensation samples in tubes each day and apply the same compensation results to the three plates that are run on that day. The current layout of the plate that is being used for new samples has wells devoted to compensation.

3. Compute the compensation matrix using plateCore and flowCore. Results have been validated against FlowJo (data not shown).

3.3.3. Morphology Gating

It is of critical importance to examine the health status of the samples used for the FACS™ CAP assay. Compromised viability of cells contributes to increased nonspecific binding of the antibodies and can confound interpretation of flow data. For special applications, a near-infrared LIVE/DEAD® fixable dead cell stain kit can be incorporated to evaluate the viability of the cells after the surface marker staining process. For most applications, however, it is sufficient to restrict data analysis to a cell population where dead cells, debris, and aggregates are excluded based on cell morphology.

1. Evaluate cell morphology based on measurements of FSC and SSC.

2. Select cells with normal morphology for further analysis using a standard gate (Fig. 1). In general, cells are very stable across

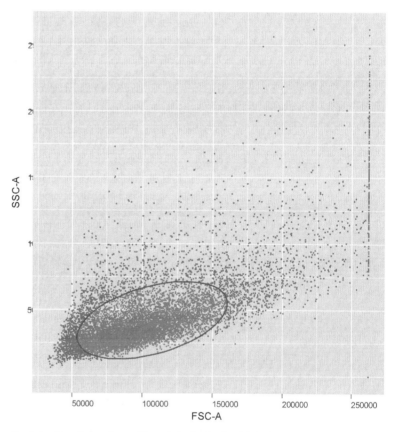

Fig. 1. A gating strategy is used to exclude dead cells, debris, and aggregates prior to data analysis. Cells with "normal morphology" are identified by use of an ellipsoid gate in the dimensions forward scatter (FSC) and side scatter (SSC). The *encircled lighter cells* (red in the electronic version) fall inside the normal morphology gate and are considered in further analyses. The *darker cells* (outside the gate, blue in the electronic version) fall outside the normal morphology gate and are excluded from further analyses.

wells and multiple plates, and a single, fixed gate is used for one FACS™ CAP run.

3.3.4. Autofluorescence Correction

Autofluorescence is one of the major concerns in flow cytometry. One characteristic of cultured adherent cells is that they have a significant level of autofluorescence (7). Thus, an additional signal is present in each channel that cannot be corrected for by compensation and confounds any attempt to identify positive cells (see Fig. 2). Additionally, the output from R has been corrected for autofluorescence using the method described by Hahne et al. (8), which facilitates automated gating for cells with high autofluorescence.

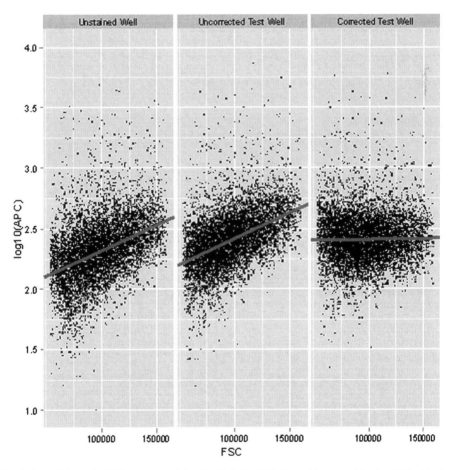

Fig. 2. The *first panel* shows log10 fluorescence intensity (FI) in the APC channel in an unstained well. The *line* (red in the electronic version) shows a linear relationship between cell size, as measured by FSC, and FI in the absence of any antibody. This FI signal is referred to as autofluorescence. The *middle panel* shows the results from a test well that has an antibody conjugated to a dye that is measured in the APC channel. A naïve analysis might classify some of these cells as having positive expression for the corresponding marker. The *rightmost panel* shows the same test well after correction for autofluorescence, and clearly removes the influence of autofluorescence from the determination of positive marker expression levels.

3.3.5. Determining Gates Based on Isotype Controls

On each BD FACS™ CAP plate, there are a variety of controls, including isotype controls for selected immunoglobulin classes and subclasses. These isotype controls are used to set thresholds (negative control gates). Any cell that has a measured fluorescence intensity (FI) above the threshold is classified as "positive."

1. For each isotype control, set the initial threshold at the 99.5%-tile (see Fig. 3). Typically, the isotype controls are very similar from plate to plate for the same cell source and typically have CVs of less than 5%.

2. Review the representative density plots and make corrections for any thresholds that appear anomalous. Once the threshold/gate has been finalized for each fluorescence channel, the

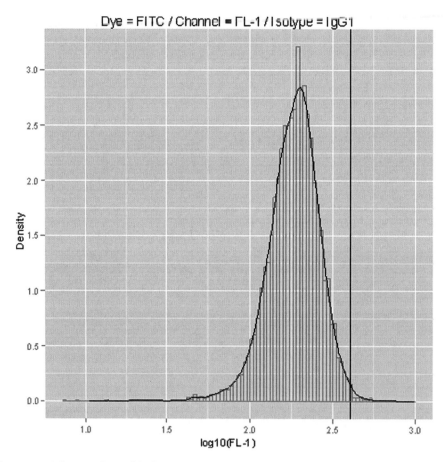

Fig. 3. Two representations are shown of the fluorescence intensity in the FL-1 channel (530/30 bandpass filter) for the FITC/IgG1 isotype control. The *x*-axis is log10 FI. The bars (red in the electronic version) show a histogram as would be commonly generated using a program such as FlowJo. In this case, the height of each bar is proportional to the number of cells with FI in the corresponding range. Flow cytometrists also often refer to this histogram as a density. The heavy black line is a smoothed density curve generated by the R package KernSmooth using the bandwidth parameter 0.03. In the rest of this article, when we refer to a density, it will be defined as in this illustration. We use the smoothed density representation as a simplified, more visually appealing representation of the distribution of fluorescent intensities that is less dependent on random fluctuations or noise in the data. The vertical line is the threshold calculated as the 99.5%-tile of the FI values.

percentage of positive cells reported reflects the fraction of cells with a fluorescent signal higher than that of 99.5% of the negative control cells.

3.3.6. Quality Control

The use of replicate plates of cells in the BD FACS™ CAP analysis primarily serves the purpose of quality control, as under normal circumstances, there is very little difference in the results among replicate plates (see Subheading 3.3.7). Although still relatively rare, the most common quality issue stems from fluidics problems that occur during the acquisition of cells by the flow cytometer.

1. To detect fluidics problems, examine empirical cumulative distribution function (ECDF) for either SSC or FSC. Figure 4a shows an example of how the ECDF plots look for two wells, one of which was normal for all three replicate plates and one of

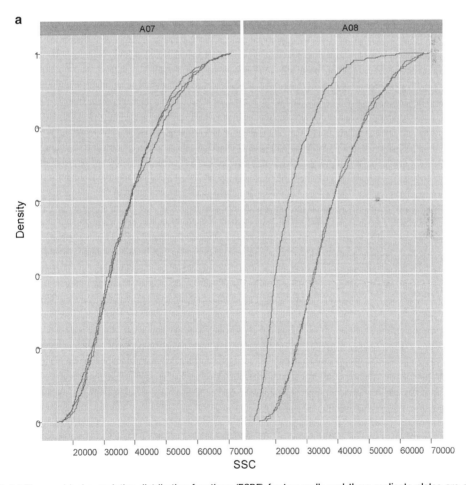

Fig. 4. (a) The empirical cumulative distribution functions (ECDF) for two wells and three replicate plates are shown represented by 3 lines. The y-value on an ECDF gives an estimate of the probability that a randomly selected event will have a value less than the corresponding x-value. Well A07 shows a normal agreement among the three replicate wells. Well A08 shows that there was a problem in one of the plates.

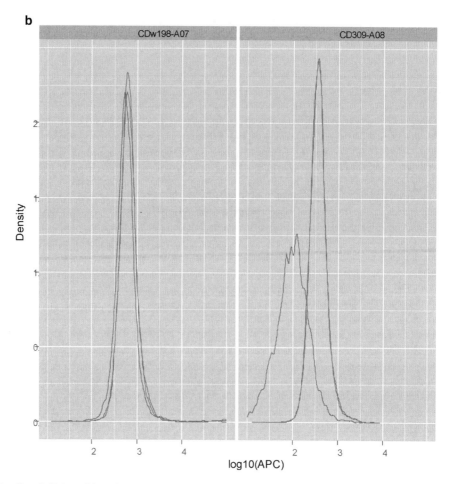

Fig. 4. (continued) (**b**) In well A07, the densities for the FI of CDw198 are very similar. In well A08, the densities of CD309 reflect the impact of the quality problem for the same plate as in (**a**). Both the antibodies shown were conjugated to dye APC. Results that are flagged as being anomalous during the QA process are marked as outliers and not included in the subsequent statistical analysis.

which had a quality problem. Figure 4b shows how the QA problem affected one of the markers measured in that well.

2. Mark all the results corresponding to the well with QA problem as outliers and exclude them in any subsequent statistical analysis.

3.3.7. Summary Statistics   The primary summary statistic is the Percent Positive; that is, the percentage of cells that have a reading greater than the value of the negative control gate, as determined from the isotype controls. Depending on the particular application, a variety of other statistics may be of interest. Due to the highly reliable nature of flow cytometry, the variability in Percent Positive from plate to plate is usually quite small. For a typical donor in the present example, the mean standard error across three plates over all markers for

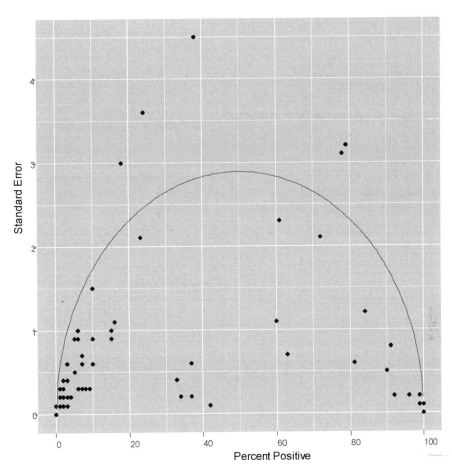

Fig. 5. For a selected donor, the mean Percent Positive from three replicate plates is shown for all markers on the x-axis. The y-axis shows the corresponding standard errors. The curve represents the expected standard error if the plates represented true biological replicates. As discussed in the text, the replicate plates should be viewed as technical replicates displaying less variability, in general, than would be expected from true biological replicates.

Percent Positive was less than 1%. From a theoretical viewpoint, we would expect the Percent Positive to follow a binomial distribution when generated from independent samples of the same cells. Figure 5 shows that the observed variability from plate to plate is generally lower than what would be expected if the plates represented true biological variability rather than simply technical replicates.

3.3.8. Normalization

For the purposes of understanding and visually comparing marker expression profiles, use the smoothed density (see Fig. 3 and (5)). For the purposes of comparing multiple marker expression across multiple donors or multiple samples, combine the data from the replicate plates to get a single density.

1. When combining the data from multiple plates, adjust for minor differences in instrument settings or cell handling conditions by first calculating the average value of the negative

control gates. Shift the values for each of the replicates by the difference between the control gate for that replicate and the average. Combine the data and calculate the final density.

2. When comparing data from multiple samples, adjust the densities in a similar fashion but do not recompute them. In this way, a single value can be used to represent the negative threshold gate across multiple samples, which greatly simplifies the display and makes it much easier to compare marker profiles across samples.

3.3.9. Categorization and Interpretation of Expression Levels

Because BD FACS™ CAP is a screening assay, the reported Percent Positive values should be interpreted with caution. In any case in which it is important to have an accurate estimate of the actual Percent Positive value for a given cell type, it is recommended that an optimized assay be used. Consequently, Percent Positive values are typically classified into one of four fairly broad categories; namely, negative, low, medium, or high.

(a) Negative corresponds to values of markers with Percent Positive ≤10%.

(b) The low category designates markers with 10% < Percent Positive ≤50%.

(c) The medium category designates markers with 50% < Percent Positive ≤85%.

(d) The high category designates markers with Percent Positive >85%.

Markers falling in the negative and high categories can be expected to stay in those categories when an optimized assay has been developed. The low and medium categories are expected to be less stable.

Summary data are presented in Fig. 6 for 79 representative cell surface proteins included in the FACS CAP screen. The data are characterized as described above and then assigned a color code and presented in a heat map. In this form, one can get a general idea of cell-specific protein expression patterns, donor-to-donor variability, and/or treatment-specific differences in cell surface protein expression. The dendograms at the top and side of Fig. 6 show the results of a cluster analysis. In particular, samples from different donors of the same cell type were always more closely related to each other than to samples from any other cell type. In addition, similarity to a different cell type could also be discovered as exemplified by the comparison of profiles of the fibroblasts (NHDF) and mesenchymal stem cells (hMSC1). Clustering also shows that there are groups of markers with similar patterns of expression levels that vary among the different cell types.

In many applications of flow cytometry in which a Percent Positive value is calculated, there are two distinct populations.

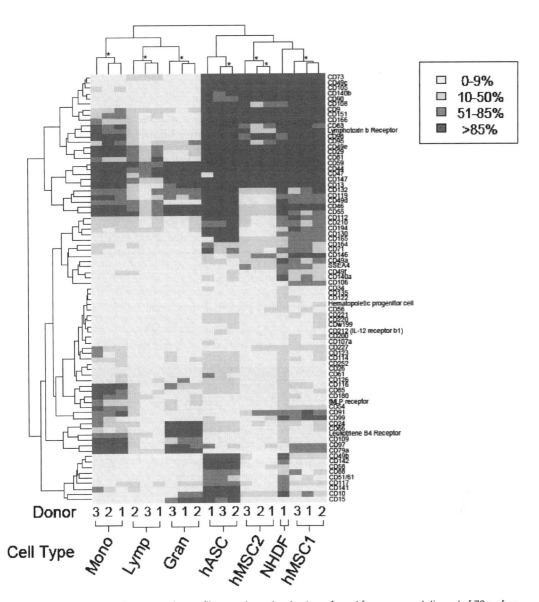

Fig. 6. Percent positive marker expression profiles are shown in a heatmap format for a representative set of 79 surface markers on varying cell types. Analysis of three donors for monocytes (Mono), lymphocytes (Lymp), granulocytes (Gran), adipose-derived stem cells (hASC), and two unrelated samples of bone marrow-derived mesenchymal stem cells (hMSC1 and hMSC2) is presented. One donor sample is shown for dermal fibroblasts (NHDF). The rows and columns are ordered based on a cluster analysis, whose results are represented in the dendrograms at the *top* and *left* of the figure. In, for example, the dendrogram at the top of the heatmap, cell samples that are most similar based on percent positive values on the markers are joined at the lowest levels of the dendrogram. Groups are merged successively based on the cluster analysis as the tree reaches the top of the dendrogram. In this case, samples from different donors of the same cell type were always more closely related to each other than to samples from any other cell type. The clustering shown on the side of the figure shows that there are groups of markers with similar patterns of expression levels that vary among the different cell types.

In this case, one population is typically negative for the marker in question and one population is positive. The Percent Positive is the percent of cells falling in the positive population. In the case of adherent cells, it is more likely that there is only a single population, and the negative control gate falls in the interior of this population. This can be seen for a representative set of markers in Fig. 7. When there is only a single population, a more accurate interpretation of the Percent Positive is "the percentage of cells with a measure intensity value greater than the negative control gate."

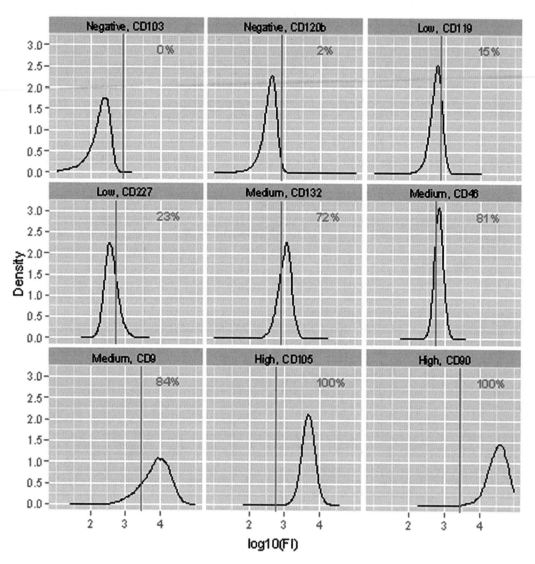

Fig. 7. Marker expression profiles (densities) are shown for a representative set of markers. The markers were chosen to span the range from 0% to 100% for Percent Positive values. The assigned expression level categories and marker name are shown in the panel titles. The negative control gate is shown as a vertical line on each panel along with the value for Percent Positive. This illustrates the interpretation of the Percent Positive as the percentage of cells greater than the negative control gate as opposed to the interpretation of the percentage of cells in a "positive" population.

3.3.10. Comparing Samples

In the present example, BD FACS™ CAP was performed on human MSCs derived from three donors and prepared as described by the Arnold Caplan Laboratory (9, 10). Three replicate plates were assayed for each donor. As discussed earlier, the replicate plates represent technical (not biological) variability; the replication is primarily used for quality purposes. Data from replicate plates are combined as discussed in Subheading 3.3.8 before comparing results among the three donors.

The test to compare the Percent Positive values among donors is based on the null hypothesis that the donors are not biologically different in regard to expression of a specific marker. In this case, the assumption is that the observed values for Percent Positive will follow a binomial distribution. The ratio of the observed variation in Percent Positive to the expected variation should follow a Chi-squared distribution. In this particular case, the distribution will only be approximate because the true Percent Positive is estimated from the data. Based on this chi-square distribution, a *P*-value is calculated for the test of the null hypothesis for each marker.

Since there are more than 200 markers in the BD FACS™ CAP screen, the multiplicity problem must be addressed before computing final *P*-values. Multiplicity problem means that the risk of making false discoveries is extremely high when making many independent tests. To control for this false discovery rate, apply the method of Hochberg and Benjamini (11). After making this correction, an adjusted *P*-value of less than 0.05 will indicate that the samples (donors in this case) have different expression profiles.

1. When the samples are not statistically different, assign a category (negative, low, medium, or high) to the marker that holds across all samples based on the discussion in Subheading 3.3.9 using the mean Percent Positive.

2. When the null hypothesis is rejected, assign the category "Different."

3. For any markers in which there are distinct subpopulations, mark these populations as "Heterogeneous" and do not consider the results of the statistical test to be meaningful because of the difficulty in interpreting the Percent Positive value.

	Negative	Low	Medium	High	Different	Hetero-geneous	Total
Frequency	167	12	8	17	7	0	211

The marker expression profiles for four representative markers that fall into the "Different" category (i.e., for which the null hypothesis is rejected) are shown in Fig. 8.

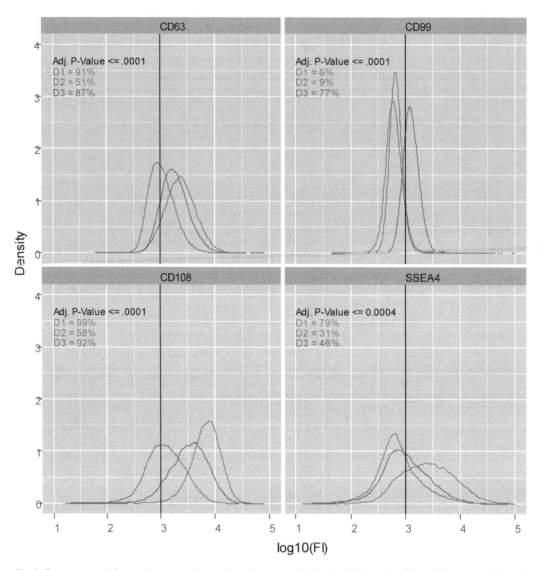

Fig. 8. Four representative markers were chosen from the seven that had statistically significant differences in Percent Positive among the three donors. The *P*-value after the Benjamini Hochberg multiplicity adjustment is shown on each panel. The Percent Positive values are also shown for each donor. For the purposes of display, the *x*-scale densities have all been normalized so that the negative control gate has the value of 3.0 for all samples.

3.3.11. Comparison of the Results to the Literature

Based on a review of the literature, the authors identified 12 markers that are commonly reported as being negative for bone marrow-derived MSCs and 15 markers that are commonly reported as being positive (see Note 4). The results for BD FACS™ CAP for these markers on the current set of donors are shown in Fig. 9. Ten of the markers reported as negative in the literature are clearly negative for these donors. One marker (CD24) shows a low expression level and one marker (SSEA4) shows an intermediate level with differences among donors (see Fig. 8 for more information on SSEA4). Of the 15 markers

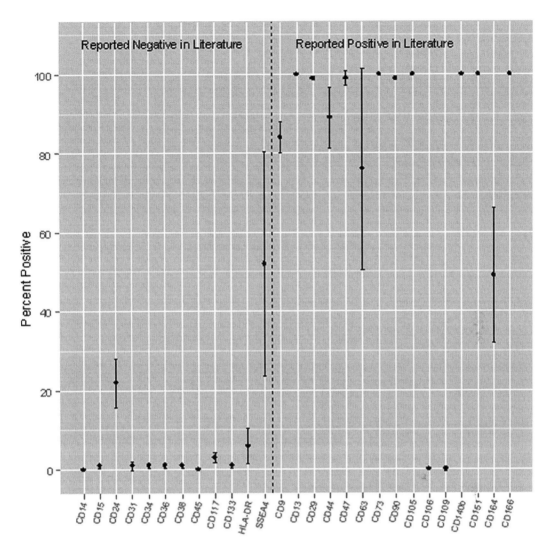

Fig. 9. Based on a review of the literature, we identified 12 markers that are commonly reported as negative and 15 markers that are commonly reported as positive on bone marrow-derived MSCs. The average Percent Positive ±2 standard errors are shown for each of these markers for the three donors in this study.

reported as positive in the literature, 9 were approximately 100% positive, while 2 markers were negative. The other four markers were expressed at varying levels.

3.3.12. Heterogeneous Populations

A population is designated as heterogeneous in regard to expression of a particular marker if there are at least two distinct populations of cells. In this case, the density function will be bimodal, with each peak representing a different population of cells.

1. To identify multiple populations, use a peak-hunting algorithm on each density.

2. Carefully review markers with apparent multiple populations before officially classifying them as heterogeneous.

Results of a validation study (data not shown) suggest that subpopulations as small as 10% of the total population can be identified, as long as they are well separated. Smaller subpopulations may be identified if more cells are analyzed. Figure 7 shows representative density functions that do not represent heterogeneous populations. For the MSC donors reported in the present example, no markers were identified as having heterogeneous expression.

4. Notes

1. Preparation of lyophilized BD FACS™ CAP plates:

    (a) The plates used in the FACS CAP screening are produced from a master plate containing 79 unique three-color antibody cocktails, each well formulated to contain saturating concentrations of each individual antibody. Row A of the master plate is intentionally left empty in order to allow the addition of unstained cells for setup as well as the addition of single stained cells for compensation controls. Wells B1 to B5 contain relevant isotype controls and wells B6 to H12 contain the three-color antibody cocktails.

    (b) The master plate is used to produce daughter lyophilized plates (Lyoplates) in an automated workflow that minimizes any possible operator-introduced variability in the final screening plates. These daughter plates are then lyophilized in order to maximize stability and to simplify the storage conditions. The lyophilized plates are stable for up to 2 years when stored at room temperature in dessicated sealed foil pouches.

    (c) On the day of the assay, three lyoplates are opened for each cell sample being analyzed and the cell suspension is immediately added to the plates to rehydrate the staining cocktails.

2. BD CompBeads may also be used for compensation if analyzing a cell type that is uncharacterized and, therefore, no known markers exist to enable cellular compensation setup.

3. Viability dye should be used when analyzing a cell sample with low viability. Typically, a FACS CAP experiment should be carried out on cells with viability greater than 90%. If this is not possible, there may be a significant false-positive rate due to the dead cells nonspecifically binding antibodies. Incorporation of a viability dye will decrease this false-positive rate.

4. The performance and sensitivity of the BD FACS™ CAP platform has been thoroughly characterized in internal validation studies, which included analyses of both adherent and suspension cell types. Sets of expressed markers found using BD FACS™ CAP on these adherent and suspension cells are generally in good agreement with published marker phenotypes. Since the BD FACS™ CAP is intended for use with many different cell types, the concentrations of the 229 antibodies employed are not necessarily optimal in every individual case. Due to inherent complexities of gating systematically across hundreds of markers, information from unstained isotype control samples is used to estimate the thresholds (negative control gates). Further antibody concentration optimization may be required for markers with negative or low expression levels to establish definitively whether or not the marker is actually expressed. Additionally, nonspecific staining introduced by a significant number of unhealthy or dead cells, as often seen in frozen samples, can be another confounding factor for data interpretation. When interpreting BD FACS™ CAP data, it is important to bear in mind that this is a profiling technology – much like microarray-based gene expression analysis – in which, to some degree, a trade-off occurs between high throughput (enabling the simultaneous interrogation of a large number of surface markers, or of genes, to yield a profile) on the one hand and sensitivity/precision on the other. That is to say, in neither FACS™ CAP nor in microarray-based gene expression analysis is each individual marker's assay optimized for a particular cell type and experimental condition, so that the results are qualitative (or, at best, semi-quantitative), providing a foundation and guide to further, more optimized, experimentation, which the end-user may wish to pursue.

Acknowledgments

We would like to acknowledge the following individuals for important contributions to the development of the FACS™ CAP technology: Keith Deluca, Megan Gottlieb, Errol Strain, Julie Leonard, Dylan Wilson, Stacy Xu, Jamal Sirriyah, John Dunne, Sharon Presnell, David Hodl, Mary Meyer, and William Busa. For the construction of the antibody plates, we would like to acknowledge the excellent technical assistance provided by Christine Chuang.

References

1. R Development Core Team (2009) R: A language and environment for statistical computing. R Foundation for Statistical Computing, Vienna, Austria (http://www-R-project.org).

2. Hahne, F., Le Meur, N., Brinkman, R. R., Ellis, B., Haaland, P., Sarkar, D., Spidlen, J., Strain, E., and Gentleman, R. (2009) flowCore: A bioconductor package for high throughput flow cytometry. *BMC Bioinformatics* **10**, 106.

3. Strain, E., Hahne, F., Brinkman, R. R., and Haaland, P. (2009) Analysis of high-throughput flow cytometry data using plateCore. *Adv. Bioinformatics* (**2009**:356141).

4. Sarkar, D., Le Meur, N., and Gentleman, R. (2008) Using flowViz to visualize flow cytometry data. *Bioinformatics* **24**, 878–9.

5. Wand, M.P. and Jones, M.C. (1995) Kernel Smoothing. Monographs of Statistics and Applied Probability. CRC Press Chapman & Hall, Boca Raton, FL.

6. Wickham, H. (2009) Ggplot2: Elegant Graphics for Data Analysis. UseR!. Springer, New York, NY.

7. Roederer, M. and Murphy, R. F. (1986) Cell-by-cell autofluorescence correction for low signal-to-noise systems: Application to epidermal growth factor endocytosis by 3 T3 fibroblasts. *Cytometry* **7**, 558–65

8. Hahne, F., Arlt, D., Sauermann, M., Majety, M., Poustka, A., Wiemann, S., and Huber, W. (2006) Statistical methods and software for the analysis of high throughput reverse genetic assays using flow cytometry readouts. *Genome Biol.* **7**, R77.

9. Lennon, D. P. and Caplan, A. I. (2006) Isolation of human marrow-derived mesenchymal stem cells. *Exp. Hematol.* **34**, 1604–5.

10. Lennon, D. P. and Caplan, A. I. (2006) Mesenchymal stem cells for tissue engineering, in: *Culture of Cells for Tissue Engineering* (Vunjak-Novakovic, G. and Freshney, R. I., eds.), John Wiley & Sons, Inc., Hoboken, NJ, pp. 23–59.

11. Hochberg, Y. and Benjamini, Y. (1990) More powerful procedures for multiple significance testing. *Stat. Med.* **9**, 811–8.

Chapter 7

Tracking Immune Cell Proliferation and Cytotoxic Potential Using Flow Cytometry

Joseph D. Tario Jr., Katharine A. Muirhead, Dalin Pan, Mark E. Munson, and Paul K. Wallace

Abstract

In the second edition of this series, we described the use of cell tracking dyes in combination with tetramer reagents and traditional phenotyping protocols to monitor levels of proliferation and cytokine production in antigen-specific CD8+ T cells. In particular, we illustrated how tracking dye fluorescence profiles could be used to ascertain the precursor frequencies of different subsets in the T-cell pool that are able to bind tetramer, synthesize cytokines, undergo antigen-driven proliferation, and/or carry out various combinations of these functional responses.

Analysis of antigen-specific proliferative responses represents just one of many functions that can be monitored using cell tracking dyes and flow cytometry. In this third edition, we address issues to be considered when combining two different tracking dyes with other phenotypic and viability probes for the assessment of cytotoxic effector activity and regulatory T-cell functions. We summarize key characteristics of and differences between general protein- and membrane-labeling dyes, discuss determination of optimal staining concentrations, and provide detailed labeling protocols for both dye types. Examples of the advantages of two-color cell tracking are provided in the form of protocols for (a) independent enumeration of viable effector and target cells in a direct cytotoxicity assay and (b) simultaneous monitoring of proliferative responses in effector and regulatory T cells.

Key words: Cell tracking, CellVue® dyes, CFSE, Cytotoxicity assay, Dye dilution proliferation assay, Flow cytometry, PKH dyes, Proliferation analysis, Regulatory T cells

1. Introduction

A multiplicity of fluorescent dyes is now commercially available for cell tracking and proliferation monitoring (reviewed in (1)). Although diverse in their chemistry and fluorescence characteristics, these reagents can be categorized into two main classes based upon their mechanism of cell labeling. Dyes of one class, here referred to as "protein dyes," permanently combine with proteins

Teresa S. Hawley and Robert G. Hawley (eds.), *Flow Cytometry Protocols*, Methods in Molecular Biology, vol. 699, DOI 10.1007/978-1-61737-950-5_7, © Springer Science+Business Media, LLC 2011

by forming a covalent bond. Dyes of the other class, here referred to as "membrane dyes," stably intercalate within cell membranes via strong hydrophobic associations. The term "proliferation dye" will be used here to refer to dyes of either class that (a) exhibit sufficiently good chemical and metabolic stability to partition approximately equally between daughter cells at mitosis, and (b) are sufficiently nontoxic that they can be used to label cells at initial intensities that are high enough to follow the resulting dye dilution through multiple rounds of cell division.

Due to their stability of cell association, cell tracking dyes of both classes have proven useful for in vivo assays of cell trafficking and recruitment in transplant and tissue repair studies (2–5) and for monitoring proliferation, differentiation, and effector functions in stem and immune cell biology (5, 6). In vivo cell tracking using fluorescent dyes also provides information complementary to that provided by MRI using paramagnetic (7) or superparamagnetic (8, 9) particles or polymers. For example, MR-active cell tracking agents offer superior in vivo 3D imaging resolution compared with whole body fluorescence imaging, but current agents either do not allow tracking of cell division history (e.g., MR-active micro- and nanoparticles do not necessarily partition equally between daughter cells) or exhibit greater toxicity than fluorescent cell tracking dyes (10). Similarly, bioluminescent reporter gene imaging is ideal for very long-term tracking studies where proliferation of the labeled population may exceed the detection limit of traditional fluorescent dyes (typically seven or eight generations), because all progeny of stably transfected parental cells will contain the reporter gene (11). Many investigators have found it advantageous to combine genetic labeling with fluorescent cell tracking dyes in order to quantify the number of cells in each generation or assess the frequency of precursors from whence they arose, something that is not possible using genetic markers alone (12, 13). Given the multitude of colors available to choose from (1), it seems likely that methods for combining fluorescent cell tracking dyes with bioluminescent markers will also be developed.

Cell tracking using fluorescent protein and membrane dyes has also proven beneficial for in vitro studies of cytotoxic effector mechanisms (see Subheading 3.3), cell membrane transfer (14, 15), and cell proliferation history (1, 4, 13, 16, 17). In vitro studies of stem/progenitor and immune cell proliferation by flow cytometry are among the most common applications of both classes of cell tracking dyes (reviewed in (1, 5)). This is true largely because of the limitations of alternate methods for proliferation monitoring. Tritiated thymidine (^{3}H-thymidine) incorporation is reproducible and sensitive. However, it presents significant safety and regulatory issues, is ill-suited for analysis of mixed populations at the single cell level, detects only cells actively synthesizing DNA at the time of the pulse, and does not allow for the isolation of daughter cells for further analyses such as immunophenotyping, gene expression, proteomics, or functional studies. Click-iT® EdU technology (available commercially from Invitrogen) detects the incorporation of a modified thymidine analog into replicating DNA under much milder conditions than labeling with bromodeoxyuridine (BrdU), can be detected using a variety of fluorochromes, and is compatible with single-cell analysis by flow cytometry (18). However, it also detects only cells actively synthesizing DNA at the time of the ethynyl-deoxyuridine pulse and, because detection requires mild permeabilization and fixation, is unsuitable for the isolation of viable daughter cells for functional studies.

In selecting fluorescent cell tracking dye(s) for a given study, it is essential to understand the advantages and limitations of different probes in order to match the probe(s) to the needs of the application. In our experience, key considerations include (1) the ability to achieve bright initial staining intensities without altering the expression or function of cellular machinery, or otherwise affecting the functional or proliferative capabilities of labeled cells relative to unlabeled controls; (2) stability of dye–cell association sufficient to ensure that probe is not lost from labeled cells due to degradation and does not transfer to unlabeled cells over the time frame of the assay; and (3) spectral compatibility with available instrument configuration(s) and other fluorochromes to be used. Ideally, the cell-labeling protocol should also be simple, rapid, and robust (i.e., readily reproducible both intra-experimentally and intra-institutionally). In this chapter, we illustrate how these considerations are addressed in the context of two immune function assays, as well as the advantages and limitations associated with combining multiple tracking dyes to increase the information available from a given assay. In particular, we discuss protocols for a direct LAK cytotoxicity assay using PKH67 and CellVue Claret, and an in vitro suppression assay that simultaneously monitors the proliferative capacities of regulatory and effector T cells using CFSE and CellVue Claret.

2. Materials

2.1. Cell Isolation and Cell Culture

1. Complete medium (CM). RPMI 1640 supplemented with 10% heat-inactivated fetal bovine serum (FBS) (Atlanta Biologicals, Lawrenceville, GA), 25 mM HEPES, 0.1 mM nonessential amino acids, 1 mM sodium pyruvate, 2 mM fresh glutamine, and 50 µg/mL gentamicin sulfate and 5×10^{-5} M β-mercaptoethanol.

2. Phosphate-buffered saline (PBS). Prepare 10× stock with 1.37 M NaCl, 27 mM KCl, 100 mM Na_2HPO_4, and 18 mM

KH_2PO_4. Adjust to pH 7.4 with HCl if necessary. Sterilize by 0.2-µm filtration and store at room temperature. Prepare working solution by diluting one part with nine parts of tissue culture grade water.

3. 10% Acid citrate dextrose (ACD, Sigma–Aldrich, St. Louis, MO) in PBS.

4. Formaldehyde, 10%, methanol free, ultra pure (Polysciences, Inc., Warrington, PA). Dilute to 2% in PBS (pH 7.4) and store at 4–8°C.

5. Hanks' balanced salt solution (HBSS) without phenol red, magnesium- and calcium- free (Invitrogen). Store at room temperature until opened, then at 4–8°C.

6. Histopaque®-1077 (Sigma–Aldrich). Store at 4–8°C and use at room temperature.

7. IL-2 (Aldesleukin, Proleukin for injection, NDC 53905-991-01; Novartis, New York, NY). Dilute stock (2.2×10^6 IU/mL) in sterile HBSS to 1×10^5 IU/mL and store at –80°C. Do not refreeze after thawing; store at 4–8°C and discard thawed product after 7 days.

8. Phytohemagglutinin-HA (PHA) (Remel, Lenexa, KS). Prepare stock of 10 mg/mL in CM from powder, sterilize by 0.2-µm filtration, and store frozen at –80°C in single-use aliquots. To induce polyclonal T-cell proliferation, incubate human peripheral blood mononuclear cells (hPBMC) at a final PHA concentration of 5 µg/mL.

9. TRIMA filters (Trima Accel Collection System; CaridianBCT, Inc., Lakewood, CO). WBC-retaining filters were obtained from the pheresis facility at Roswell Park Cancer Institute, after informed consent from donors, and used to isolate hPBMC in quantity for some of the studies described here.

10. Versene, 0.48 mM EDTA·4Na in PBS (Mediatech, Manassas, VA). Adjust to pH 7.4 with HCl if necessary.

11. K562 cell line. Kind gift of Dr. Myron S. Czuczman, Roswell Park Cancer Institute, Buffalo, New York. Also available for purchase, #CCL-243; American Type Culture Collection, Manassas, VA.

2.2. Antibodies

1. Anti-CD3 (clone OKT3) and anti-CD28 (clone 28.2). Azide-free, unconjugated preparations, each at a concentration of 1.0 mg/mL (eBioscience, San Diego, CA).

2. Fluorochrome-conjugated monoclonal antibodies (mAbs). CD4 phycoerythrin cyanine 7 (PECy7, clone SK3), CD25 allophycocyanin (APC, clone 2A3), and CD127 phycoerythrin (PE, clone hIL-7R-M21) (all from BD Biosciences, San Jose, CA); CD45 Pacific Blue (PacBlue, clone HI30) (BioLegend, San Diego, CA).

2.3. Flow Cytometry Reagents

1. 7-Aminoactinomycin D (7-AAD) (Invitrogen, Carlsbad, CA). Reconstitute powdered solid to 1 mg/mL in PBS and store at –20°C. Make a weekly working stock by diluting thawed 1-mg/mL stock to 100 μg/mL in PBS and store at 4–8°C. Add 4 μL of working stock to each 100 μL of cells (4 μg/mL final) and allow the cells to stand on ice for 30 min prior to data acquisition.

2. FCM buffer. PBS (pH 7.2) supplemented with 1% BSA, 0.1% sodium azide, and 40 μg/mL tetrasodium ethylenediaminetetraacetic acid.

3. Human IgG block. Reconstitute human IgG Cohn fraction II and III globulins (Sigma–Aldrich) to 12 mg/mL in RPMI 1640 supplemented with 25 mM HEPES, 20 μg/mL gentamicin sulfate, and 2 mg/mL BSA (Sigma–Aldrich). Store frozen at –20°C until use. Once thawed, store at 4–8°C for no longer than 1 month.

4. LIVE/DEAD® Fixable Violet (Invitrogen). Reconstitute with DMSO according to the manufacturer's instructions. Store frozen at –20°C for no longer than 6 months. Thaw a fresh aliquot daily and dilute 1:50 in PBS. Add 5 μL per test and incubate for 30 min in a buffer free of exogenous protein before washing and fixing in 2% formaldehyde for assessment of viability by flow cytometry.

5. Spherotech AccuCount Ultra-Rainbow Fluorescent Particles (Spherotech, Lake Forest, IL). These are used for single platform enumeration of absolute cell numbers by flow cytometry as described in Subheading 3.3.

2.4. Cell Tracking Dyes

1. CellVue® Claret, PKH26, and PKH67 fluorescent cell linker kits (Sigma–Aldrich). The kits contain 1 mM of stock solutions in ethanol and cell-labeling diluent for general cell membrane labeling (Diluent C). CellVue Claret is also available from Molecular Targeting Technologies, Inc. (West Chester, PA). Store tightly capped at room temperature to avoid evaporation of ethanol and associated increases in dye concentration. If any dye solids are visible, sonicate the dye stocks to resolubilize before use and verify that dye absorbance remains within the range specified in the Certificate of Analysis available for each kit.

2. 5-(and-6)-carboxyfluorescein diacetate, succinimidyl ester (CFDA-SE; Invitrogen). Reconstitute for cell labeling as described in Subheading 3.1. As the nonfluorescent CFDA precursor diffuses across the plasma membrane into the cytoplasm, its acetate substituents are cleaved by nonspecific esterases, forming the fluorescent amino-reactive product, carboxyfluorescein succinimidyl ester (CFSE).

2.5. Flow Cytometer and Other Equipment

1. For routine data acquisition, any flow cytometer capable of acquiring forward and side scatter, PacBlue, FITC, PE, PerCP/PECy5, PECy7, and APC would be appropriate. All of the data except for those presented in Figs. 2b, 4b, and 7 were acquired using an LSRII flow cytometer (BD Biosciences) fitted with a 25-mW 407-nm violet diode laser, a 20-mW 488-nm blue optically pumped semiconductor laser, and a 20-mW 635-nm HeNe laser. From the 407-nm laser, PacBlue or LIVE/DEAD Fixable Violet was detected using a 450/50-nm bandpass filter; from the 488-nm laser, CFSE or PKH67, PKH26, 7-AAD, and CD4 PECy7 fluorescence were detected using 530/30-, 575/26-, 685/35-, and 780/60-nm bandpass filters, respectively. CellVue Claret fluorescence was excited using the 633-nm line and collected using a 660/20-nm bandpass filter.

2. For sorting experiments, any flow cytometer capable of collecting peak area, width, and height and sorting based on forward and side scatter, PE, PECy7, and APC would be appropriate. The data shown in Fig. 7 were acquired on a FACSAria II flow cytometer (BD Biosciences) fitted with a 100-mW 355-nm solid state UV laser, a 100-mW 405-nm violet diode laser, a 50-mW 488-nm blue optically pumped semiconductor laser, and a 40-mW 639-nm red diode laser. From the 488-nm laser, CD127 PE and CD4 PECy7 fluorescence were detected using 575/26-nm and 780/60-nm bandpass filters, respectively. CD25 APC was excited using the 639-nm line and collected using a 660/20-nm bandpass filter.

3. A tube rotator (#13916-822; VWR, West Chester, PA) was used for monocyte depletion of hPBMC and preparation of accessory cells for the studies described in Subheading 3.4.

3. Methods

Virtually any eukaryotic cell can be stained with either class of tracking dye after a single cell suspension has been obtained (see Notes 1 and 2). The labeling conditions described below have been successfully used to stain hPBMC and cultured cell targets used for the immune function assays discussed here, but are likely to require modification for other cell types, assay systems, or dye combinations (see Notes 3–5). Although CFSE is used herein to represent a typical protein-labeling dye, and PKH26, PKH67, and CellVue® Claret to represent typical membrane-labeling dyes, many other tracking dyes are available (see Note 6) and the principles described in Subheadings 3.1 and 3.2 also apply to optimization of conditions for use of those dyes.

3.1. hPBMC Staining with CFSE

1. Prepare a 5-mM stock solution of CFSE (MW 557.47 g/mol) in anhydrous DMSO (see Notes 7–9).

2. Wash cells to be labeled twice in serum-free PBS (or HBSS) and resuspend in serum-free buffer at a final concentration of 5×10^7 cells per mL (range $0.5–50 \times 10^6$ cells/mL; see Notes 5, 10, and 11), using a tube that will hold at least six times the volume of the cell suspension.

3. Immediately prior to cell labeling, prepare a 50-µM working CFSE solution by diluting the 5-mM stock solution of CFSE in DMSO from step 1 into PBS (see Note 12).

4. For a final staining concentration of 5 µM CFSE, add 100 µL of working CFSE solution per mL of cell suspension (e.g., for 2 mL of cells at 5×10^7 cells/mL, add 200 µL of 50 µM CFSE; see Notes 13–15).

5. Immediately vortex the tube briefly to disperse CFSE throughout the cell suspension. Incubate at ambient temperature (~21°C) for 5 min, with occasional mixing either manually or on a rotator (see Notes 16 and 17).

6. Stop the reaction by adding a 5× volume of CM (containing 10% FBS) or a 1× volume of FBS, and mixing well (see Note 18).

7. Wash the cells twice with 5–10 volumes of CM, centrifuge at $400 \times g$ for 5 min at ~21°C, and discard the supernatant. After resuspension of the cell pellet for the second wash, remove an aliquot for cell counting. After the final wash, adjust the cell concentration to 5×10^5 cells/mL during the final resuspension in CM.

8. Assess recovery, viability, and fluorescence intensity profile of labeled cells immediately post-staining to determine whether to proceed with the assay setup (see Note 19 and Figs. 1 and 2).

9. At 24-h post-labeling, verify that labeled but non-proliferating cells (e.g., unstimulated control) are resolved well enough from unstained cells for purposes of the assay to be performed (Figs. 1 and 2) and that CFSE fluorescence can be adequately compensated in adjacent spectral windows used for measurement of other probes such as PE and RFP (see Notes 6 and 20–22). If samples are to be fixed and analyzed in batch mode, verify that loss of intensity due to fixation does not compromise the ability to distinguish desired number of daughter generations (see Note 23 and Fig. 2).

10. Verify that labeled cells are functionally equivalent to unlabeled cells (see Note 24).

3.2. hPBMC Staining with PKH26, PKH67, or CellVue Claret® Membrane Dyes

1. Wash cells to be labeled twice in serum-free PBS or HBSS (see Note 5), using a conical polypropylene tube (see Note 25) sufficient to hold at least six times the final staining volume in step 5. After resuspension of the cell pellet for the second

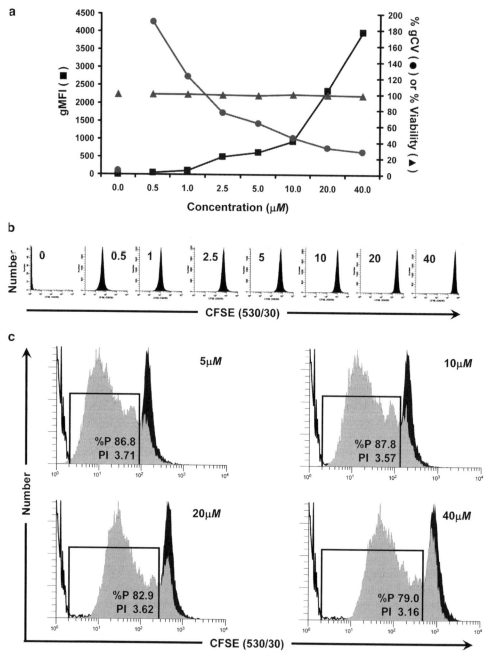

Fig. 1. Considerations for optimization of hPBMC staining with CFSE. The optimal concentration for any tracking dye is that which yields cells that are as brightly and homogeneously stained as possible, while also exhibiting good viability, unaltered cell function, and the ability to compensate adequately for color overlap with other probes to be used. In this study, the maximum tolerated concentration of CFSE was determined for hPBMC isolated from peripheral blood, labeled as described in Subheading 3.1 (final concentrations: 5×10^7 cells/mL and 0.5–40 μM CFSE) and analyzed by flow cytometry immediately upon completion of staining (T_0). (a) The relationships between dye concentration, initial staining intensity (geometric mean fluorescence intensity; gMFI), peak width (calculated as % gCV = geometric SD/gMFI × 100), and viability (% of cells able to exclude trypan blue) are shown. Viability was minimally affected at all concentrations of CFSE tested. Increasing CFSE concentrations led to increasing intensities (gMFI) and decreasing peak widths (gCV). (b) Individual histograms for each test sample shown in (a) were collected at a constant CFSE detector voltage, which was set so that the 40-μM

wash, remove an aliquot for cell counting (see Note 15) and determine the volume needed to prepare a 2× working cell solution at a concentration of 2×10^6 cells per mL in step 3 (range $2{-}100 \times 10^6$ cells/mL; see Table 1 and Note 26).

2. Following the second wash in step 1, aspirate the supernatant, taking care to minimize the amount of remaining buffer (no more than 15–25 μL) while avoiding aspiration of cells from the pellet (see Notes 27 and 28). Flick the tip of the conical tube once or twice with a finger to loosen/resuspend the cell pellet in the small amount of fluid remaining, but avoid significant aeration since this reduces cell viability.

3. To a second conical polyproplene tube (see Note 25), add a volume of Diluent C staining vehicle (provided with each membrane dye kit) equal to that calculated in step 1 for the preparation of the 2× cell solution. Prepare a 2× PKH26 or CellVue Claret working dye solution by adding the appropriate amount of 1 mM ethanolic dye stock to the Diluent C (e.g., add 2 μL of dye to 1.0 mL of Diluent C for a 2× working dye solution for a 2-μM working stock and a final dye concentration of 1 μM after admixture with 2× cells in step 5). Immediately triturate several times, then flick or gently vortex the tube to ensure complete dispersion of dye in the diluent, avoiding deposition of fluid in cap or as droplets on walls. Proceed with steps 4 and 5 as rapidly as possible (see Notes 29 and 30).

4. Prepare a 2× cell suspension by adding the volume of Diluent C calculated in step 1 to the partially resuspended cell pellet from step 2. Triturate three to four times to obtain a single-cell suspension and proceed immediately to step 5. Excessive mixing should be avoided since this reduces cell viability.

5. Rapidly admix the 2× cell suspension prepared in step 4 into the 2× working dye solution prepared in step 3, triturating three to four times immediately upon completion of addition in order to achieve as nearly instantaneous exposure of all cells to the same amount of dye as is possible (see Note 31).

Fig. 1. (continued) sample remained fully on scale in the last decade. Note that at this voltage, unstained cells were not fully on scale in the first decade. (c) Samples stained with the indicated concentrations of CFSE were cultured in the presence of anti-CD3 and anti-CD28 for 4 days (gray histograms) and compared to controls that were CFSE stained but unstimulated (black histograms) or unstained (unfilled histograms) to determine the effect of CFSE concentration on proliferative potential (see Subheading 3.4.5 for description of methods for quantifying the extent of proliferation). Although post-staining viabilities were similar at all concentrations (Fig. 1a), both the proliferative fraction (% proliferating cells; %P) and the proliferative index (fold increase in cell number; PI) decreased at the highest concentration (40 μM) and %P also decreased at 20 μM, indicating that some inhibition of proliferation was occurring at higher concentrations of CFSE.

a

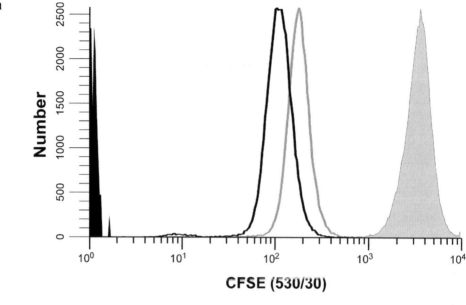

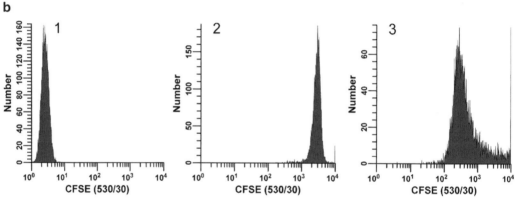

Fig. 2. Additional factors affecting intensity and heterogeneity of CFSE labeling. (**a**) Proliferation-independent sources of intensity loss. hPBMC isolated from peripheral blood were stained with CFSE as described in Subheading 3.1 (final concentrations: 1×10^7 cells/mL, 5 μM CFSE). Unfixed samples were analyzed immediately post-staining (filled gray histogram; gMFI = 3,350, gCV = 35.2%) and after 24 h in culture at 37°C without stimulus; either unfixed (unfilled histogram, *gray line*; gMFI = 182, gCV = 33.0%) or fixed in 2% methanol-free formaldehyde (unfilled histogram, *black line*; gMFI = 109, gCV = 45.0%). Data were collected on an LSR II flow cytometer using a lymphocyte scatter gate and with CFSE detector voltage set so that the unfixed $T = 0$ day (T_0) sample remained fully on scale in the last decade, a voltage at which the unstained T_0 sample (filled black histogram) was not fully on scale in the first decade. Due to the substantial (>tenfold) proliferation-independent intensity loss characteristically seen during the first 24 h, T_0 samples should not be used as compensation controls, and a stabilization period in culture is required before labeled cells are used for in vitro cytotoxicity or proliferation assays (see Notes 20–22). Further intensity losses due to fixation are less pronounced but must be taken into account when selecting optimal staining conditions if fixed samples are to be analyzed in batch mode (see Note 23). (**b**) Effect of staining conditions on CFSE fluorescence distributions. Replicate samples of logarithmically growing cultured U937 cells were stained at 1×10^7 cells/mL with 0.5 μM CFSE for 5 min at either 37°C with occasional mixing after dye addition or at ambient temperature without further mixing (see Subheading 3.1 and Note 16). Stained cells were washed and then analyzed on a CyAn flow cytometer (Beckman Coulter, Miami, FL) using constant instrument settings (HV = 351). *Histogram 1*: unstained control with CFSE detector voltage adjusted to place all cells on scale in the first decade, with few/no cells accumulating in the first channel. *Histogram 2*: cells labeled at 37°C with immediate mixing gave an ideal staining profile, with a bright, homogenously stained, symmetrical population falling in the fourth decade and very few cells accumulating in the last channel (gMFI = 2,817, gCV = 23.4%). *Histogram 3*: Cells labeled at ambient temperature without further mixing were suboptimally labeled at this relatively low concentration as evidenced by their dim, asymmetric, right skewed, staining pattern (gMFI = 468, gCV = 274%).

Table 1
Non-Perturbing Membrane-Dye Staining Conditions for Selected Cell Types

Cell type	Final cell concentration	Final dye concentration	Reference(s)
hPBMC[a] (high cell #)	1×10^7 cells/mL	2 µM PKH67	(49)
	3×10^7 cells/mL	4 µM CellVue Claret	(45)
	5×10^7 cells/mL	5 µM CellVue Claret	Fig. 6
hPBMC[a] (low cell #)	5×10^6 cells/mL	2 µM PKH26	(50)
	1×10^6 cells/mL	1 µM CellVue Claret	Figs. 8 and 10
Cultured cell lines	1×10^7/mL	15 µM PKH26 (U937)	Fig. 4b
	1×10^7/mL	12.5–15 µM PKH26 (U937)	(41)
	1×10^7/mL	1 µM PKH67 (K562)	Fig. 5
	1×10^7/mL	1 µM PKH67 (polyclonal T cell lines)	(15)
	1×10^7/mL	10 µM CellVue Claret (YAC-1)	E. Breslin, personal communication

[a]A low-speed wash ($300 \times g$) post-Ficoll–Hypaque was used to minimize platelet contamination (see Note 11)

6. After 3 min, stop the labeling by adding a 5× volume of CM (containing 10% FBS) or a 1× volume of FBS or other cell-compatible protein, and mixing well (i.e., if 1 mL of cells was combined with 1 mL of dye, then add 10 mL of CM or 2 mL of FBS) (see Note 32).

7. Centrifuge the stained cells ($400 \times g$ for 5 min at ~21°C) and then wash twice in CM. After the first wash, resuspend the cells and transfer them to a clean conical polypropylene tube (see Note 33). After the final wash, count and resuspend the cells to 1.5×10^6 cells/mL in CM.

8. Assess recovery, viability, and fluorescence intensity profile of labeled cells immediately post-staining to determine whether to proceed with assay setup (see Note 19 and Figs. 3 and 4).

9. Verify that labeled but non-proliferating cells (e.g., unstimulated control) are resolved well enough from unstained cells for purposes of the assay to be performed (see Figs. 3 and 4) and that membrane dye fluorescence can be adequately compensated in adjacent spectral windows used for measurement of other probes (see Notes 6, 34, and 35).

10. Verify that labeled cells are functionally equivalent to unlabeled cells (see Note 24).

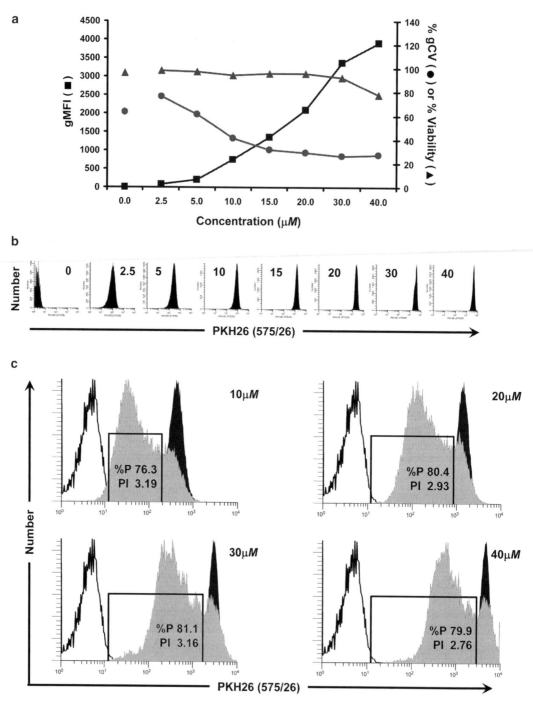

Fig. 3. Considerations for optimization of hPBMC staining with PKH26. In this study, monocyte-depleted lymphocytes isolated from TRIMA filters (Subheading 3.4.1) were labeled with the indicated concentrations of PKH26 in Diluent C as described in Subheading 3.2 (final cell concentration: 5×10^{7}/mL) to determine the maximum tolerated concentration. Cells were analyzed by flow cytometry immediately upon completion of staining (T_0). (a) The effect of PKH26 concentration on staining intensity, CV, and viability was determined in a titration study similar to that described in Fig. 1a. Viability was minimally affected at concentrations up to 30 μM but decreased at 40 μM, most likely due to the ethanol vehicle present in the 1-mM PKH26 dye stock (final concentration in staining step: 4% at 40 μM). As with CFSE, increasing

**3.3. Total Cytotoxicity:
Quantitation
of Cell-Mediated
Killing Using Multiple
Tracking Dyes**

The radioactive chromium (^{51}Cr)-release assay has traditionally been considered the gold standard for determining the cytolytic potential of effector cells (19–21). Although the assay is reliable, it has a number of disadvantages and functional limitations. The major disadvantage is the use of radioactivity, which is potentially hazardous and impractical for some laboratories. Other limitations include difficulty in labeling targets with ^{51}Cr and the spontaneous release of ^{51}Cr from targets, causing extremely high background levels, which makes resolution of effector-mediated lysis difficult. High backgrounds can be particularly problematic for longer term assays (18 h–10 days), which are sometimes required to detect low-frequency effectors (22) or to measure antibody-dependent cytotoxicity (23). The use of flow cytometry and cell tracking dyes to measure cytotoxicity does not require radioactivity and has the distinct advantage of being able to measure killing at the single cell level even when targets and effectors cannot be distinguished on the basis of light scatter. In the simplest format, target cells are labeled with a tracking dye and incubated for a period of time, after which viability is assessed by flow cytometry. However, a wide variety of in vitro and in vivo cytotoxicity assays have been described, in which different combinations of tracking dyes, viability probes, and antibody reagents are used to further characterize effectors, targets, and mechanisms of killing (6, 21–29). The protocol described here uses killing of a cultured cell line (K562) by lymphokine (IL-2) activated killer (LAK) cells as a model system, but the principles and general procedures are applicable to virtually any effector–target combination. In addition to illustrating that a new far-red cell tracking dye (CellVue Claret) does not alter LAK functionality, we discuss two different methods for measuring target cell death: (a) on a relative basis by determining percentage of targets deemed dead based on their inability to exclude 7-AAD, and (b) on an absolute basis by using counting beads to enumerate the number of viable target cells that remain when effectors are present versus when they are absent. The latter method is unaffected by cells lost due to complete lysis and, therefore, is particularly useful for longer term cytotoxicity assays.

Fig. 3. (continued) PKH26 concentrations yielded increasing fluorescence intensities and decreasing peak widths. (**b**) Individual histograms for each test sample shown in (**a**) were collected at a constant PKH26 detector voltage, which was set so that the 40-µM sample remained fully on scale in the last decade. Note that at this voltage, unstained cells were mostly on scale in the first decade. (**c**) Samples stained with the indicated concentrations of PKH26 were cultured in the presence of anti-CD3 and anti-CD28 for 4 days (*gray histograms*) and compared to unstimulated stained (*black histograms*) and unstained controls (*unfilled histograms*). The proliferative fraction (%P) was essentially identical at the three highest concentrations but slightly lower at the lowest concentration (10 µM) due to the overlap of highly proliferated cells with the unstained cell region. Proliferative index (PI) was somewhat reduced at the highest PKH26 concentration (40 µM), suggesting that the rate of expansion had decreased and that the proliferative potential as well as viability had been compromised.

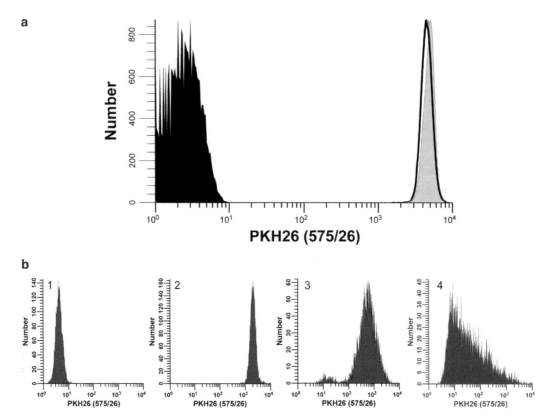

Fig. 4. Additional factors affecting intensity and heterogeneity of PKH26 staining. (**a**) Stability of PKH26 intensity during the first 24 h of culture and after fixation. hPBMC isolated from peripheral blood were prepared and labeled with PKH26 in Diluent C as described in Subheading 3.2 (final concentrations: 5×10^7 cells/mL, 10 μM PKH26). Unfixed samples were analyzed immediately post-staining (T_0) (filled gray histogram; gMFI = 4,874, gCV = 16.0%) and after 24 h in culture (T24) at 37°C without stimulus; either unfixed (unfilled histogram, gray line; gMFI = 4,557, gCV = 17.9%) or fixed in 2% methanol-free formaldehyde (unfilled histogram, black line; gMFI = 4,536, gCV = 17.6%). Data were collected on an LSR II flow cytometer using a lymphocyte scatter gate, and with PKH26 detector voltage set so that the unfixed T_0 sample remained fully on scale in the last decade and the unstained control mostly on scale in the first decade. In contrast to CFSE (Fig. 2a), intensity and distribution differences between T_0 and T24 fixed and unfixed samples were minimal. (**b**) Effect of staining conditions on PKH26 fluorescence distributions. Replicate samples of logarithmically growing, cultured U937 cells were stained with PKH26 (final concentrations: 1×10^7 cells/mL, 12.5–15 μM PKH26) for 3 min at ambient temperature, with or without trituration after addition of 2× cells to 2× dye (see Subheading 3.2 and Notes 29–31). Stained cells were washed and then analyzed on a CyAn flow cytometer using constant instrument settings (HV = 547). *Histogram 1*: unstained control with PKH26 detector voltage adjusted to place all cells on scale in the first decade, with few/no cells accumulating in the first channel. *Histogram 2*: staining with 15 μM dye by addition of 2× cells to 2× dye *with* immediate trituration resulted in a bright, homogenously stained symmetrical population of cells placed in the fourth decade, with no cells accumulating in the last channel (gMFI = 2,548, gCV = 26.2%). *Histogram 3*: staining with 15 μM dye by addition of 2× cells to 2× dye *without* immediate trituration resulted in a reduced intensity and a broader CV (gMFI = 505, gCV = 116%) as well as a dimly stained subpopulation, possibly due to a drop of cells dispensed down the wall of the tube and not well-mixed with the final staining solution. *Histogram 4*: a staining error led to 3 μL of concentrated ethanolic dye stock being added directly to 2× cells in Diluent C without further trituration rather than being used to prepare a 2× dye solution in Diluent C. This resulted in a final dye concentration of 12 μM but yielded extremely dim and heterogenous staining (gMFI = 32.9, gCV = 1,020%). The observed right skewing most likely reflects poor mixing due to the combined effects of widely disparate cell and dye volumes, lack of trituration, and the fact that cells closest to the dye-dispensing point would be exposed to a higher concentration of dye than those farther away.

3.3.1. Generation of Stained LAK Effector Cells

1. Prepare hPBMC from heparinized peripheral blood using the laboratory's standard density gradient fractionation protocol, with the addition of a final low-speed wash step ($300 \times g$) to minimize platelet contamination (see Note 11). Count and adjust to 1×10^8 hPBMC/mL.

2. Stain hPBMC with CellVue Claret at a final dye concentration of 5 µM and a final cell concentration of 5×10^7 cells/mL, according to the procedures described in Subheading 3.2 (see Note 19).

3. Assess recovery, viability, and fluorescence intensity profile of labeled cells immediately post-staining to determine whether to proceed with assay setup (see Note 19).

4. Resuspend labeled hPBMC in CM at 3×10^6 cells/mL (typically 5–10 mL total volume) and incubate upright in a T25 flask with 1,000 IU/mL of IL-2 at 37°C for 4 days to generate LAK effector cells. Set up a parallel flask of unstained hPBMC for use as assay and instrument setup controls (see Subheadings 3.3.3 and 3.3.4).

5. On day 4, harvest LAK effector cells, triturating to disperse any cell clusters into a single cell suspension. Wash once with 50 mL of CM, count, and resuspend at 1×10^7 cells/mL in CM.

3.3.2. Labeling K562 Target Cells

1. On day 4 of the LAK induction period, harvest logarithmically growing K562 targets (see Note 37). Wash twice with 50 mL of HBSS, count, and adjust to 2×10^7 cells/mL in Diluent C for staining.

2. Stain K562 cells with PKH67 at a final dye concentration of 1 µM and a final cell concentration of 1×10^7 cells/mL, according to the procedures described in Subheading 3.2 (see Note 38).

3. Assess recovery, viability, and fluorescence intensity profile of labeled cells immediately post-staining to determine whether to proceed with assay setup (see Note 19).

4. Wash PKH67-labeled K562 targets twice in 15 mL of CM. Count and adjust to 1×10^5 cells/mL in CM.

3.3.3. Cytotoxicity Assay

1. In a 96-well round-bottom plate, make triplicate serial 1:2 dilutions of the LAK effectors as follows: Pipet 200 µL of the stained LAK cell suspension into the first well, and 100 µL of CM into each of seven adjacent wells. Serially transfer 100 µL of LAK cells from the first well to the second, then from the second to the third, etc., ending with a transfer of 100 µL from the seventh well to the eighth well and removal of 100 µL of cell suspension from the eighth well.

2. Add 100 µL of stained K562 targets to each well, creating effector-to-target ratios of 100:1, 50:1, 25:1, 12.5:1, 6.2:1, 3.1:1, 1.6:1, and 0.8:1 (total volume per well: 200 µL).

3. Add 100 µL of targets and 100 µL of effectors to the target-only and effector-only wells, respectively, followed by 100 µL of CM (see Note 39). Incubate the plate at 37°C for 4 h (see Note 40).

4. After the incubation period has elapsed, label test wells directly in the 96-well plate with a saturating amount of anti-CD45 PacBlue on ice for 30 min (see Note 41).

5. Transfer the contents of each well into individually labeled 12 × 75 mm round-bottom tubes compatible with the laboratory's flow cytometer. Wash each well with 200 µL of cold FCM buffer and transfer the wash fluid to the appropriate tube.

6. Wash each sample once with 3 mL of cold FCM buffer and resuspend in 150 µL of FCM buffer.

7. Add 8 µL of 7-AAD (100 µg/mL stock) and 50 µL of Spherotech enumeration beads (stock concentration ~1 × 10^6 beads/mL; final concentration in tube ~2.4 × 10^5 beads/mL) using reverse pipetting technique. Let the setup stand for 30 min on ice so 7-AAD can equilibrate before initiating acquisition of flow cytometric data.

3.3.4. Flow Cytometric Acquisition and Analysis

1. Establish appropriate voltage settings using autofluorescence and single color controls from Table 2 (see Notes 39, 41, and 42).

2. Using the single color controls from Table 2, adjust compensation settings according to your laboratory and/or instrument manufacturer's standard procedures (see Note 39).

3. Acquire data on the flow cytometer using the gating strategy summarized in Fig. 5.

4. Calculate cytotoxicity using the method described in step 5 or 6 (see Fig. 6 and Notes 43 and 44).

5. *Method 1*: Percent cytotoxicity based on 7-AAD uptake by target cells is calculated from Fig. 5, plot 4 as

$$\% \text{cytotoxicity} = \frac{(\text{PKH67}^+ - 7\text{-AAD}^+ \text{ events})}{(\text{total number of PKH67}^+ \text{ events})}$$

Similarly, % viable LAK effector cells is calculated from Fig. 5, plot 6 as

$$\% \text{viable effectors} = \frac{(\text{CellVue Claret}^+ - 7\text{-AAD}^- \text{ events})}{(\text{total number of CellVue Claret}^+ \text{ events})}$$

Table 2
Recommended Assay and Instrument Controls for Measuring Cytotoxicity

	Cells	Label	Treatment	Comments
Assay controls[a]	LAK effectors only	Claret	Incubate with assay samples	*Negative control*: used to calculate spontaneous LAK cell death
	K562 targets only	PKH67	Incubate with assay samples	*Negative control*: used to calculate spontaneous K562 cell death
	K562 (heat killed) + LAK effectors	PKH67 Claret	Incubate with assay samples	*Positive control*: only appropriate for assessment by 7-AAD exclusion, not by bead enumeration
	LAK cells	None	Same E:T ratios as test samples	*Staining control*: used to verify that tracking dye-labeled cells kill equivalently to unlabeled cells[e]
Instrument controls[b]	K562 cells	None	None	Select voltage for LAK/Claret channel[f]
	K562 cells	PKH67	None	Select voltage for K562/PKH67 channel; set color compensation for all other channels; set negative region in LAK/Claret channel (Fig. 5, plot 3)
	K562 cells	CD45 PacBlue	None	Set CD45 threshold or gate to include both targets and effectors (K562 MFI < LAK MFI)
	LAK cells	None[c]	None	Select voltage for K562/PKH67 channel[g]
	LAK cells	Claret	None	Select voltage for LAK/Claret channel; set color compensation in all other channels; set negative region in K562/PKH67 channel (Fig. 5, plot 5)
	LAK cells	CD45 PacBlue	None	Color compensation
	LAK cells	7-AAD	Heat killed[d]	Color compensation

[a] Negative and positive assay controls are included in the experimental plate with test samples, or set up in parallel with the experimental plate, to verify that the expected biological outcomes can be recognized using the chosen instrument conditions

[b] Instrument controls are used to establish instrument voltages and compensation settings

[c] For unstained LAK cell controls, it will be necessary to set up a separate culture of unstained PBMC with IL-2 at the same time as the CellVue Claret-stained PBMC

[d] To heat kill, incubate at 56°C for 30 min. K562 cells could also be used but light scatter properties after heat killing differ substantially from those seen after LAK killing

[e] Needed only to establish optimized staining conditions for tracking dye when assay is first being implemented in the laboratory; not required on a routine basis

[f] Use of unstained LAK to select high voltage for CellVue Claret detector would place unstained K562 cells midscale due to their much greater autofluorescence. Therefore, unstained K562 cells were used instead to maximize dynamic range

[g] Use of unstained K562 to select high voltage for the PKH67 detector would place unstained LAK cells offscale low due to their much lower autofluorescence. Therefore, unstained LAK cells were used instead to maximize dynamic range

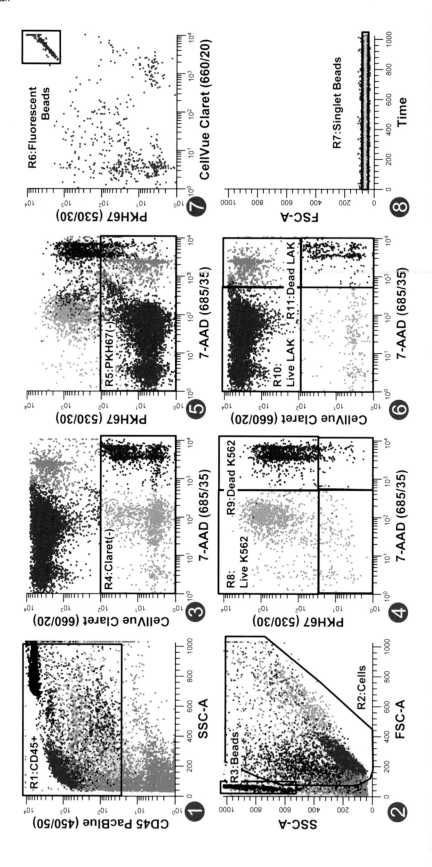

Fig. 5. Analysis strategy for simultaneous quantitation of target and effector viability in a direct cytotoxicity assay. In this multicolor cytotoxicity assay, LAK effector cells stained with CellVue Claret (final concentrations: 5×10^7 cells/mL, 5 μM dye) and K562 targets stained with PKH67 (final concentrations: 1×10^7 cells/mL, 1 μM dye) were co-cultured at 37°C to assess LAK cell-mediated killing. Killing was assessed using two different metrics: (1) as % of targets able to exclude a viability probe (7-AAD) and (2) using counting beads to enumerate the number of viable target cells that remained when effectors were present versus when they were absent. Because these beads have very low forward scatter (R3, plot 2), it was not possible to set an acquisition threshold on this parameter and reliably ensure that all bead events were collected. Side scatter was, therefore, used as the thresholding parameter, with anti-CD45 PacBlue being used to include all leukocytes and beads (plot 1; see Note 41). A reciprocal gating strategy was applied to assess target and effector cell numbers and viability. A two-parameter plot of CellVue Claret versus 7-AAD fluorescence, gated on CD45+ events (R1, plot 1) with cell-like light scatter (R2, plot 2) that were not beads (not R6, plot 7), was used to identify target cells as events that were not CellVue Claret positive (R4, plot 3). Live versus dead target cells were then enumerated on a two-parameter plot of PKH67 versus 7-AAD fluorescence (plot 4, gated on R1&R2&R4 and not R6). Similarly, CellVue Claret-stained LAK cells were identified as PKH67 negative (R5) on a plot of CellVue Claret versus 7-AAD fluorescence (plot 5, gated on R1&R2 and not R6. This strategy was used because substantial differences in autofluorescence between targets and effectors (see Table 2) made it difficult to establish instrument settings that gave complete resolution between PKH67+ K562 targets and PKH67 negative (CellVue Claret+) LAK cells (plot 5), whereas much better resolution was possible between CellVue Claret + LAK cells and CellVue Claret negative (PKH67+) K562 cells (plot 3). Counting beads (R6), which exhibit broad-spectrum fluorescence (plot 7, gated on R3), were enumerated in plot 8 (gated on R3&R6) after gating on R7 to exclude doublets and larger aggregates. Extent of cytotoxic killing was assessed as described in Subheading 3.3.4, steps 5 and 6. Color codes for plots 1–6: *blue* = viable LAK effectors; *green* = nonviable LAK effectors; *light blue* = viable K562 targets; *red-brown* = dead K562 targets; *black* = enumeration beads; *gray* = noise/debris; *yellow* = very low forward scatter events.

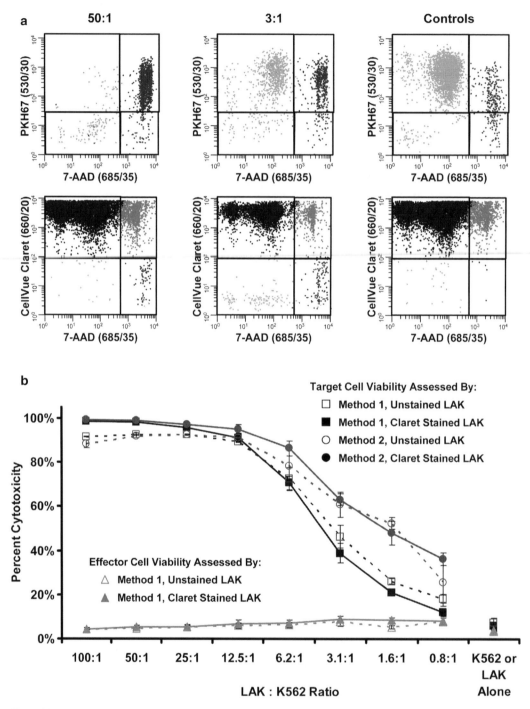

Fig. 6. LAK cell-mediated killing of K562 targets is unaffected by staining with CellVue Claret. LAK cells were labeled with CellVue Claret and incubated with PKH67-labeled K562 cells at effector to target (E:T) ratios ranging from 100:1 to 0.8:1 for 4 h at 37°C. Test samples and controls (Table 2) were analyzed using the gating strategies described in Fig. 5. (a) Representative plots from test samples with 50:1 and 3:1 E:T ratios and from the K562 target only and LAK cell only controls. (b) LAK-induced cytotoxicity of K562 cells was assessed for each condition as described in Fig. 5 using either Method 1 (based on percent of target cells that took up 7-AAD, Subheading 3.3.4, step 5; *squares*) or Method 2 (based on number of viable target cells remaining in the presence versus absence of effectors, Subheading 3.3.4, step 6; *circles*).

6. *Method 2*: This alternative method uses a calculation comparable to the approach used in a standard ^{51}Cr release assay, using regions R7 (singlet beads) and R8 (live K562 targets) defined on plots 4 and 8 of Fig. 5.

$$\%\text{cytotoxicity} = \left(1 - \left[\frac{(R8(\text{test}) \: / \: R7(\text{test}))}{(R8(\text{target only}) \: / \: R7(\text{target only}))}\right]\right) \times 100\%$$

For example, using the numbers of events obtained from the data shown in Fig. 5 gives the following result:

$$\% \text{ cytotoxicity} = \left(1 - \left[\frac{(935 \: / \: 23,276)}{(2801 \: / \: 22,639)}\right]\right) \times 100\%$$
$$= (1 - 0.322) \times 100\%$$
$$= 68\%$$

3.4. Tracking Proliferation: Inhibitory and Enhancing Effects of Treg and Teff Cell Interactions

Regulatory T cells (Treg) exert potent immunosuppressive effects in autoimmune diseases, transplantation, and graft-versus-host disease (30), inhibiting proliferation of effector T cells (Teff) primarily by downregulating induction of their IL-2 mRNA (31). Phenotypically, Treg are defined by their co-expression of CD3, CD4, CD25, the transcription factor FOXP3, and dim expression of CD127, along with several other surface markers shared with activated T cells such as GITR and CTLA-4 (30, 32, 33). As reviewed by Brusko et al. (34), assays that use tracking dyes to monitor Treg suppression of anti-CD3 plus IL-2-induced effector T-cell (Teff) proliferation have significant advantages over in vitro suppression assays using ^{3}H-thymidine, a standard measure of Treg activity. In particular, although they require approximately tenfold more cells, tracking dye-based assays reflect total Teff proliferation throughout the 4-day culture period rather than simply measuring DNA synthesis during the final hours of the response and can be extended to enable simultaneous monitoring of low level Treg proliferation as well (34). Our experience with a single-color in vitro suppression assay has been that it can be difficult to reliably distinguish highly proliferated CFSE^{dim} Teff from unlabeled Treg, since both populations express similar levels of CD4. Use of a second tracking dye has the advantage of not only simplifying discrimination between Treg and CFSE^{dim} Teff, but

Fig. 6. (continued) As an internal control, percentage of dead LAK effectors (*triangles*) was assessed by Method 1 at each E:T ratio and verified to be acceptably low and relatively constant. To determine whether CellVue Claret staining affected their cytolytic potential, parallel studies were performed using CellVue Claret-stained (*solid lines*) and unstained (*dashes*) LAK effectors. The data indicate that LAK cells kill K562 cells in a concentration-dependent manner and that the tracking dye did not affect function. Interestingly, Method 2 was slightly more sensitive at detecting target cell loss (Note 43). Representative data from one of two replicate experiments are shown; data points signify the mean ± 1 standard deviation of triplicate samples.

also allowing assessment of whether increasing numbers of Teff in the assay have any effect on Treg proliferation. In the variation described here, isolation of CD4+ Treg and Teff by sorting was combined with CFSE labeling of Teff and CellVue Claret labeling of Treg to ascertain the effect of Treg:Teff ratio on the proliferative response of each cell type (see Note 45). Parallel studies using unstained Treg confirmed that their ability to suppress Teff proliferation was unaltered by labeling with CellVue Claret.

3.4.1. Preparation of Monocyte-Depleted Lymphocytes (see Note 46)

1. Prepare TRIMA filtrate by draining TRIMA filter into a 50-mL conical tube, followed immediately by rinsing the filter with 40 mL of 10% ACD in PBS to dislodge trapped cells (35) (see Note 2).

2. Isolate hPBMC from the TRIMA filtrate using the laboratory's standard density gradient fractionation protocol, with the addition of a final low-speed wash ($300 \times g$) to minimize platelet contamination (see Note 11).

3. To separate lymphocytes from monocytes via cold aggregation (36, 37), resuspend hPBMC in 50 mL of cold CM and dispense 12.5 mL each into four 15-mL conical polypropylene tubes. Affix the tubes onto the fins of tube rotator and rotate along their horizontal axis, parallel to the benchtop, at 18 rpm at 4°C to induce monocyte aggregation (see Note 47). After 30–45 min, visible 1–3 mm aggregates will form that contain primarily monocytes.

4. Remove the tubes from the rotator and place vertically on ice for 15 min, permitting aggregated cells to precipitate at $1 \times g$ to the bottom of each tube.

5. Harvest supernatant containing the monocyte-depleted lymphocytes, wash twice with cold HBSS, and use for isolation of Treg, Teff, and accessory cells (see Subheading 3.4.2 and Notes 48 and 49).

3.4.2. Isolation of Treg, Teff, and Accessory Cells by Flow Cytometry and Sorting (see Note 50)

1. Adjust monocyte-depleted lymphocytes from Subheading 3.4.1, step 5, to 5×10^7 cells/mL in HBSS and incubate for 10 min with 600 μg/mL of human IgG to block Fc receptor binding.

2. Add a mAb cocktail containing anti-CD127 PE, anti-CD4 PECy7, and anti-CD25 APC to the IgG-blocked lymphocytes and incubate on ice for 30 min (see Note 51).

3. Wash the cells twice with HBSS and resuspend at 1.5×10^7 cells/mL in HBSS.

4. Sort antibody-labeled cells on a fluorescence-activated cell sorter (e.g., FACSAria II or equivalent) into glass tubes containing CM at a rate that provides for purities of 95% or greater (see Note 52). The gating logic used to sort Treg, Teff, and accessory cells is illustrated in Fig. 7.

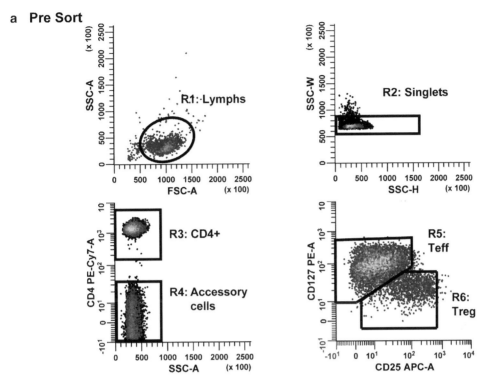

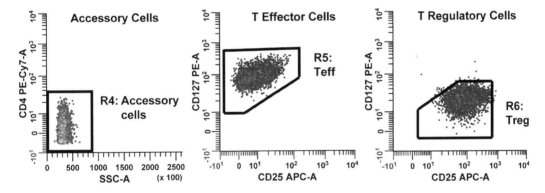

Fig. 7. Sorting logic for Teff, Treg, and accessory cells. Monocyte-depleted lymphocytes prepared from TRIMA filters were stained with CD127 PE, CD4 PE-Cy7, and CD25 APC and then sorted into populations of Teff, Treg, and accessory cells to study the effects of Teff and Treg on each other's proliferative response. (**a**) Histograms were sequentially gated by applying the lymphocyte scatter region (R1 on FSC vs. SSC plot) and side scatter-based multiplet exclusion criteria (R2 on SSC-H vs. SSC-W plot); R1 and R2 to the SSC-A versus CD4 PECy7-A plot; and R1, R2, and R3 to the CD25 APC-A versus CD127 PE-A plot. Teff cells were defined as CD4+ CD127^bright CD25^dim singlet lymphocytes. Treg cells were defined as CD4+ CD127^dim CD25+ singlet lymphocytes. Accessory cells were defined as CD4– singlet lymphocytes. (**b**) After sorting, purity was assessed for each population and found to be greater than 97% in all cases.

3.4.3. Proliferation Protocol   1. Stain sorted Treg with CellVue Claret (final cell concentration of 1×10^6/mL; final dye concentration, 1 μM) according to the procedures described in Subheading 3.2. Wash in CM, count, and adjust to 1×10^6 cells/mL.

2. Stain sorted Teff cells with CFSE (final cell concentration of 5×10^7/mL; final dye concentration, 5 μM) according to the procedures described in Subheading 3.1. Wash in CM, count, and adjust to 5×10^5 cells/mL.

3. In a 96-well round-bottom plate, make triplicate serial 1:2 dilutions of the Treg as follows: Pipet 200 μL of the stained Treg suspension into the first well and 100 μL of CM into an adjacent set of four wells. Serially transfer 100 μL of Treg from the first well to the second, then from the second to the third, etc., ending with the transfer of 100 μL from the fourth well to the fifth well and removal of 100 μL of cell suspension from the fifth well (see Note 40).

4. Add 100 μL of stained Teff to each well, creating Treg-to-Teff ratios of 2:1, 1:1, 0.5:1, 0.25:1, and 0.125:1 (see Note 53).

5. Add 50 μL of Treg and 100 μL of Teff cell to the Treg-only and Teff-only wells, respectively (see Notes 54 and 55).

6. Centrifuge sorted accessory cells ($400 \times g$ for 5 min at ~21°C), pool into a 50-mL conical tube, adjust to 1×10^6 cells/mL with CM, and irradiate with 3,000 rad of gamma irradiation to inhibit proliferation. After irradiation, adjust the concentration to 5×10^5 cells/mL in CM.

7. To an aliquot of accessory cells commensurate with the size of the experiment, add azide-free anti-CD3 (clone OKT3) to a final concentration of 3 μg/mL and anti-CD28 (clone 28.2) to a final concentration of 1.5 μg/mL. Add 0.1 mL of this preparation to each test well from step 4, yielding a final concentration of 1 μg/mL of anti-CD3 and 0.5 μg/mL of anti-CD28 in a final volume of 0.3 mL/well.

8. Add CM to bring each well to a final volume of 0.3 mL and incubate in a humidified 37°C incubator with 5% CO_2 for 96 h (see Note 56).

9. After the 96-h incubation, remove the plate from the incubator and harvest cells from each well into individually labeled 12×75 mm round-bottom tubes compatible with the laboratory's flow cytometer and place on ice. Rinse each well with 200 μL of cold HBSS, adding with the appropriate tube. QS each tube to 3 mL with HBSS.

10. Centrifuge at $400 \times g$ for 5 min at ~21°C and resuspend each pellet in 100 μL of cold HBSS buffer, adding 10 μL of human IgG to block Fc receptor binding.

11. Incubate for 10 min on ice and then label with anti-CD4 PECy7 (clone SK3) and 5 μL of LIVE/DEAD® Fixable Violet reagent, diluted 1:50 from frozen DMSO stock (see Note 57).

12. Incubate for 30 min on ice and then wash two times with FCM buffer. Resuspend the cells in 300 μL of FCM buffer for flow cytometric analysis.

3.4.4. Flow Cytometric
Acquisition and Analysis

1. Establish appropriate voltage settings using autofluorescence and single-color controls from Table 3 (see Notes 54 and 58).

2. Using the single-color controls from Table 3, adjust compensation settings according to your laboratory and/or

Table 3
Recommended Assay and Instrument Controls for Measuring Immune Cell Proliferation

	Cells	Label	Treatment	Notes
Assay controls[a]	Teff	CFSE	+acc, no stimulus[d]	*Negative control:* spontaneous Teff proliferation
	Teff	CFSE	+acc + stimulus	*Positive control:* maximum Teff proliferation
	Treg	Claret	+acc, no stimulus	*Negative control:* spontaneous T regulatory proliferation
	Treg	Claret	+acc + stimulus	Define Treg proliferation
	Teff	CFSE	PHA	*Positive control:* verify that Teff are able to proliferate when cell surface receptors are bypassed
	acc	None	None	*Optional:* confirm that they remain CD4– throughout assay and will not be confused with highly divided Teff
	Teff	Varying CFSE	+acc + stimulus	*Staining control:* used during assay setup to verify that concentration of tracking dye chosen does not affect Teff proliferation (Fig. 1c)[e]
	Treg	None	+acc + stimulus	*Staining control:* used during assay setup to verify that tracking dye labeled cells proliferate equivalently to unlabeled cells (Fig. 9)[e]
Instrument controls[b]	Teff	None	Unstimulated, accessory only	Set voltage and negative region in CFSE channel
	Teff	CFSE	Unstimulated	Set voltage in CFSE channel; set color compensation in other channels; estimate location of undivided Teff
	Treg	None	Unstimulated	Set voltage and negative region in Claret channel
	Treg	Claret	Unstimulated	Set voltage in Claret channel; set color compensation in other channels; estimate location of undivided Treg
	Teff (no Claret)	CD4 PE-Cy7	Unstimulated	Set compensation
	acc + Teff (no CFSE)[c]	LIVE /DEAD Fixable Violet	Unstimulated	Set compensation

[a]Assay controls are included in the experimental plate with test samples to verify that the expected biological outcomes can be recognized using the chosen instrument conditions
[b]Instrument controls are used to establish instrument voltages and settings
[c]Irradiated accessory cells will be nonviable and should be 100% positive for this viability probe
[d]Acc = accessory cells (CD4 negative lymphocytes); stimulus = anti-CD3 plus anti-CD28
[e]Needed only to establish optimized staining conditions for tracking dye when assay is first being implemented in the laboratory; not required on a routine basis

instrument manufacturer's standard procedures (see Notes 55 and 58).

3. Acquire data on the flow cytometer using the gating strategy shown in Fig. 8 (see Note 59).

3.4.5. Calculation of Proliferative Fraction and Proliferative Index

Either Proliferative Fraction (%P), a semi-quantitative estimate of percent proliferating cells, or Proliferative Index (PI), a more quantitative estimate of fold population expansion, may be used to analyze the extent of proliferation. In either approach, the starting point is a single parameter tracking dye dilution profile for the appropriate subpopulation of viable lymphocytes (here CFSE for Teff and CellVue Claret for Treg), created using the gating strategy described in Fig. 8.

1. Calculation of %P. To calculate %P, a stained, unstimulated control is used to set the upper boundary for enumeration of daughter cells, selecting an intensity that gives an acceptably low value for dividing cells in the absence of stimulus (e.g., 1–5%; *see* Figs. 9 and 10). An unstained control is used to define the lower boundary for the enumeration of proliferating cells, selecting an intensity that gives an acceptably low value for dividing cells in the absence of proliferation dye. %P is then defined as the percentage of proliferating cells with fluorescence intensity less than that of the stained but unstimulated control and more than that of the unstained control.

2. Calculation of PI. To calculate PI, a specifically designed peak-modeling software such as ModFit LT (Verity Software House, Topsham, ME), FCS Express (De Novo Software, Los Angeles, CA), or FlowJo (TreeStar, Ashland, OR) is used to fit the viable, lymphocyte-gated, single-parameter CFSE and CellVue

---

Fig. 8. Simultaneous analysis of Teff and Treg proliferation during an in vitro suppression assay. After isolation by sorting as described in Fig. 7, Teff cells were stained with CFSE (final concentrations: 5×10^7 cells/mL, 5 µM dye) and co-incubated for 4 days with sorted Treg stained with CellVue Claret (final concentrations: 1×10^6 cells/mL, 1 µM dye) in the presence of anti-CD3, anti-CD28, and sorted, irradiated accessory cells (see Subheading 3.4.3 for details). Representative data are shown for one of three triplicate samples at a Treg:Teff ratio of 0.25:1. LIVE/DEAD Fixable Violet reagent was used to exclude dead cells (R1, *upper left plot*; in the electronic version: accessory cells = *red-brown*, nonviable Teff = *gray* and nonviable Treg = *red*) from all other data plots. CellVue Claret staining was used to distinguish viable Treg (R4, center right plot) from viable but highly proliferated Teff (R5, center right plot). A single parameter CFSE (530/30) proliferation profile for Teff (*lower left plot*) was generated by gating on cells that were CFSE + (R5), CD4+ (R3), viable (not R1), and had lymphocyte scatter properties (R2). A single parameter CellVue Claret (660/20) proliferation profile for Treg was generated by gating on cells that were CellVue Claret + (R4), CD4+ (R3), viable (not R1), and had lymphocyte scatter properties (R2). Note the generous lymphocyte region (R2) defined to include lymphocyte blasts. Proliferative fractions, representing the percent of cells that have undergone one or more divisions, were calculated as described in Subheading 3.4.5 (R6 = 78.6% and R7 = 37.6% for Teff and Treg, respectively). Proliferative indices (PI), representing fold expansion of Teff and Treg populations during the culture period, were calculated as described in Subheading 3.4.5 using Gaussians to model each of the generational peaks (e.g., in the electronic version: *blue* = parental generation, *orange* = first daughter generation, etc.).

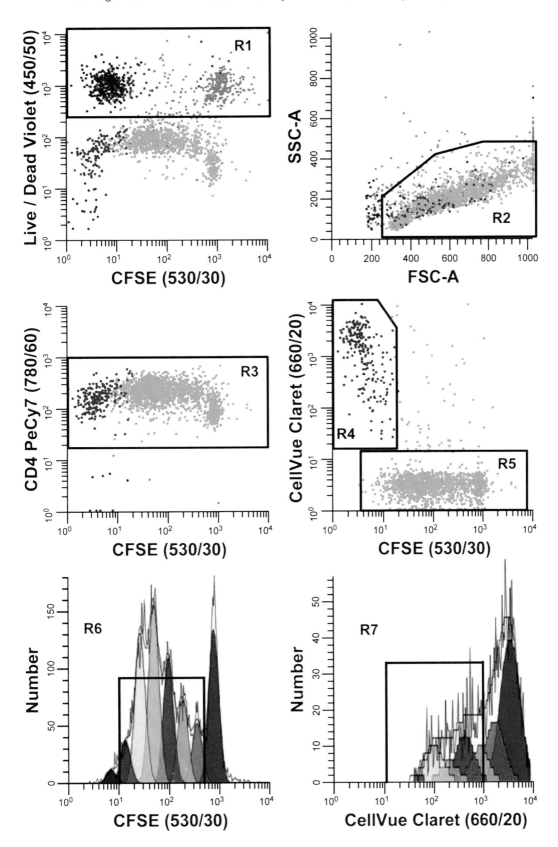

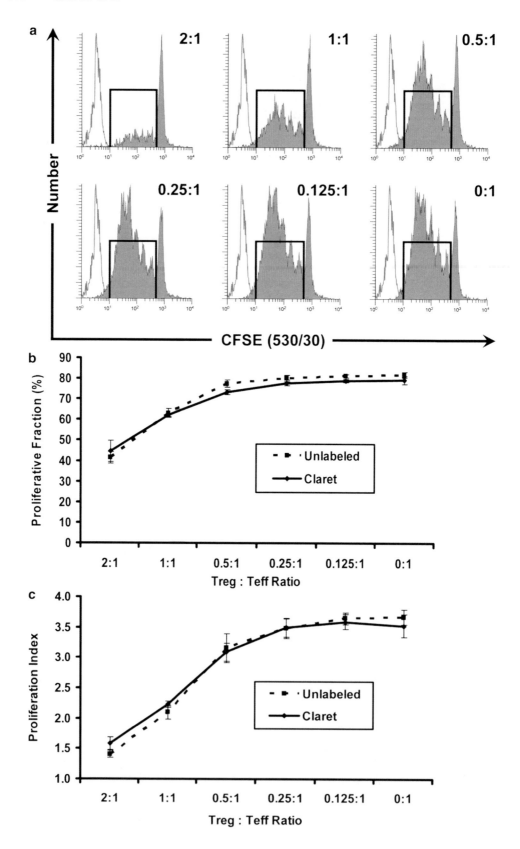

Claret data. These programs use a nonlinear least squares analysis to find iteratively the best fit to the raw data by changing the position, height, and CV of each generational Gaussian. After loading and gating the histogram, users define the location of the parental generation, its spacing, and if necessary its SD. When modeling lipophilic dyes, an equal spacing between generations is assumed, whereas when modeling CFSE, an unequal spacing must be assumed to adjust for observed nonlinearities in peak spacing (possibly due to continued slow dye loss even after 24 h). The area under each Gaussian is taken as a measure of the relative number of cells in that generation and the sum of all Gaussians corresponds to the relative number of cells in the total population. These values are then used internally by the software to calculate the PI.

3. Calculation of percent suppression. The degree of suppression observed when Treg cells are co-cultured with Teff cells is calculated using one of the two following methods:

Method 1:

$$\%\text{Suppression} = \left(1 - \left[\frac{\text{Px}_{\text{Treg + Teff}}}{\text{Px}_{\text{Teff}}}\right]\right) \times 100\%$$

This method is appropriate when the proliferation metric (Px) is %P or any other measure for which the value goes to 0 when the proliferative response is fully suppressed.

Method 2:

$$\%\text{Suppression} = \frac{\left(\text{Px}_{\text{Teff}} - \text{Px}_{\text{Treg + Teff}}\right)}{\left(\text{Px}_{\text{Teff}} - 1\right)} \times 100\%$$

This method is appropriate when the Px is PI or any other measure for which the value goes to 1 when the proliferative response is fully suppressed (51).

Fig. 9. Inhibition of Teff proliferation by Treg cells. Teff cells were stained with CFSE and incubated with graded ratios of Treg cells in the presence of anti-CD3, anti-CD28, and accessory cells, as described in Fig. 8. The maximum proliferative potential of Teff was assessed in the absence of Treg cells (Treg:Teff ratio of 0:1). As increasing numbers of Tregs were added to the culture system, increasing inhibition of Teff cell proliferation was observed, as expected. Similar results were obtained with both CellVue Claret-stained (*solid line*) or unstained (*dashed line*) Treg, indicating that staining with the CellVue Claret tracking dye did not affect the potency of inhibition by Treg cells. Proliferative fraction (%P) and Proliferative Index (PI) were determined as described in Fig. 8 and Subheading 3.4.5. Data points in (**b**) and (**c**) represent the mean ± 1 standard deviation of triplicate samples. (**a**) Representative proliferation profiles for Teff at varying Treg:Teff ratios (*filled histograms*), showing increasing inhibition of proliferation at higher ratios. Unstained, unstimulated cells (*unfilled histograms*) are overlaid for reference. Stained, unstimulated cells largely overlapped with the stimulated parental population (not shown in this figure; see Fig. 1c). (**b**) Effect of Treg:Teff ratio on Proliferative Fraction of Teff. (**c**) Effect of Treg:Teff ratio on Proliferative Index of Teff.

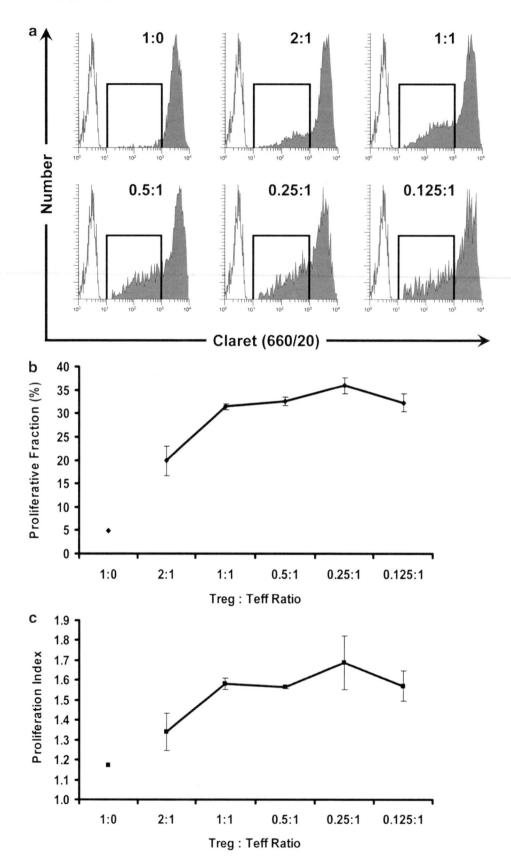

4. Notes

1. Lymphocytes and monocytes are typically isolated from anticoagulated blood using standard Ficoll-Hypaque density centrifugation techniques prior to labeling, but cryopreserved PBMCs, adherent cell lines (harvested using trypsinization), and non-adherent lines are also suitable for staining. Cells may be labeled while adherent by flooding the culture dish or flask with dye solution. However, this typically gives considerably more heterogeneous intensity distributions, especially for membrane dyes (38), and makes their interpretation in dye dilution proliferation assays more complex. Labeling of single cell suspensions is, therefore, generally preferred.

2. Labeled cells are typically placed back into culture for in vitro assays or injected into animal models for in vivo functional studies. Standard sterile technique should, therefore, be followed throughout the labeling protocols described in Subheadings 3.1–3.4.

3. Amount of dye required for bright but nontoxic staining will in general increase as total number and/or size of cells to be stained increases. However, exact concentrations resulting in over-labeling and loss of function will vary depending on cell type and class of tracking dye used (e.g., Table S1 in (1)). Therefore, appropriateness of final cell concentration and final tracking dye concentration used for labeling should always be verified by comparing viability and functionality of labeled versus unlabeled cells. Similarly, both cell and dye concentrations used for labeling should be reported in any publication.

4. Total number of cells to be stained will depend on the number of replicates and controls required by the experimental protocol. Staining intensities are most easily reproduced when

Fig. 10. Enhanced Proliferation of Treg cells at low Treg:Teff ratios. The use of a two-dye system allows for the discrimination of Treg and Teff from accessory cells and from each other, and the simultaneous measurement of proliferation in both subsets. In the same samples as shown in Fig. 9, the proliferation of CellVue Claret-stained Treg cells was monitored at different Treg:Teff ratios. Treg are generally anergic and, as expected, did not proliferate when incubated with anti-CD3, anti-CD28, and accessory cells in the absence of Teff cells (Treg:Teff ratio of 1:0). However, as the proportion of Teff in the cultures was increased (i.e., as the Treg:Teff ratio decreased), the extent of Treg proliferation also increased. Proliferative fraction (%P) and Proliferative Index (PI) were determined as described in Fig. 8 and Subheading 3.4.5. Data points in (b) and (c) represent the mean ± 1 standard deviation of triplicate samples. (a) Representative proliferation profiles for Treg at varying Treg:Teff ratios showing increasing proliferation at lower ratios (increased proportion of Teff in cultures) (*filled histograms*). Unstained, unstimulated cells (*unfilled histograms*) are overlaid for reference. (b) Effect of Treg:Teff ratio on Proliferative Fraction of Treg. (c) Effect of Treg:Teff ratio on Proliferative Index of Treg.

staining is done in volumes ranging from 0.5 to several milliliters. Once an approximate cell concentration has been established based on these factors, a preliminary dye titration experiment is recommended to determine (or verify) the optimal concentration of the tracking dye (39, 40).

5. Exogenous protein reduces labeling efficiency for both protein and membrane dyes and is, therefore, normally removed by washing the cells with a protein-free buffer such as PBS or HBSS prior to staining. However, when labeling must be done at relatively low cell concentrations due to limited number of cells or other experimental concerns, addition of exogenous protein may aid in avoiding over-labeling and resultant loss of cell viability or functionality (see Note 10 and (4)).

6. Selection of proliferation tracking dye(s) for a given study is typically based on spectral compatibility with other fluorochromes to be used in the study (1, 41), and the ability to achieve acceptable starting intensity without adverse effect on function (1, 4, 42–44). CFSE, PKH26, and PKH67 can be excited using a 488-nm laser line (absorption maxima: 492, 551, and 490 nm, respectively) and have emission maxima of 516, 565, and 504 nm, respectively. CellVue Claret (absorption maximum: 654 nm) can be excited using red diode or HeNe laser lines at 633–635 nm or a 647-nm Kr laser line and has an emission maximum of 677 nm.

7. CFSE is highly lipophilic but poorly soluble in ethanol or other polar organic solvents. If CFSE is purchased as a bulk powder, it should be accurately weighed out and made up as a 5-mM stock solution (MW 557.47 g/mol) in freshly opened anhydrous DMSO. Although the entire contents of dye vial can be dissolved in a calculated volume of DMSO (4), the final dye concentration should be confirmed spectrophotometrically (e.g., by absorption at 490 nm) and adjusted as needed for consistency, since exact weights contained may otherwise vary sufficiently from vial to vial to require re-titration of new versus old dye stocks in order to avoid toxicity (E. Tenorio, personal communication).

8. Aliquots of 5 mM CFSE dye stock in DMSO can be stored in a dessicator at –20°C for several months but repeated freezing and thawing of a given aliquot should be avoided since DMSO is hygroscopic and takes up moisture from the air. The presence of water leads to reduced labeling efficiency due to hydrolysis of both the diacetate ester moieties required for entry into cells and the succinimidyl ester moieties required for covalent reaction with amino groups under physiologic conditions.

9. If the necessary weighing or spectrophotometric instrumentation is not available in the laboratory, fresh 5 mM stock

may be prepared for use in step 1 of Subheading 3.1, by adding anhydrous DMSO to commercially available single-use vials containing pre-weighed amounts of CFSE. However, it should be noted that cost per milligram of dye is greater for such vials than for dye purchased in bulk powder form.

10. When staining cells at concentrations $<1 \times 10^7$/mL, inclusion of exogenous protein (e.g., 5% v/v FBS or 1% v/v serum albumin) in the resuspension buffer is suggested to avoid over-labeling and loss of cell function (see Notes 5 and 14). Resuspension in a serum-free culture medium will also reduce labeling efficiency and potential for over-labeling, due to the presence of free amino acids that compete for reaction with CFSE. Alternatively, if addition of exogenous protein must be avoided due to other experimental considerations, the working stock of CFSE prepared in Subheading 3.1, step 3 may be further diluted in buffer prior to initiation of cell labeling in Subheading 3.1, step 4. The time between initial dilution and initiation of cell labeling should be minimized since hydrolysis begins immediately upon dilution of the DMSO stock into aqueous solution and proceeds very rapidly.

11. Platelets present in variable amounts act as "hidden" sources of added protein or membrane that can affect labeling efficiency even when hPBMC and dye concentrations are carefully reproduced. Addition of a final low-speed wash step (5 min at $300 \times g$) minimizes platelet contamination of hPBMC and improves consistency of staining with both protein- and membrane-labeling dyes.

12. Ensure that the 5 mM CFSE stock in DMSO is completely thawed prior to preparation of the working stock, but minimize the length of time that the DMSO stock is exposed to ambient conditions to limit the uptake of moisture. The CFSE working stock solution should be clear and colorless. If there is any sign of yellowing, it should not be used, since this indicates conversion to carboxyfluorescein, the charged fluorescent hydrolysis product which will not enter cells.

13. This concentration was chosen such that following a 24-h stabilization period, the fluorescence intensity of non-dividing lymphocytes should fall within the third and fourth decade of a four-decade log amplifier when unstained cells are placed in the first decade (see also Notes 20–22).

14. For an hPBMC concentration of 1×10^7/mL, staining at a final concentration of 0.5–1.0 μM CFSE is recommended to avoid over-labeling. CFSE labels proteins indiscriminately and if the function of critical residues is modified by labeling,

it can interfere with signal transduction pathways, proliferation, and other cell functions even when cell viability remains acceptable (see Table S1 in (1)). More extensive labeling increases the likelihood of altered cell function(s) and the extent of labeling is a function of dye concentration, cell concentration, labeling time, and labeling conditions (temperature, mixing, etc.). Concentrations given here for both cells and dye should, therefore, be taken only as a starting point and verified in each user's experimental system.

15. Obtaining reproducible starting intensities from study to study requires accurately reproducing both dye and cell concentrations. Therefore, cell counting using a Coulter Counter or other automated cell counter rather than manual counting using a hemocytometer is recommended, since results of replicate hemocytometer counts often vary by as much as 15–20%.

16. Since uptake of CFSE into cells and reaction with free amino groups occur very rapidly, it is critical to disperse the dye solution quickly and evenly throughout the cell suspension immediately after addition.

17. Once formed by hydrolysis, carboxyfluorescein is sensitive to photobleaching. Therefore, covering with aluminum foil or placing in a dark location is recommended to protect tubes or wells containing CFSE-labeled cells from exposure to high intensity light or prolonged exposure to room light.

18. Inclusion of protein in the stop solution is essential, since it reacts with and inactivates free CFSE. Free amino acids in culture medium further aid in the inactivation. Alternatively, PBS or HBSS containing 1–2% serum albumin may be used as a stop solution.

19. For starting cell numbers of 10^7 or more, recoveries of at least 85% and viabilities of at least 90% should be obtained for freshly drawn hPBMC (e.g., Fig. 1a and Table 3 in (45)). However, recoveries typically decrease at lower cell numbers and may also be lower for preparations in which the cells are older or have been subjected to other stresses (e.g., pheresis, elutriation, or cryopreservation and thawing). Staining intensity and CV will vary for different cell types, but a bright symmetrical fluorescence intensity profile coupled with poor recovery and/or viability usually indicates substantial overlabeling and the need to increase cell concentration, decrease dye concentration, or both. Conversely, heterogeneous and/or dim staining (<2 log separation from unstained control) coupled with good recovery and viability suggests underlabeling and the need to decrease cell concentration, increase dye concentration, or both.

20. The great majority of cell-associated CFSE is lost within the first 24–48 h as cells export and/or degrade unreacted dye and labeled proteins/peptides that are short-lived, damaged by the covalent labeling process, destined for secretion, etc. This typically results in (a) up to a 1-log decrease in mean fluorescence intensity and (b) an approximately 5–10% decrease in CV, since T_0 intensities reflect primarily variations in cell size, whereas T24–48 intensities reflect primarily variations in cellular content of stable long-lived proteins (see Fig. 2a).

21. Some labeling protocols recommend a washout period in which freshly stained cells are incubated in CM at 37°C for 30–60 min to hasten fluorescence stabilization. Our experience has been that a relatively short washout period of this type is not sufficient to prevent transfer of exported dye to unlabeled cells present in a co-culture, suggesting that labeled cells continue to export labeled proteins beyond the washout period (41). Therefore, at least a 24-h stabilization period in culture is recommended before labeled cells are used for in vitro cytotoxicity or proliferation assays. For in vivo cytotoxicity or proliferation assays, if labeled cells are to be reinfused within less than 24 h post-staining, it is critical to include control animals in which minimal killing or proliferation is expected in order to verify that ex vivo labeling conditions resulted in adequate resolution between stably labeled cells and unlabeled cells.

22. T_0 CFSE-labeled samples are NOT appropriate for use as compensation controls because they are so much brighter than T24 samples that they typically cannot be run on the same intensity scale when unstained cells are placed within the first decade (see Note 20 and Fig. 2a). For proliferation assays based on CFSE dye dilution, it is, therefore, critical to select a staining concentration that gives adequate separation from unstained cells at 24 h without unacceptable fluorescence overlap into spectral channels used to detect other reagents. It is also essential to ensure that sufficient cells are prepared to provide unstimulated T24 samples for use as compensation controls. For an example of typical controls to set up for a proliferation assay, see Table 2 in ref. (45). Inability to achieve adequate color compensation for a CFSE-labeled but unstimulated (i.e., non-proliferating) T24 sample indicates the need to reduce CFSE concentration, increase cell concentration, or both during the staining step, while recognizing that this may reduce the number of daughter cell generations that can be resolved from unstained cells.

23. Fixation of CFSE-labeled cells in EtOH (46) or methanol-free formaldehyde (Fig. 2a) leads to further loss of cell-asso-

ciated CFSE (30–50% decrease in fluorescence intensity), most likely due to leakage of small but stably labeled peptides or proteins out of the cells as membrane permeabilization occurs. Fortunately, the decrease from fresh to fixed cells does not appear to affect the shape of dye dilution profiles, which can therefore still be used to deduce cell proliferation history so long as the weaker fluorescence of fixed cells does not compromise the ability to resolve the desired number of generations from unstained cells.

24. For dye dilution proliferation assays using hPBMC (or PBMC from other species), it may be necessary to use an independent method such as ^{3}H-thymidine incorporation (see Table S1 in (1)) to verify that cell function is unaltered by labeling with tracking dye at the chosen concentration.

25. Like most membrane-intercalating dyes, PKH26, PKH67, and CellVue Claret contain not only aromatic chromophores but also lipophilic alkyl tails, all of which readily adsorb to the walls of polystyrene tubes or plates. This can substantially reduce labeling efficiency, particularly when working at relatively low dye concentrations (2 μM or less). Use of polypropylene tubes is strongly recommended to minimize adsorptive dye loss and maximize reproducibility of labeling.

26. The PKH and CellVue dyes are incorporated into membranes based on strong hydrophobic forces that drive partitioning from the aqueous phase, in which the dyes are highly insoluble, into cell membranes, where they remain stably intercalated due to noncovalent interactions with the lipid bilayer. This means the final staining intensity is strongly affected by both the amount of dye and the amount of membrane present in the staining step. As shown in Table 1, dye concentrations required for bright but nonperturbing staining, therefore, vary with cell type/size as well as cell concentration, making it important to publish and accurately reproduce both dye and cell concentrations in order to obtain reproducible results.

27. Membrane-intercalating dyes are even more lipophilic than CFSE due to long chain alkyl moieties that mediate stable insertion into the membrane bilayer. Although the dyes are modestly soluble in polar organic solvents such as ethanol, the alkyl chains tend to self-associate in aqueous solutions, reducing membrane-labeling efficiency. The presence of salts substantially increases the extent of micelle and aggregate formation and decreases general membrane-labeling efficiency, although phagocytic cells can become differentially labeled by taking up dye aggregates. Use of conical rather than round-bottom tubes is highly recommended, since this facilitates more complete removal of salt-containing media or buffers prior to cell labeling.

28. The technique used to remove supernatant prior to resuspension in Diluent C or during wash steps can, in our experience, substantially impact both quality of staining and cell recovery when labeling with membrane dyes. Tube inversion and blotting typically leaves ~100 μL of supernatant, leading to significant salt remaining in Subheading 3.2, step 2 and reduced labeling efficiency in Subheading 3.2, step 5. Subsequent aspiration to reduce the amount of fluid risks loss of cells at the top of the pellet that have been loosened as the fluid drains back down the side of the tube. The best method that we have observed for maximizing fluid removal while minimizing cell loss is to fit a sterile disposable 200-μL pipette tip at the end of a vacuum aspirator, which reduces the aperture size and makes it easier to position the point of suction accurately relative to the cell pellet.

29. The fact that the PKH and CellVue dyes label cells by partitioning into membranes has two other significant consequences beyond the negative effect of salts on staining efficiency. First, staining occurs even more rapidly than for CFSE, with the partitioning of dye into membranes being essentially complete within 15–30 s after admixing 2× cells and 2× dye solutions. Second, to obtain bright, homogeneous labeling, it is critically important that admixing of cells and dye be done so that all cells are exposed to the same concentration of dye at the same time. This is much more easily achieved by admixing similar volumes of dye and cells than by trying to disperse a small volume of ethanolic dye into a much larger volume of cell suspension (see Fig. 4b).

30. Diluent C is an aqueous isotonic, iso-osmotic, salt-free staining vehicle that contains neither organic solvents nor physiologic salts. Although Diluent C is designed to maximize dye solubility and minimize cell toxicity for short periods (5–15 min), the longer cells and dye stand in Diluent C, the more likely that (a) staining efficiency will decrease due to dye aggregation and (b) decreased cell viability or function may result from lack of physiologic salts. Subheading 3.2, steps 2–5 should, therefore, be completed in as short a period as possible, preferably 1–2 min. When multiple samples are to be labeled, it is highly recommended that processing through the first wash of Subheading 3.2, step 7 be completed for each sample before the next sample is stained. However, remaining steps may then be carried out in parallel for all samples.

31. In theory, results should be the same whether 2× cells are admixed with 2× dye or vice versa. However, our experience has been that adding cells to dye makes it easier to reliably obtain homogeneous staining. Adding dye to cells runs the

risk that any cells on the wall of the tube may remain unstained if they are above the level of admixed solution or stain at lower intensity if they do not mix with bulk staining solution until after trituration is completed (see Fig. 4b).

32. Some literature protocols suggest the use of CM, which contains 10% FBS, as the stop reagent. However, our experience has been that use of neat FBS results in more efficient removal of unbound dye (due to its higher protein concentration) and also reduced likelihood of forming dye aggregates large enough to sediment with cells during subsequent wash steps (due to its lower ionic strength). If CM is used, a larger volume (5× rather than 1×) is recommended to ensure sufficient protein is present to adsorb all unbound dye.

33. Even when polypropylene tubes are used (see Note 25), some dye adsorption to tube walls may occur. Therefore, washing efficiency is improved if cells are transferred to a fresh polypropylene tube after aspiration of the stop solution and resuspension of the cell pellet for the first wash in Subheading 3.2, step 6. This is particularly important if CM is used as the stop reagent (see Note 32), since carryover of dye particles may result in inadvertent labeling of other cell types present in the culture.

34. In contrast to CFSE, the fluorescence intensity of PKH26, PKH67, or CellVue Claret labeled cells does not decrease significantly in the absence of cell proliferation, and the intensity is also stable to fixation with neutral methanol-free formaldehyde (e.g., 1–2% paraformaldehyde dissolved in PBS). A small aliquot of cells fixed at T_0 can, therefore, be used as a compensation control to evaluate the overlap of membrane dye signal into spectral regions used to detect other reagents. For an example of typical controls to set up for a lymphocyte proliferation assay, see Table 2 in ref. (45). As with CFSE, the inability to achieve adequate color compensation indicates the need to reduce dye concentration, increase cell concentration, or both, during the staining step (see Note 22).

35. Paradoxically, our experience and that of others suggest that while over-staining with membrane dyes may not reduce viability or functionality, it can sometimes result in increased T24 fluorescence intensity compared with T_0 intensity, either in the absence (see Fig. 1 in (1) and Fig. 1 in (45)) or presence (B. Lahti, personal communication) of cell proliferation. The most likely explanation is stacking and self-quenching of dye in the plasma membrane at T_0 that is relieved as the dye redistributes into intracellular membranes via normal membrane trafficking mechanisms (38). Dye dilution appears to proceed linearly with cell division once stacking/quenching is

relieved, presumably because total number of molecules per cell does not change, and membrane dyes are typically pH insensitive in physiologic ranges. However, for assay systems where proliferation monitoring is to be initiated immediately (e.g., continuously growing cell lines or transfectants), cell and/or dye concentration(s) in the staining step should be chosen to ensure that dye dilution proceeds linearly from T_0. Finally, it should be noted that although CFSE over-labeling may also result in stacking/quenching, any modest increase from T_0 to T24 would be completely undetectable in the face of the substantial division-independent dye loss seen during the same period.

36. Regardless of which class of tracking dyes is used, labeling prior to LAK induction with IL-2 is preferable to post-induction labeling. This is particularly true if CFSE or other protein dyes are used, since staining immediately before assay initiation runs the risk that early dye efflux could result in unintended transfer of label to target cells (41). It is also recommended when labeling with membrane dyes, since it gives the cells time to recover from thermal or other stresses incurred during the staining procedure.

37. Use of target cells obtained from high-density cultures should be avoided, due to the presence of significant numbers of apoptotic and/or dead cells. Such cells will readily stain with tracking dyes but will give highly variable staining with 7-AAD or other dyes used to determine viability by dye exclusion, making it much more difficult to establish the intensity limit above which a target cell is to be considered nonviable.

38. Cell and dye concentrations given in Subheading 3.3.1, step 2 and Subheading 3.3.2, step 2 were selected to give staining intensities that (a) were on scale in the last decade of a 4-log intensity scale when unstained cells were placed in the first decade, and (b) could readily be compensated in adjacent spectral channels. It should be noted that a slight peak asymmetry is of somewhat less concern when labeling targets or effectors for a cytotoxicity assay than when labeling hPBMC for a proliferation assay. It is however, important to ensure that 100% of cells are labeled with the chosen tracking dye so that a clear distinction can be made between cells that are labeled, and cells that are unlabeled or those that are labeled with a different dye.

39. Table 2 summarizes recommended assay and staining/instrument setup controls, as well as additional controls required at the time of initial assay setup or useful for troubleshooting (if sufficient cells are available to accommodate their inclusion).

40. For longer term assays, it may be desirable to set up the plate with a border of CM-filled wells around the periphery in

order to minimize evaporation-associated variability in cell and cytokine concentrations in the assay wells.

41. CD45 was used here as a gating parameter because both LAK and K562 are CD45+ (although K562 are dimmer than LAK). The PacBlue conjugate was selected here for spectral compatibility with the other probes used in the assay. Because the enumeration beads also fluoresce in the PacBlue channel (450/50), they also will be included in the "CD45" gate (Fig. 5, plot 1). A low SSC threshold was used to reduce debris but required substantial care to ensure that all events corresponding to both dead and live targets and effectors were included above the threshold. An alternative strategy would be to set a threshold on CD45 PacBlue that included all bead and cellular events (live and dead), avoiding the necessity for a side scatter threshold.

42. Choice of optimized labeling conditions for tracking dye(s) should have already established that stained cells will be on scale, without events accumulating in the highest channel. This condition should be established when compensation is set to 0% and voltages are adjusted to place unstained control cells in the first decade, but sufficiently above the left axis such that events do not accumulate in the lowest channel. In addition, it is critical to recognize that cells labeled with tracking dyes can be extremely bright, requiring substantially decreased detector voltage settings compared with those used for the detection of immunofluorescence. If the tracking dye signal in a secondary channel (set to a higher detector voltage) is greater than its intensity in the primary channel, it will be impossible to compensate properly for tracking dye overlap in the secondary channel (41). In this case, staining with concentrations of tracking dye(s) may be required.

43. As shown in Fig. 6, CellVue Claret staining does not affect the cytolytic potential of LAK cells. Somewhat surprisingly for a 4-h assay, the use of 7-AAD to quantify dead target cells (Method 1) appears to be less sensitive than the use of enumeration beads to count remaining viable target cells (Method 2). To test the hypothesis that a too-stringent CD45 region or scatter region (R1 or R2 of Fig. 5, respectively) might have excluded some dead K562 targets from the PKH67 + 7-AAD + gate in Method 1, we evaluated the effect of extending region R1 to include events with lower CD45 intensity, or expanding region R2 to incorporate events with lower forward scatter. These manipulations (especially in combination) gave decreased rather than increased values for percent cytotoxicity. This was true using either Method 1 or 2 and did not reduce the differential between the two. In order to enumerate dead cells with a viability dye, Method 1 requires

that cells remain intact enough to contain a nucleus. It has been demonstrated in longer term assays that killed cells may fragment, enucleate, and thus become uncountable by this method (23). The data of Fig. 6 suggest that even in a relatively short-term assay, Method 2 – which relies on enumeration of live cells remaining in the sample – is inherently a more accurate way to quantify cytotoxicity. An alternate hypothesis considered was that Method 1 might be more sensitive to the quadstat region used to determine % dead targets than Method 2, due to a consistent decrease in PKH67 intensity observed for dead versus live K562 targets, which was most pronounced at low E:T ratios. For example, in the *center panel* of Fig. 6a (3:1 E:T ratio), gMFI = 299 for PKH67 + 7-AAD+ events, gMFI = 498 for PKH67 + 7-AAD– events, and approximately 8% of PKH67 + 7-AAD+ events fall in the lower right quadrant (i.e., are PKH67^{dim}). Lowering the quadstat boundary to include ≥98% of PKH67^{dim} 7-AAD+, however, actually resulted in decreased values for percent cytotoxicity, not increased values, due to increased levels of noise in the live K562 region. Further analysis of the reduced PKH67 intensity observed for 7-AAD+ targets suggested that it was at least partially due to membrane transfer from K562 targets to LAK effectors, as evidenced by increases in both PKH67 and CD45 intensity observed for live CellVue Claret-positive LAK cells at low E:T ratios (data not shown).

44. Number of non-viable effectors present in the assay can be calculated similarly, using a CellVue Claret versus 7-AAD histogram as gated in Fig. 5, plot 6 (gated on R1&R2&R5 and not R6) and then dividing the number of dead effectors (CellVue Claret and 7-AAD dual positive; R11) by the total number of effectors (CellVue Claret positive; R10 + R11). This may be helpful in troubleshooting if target cell killing is lower than expected and/or in longer term assays in which some effector cell death is expected.

45. During the analysis of suppression assays, fluorescent monoclonal antibodies (mAb) such as anti-CD4 can be used to discriminate Teff and Treg from accessory cells, which are required to maximize Teff proliferation if soluble anti-CD3 is used for stimulation rather than plate-bound anti-CD3 (data not shown).

46. Anti-coagulated peripheral blood or apheresis packs used for stem or mononuclear cell collections (47) are suitable sources of both lymphocytes and accessory cells. WBC-retaining TRIMA filters were used as a starting point for the studies described here because they represent a ready source of large numbers of hPBMC (average recovery: $1.2 \pm 0.4 \times 10^9$ cells; $n = 39$), an important consideration when using multiparameter flow cytometry and sorting to isolate low-frequency subpopulations such as Treg cells.

47. Be sure to rotate tubes on their horizontal axis to avoid foaming, sloshing from side to side, or end-over-end mixing, which will inhibit monocyte aggregation.

48. Typical lymphocyte recoveries from a TRIMA filter are $7.1 \pm 3.1 \times 10^8$ cells. Typical composition is $71.8 \pm 7.1\%$ T cells, $3.3 \pm 3.6\%$ monocytes, and $24.7 \pm 6.4\%$ B/NK cells as determined by CD45 FITC/CD14 PE/CD3 PerCP staining and flow cytometric analysis ($n = 39$).

49. Though not further used in the studies described here, the cell pellet obtained after cold aggregation and $1 \times g$ fractionation is an excellent source of relatively pure monocytes. These may be harvested with a 10-mL pipette, being careful not to disrupt the aggregates. To remove platelets, overlay the aggregates onto 5 mL of FBS and let them to sediment at $1 \times g$ on ice for 15 min. Recover pellet, and wash twice with 15 mL of Versene to remove platelets and once with 15 mL of CM. Typical recoveries are $2.1 \pm 1.2 \times 10^8$ monocytes with $74 \pm 10\%$ purity ($n = 39$).

50. All procedures should be carried out on ice with cold reagents.

51. All mAbs should be used at concentrations previously determined to be saturating. In general, we have found that the concentration of mAb determined to be saturating for 100 μL of whole blood will also be saturating when staining up to $3–5 \times 10^6$ lymphocytes in a volume of 100 μL.

52. After sorting, we routinely obtain $1–4 \times 10^6$ Treg and $1–2 \times 10^7$ Teff cells from a starting preparation of 1.5×10^8 monocyte-depleted lymphocytes. Accessory cells were simultaneously sorted with the Treg and Teff cells until 1.5×10^7 cells had been recovered.

53. Ideally, given sufficient cell numbers, samples should be prepared in triplicate.

54. Table 3 summarizes recommended assay and staining/instrument setup controls, as well as additional controls required at the time of initial assay setup or useful for troubleshooting (if sufficient cells are available to accommodate their inclusion).

55. Single-color compensation controls are best prepared from cells treated in the same fashion as the test samples. This is particularly important for CFSE, which exhibits rapid non-proliferation-related intensity losses during the first 24 h and more slowly thereafter (see Fig. 2 and (1, 48)). If insufficient numbers of CellVue Claret-stained Treg cells are available for both test and control samples, it is possible, although not ideal, to stain other lymphocytes (e.g., excess Teff) under identical cell concentration and dye conditions for use as Treg surrogates in establishing appropriate detector voltages and compensation settings.

56. Since many wells in the 96-well plate will remain empty, it is recommended that wells surrounding test and control cells for the assay be filled with CM to minimize evaporation in the assay wells.

57. Although 7-AAD can be used as an alternative to LIVE/ DEAD Fixable Violet reagent (45), color compensation for CellVue Claret is simpler with the latter since there is greater spectral separation than with 7-AAD. In our experience, labeling cells with LIVE/DEAD Fixable Violet reagent can successfully be performed in the presence of low levels of exogenous protein, although this will reduce the separation between viable and nonviable populations. Here we chose to incubate simultaneously with human IgG, anti-CD4 PECy7, and LIVE/DEAD reagent in order to minimize cell losses associated with sequential staining and additional wash steps.

58. Choice of optimized labeling conditions for both Treg (here CellVue Claret stained) and Teff (here CFSE stained) should have already established that stained cells will be on scale, without events accumulating in the highest channel. This condition should be established when compensation is set to 0% and voltages are adjusted to place unstained control cells in the first decade, but sufficiently above the left axis so that events do not accumulate in the lowest channel.

59. Total number of cells to be collected depends upon the actual population of interest. In a cell proliferation experiment, the population of interest may be distributed over a broad range of intensities representing up to seven or eight generations. Therefore, to model and calculate the number of cells in each generation accurately, a large number of cells may need to be collected. It is most appropriate to set a stop region on the specific population of interest and collect sufficient cells for analysis within this region. In this example, if the population of interest is Treg cells, it is recommended that a stop region be set on viable Treg cells and a minimum of 5,000 events be collected within this region. When studying rare cells, it may be necessary to simply run the sample tube nearly dry in order to collect the maximum possible number of events.

Acknowledgments

The authors have had the opportunity to work with many wonderful people on the development of these techniques over the years. In particular, they would like to acknowledge the technical and intellectual contributions of Bruce Bagwell (Verity Software House), Drew Bantly (University of Pennsylvania),

Nadège Bercovici (IDM), Lizanne Breslin (PTI Research and SciGro), Jan Fisher (Dartmouth Medical School), Alice Givan (Dartmouth Medical School), Brian Gray (Molecular Targeting Technologies), Jonni Moore (University of Pennsylvania), Betsy Ohlsson-Wilhelm (SciGro), Feng Qian (Roswell Park), Earl Timm, Jr. (Roswell Park), and Mary Waugh (Dartmouth Medical School). They would also like to thank the Bowdoin class of 2006 from the Annual Courses in Flow Cytometry (Research Methods and Applications) who generated the data for Figs. 2b and 4b.

Flow cytometry was performed at Roswell Park Cancer Institute's Flow Cytometry Laboratory, which was established in part by equipment grants from the NIH Shared Instrument Program, and receives support from the Core Grant (5 P30 CA016056-29) from the National Cancer Institute to the Roswell Park Cancer Institute.

References

1. Wallace, P. K., Tario, J. D., Jr., Fisher, J. L., Wallace, S. S., Ernstoff, M. S., and Muirhead, K. A. (2008) Tracking antigen-driven responses by flow cytometry: monitoring proliferation by dye dilution. *Cytometry A* **73**, 1019–34.

2. Parish, C. R. (1999) Fluorescent dyes for lymphocyte migration and proliferation studies. *Immunol Cell Biol* **77**, 499–508.

3. Poon, R. Y., Ohlsson-Wilhelm, B. M., Bagwell, C. B., and Muirhead, K. A. Use of PKH membrane intercalating dyes to monitor cell trafficking and function. In: R. A. Diamond and S. DeMagio (eds.), In Living Color: Flow Cytometry and Cell Sorting Protocols, pp. 302–52. New York: Springer, 2000.

4. Quah, B. J., Warren, H. S., and Parish, C. R. (2007) Monitoring lymphocyte proliferation in vitro and in vivo with the intracellular fluorescent dye carboxyfluorescein diacetate succinimidyl ester. *Nat Protoc* **2**, 2049–56.

5. Wallace, P. K. and Muirhead, K. A. (2007) Cell tracking 2007: a proliferation of probes and applications. *Immunol Invest* **36**, 527–561.

6. Fuse, S. and Underwood, E. (2007) Simultaneous analysis of in vivo CD8+ T cell cytotoxicity against multiple epitopes using multicolor flow cytometry. *Immunol Invest* **36**, 829–845.

7. Schafer, R., Wiskirchen, J., Guo, K., Neumann, B., Kehlbach, R., Pintaske, J., Voth, V., Walker, T., Scheule, A. M., Greiner, T. O., Hermanutz-Klein, U., Claussen, C. D., Northoff, H., Ziemer, G., and Wendel, H. P.

(2007) Aptamer-based isolation and subsequent imaging of mesenchymal stem cells in ischemic myocard by magnetic resonance imaging. *Rofo* **179**, 1009–15.

8. Flexman, J. A., Minoshima, S., Kim, Y., and Cross, D. J. (2006) Magneto-optical labeling of fetal neural stem cells for in vivo MRI tracking. *Conf Proc IEEE Eng Med Biol Soc* **1**, 5631–4.

9. Stroh, A., Boltze, J., Sieland, K., Hild, K., Gutzeit, C., Jung, T., Kressel, J., Hau, S., Reich, D., Grune, T., and Zimmer, C. (2009) Impact of magnetic labeling on human and mouse stem cells and their long-term magnetic resonance tracking in a rat model of Parkinson disease. *Mol Imaging* **8**, 166–78.

10. Modo, M., Beech, J. S., Meade, T. J., Williams, S. C., and Price, J. (2009) A chronic 1 year assessment of MRI contrast agent-labelled neural stem cell transplants in stroke. *Neuroimage* **47 Suppl 2**, T133–42.

11. Sun, N., Lee, A., and Wu, J. C. (2009) Long term non-invasive imaging of embryonic stem cells using reporter genes. *Nat Protoc* **4**, 1192–201.

12. Schierling, W., Kunz-Schughart, L. A., Muders, F., Riegger, G. A., and Griese, D. P. (2008) Fates of genetically engineered haematopoietic and mesenchymal stem cell grafts in normal and injured rat hearts. *J Tissue Eng Regen Med* **2**, 354–64.

13. Cicalese, A., Bonizzi, G., Pasi, C. E., Faretta, M., Ronzoni, S., Giulini, B., Brisken, C., Minucci, S., Di Fiore, P. P., and Pelicci, P. G. (2009) The tumor suppressor p53 regulates

polarity of self-renewing divisions in mammary stem cells. *Cell* **138**, 1083–95.

14. Daubeuf, S., Aucher, A., Sampathkumar, S.-G., Preville, X., Yarema, K. J., and Hudrisier, D. (2007) Chemical labels metabolically installed into the glycoconjugates of the target cell surface can be used to track lymphocyte/target cell interplay via trogocytosis: comparisons with lipophilic dyes and biotin. *Immunol Invest* **36**, 687–712.

15. Gertner-Dardenne, J., Poupot, M., Gray, B. D., and Fournié, J.-J. (2007) Lipophilic fluorochrome trackers of membrane transfers between immune cells. *Immunol Invest* **36**, 665–685.

16. Hawkins, E. D., Hommel, M., Turner, M. L., Battye, F. L., Markham, J. F., and Hodgkin, P. D. (2007) Measuring lymphocyte proliferation, survival and differentiation using CFSE time-series data. *Nat Protoc* **2**, 2057–67.

17. Kusumbe, A. P. and Bapat, S. A. (2009) Cancer stem cells and aneuploid populations within developing tumors are the major determinants of tumor dormancy. *Cancer Res* **69**, 9245–53.

18. Hamelik, R. M. and Krishan, A. (2009) Click-iT assay with improved DNA distribution histograms. *Cytometry A* **75**, 862–5.

19. Roder, J. C., Haliotis, T., Klein, M., Korec, S., Jett, J. R., Ortaldo, J., Heberman, R. B., Katz, P., and Fauci, A. S. (1980) A new immunodeficiency disorder in humans involving NK cells. *Nature* **284**, 553–5.

20. Morales, A. and Ottenhof, P. C. (1983) Clinical application of a whole blood assay for human natural killer (NK) cell activity. *Cancer* **52**, 667–70.

21. Zaritskaya, L., Shurin, M. R., Sayers, T. J., and Malyguine, A. L. (2010) New flow cytometric assays for monitoring cell-mediated cytotoxicity. Expert Rev. Vaccines **9**, 601–616.

22. Sheehy, M. E., McDermott, A. B., Furlan, S. N., Klenerman, P., and Nixon, D. F. (2001) A novel technique for the fluorometric assessment of T lymphocyte antigen specific lysis. *J Immunol Methods* **249**, 99–110.

23. Flieger, D., Spengler, U., Beier, I., Sauerbruch, T., and Schmidt-Wolf, I. G. (2000) Prestimulation of monocytes by the cytokines GM-CSF or IL-2 enhances the antibody dependent cellular cytotoxicity of monoclonal antibody 17-1A. *Z Gastroenterol* **38**, 615–22.

24. Aubry, J. P., Blaecke, A., Lecoanet-Henchoz, S., Jeannin, P., Herbault, N., Caron, G., Moine, V., and Bonnefoy, J. Y. (1999) Annexin V used for measuring apoptosis in the early events of cellular cytotoxicity. *Cytometry* **37**, 197–204.

25. Ely, P., Wallace, P. K., Givan, A. L., Graziano, R. F., Guyre, P. M., and Fanger, M. W. (1996) Bispecific-armed, interferon gamma-primed macrophage-mediated phagocytosis of malignant non-Hodgkin's lymphoma. *Blood* **87**, 3813–21.

26. Lee-MacAry, A. E., Ross, E. L., Davies, D., Laylor, R., Honeychurch, J., Glennie, M. J., Snary, D., and Wilkinson, R. W. (2001) Development of a novel flow cytometric cell-mediated cytotoxicity assay using the fluorophores PKH-26 and TO-PRO-3 iodide. *J Immunol Methods* **252**, 83–92.

27. Schutz, C., Fischer, K., Volkl, S., Hoves, S., Halbritter, D., Mackensen, A., and Fleck, M. (2009) A new flow cytometric assay for the simultaneous analysis of antigen-specific elimination of T cells in heterogeneous T cell populations. *J Immunol Methods* **344**, 98–108.

28. Wilkinson, R. W., Lee-MacAry, A. E., Davies, D., Snary, D., and Ross, E. L. (2001) Antibody-dependent cell-mediated cytotoxicity: a flow cytometry-based assay using fluorophores. *J Immunol Methods* **258**, 183–91.

29. Chen, X., Wang, B., and Chang, L. J. (2006) Induction of primary anti-HIV CD4 and CD8 T cell responses by dendritic cells transduced with self-inactivating lentiviral vectors. *Cell Immunol* **243**, 10–8.

30. Liu, W., Putnam, A. L., Xu-Yu, Z., Szot, G. L., Lee, M. R., Zhu, S., Gottlieb, P. A., Kapranov, P., Gingeras, T. R., Fazekas de St Groth, B., Clayberger, C., Soper, D. M., Ziegler, S. F., and Bluestone, J. A. (2006) CD127 expression inversely correlates with FoxP3 and suppressive function of human CD4+ T reg cells. *J Exp Med* **203**, 1701–11.

31. Shevach, E. M. (2009) Mechanisms of FOXP3+ T regulatory cell-mediated suppression. *Immunity* **30**, 636–45.

32. Boros, P. and Bromberg, J. S. (2009) Human FOXP3+ regulatory T cells in transplantation. *Am J Transplant* **9**, 1719–24.

33. Corthay, A. (2009) How do regulatory T cells work? *Scand J Immunol* **70**, 326–36.

34. Brusko, T. M., Hulme, M. A., Myhr, C. B., Haller, M. J., and Atkinson, M. A. (2007) Assessing the in vitro suppressive capacity of regulatory T cells. *Immunol Invest* **36**, 607–628.

35. Neron, S., Thibault, L., Dussault, N., Cote, G., Ducas, E., Pineault, N., and Roy, A. (2007) Characterization of mononuclear cells remaining in the leukoreduction system chambers of apheresis instruments after routine

platelet collection: a new source of viable human blood cells. *Transfusion* **47**, 1042–9.

36. Wallace, P. K., Keler, T., Coleman, K., Fisher, J., Vitale, L., Graziano, R. F., Guyre, P. M., and Fanger, M. W. (1997) Humanized mAb H22 binds the human high affinity Fc receptor for IgG (FcgammaRI), blocks phagocytosis, and modulates receptor expression. *J Leukoc Biol* **62**, 469–79.

37. Mentzer, S. J., Guyre, P. M., Burakoff, S. J., and Faller, D. V. (1986) Spontaneous aggregation as a mechanism for human monocyte purification. *Cell Immunol* **101**, 312–9.

38. Rousselle, C., Barbier, M., Comte, V. V., Alcouffe, C., Clement-Lacroix, J., Chancel, G., and Ronot, X. (2001) Innocuousness and intracellular distribution of PKH67: a fluorescent probe for cell proliferation assessment. *In Vitro Cell Dev Biol Anim* **37**, 646–55.

39. Horan, P. K., Melnicoff, M. J., Jensen, B. D., and Slezak, S. E. (1990) Fluorescent cell labeling for in vivo and in vitro cell tracking. *Methods Cell Biol* **33**, 469–90.

40. Wallace, P. K., Palmer, L. D., Perry-Lalley, D., Bolton, E. S., Alexander, R. B., Horan, P. K., Yang, J. C., and Muirhead, K. A. (1993) Mechanisms of adoptive immunotherapy: improved methods for in vivo tracking of tumor-infiltrating lymphocytes and lymphokine-activated killer cells. *Cancer Res* **53**, 2358–67.

41. Tario, J. D., Jr., Gray, B. D., Wallace, S. S., Muirhead, K. A., Ohlsson-Wilhelm, B. M., and Wallace, P. K. (2007) Novel lipophilic tracking dyes for monitoring cell proliferation. *Immunol Invest* **36**, 861–85.

42. Givan, A. L. (2007) A flow cytometric assay for quantitation of rare antigen-specific T-cells: using cell-tracking dyes to calculate precursor frequencies for proliferation. *Immunol Invest* **36**, 563–80.

43. Givan, A. L., Fisher, J. L., Waugh, M. G., Bercovici, N., and Wallace, P. K. (2004) Use of cell-tracking dyes to determine proliferation precursor frequencies of antigen-specific T cells. *Methods Mol Biol* **263**, 109–24.

44. Lyons, A. B. and Doherty, K. V. Flow cytometric analysis of cell division by dye dilution. In: J. P. Robinson, Z. Darzynkiewicz, R. Hoffman, J. P. Nolan, A. Orfao, P. S. Rabinovitch, and S. Watkins (eds.), Current Protocols in Cytometry, Vol. 9.11. New York: John Wiley & Sons, Inc., 2004.

45. Bantly, A. D., Gray, B. D., Breslin, E., Weinstein, E. G., Muirhead, K. A., Ohlsson-Wilhelm, B. M., and Moore, J. S. (2007) Cellvue Claret, a new far-red dye, facilitates polychromatic assessment of immune cell proliferation. *Immunol Invest* **36**, 861–85.

46. Matera, G., Lupi, M., and Ubezio, P. (2004) Heterogeneous cell response to topotecan in a CFSE-based proliferation test. *Cytometry A* **62**, 118–28.

47. Wallace, P. K., Romet-Lemonne, J. L., Chokri, M., Kasper, L. H., Fanger, M. W., and Fadul, C. E. (2000) Production of macrophage-activated killer cells for targeting of glioblastoma cells with bispecific antibody to FcgammaRI and the epidermal growth factor receptor. *Cancer Immunol Immunother* **49**, 493–503.

48. Lyons, A. B. and Parish, C. R. (1994) Determination of lymphocyte division by flow cytometry. *J Immunol Methods* **171**, 131–7.

49. Bercovici, N., Givan, A. L., Waugh, M. G., Fisher, J. L., Vernel-Pauillac, F., Ernstoff, M. S., Abastado, J. P., and Wallace, P. K. (2003) Multiparameter precursor analysis of T-cell responses to antigen. *J Immunol Methods* **276**, 5–17.

50. Givan, A. L., Fisher, J. L., Waugh, M., Ernstoff, M. S., and Wallace, P. K. (1999) A flow cytometric method to estimate the precursor frequencies of cells proliferating in response to specific antigens. *J Immunol Methods* **230**, 99–112.

51. Munson, M.E. (2010) An improved technique for calculating relative response in cellular proliferation experiments. *Cytometry* **77A**, 909–910.

Chapter 8

Multiparameter Intracellular Cytokine Staining

Patricia Lovelace and Holden T. Maecker

Abstract

Intracellular cytokine staining (ICS) is a popular method for visualizing cellular responses, most often T-cell responses to antigenic or mitogenic stimulation. It can be coupled with staining for other functional markers, such as upregulation of CD107 or CD154, as well as phenotypic markers that define specific cellular subsets, e.g. effector and memory T-cell compartments. Recent advances in multicolor flow cytometry instrumentation and software have allowed the routine combination of 8–12 (or more) markers in combination, creating technical and analytical challenges along the way, and exposing a need for standardization in the field. Here, we will review best practices for antibody panel design and procedural variables for multicolor ICS, and present an optimized protocol with variations designed for use with specific markers and sample types.

Key words: Antigen specific, Intracellular staining, Multicolor, Polychromatic, Fixation, Permeabilization, AIDS vaccine research, T cells

1. Introduction

With the use of secretion inhibitors such as monensin or brefeldin A, secreted cytokines and other proteins can be retained intracellularly. These proteins thus become available for antibody staining, upon fixation and permeabilization of the cells (1, 2). In general, short-term stimulation of cells with mitogen or antigen is required to induce cellular activation and production of cytokines. One common application of this technique is the visualization of antigen-specific T cells in PBMC (3) or whole blood (4). This requires stimulation with protein antigens or, commonly, pools of overlapping peptides spanning a protein sequence of interest (5). The latter, when designed with sufficient length and overlap between peptides, can efficiently stimulate both CD4 and CD8 T-cell responses.

Teresa S. Hawley and Robert G. Hawley (eds.), *Flow Cytometry Protocols*, Methods in Molecular Biology, vol. 699, DOI 10.1007/978-1-61737-950-5_8, © Springer Science+Business Media, LLC 2011

A common protocol for antigen-specific stimulation of T cells for intracellular cytokine staining (ICS) is as follows. Whole blood or PBMC are incubated with antigen or peptide mixtures for 6–16 h. Brefeldin A and/or monensin is added at the time of stimulation (for peptides) or after 2 h (for proteins, to allow for intracellular antigen processing, which is compromised by the secretion inhibitor). At the end of the stimulation period, cells can be held at 4–18°C until ready to process. They are then treated with EDTA to remove adherent cells, fixed (usually with formaldehyde), permeabilized (usually with a detergent), and stained for intracellular determinants. In some cases, surface marker staining is done in conjunction with intracellular staining (this usually works well for CD3, CD4, and CD8). However, most other cell-surface markers require staining prior to fixation, because the epitopes recognized by staining antibodies are sensitive to fixation and/or permeabilization.

Intracellular staining for multiple cytokines is now often combined with staining for other functional and phenotypic markers as well. This has been made possible by the availability of flow cytometers with digital signal processing, and detectors for up to 18 colors. Along with this instrumentation, software for automated calculation of compensation between colors is now routinely used, often in combination with single-stained capture beads that make construction of compensation controls easier and more precise (since the actual experimental antibodies can be used for compensation, an important consideration for some tandem dye conjugates). Finally, software and fluorescent beads to automate instrument setup and track performance over time are now available, making longitudinal standardization of experiments, at least for a single instrument, much easier. Standardization across instruments, especially given the degree of instrument customization seen in the field, can still be difficult, however.

Despite the advances in tools for multicolor flow cytometry, designing optimal antibody panels of 8 or more colors can be a challenge. The optical spectrum is limited, such that addition of new fluorescent reagents tends to create more spillover into existing detectors. In some cases, this can severely compromise the ability to use those detectors for measurements requiring high-resolution sensitivity.

A general discussion of rules for antibody panel design, along with suggestions for specific fluorochrome combinations and panels, is given in (6, 7). These rules are very briefly summarized here.

1. Use the dimmest fluorochromes for brightly staining antibodies (CD45, CD4, CD8, CD3, etc.), while reserving the brightest fluorochromes for dimly staining antibodies (see Table 1).

2. Avoid spillover from a bright cell population into a detector requiring high-resolution sensitivity (see Fig. 1). In some cases, two fluorochromes with high spillover between them

Table 1
Fluorochrome brightness

	Bright	Medium	Dim
Fluorochromes:	Qdot 655 PE AlexaFluor 647 or APC PE-Cy5 PE-Cy7	Qdot 705 Qdot 605 PerCP-Cy5.5 APC-Cy7 AlexaFluor 488 or FITC AlexaFluor 700	Qdot 800 Pacific Orange Pacific Blue or V450 AmCyan
Antibodies to use:	IL-2, CD25, CD127, PD-1, CTLA-4, or other markers requiring high-resolution sensitivity	CD27, CD28, TNFα, IFNγ, or other markers with intermediate separation between positive and negative populations.	CD45, CD4, CD8, CD3, or other markers with good separation between positive and negative populations

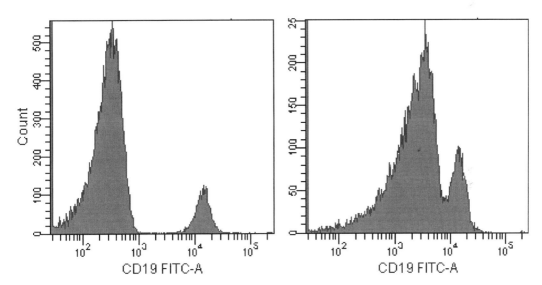

Fig. 1. Effect of a bright signal in AmCyan on resolution sensitivity in FITC. *Left panel:* staining with CD19 FITC alone. *Right panel:* combined staining of CD45 AmCyan and CD19 FITC. Note the spread of the CD19-negative peak, causing loss of resolution sensitivity. This is only a problem when both markers are present on the same cell population (e.g. use of CD3 AmCyan and CD19 FITC would not pose a problem). Similar issues arise whenever there is significant spillover from a bright population in one detector to another detector where high-resolution sensitivity is required.

can be used for markers that identify distinct, non-overlapping cell populations, thereby negating spillover concerns.

3. Avoid potential artifacts of tandem dye degradation, either by avoiding tandem dyes that are particularly susceptible to breakdown (APC-Cy7 and PE-Cy7), or by choosing reagent combinations such that small amounts of tandem breakdown

will not compromise readouts in the parent dye channels (APC or PE). See Maecker et al. (7) for further discussion and examples of this issue.

There are also practical considerations to panel design, such as what antibody conjugates are commercially available. In general, it is best to use direct fluorochrome conjugates for multicolor work and for intracellular staining, since nonspecific binding can be a significant concern in these situations.

Even direct antibody conjugates can be optimized by titration for a particular application. Optimal titers should be picked on the basis of maximal signal:noise, which is often obtained below the titer recommended by the manufacturer.

While panels will constantly be refined to include new markers of interest, a degree of standardization is helpful, to avoid extensive re-optimization of new panels and provide for some degree of longitudinal comparisons. In Table 2, we provide some suggested panels currently in use in our laboratory.

Finally, optimal detection of certain marker combinations requires modification of stimulation and processing steps (e.g. CD107 or CD154). These variables are summarized in Table 3.

Table 2
Some suggested multicolor antibody panels

	8-Color T cells 3 Memory markers + 2 functions[a]	10-Color T cells 1 Memory marker + 5 functions[b]	10-Color T cells 3 memory markers + 3 functions[c]
FITC or AlexaFluor 488	Anti-IFNγ	CD27	Anti-IFNγ
PE	Anti-IL-2	CD154	Anti-IL-2
PE-Texas Red or PE-AlexaFluor 610		CD107	Anti-TNFα
PerCP-Cy5.5	CD28	CD4	CD28
PE-Cy7	CD45RA	Anti-IFNγ	CD45RA
APC or AlexaFluor 647	CD27	Anti-IL-2	CD27
AlexaFluor 700		Anti-TNFα	CD3
APC-H7	CD8	CD8	CD8
Pacific Blue or V450	CD3	CD3	CD4
AmCyan or Pacific Orange	CD4	CD14	CD14

[a]Used in Ref. (15)
[b]Modified from panel used for International Flow Cytometry School (IFCS) 2006, Florence, Italy
[c]Used for Multicolor ICS Users Group standardization studies (see also http://maeckerlab.typepad.com for additional panel suggestions and details)

Table 3
Procedural variables for different functional markers

Variable	IL-2, IL-4, IL-5, IL-13, IFNγ, MIP-1β, TNFα	CD107, CD154	IL-10, TGFβ
Stimulation conditions	6–12 h	5–6 h in the presence of staining antibodies for these markers	12–24 h (serum-free medium for TGFβ[a])
Secretion inhibitor	Brefeldin A	Monensin[b]	Monensin
Fixation/ permeabilization system	FACS lysing solution, FACS permeabilizing solution 2[c]	FACS lysing solution, FACS permeabilizing solution 2[c]	Cytofix, Cytoperm[d]

[a]Serum-free medium (e.g. AIM V; Invitrogen, Carlsbad, CA) produces much stronger TGFβ responses, presumably because serum contains free TGFβ that blocks staining for this marker

[b]CD107 and CD154 are taken up by endocytic vesicles and degraded. Monensin blocks this degradation by preventing acidification of these vesicles. When performing combined assays with cytokines, a monensin plus brefeldin A combination is recommended (see Note 5)

[c]BD Biosciences, San Jose, CA. Note that the Cytofix/Cytoperm system (BD Biosciences, San Diego, CA) is also used successfully for these markers by many investigators

[d]BD Biosciences, San Diego, CA

2. Materials

2.1. Reagents

1. Test specimens: heparinized whole blood, freshly isolated PBMC, or cryopreserved PBMC, isolated by Ficoll gradient centrifugation or via cell preparation tubes (CPT) (BD Vacutainer; BD Diagnostics, Franklin Lakes, NJ, or equivalent).

2. RPMI-1640 medium with 20 mM HEPES, 10% fetal bovine serum, and antibiotic/antimycotic solution (cRPMI-10; components from Sigma Chemical Co., St. Louis, MO, or equivalent supplier).

3. Stimulation antigens, e.g. peptide mixes (5). To prepare aliquots of lyophilized peptides, resuspend peptides or peptide mixes in DMSO at a concentration of 500 µg/mL/peptide or greater. Store resuspended peptides in aliquots at –80°C. Optional: Preconfigured plates containing lyophilized stimulation reagents, costimulatory antibodies, and secretion inhibitor(s) can be purchased (BD Lyoplate; BD Biosciences, San Jose, CA) (8).

4. Staphylococcal enterotoxin B (SEB), 50 µg/mL in sterile PBS, to use as a positive control for stimulation.

5. Recommended: costimulatory antibodies to CD28 and CD49d, 0.1 mg/mL each in sterile PBS (BD FastImmune; BD Biosciences).

6. Brefeldin A, 5 mg/mL in DMSO; or brefeldin A plus monensin, 2.5 mg/mL each in 50% DMSO plus 50% methanol (Sigma).

7. 20 mM EDTA in PBS, pH 7.4 (BD FastImmune or equivalent).

8. Erythrocyte lysis and cell fixation reagent, e.g. BD FACS lysing solution (BD Biosciences) or equivalent.

9. Cell permeabilizing reagent, e.g. BD FACS permeabilizing solution 2 (BD Biosciences) or equivalent.

10. Fluorescent-labeled antibodies (see for example Table 2). Optional: Preconfigured plates containing lyophilized antibody cocktails (BD Lyoplate) (8).

11. Wash buffer: 0.5% bovine serum albumin plus 0.1% NaN$_3$ in PBS.

12. Recommended: BD CompBeads (anti-mouse Igκ, anti-rat Igκ, or anti-rat/hamster Igκ; BD Biosciences), for creating single-color compensation controls.

13. (Optional) To reduce biohazard potential, or if samples will be stored for >24 h prior to acquisition: 1% paraformaldehyde in PBS [dilute 10% paraformaldehyde (EM Science, Gibbstown, NJ) 1:10 in PBS], or BD Stabilizing Fixative (BD Biosciences).

2.2. Equipment

1. For whole blood assays, 96-well conical bottom deep-well polypropylene plates and lids. For fresh or frozen PBMC, 96-well conical bottom polypropylene plates with lids (e.g. BD Falcon, BD Biosciences, or equivalent) (see Note 1).

2. 12-channel aspiration manifold with 35 mm prongs for deep-well plates and 7 mm prongs for regular plates (V&P Scientific, San Diego, CA).

3. Plate holders for table-top centrifuge (e.g. Sorvall Instruments, Newtown, CT).

4. Polychromatic flow cytometer with digital signal processing, e.g. BD LSR II equipped with Cytometer Setup and Tracking beads and software (BD Biosciences) or Dako Cyan ADP (Dako Corporation, Fort Collins, CO).

5. Optional: 96-well plate loader for flow cytometer.

3. Methods

3.1. Sample Collection

1. (a) For fresh PBMC (see Note 2): Resuspend at 5×10^6 to 1×10^7 viable cells/mL in warm (37°C) cRPMI-10 (see Note 3).

   (b) For cryopreserved PBMC: Thaw briefly in a 37°C water bath, then slowly dilute up to 10 mL with warm (37°C) cRPMI-10 and centrifuge for approximately 7 min at $250 \times g$. Resuspend in a small volume of warm cRPMI-10, perform a viable cell count, and dilute to a concentration of about 1.5×10^7 viable cells/mL for resting overnight. Incubate at 37°C for 6–18 h prior to stimulation in a conical centrifuge tube, tilted (see Note 4). Resuspend and recount cells after the rest. Using this new count, prepare the final suspension at 5×10^6 to 1×10^7 viable cells/mL in warm (37°C) cRPMI-10.

(c) For whole blood assays: Collect whole blood in sodium heparin and store at room temperature for not more than 8 h prior to use.

2. Add 200 µL of cell suspension per well to an appropriate 96-well plate.

3. For assays using preconfigured lyophilized stimulation reagents in plates: Add 200 µL of cell suspension directly to the appropriate wells in the stimulation lyoplate, allow the cells to stand for a few minutes, and then pipet up and down thoroughly to mix. Skip to Subheading 3.2, step 6.

3.2. Cell Activation

1. Prepare the secretion block reagents.

(a) For assays not involving CD107 or CD154: Thaw an aliquot of 5 mg/mL of brefeldin A stock (see Note 5). Dilute 1:10 in sterile PBS to make a 50× working stock.

(b) For assays measuring CD107 and/or CD154: Thaw an aliquot of 2.5 mg/mL of brefeldin A plus 2.5 mg/mL of monensin stock (see Note 5). Dilute 1:10 in sterile PBS to make a 50× working stock.

2. Thaw and dilute peptide stock aliquots (see Note 6) in sterile PBS, if necessary, to achieve a 50× working stock that is between 50 and 100 µg/mL/peptide (when diluted 1:50, this will yield a final concentration of 1–2 µg/mL/peptide).

3. For each stimulation condition, prepare a "master mix" of the 50× working stocks and costimulatory antibodies as follows:

(a) 4 µL/well of peptides, SEB (positive control), or PBS (negative control).

(b) 4 µL/well of brefeldin A or brefeldin A plus monensin.

(c) 4 µL/well of CD28 plus CD49d Ab stock (see Note 7).

(d) Always prepare a slight excess of each master mix.

4. Pipet 12 µL of the appropriate master mix into each well containing cells. Mix by pipetting gently.

5. For assays involving CD107 and/or CD154, also add the recommended titer of the antibody conjugate(s) to each well. Minimize exposure to light, particularly for tandem dye conjugates (see Note 8).

6. Cover the plate and incubate for 6–12 h at 37°C (see Notes 9 and 10).

*3.3. Sample
Processing*

1. To halt activation and detach adherent cells, add 20 μL per well of 20 mM EDTA in PBS and mix by pipetting.

2. Incubate for 15 min at room temperature, then mix again by vigorous pipetting to resuspend adhered cells fully.

3. For PBMC, centrifuge plate at $250 \times g$ for 5 min. Aspirate the supernatant with the appropriate vacuum manifold for the plate (see Note 11).

4. For assays using amine-reactive dye for staining nonviable cells: Resuspend the amine dye at optimum concentration in PBS (usually around 2.5 μg/mL, but this should be determined for individual lots of dye). Resuspend each well with 100 μL of this solution, incubate for 20 min at room temperature, then add 100 μL of wash buffer, and wash as in step 3 above. Amine dyes can be used with whole blood, but higher concentrations will be required because the blood is not washed into PBS prior to dye staining. Therefore, the staining intensity may be reduced.

5. For assays using liquid reagents and cell-surface markers other than CD3, CD4, and CD8: Resuspend each well in 100 μL of wash buffer (for PBMC) and add optimal titers of all Abs to cell-surface markers (see Note 12). Incubate for 30–60 min at room temperature, then add 100 μL of wash buffer (for PBMC), and wash as in step 3 above.

   (a) For assays using preconfigured lyophilized staining reagents and cell-surface staining Abs, resuspend the appropriate wells of the surface Ab plate with 50 μL of wash buffer. Let sit for a few minutes, then pipet up and down thoroughly to mix. Transfer the solution to appropriate wells of the cell plate, incubate for 30–60 min at room temperature in the dark, then add 100 μL of wash buffer, and wash as in step 3 above.

6. For PBMC, resuspend cell pellets with 100 μL of 1× BD FACS lysing solution per well. For whole blood, add 2 mL of room temperature 1× BD FACS lysing solution per well, pipetting up and down to mix. Incubate both types of assay at room temperature for 10 min in FACS lysing solution (*see* Notes 13 and 14).

7. For PBMC, add 100 μL of wash buffer to each well, then centrifuge the plate at $500 \times g$ for 5 min (see Note 15). For whole blood, simply centrifuge the plate at $500 \times g$. Aspirate the supernatant for both with the appropriate vacuum manifold for the plate.

8. For PBMC, resuspend cell pellets with 200 μL of 1× BD FACS permeabilizing solution 2 per well. For whole blood, resuspend cell pellets in 1 mL of 1× BD FACS permeabilizing solution 2 per well. Incubate both types of assay at room temperature for 10 min (see Note 14).

9. Centrifuge the plate at $500 \times g$ for 5 min (see Note 15). Aspirate the supernatant with appropriate vacuum manifold for the plate.

10. For standard plates, add 200 µL of wash buffer to each well and wash as in step 9 above. For deep-well plates, add 1.5 mL of wash buffer to each well and wash as in step 9 above.

11. For standard plates, add 200 µL of wash buffer to each well and wash a second time as in step 9 above.

12. For assays using liquid reagents: Resuspend the pellet in 100 µL of wash buffer and add optimal titers of all Abs to intracellular markers and surface markers not already stained. Incubate in the dark at room temperature for 60 min, mixing by pipetting or gentle agitation every 15–20 min.

    (a) For assays using preconfigured lyophilized intracellular stain- ing reagents, resuspend the appropriate wells of the intracel- lular antibody plate with 50 µL of wash buffer. Let sit for a few minutes, then pipette up and down thoroughly to mix. Transfer the solution to the appropriate wells of the cell plate and incubate at room temperature in the dark for 60 min, mixing by pipetting or gentle agitation every 15–20 min.

13. Wash again as in steps 10 and 11 above.

14. Resuspend pellets with 150 µL of wash buffer. Store at 4°C in the dark until ready for data acquisition, which should be per- formed within 24 h. Optional: resuspend pellets with 150 µL of 1% paraformaldehyde in PBS or BD Stabilizing Fixative (see Note 16).

3.4. Data Acquisition and Analysis

1. First determine optimal PMT settings for the instrument and reagent panel in question. Using CS&T beads and software on a BD LSR-II, start with CS&T baseline voltages and then perform the following:

    (a) Run single-stained compensation controls (see Note 17) and decrease PMT voltage gain, if needed, to ensure that no events are in the highest fluorescence channel. Increase PMT voltage gain, if needed, to ensure that positive peaks are at least twofold brighter in their primary detec- tor compared to other detectors.

    (b) Run a fully stained positive control sample and decrease PMT voltage gain, if needed, to ensure that no events are in the highest fluorescence channel. If changes are made, repeat steps (a) and (b) until no more changes are required. Save the resulting settings as an Application Setting in FACS Diva software (BD Biosciences).

2. Acquire the single-stained compensation controls and use the software's automated algorithm to calculate compensation (see Note 18).

3. Create a template for acquisition that displays the relevant parameters in the test samples in the form of dot plots. This template need not be the same as that used for analysis, i.e., it does not need to specify all gates or regions of interest. In fact, a simplified acquisition template will allow faster processing of data. However, the template should show any gates used to define the saved population of cells or the stopping criteria (e.g. CD3+ cells).

4. Set an appropriate threshold, usually on FSC or CD45+ events, to eliminate debris and unwanted events. Set the stopping and storage criteria to obtain sufficient events for analysis. It is usually safest to store all events (rather than a gated subset) to allow for re-gating and exploration of other subsets. However, sometimes a threshold or gate on CD3+ cells may be employed in order to reduce file sizes (see Note 19).

5. Record data from samples.

6. Analyze data using the acquisition software or compatible third-party software. Be sure to define all regions of interest and report the desired statistics on these (see Note 20 and Fig. 1). Where possible, use a batch analysis function to analyze all samples from a given experiment or study, and export the statistical data to a spreadsheet (see Note 21).

7. For large studies, it is helpful to create a database to accept the statistical output files from batch analysis. This database can then be queried to create data tables from subsets of the data, allowing rapid graphing, statistical analysis, background subtraction, conversion to absolute counts, etc.

4. Notes

1. *Plates versus tubes*: Cells can also be stimulated in 15-mL conical polypropylene tubes, with staining in 12×75 mm polystyrene tubes (BD Falcon). However, plates are preferred for ease of handling multiple samples, and results for human PBMC are equivalent to those in tubes (9).

2. *Fresh PBMC*: If PBMC are not to be cryopreserved, they should ideally be prepared on the day of blood draw and then either stimulated the same day, or rested at 37°C in cRPMI-10 overnight and stimulated the following day. Overnight resting at 37°C increases the staining intensity of cytokines, but the effect is more pronounced with cryopreserved samples. Overnight shipping of whole blood or PBMC at ambient temperatures can cause a variable decrease in cell function and should be avoided if possible, although shipping PBMC is preferable to shipping whole blood.

3. Higher cell concentrations (1×10^7/mL, 2×10^6/well) should be used when possible, especially when response levels are low and/or there are many cell subsets to enumerate.

4. *Cryopreserved PBMC*: If cells cannot be stimulated within 24 h of blood draw, they should be cryopreserved by a validated protocol (10). Upon thawing, recoveries of >60% and viabilities of >80% should be obtained to minimize loss of functional responses. The method of thawing is equally as important as that of cryopreservation (10). Thawed cells should be rested in cRPMI-10 for 6–18 h at 37°C to maximize cytokine staining intensity (9). Some cell loss may occur during this period, so rest the cells at a higher concentration than you will ultimately use. Recount and resuspend at the desired final concentration after resting.

5. *Brefeldin A versus monensin*: Secretion of most cytokines of interest (IFNγ, IL-2, etc.) is best inhibited by brefeldin A at 10 µg/mL cells. However, CD107 and CD154 are transiently expressed on the cell surface. Therefore, staining Abs to CD107 and/or CD154 is added to the stimulation culture to bind the antigen(s) as soon as they are expressed. Monensin increases the intensity of staining under these conditions by preventing the acidification and degradation of lysosomal vesicles that contain the recycled CD107 and CD154. Thus, for combined cytokine and CD107 or CD154 detection, 5 µg/mL each of brefeldin A and monensin is recommended.

6. *Peptide mixes*: Peptide mixes can be prepared and lyophilized as premixed pools of up to several hundred peptides (5). These can then be resuspended in DMSO at high concentration per peptide, avoiding DMSO toxicity. The total concentration of DMSO in the assay should be kept at <0.5%.

7. *Costimulatory antibodies*: Antibodies to CD28 and CD49d can increase the cytokine response to protein antigens, peptides, and SEB by amplifying the signal for low-affinity T cells (11). In occasional donors, they increase cytokine production in the absence of antigen (TNFα is usually the most affected).

8. *Adding staining Abs during stimulation*: As described in Note 5, staining Abs to CD107 and CD154 are best added during stimulation, to capture the transiently expressed antigen. Fluorochrome-conjugated Abs are sensitive to light exposure, so they should be handled in low light and, once added, the samples should be incubated in the dark. Certain tandem dyes such as APC-Cy7 and PE-Cy7 are particularly sensitive to light and temperature (7) and are not optimal choices for use in stimulation cultures.

9. *Stimulation time*: A minimum of 5–6 h allows adequate detection of most proinflammatory cytokines such as IFNγ,

TNFα, and IL-2 (12). Increasing the time of incubation (in the presence of brefeldin A) increases cytokine staining intensity, but is not recommended for CD107 or CD154. For whole proteins requiring intracellular processing, a pre-incubation of 2 h prior to adding brefeldin A and/or monensin is recommended (12). CD8 responses to whole protein antigens can sometimes be detected and are increased with longer incubation in antigen alone, but not in all donors (13).

10. *Automating incubation times.* A programmable heat block, incubator, or water bath can be used for time activation, cooling the samples to 4–18°C at the end of a specified period at 37°C, and holding them for later processing.

11. A fixed-length vacuum manifold helps achieve consistent washing without undue cell loss in microtiter plates. Because of the small wash volume, a sufficient number of washes and efficient removal of supernatant are essential.

12. CD3, CD4, and CD8 can be stained either before or after fixation and permeabilization. Down-modulation of these antigens occurs to a variable degree depending upon the stimulus. Cells that have down-modulated these antigens can be better detected by intracellular staining (post-fixation and permeabilization) (5), although the overall staining intensity is usually decreased. Most other cell-surface antigens are optimally stained before fixation.

13. *Freezing of activated samples.* Samples can be frozen at –80°C directly in FACS lysing solution (12, 14). This allows for samples to be sent to another laboratory for processing, or for longitudinal samples to be accumulated for batch processing.

14. *Fixation and permeabilization steps.* Solutions for these steps should be stored and used at 22–25°C. FACS lysing solution simultaneously lyses erythrocytes and fixes leukocytes. While erythrocyte lysis is not required for PBMC samples, fixation is still helpful to prevent cell loss prior to permeabilization.

15. *Centrifugation speed.* All centrifugation post-fixation should be done at higher g force ($500 \times g$) due to increased cell buoyancy.

16. Use of paraformaldehyde is only helpful when samples are stored for more than 24 h prior to acquisition, or to ensure neutralization of potentially biohazardous samples. In addition to subtle effects on cell scatter and fluorescence, storage in paraformaldehyde can cause degradation of tandem dyes such as APC-Cy7 and PE-Cy7. An alternative fixative is available that protects these tandems from degradation (BD Stabilizing Fixative), but it is not compatible with AmCyan staining.

17. *Compensation controls*: Where possible, anti-immunoglobulin-coated capture beads (BD Biosciences) are preferred as compensation controls, because they provide a bright and homogeneous population of events stained with the antibody conjugate of interest. Ideally, the same lot of antibody should be used for compensation as is used in the experiment. In practice, however, this is only important for certain tandem conjugates, such as APC-Cy7 and PE-Cy7. The compensation controls should ideally be treated identically to the experimental samples in terms of fixation, etc., although this too is important only for the above tandem dyes.

18. *When to apply compensation*: While compensation can be calculated and changed at any time by software packages such as FlowJo (TreeStar, Ashland, OR) or FACS Diva, it is helpful to perform compensation before sample acquisition, so that any setup problems can be more readily detected.

19. *Number of events to collect*: Because multiparameter ICS assays tend to divide responding populations of cells into ever-smaller subsets, it is important to process and collect enough cells per sample to allow statistically significant differences between samples to be detected. The number of events required will depend upon the anticipated levels of responses and background, as well as the number of subsets of responding cells being identified. Statistical tools for sample size calculation can be found at http://maeckerlab.typepad.com.

20. *Gating of down-modulated cells*: Be sure that gates set on CD3, CD4, and CD8 parameters include dim-positive cells, since down-modulation of these markers occurs with activation. When using dynamic gating (see Note 21), set the region size to the maximum value possible without causing inclusion of neighboring populations. Some donors have a significant population of CD4 plus CD8dim T cells. This population contains a disproportionate number of cells specific for chronic antigens such as CMV and HIV, and should be included in the CD4+ T-cell gate to avoid under-reporting of responses.

21. *Batch analysis*: Dynamic gating tools such "Snap-To" gates in FACS Diva can be used to accommodate staining differences between samples for populations such as CD3+, CD4+, and CD8+ cells (see Fig. 1). This in turn allows the use of a single analysis template and batch analysis across multiple samples in an experiment or study. However, dynamic gates are not always useful for rare populations, and their specifications (size and movement) need to be adjusted for the data set being analyzed. Batch analysis and dynamic gating thus do not replace the need for visual inspection of all data.

Acknowledgments

Details of this protocol were optimized by Laurel Nomura and Maria Suni (BD Biosciences).

References

1. Jung, T., Schauer, U., Heusser, C., Neumann, C., and Rieger, C. (1993) Detection of intracellular cytokines by flow cytometry, *J Immunol Methods* 159, 197–207.

2. Prussin, C. and Metcalfe, D. D. (1995) Detection of intracytoplasmic cytokine using flow cytometry and directly conjugated anticytokine antibodies, *J Immunol Methods* 188, 117–128.

3. Waldrop, S. L., Pitcher, C. J., Peterson, D. M., Maino, V. C., and Picker, L. J. (1997) Determination of antigen-specific memory/effector CD4+ T cell frequencies by flow cytometry: evidence for a novel, antigen-specific homeostatic mechanism in HIV-associated immunodeficiency, *J Clin Invest* 99, 1739–1750.

4. Suni, M. A., Picker, L. J., and Maino, V. C. (1998) Detection of antigen-specific T cell cytokine expression in whole blood by flow cytometry, *J Immunol Methods* 212, 89–98.

5. Maecker, H. T., Dunn, H. S., Suni, M. A., Khatamzas, E., Pitcher, C. J., Bunde, T., Persaud, N., Trigona, W., Fu, T. M., Sinclair, E., Bredt, B. M., McCune, J. M., Maino, V. C., Kern, F., and Picker, L. J. (2001) Use of overlapping peptide mixtures as antigens for cytokine flow cytometry, *J Immunol Methods* 255, 27–40.

6. Maecker, H. T. (2009) Multiparameter flow cytometry monitoring of T cell responses, *Methods Mol Biol* 485, 375–391.

7. Maecker, H. T., Frey, T., Nomura, L. E., and Trotter, J. (2004) Selecting fluorochrome conjugates for maximum sensitivity, *Cytometry A* 62, 169–173.

8. Maecker, H. T., Rinfret, A., D'Souza, P., Darden, J., Roig, E., Landry, C., Hayes, P., Birungi, J., Anzala, O., Garcia, M., Harari, A., Frank, I., Baydo, R., Baker, M., Holbrook, J., Ottinger, J., Lamoreaux, L., Epling, C. L., Sinclair, E., Suni, M. A., Punt, K., Calarota, S., El-Bahi, S., Alter, G., Maila, H., Kuta, E., Cox, J., Gray, C., Altfeld, M., Nougarede, N., Boyer, J., Tussey, L., Tobery, T., Bredt, B., Roederer, M., Koup, R., Maino, V. C., Weinhold, K., Pantaleo, G., Gilmour, J., Horton, H., and Sekaly, R. P. (2005) Standardization of cytokine flow cytometry assays, *BMC Immunol* 6, 13.

9. Suni, M. A., Dunn, H. S., Orr, P. L., deLaat, R., Sinclair, E., Ghanekar, S. A., Bredt, B. M., Dunne, J. F., Maino, V. C., and Maecker, H. T. (2003) Performance of plate-based cytokine flow cytometry with automated data analysis, *BMC Immunol* 4, 9.

10. Disis, M. L., Dela Rosa, C., Goodell, V., Kuan, L. Y., Chang, J. C., Kuus-Reichel, K., Clay, T. M., Kim Lyerly, H., Bhatia, S., Ghanekar, S. A., Maino, V. C., and Maecker, H. T. (2006) Maximizing the retention of antigen specific lymphocyte function after cryopreservation, *J Immunol Methods* 308, 13–18.

11. Waldrop, S. L., Davis, K. A., Maino, V. C., and Picker, L. J. (1998) Normal human CD4+ memory T cells display broad heterogeneity in their activation threshold for cytokine synthesis, *J Immunol* 161, 5284–5295.

12. Nomura, L. E., Walker, J. M., and Maecker, H. T. (2000) Optimization of whole blood antigen-specific cytokine assays for CD4(+) T cells, *Cytometry* 40, 60–68.

13. Maecker, H. T., Ghanekar, S. A., Suni, M. A., He, X. S., Picker, L. J., and Maino, V. C. (2001) Factors affecting the efficiency of CD8+ T cell cross-priming with exogenous antigens, *J Immunol* 166, 7268–7275.

14. Nomura, L. E., DeHaro, E. D., Martin, L. N., and Maecker, H. T. (2003) Optimal preparation of rhesus macaque blood for cytokine flow cytometric analysis, *Cytometry* 53A, 28–38.

15. Nomura, L. E., Emu, B., Hoh, R., Haaland, P., Deeks, S. G., Martin, J. N., McCune, J. M., Nixon, D. F., and Maecker, H. T. (2006) IL-2 production correlates with effector cell differentiation in HIV-specific CD8+ T cells, *AIDS Res Ther* 3, 18.

Chapter 9

Phospho Flow Cytometry Methods for the Analysis of Kinase Signaling in Cell Lines and Primary Human Blood Samples

Peter O. Krutzik, Angelica Trejo, Kenneth R. Schulz, and Garry P. Nolan

Abstract

Phospho-specific flow cytometry, or phospho flow, measures the phosphorylation state of intracellular proteins at the single cell level. Many phosphorylation events can be analyzed simultaneously in each cell, along with cell surface markers, enabling complex biochemical signaling networks to be resolved in heterogeneous cell populations. The method has been applied to many diverse areas of biology, including the characterization of signaling pathways in normal immune responses to antigenic stimulation and microbial challenge, alteration of signaling networks that occur in cancer and autoimmune diseases, and high-throughput, high-content drug discovery. In this chapter, we provide detailed experimental protocols for performing phospho flow in cell lines, Ficoll-purified peripheral blood mononuclear cells, and whole blood. These protocols are applicable to both human and murine samples. We also provide methods for the validation of surface marker antibodies for use in phospho flow. Finally, we discuss data analysis methods, in particular, how to quantify changes in phosphorylation and how to visualize the large data sets that can result from experiments in primary cells.

Key words: Phospho flow cytometry, Signaling, Cytokine, Cell-based, Primary cells, Drug discovery

1. Introduction

Phospho-specific flow cytometry, or phospho flow, takes advantage of the two key traits of flow cytometry: multiparameter measurements and single-cell resolution (1, 2). By measuring ten or more fluorescent parameters for each individual cell that runs through the cytometer, phospho flow enables researchers to measure multiple kinase signaling pathways in heterogeneous cell populations such as peripheral blood. This has allowed our laboratory and others to explore the normal signaling responses of the immune

Teresa S. Hawley and Robert G. Hawley (eds.), *Flow Cytometry Protocols*, Methods in Molecular Biology, vol. 699,
DOI 10.1007/978-1-61737-950-5_9, © Springer Science+Business Media, LLC 2011

system in response to foreign challenges (3–5), the signaling changes that occur in disease states such as autoimmunity and cancer (6–9), and to perform high-content drug screening in both cell lines and primary cells (10, 11).

Like any other method, phospho flow cytometry requires practice to perform it accurately and reproducibly. In this chapter, we progress from a simple cell line experiment to more complex experiments in blood samples. After having taught hundreds of researchers how to perform phospho flow, we have learned that the key to success is in following this progression from simple to complex, without skipping steps.

The largest difficulty faced by novice users is attempting to do too much in their first experiments (for instance, trying to analyze phospho-protein levels in a cell population that requires five surface markers to define and only comprises 0.3% of the total population of cells). Therefore, in our laboratory, all new researchers are taught to perform a simple phospho flow experiment using the U937 cell line (see Subheading 3.1). The basic steps include cell stimulation, fixation, permeabilization, and staining (12). The U937 cell line is easy to grow in suspension, responds robustly to stimulation, and provides consistent results. In this experiment, only phospho-proteins are analyzed, and compensation is not required on the flow cytometer. This simplifies the experiment and allows the researcher to focus on performing the phospho flow method rather than setting up the cytometer. Once mastered, the U937 cell line experiment gives the researcher confidence that they can measure intracellular signaling events by flow cytometry, and provides the basis for performing more advanced experiments in primary samples.

In the subsequent sections, we add levels of complexity to the simple cell line experiment. In the PBMC experiment (see Subheading 3.2), surface markers are analyzed in addition to the intracellular proteins. In this way, cell types of interest (such as T cells, B cells, and monocytes) can be identified and analyzed biochemically. It is important to note that the choice of surface marker antibodies, as well as the titration and validation for use in this platform, must be considered carefully. In the whole blood experiment (see Subheading 3.3), signaling is measured directly in whole blood without first purifying the mononuclear cells by Ficoll separation. This enables measurements in the most physiologically relevant context, and is particularly appealing for pharmacodynamic monitoring of drugs directly in patient samples, or for diagnostic stratification of disease states.

We then present a slightly modified method, which we term "sequential staining," whereby surface markers that are difficult to analyze after permeabilization are stained prior to cell permeabilization (see Subheading 3.4). Although useful for staining surface antigens, this method has limitations as to the fluorophores that can be used.

Finally, we present a basic outline of how to analyze the data obtained with phospho flow (see Subheading 3.5). Unlike traditional flow cytometry, which typically compares percentages of cells in a particular gate, phospho flow is a quantitative method that compares the fluorescence intensity of a population before and after stimulation with a cytokine or other molecule. Here, we provide simple equations and ways to visualize the data, which can quickly become overwhelming in the light of the number of cell types and phospho-proteins that can be analyzed simultaneously in one sample.

2. Materials

1. Cells of interest: cell line (e.g. U937, Jurkat, THP-1, and Ramos), Ficoll-purified human peripheral blood mononuclear cells (PBMCs), and human whole blood.

2. Culture media for cells of interest: typically RPMI containing 10% FBS, 100 U/mL penicillin, 100 µg/mL streptomycin, and 2 mM L-glutamine (RPMI-10).

3. Stimuli: recombinant human IFN-γ, IL-4, IL-6, IL-10 (BD Biosciences, San Jose, CA); lipopolysaccharide (LPS) (Sigma–Aldrich, St. Louis, MO).

4. Fixative:

    (a) For experiments using cell lines or purified PBMCs: 16% formaldehyde in water in sealed ampules (EM grade from Electron Microscopy Sciences, Hatfield, PA).

    (b) For experiments in whole blood: Lyse/Fix buffer (#558049; BD Biosciences).

5. Permeabilization reagent: 100% methanol.

6. Staining medium: PBS with 0.5% BSA and 0.02% sodium azide.

7. Antibodies:

    (a) Phospho-specific antibodies: pStat1 Alexa Flour 647 (Ax647) (clone 4a), pStat6 Ax488 (clone J71-773.58.11), pStat3 Ax488 (clone 49), and p-p38 Ax647 (clone 36) (BD Biosciences).

    (b) Surface marker antibodies: CD3 PE-Cy7 (UCHT-1), CD4 Pacific Blue (RPA-T4), CD20 PerCP-Cy5.5 (H1), and CD33 PE (P67.6) (BD Biosciences).

8. Tubes or plates:

    (a) When processing smaller numbers of samples, 12 × 75 mm polystyrene (5 mL capacity) BD Falcon tubes (FACS tubes) are suggested.

    (b) When processing large numbers of samples or performing experiments in volumes less than 200 µL, 96-well V-bottom deep block plates (2 mL capacity, polypropylene) (#40002-012; VWR Scientific, West Chester, PA) are more convenient.

9. CO_2 incubator, 37°C.

10. Swinging bucket centrifuge with tube and plate carriers (capable of ~500× g force).

11. Flow cytometer equipped with 488- and 633-nm lasers (e.g. BD FACSCalibur; BD Biosciences) for U937 experiment; and 405-, 488-, and 633-nm lasers for PBMC and whole blood experiments (e.g. BD LSRII; BD Biosciences).

12. Analysis software: we suggest using Cytobank, web-based software for storing, sharing, analyzing, and visualizing flow cytometry data sets (accessible at http://www.cytobank.org).

3. Methods

The phospho flow method can be broken down into the following six steps:

1. Stimulation
   Cells of interest, which can be a cell line, or primary cells from a mouse or human, are treated with various molecules that might affect cell signaling such as cytokines, small molecule drugs, or growth factors. Cells can also be obtained directly from patients and tested without further treatment to examine "basal" signaling, but we have found that probing the cells with stimuli better reveals signaling changes in disease states.

2. Fixation
   Cells are fixed with formaldehyde in order to stop signaling and phosphorylation events as rapidly as possible. Formaldehyde is a cross-linking agent that enters the cell and reacts with amine groups on proteins. This binding can lead to protein–protein cross-linking but more importantly arrests enzymatic activity and halts cellular metabolism. Some signaling events, particularly on proteins that are proximal to the cell membrane such as Syk, ZAP70, and PLCγ, can decay within 15 min of induction. Therefore, cells must be fixed without washing, in order to preserve the most accurate representation of the phosphorylation state.

3. Permeabilization
   Cells are permeabilized with methanol so that antibodies against intracellular proteins can enter the cells and stain their

target antigens. Although different permeabilization reagents are available, methanol, or other similar denaturing agents, is required for staining the Signal Transducer and Activator of Transcription (Stat) proteins which are critical to nearly all cytokine-mediated signaling pathways. In addition, we have found methanol to be the most universal reagent, working for the largest number of phospho-specific antibodies.

4. Staining

Cells are stained with antibodies specific to the phosphorylated form of intracellular signaling proteins. These antibodies do not bind to the non-phosphorylated form, and therefore, the amount of antibody binding to a cell directly correlates to the amount of phosphorylation on the target protein. Primary blood samples must also be stained with antibodies against surface antigens to identify cell types of interest. These surface antibodies must be carefully validated for use in permeabilized cells, as many show greatly increased background staining when exposed to the massive number of protein and nucleic acid epitopes that are present inside the cell, but not on the cell surface. If necessary, cells can be surface stained after fixation, but prior to permeabilization, in a method called sequential staining, outlined below.

5. Acquisition

Cells are acquired on a flow cytometer, and the fluorescence intensity of each antibody binding to each cell is measured.

6. Analysis

Phospho flow data analysis requires quantitative comparisons between samples to determine the amount of protein phosphorylation induced after stimulation, or in a particular disease state. This differs from typical surface phenotyping experiments done with flow cytometry, where the percentage of positive cells is the main metric. Therefore, in phospho flow analysis, the fluorescence intensity of the phospho-antibody staining is compared to a control, and a "fold change" value is calculated. Data sets produced with phospho flow can become extremely large due to the multiparametric nature of the experiments. For instance, a typical experiment might measure five stimulation conditions in five cell populations, across five different signaling proteins. This experiment, which might only require about ten samples to be run on the cytometer, would yield 125 data points. Therefore, we often display data in heatmap format, as a tool to summarize the data for overall visualization. It is important to remember that each point in a heatmap is representative of thousands of individual cells, requiring the researcher to examine samples with interesting signaling phenotypes more closely.

3.1. U937 Cell Line

The U937 cell line experiment is the simplest phospho flow experiment that all new users should perform (12). In this experiment, cells are stimulated with two cytokines, IFN-γ and IL-4. The cells are then fixed with formaldehyde, permeabilized with methanol, washed, and stained with phospho-specific antibodies. Because the cell line is extremely consistent in its response profile, users can optimize their technique by repeating the experiment until the expected pattern of phosphorylation is observed, and the shifts in phospho-protein levels are adequate.

1. Grow U937 cell line to ~0.5 to 1×10^6 cells/mL in RPMI-10 (5 mL is required for this experiment).

2. Quickly place 1 mL of cells in five FACS tubes, numbered 1–5 (see Note 1).

3. Quickly add the cytokines IFN γ and IL 4 at 10 ng/mL final concentration, in the order listed below (see Notes 2 and 3).

Tube 1	2	3	4	5
None	None	IFN-γ	IL-4	IFN-γ+IL-4

4. Incubate the cells for 15 min at 37°C in a 5% CO_2 incubator.

5. Quickly add 100 µL of 16% formaldehyde (final concentration of ~1.5%) (see Note 4).

6. Keep the tubes at room temperature for 10 min.

7. Centrifuge the samples ($500 \times g$, 4°C, 5 min).

8. Decant the supernatant by inverting and flicking the tubes, leaving the pellet of cells at the bottom.

9. Resuspend the cell pellet in the residual medium (normally 30–70 µL) by vortexing briefly for 3–5 s or shaking the tubes vigorously by hand.

10. Add 1 mL of ice cold methanol to the resuspended cells.

11. Vortex the cells for 3–5 s.

12. Place the tubes on ice for 15–30 min (see Note 5).

13. Add 3 mL of staining medium on top of the methanol.

14. Repeat steps 7–9.

15. Add 4 mL of staining medium to the cell pellet.

16. Repeat steps 7–9.

17. Add 80 µL of staining medium to the pellet. With the residual volume typically left after decanting the supernatant, each tube should have 110–150 µL of suspension at this point.

18. Transfer 80 µL of the cell suspension into a fresh set of five FACS tubes, numbered A1–5 (see Note 6).

19. Stain the cells by adding 20 µL each of pStat1 Ax647 and pStat6 Ax488 phospho-specific antibodies to tubes A2–5, and add 40 µL of staining medium to tube A1 as a control. The total volume will be 120 µL. Tube A1 serves as an unstained control useful for instrument setup and comparison of phospho-protein staining intensities.

20. Vortex to mix.

21. Stain for 30 min to 1 h at room temperature (see Note 7).

22. Add 3 mL of staining medium.

23. Repeat steps 7–9.

24. Add 100 µL of staining medium. Keep the samples cold until acquisition (see Note 8).

25. Run the samples on a flow cytometer. Use the unstained control to set the instrument PMTs for Ax488 (FL1 with 530/30 bandpass filter on FACSCalibur) and Ax647 (FL4 with 661/16 bandpass filter) so that the cells appear in the lower quadrant of each parameter. Because Ax488 and Ax647 are spectrally distinct and are excited by the 488- and 633-nm lasers, respectively, compensation is not required.

26. Acquire 10,000 events for each sample. Expected results are shown in Fig. 1. IFN-γ induces pStat1, while IL-4 induces pStat6. The combination stimulation should induce both phospho-proteins.

3.2. Primary Human PBMC

Experiments in primary cells are more complex than the U937 cell line experiment in that surface markers must also be used to identify cell types within the heterogeneous PBMC sample. It is absolutely critical to validate the surface marker antibodies for use in permeabilized cells. Many antibodies that work well in live cell staining do not separate the appropriate cell populations once the cells have been fixed and permeabilized. See Subheading 3.4 for more information regarding surface marker antibodies used in phospho flow.

In addition, primary samples first require isolation, freezing, and thawing of the PBMCs, all of which can affect signaling responses. The method outlined below is used routinely in our laboratory for clinical samples.

3.2.1. Preparation and Freezing of PBMCs

Note: If PBMCs are to be prepared immediately before the experiment, skip to Subheading 3.2.2, step 9.

1. Prepare PBMCs from whole blood by Ficoll purification.

2. Resuspend the cells at approximately ten million cells per mL in ice-cold FBS containing 10% DMSO (the freezing medium).

3. Aliquot 1 mL of cells per cryovial.

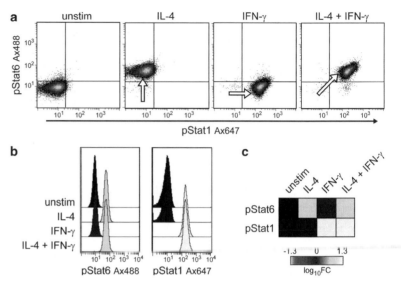

Fig. 1. Sample data from U937 cell line experiment. U937 cells were left unstimulated, or were treated with 10 ng/mL of IFN-γ, IL-4, or the combination of both IFN-γ and IL-4. Cells were fixed, permeabilized, and stained with phospho-specific antibodies against Stat1 (pY701) and Stat6 (pY641). The pStat6 antibody was conjugated to Ax488, while the pStat1 antibody was conjugated to Ax647, enabling simultaneous detection by flow cytometry. Cells were first gated based on forward and side scatter characteristics to eliminate cellular debris from the analysis. (a) Two-dimensional density plot showing stimulation of Stat6 phosphorylation with IL-4, and Stat1 phosphorylation with IFN-γ. Adding both cytokines induced both phosphorylation events. (b) Histogram overlay representation of the data. (c) Heatmap representation of the data. Histograms and heatmaps are colored (in the electronic version) according to the fold change (FC) of phosphorylation relative to the unstimulated sample, with lighter shades indicating a positive fold induction. Data analysis (including heatmap generation) was performed with Cytobank software.

4. Freeze slowly (1°C/min) in –80°C freezer (see Note 9).

5. Transfer to liquid nitrogen the next day for long-term storage.

3.2.2. Phospho Flow in PBMCs

1. Remove a vial of frozen PBMCs from liquid nitrogen.

2. Thaw quickly (1 min) in 37°C water bath.

3. Pipette the cells into 25 mL of RPMI-10 at room temperature in a 50-mL conical tube.

4. Centrifuge the samples (500×*g*, room temperature, 5 min).

5. Aspirate or decant the supernatant into 10% bleach solution (see Note 10).

6. Tap the tube onto the benchtop to dislodge the cell pellet.

7. Add 25 mL of RPMI-10 and pipette up and down to break any cell clumps (see Note 11).

8. Repeat steps 4–6.

9. Add 5 mL of RPMI-10 and pipette up and down to break any cell clumps. The cell concentration should be at ~2×10^6/mL at this point.

10. Place the cells in 37°C 5% CO_2 incubator for 1 h (see Note 12).

11. (Perform steps 11 and 12 as quickly as possible to avoid cooling the cells.) Place 1 mL of cells into five FACS tubes labeled 1–5.

12. Add the following stimuli.

Tube 1	2	3	4	5
None	None	IL-6 (50 ng/mL)	IL-10 (50 ng/mL)	LPS (1 μg/mL)

Note: At this point in the experiment, the protocol of the U937 experiment in Subheading 3.1, from step 4 onward, is followed. The only change in the PBMC experiment is that regarding the choice of antibodies added during the staining step.

13. Incubate the cells for 15 min at 37°C in a 5% CO_2 incubator.

14. Quickly add 100 μL of 16% formaldehyde (final concentration of ~1.5%).

15. Keep the tubes at room temperature for 10 min.

16. Centrifuge the samples ($500 \times g$, 4°C, 5 min).

17. Decant the supernatant by inverting and flicking the tubes, leaving the pellet of cells at the bottom.

18. Resuspend the cell pellet in the residual medium (normally 30–70 μL) by vortexing for a few seconds or shaking the tubes vigorously by hand.

19. Add 1 mL of ice-cold methanol to the pellet.

20. Vortex the cells for 3–5 s to suspend them in the methanol.

21. Place the tubes on ice for 15–30 min (see Note 5).

22. Add 3 mL of staining medium on top of the methanol.

23. Repeat steps 16–18.

24. Add 4 mL of staining medium to the cell pellet.

25. Repeat steps 16–18.

26. Add 50 μL of staining medium to the pellet. With the residual volume typically left after decanting the supernatant, each tube should have 80–120 μL of the suspension at this point.

27. Transfer 50 μL of the cell suspension into a fresh set of five FACS tubes, numbered A1–5 (see Note 6).

28. Create the antibody cocktail below and add 80 μL to tubes A2–5 (total volume will be 130 μL). Add 80 μL of staining medium to tube A1. Antibody cocktail should be made such that there is enough for at least one extra sample, to ensure having enough for all samples and accommodating small pipetting errors. Here, antibody cocktail is prepared for five samples, although only four are stained.

Antibody	Volume per sample (μL)	Total volume (μL)
CD3 PE-Cy7	10	50
CD4 PacBlu	5	25
CD20 PerCP-Cy5.5	20	100
CD33 PE	5	25
pStat3 Ax488	20	100
p-p38 Ax647	20	100
Total	80	400

29. Vortex to mix.

30. Stain for 1 h at room temperature (see Note 7).

31. Add 3 mL of staining medium.

32. Repeat steps 16–18.

33. Add 100 μL of staining medium. Keep the samples cold until acquisition (see Note 8).

34. Prepare proper compensation controls for each fluorophore being used in the experiment.

35. Acquire compensation controls and samples on a flow cytometer. Acquire at least 50,000 events per sample. Expected results are shown in Fig. 2. IL-6 and IL-10 induce Stat3 phosphorylation in many cell types, while LPS induces p38 phosphorylation in monocytes only.

3.3. Primary Human Whole Blood

Experiments in whole blood enable measurements to be made in the most physiologically relevant environment for human samples. Unlike Ficoll-purified PBMC samples, whole blood contains neutrophils and granulocytes (creating a much different forward vs. side scatter plot), red blood cells, and all of the protein factors present in the serum. This provides a more "normal" context for cell signaling to occur. However, the presence of the massive number of red blood cells complicates flow cytometry and requires that the cells be lysed prior to analysis.

In this method, the red blood cells are lysed simultaneously as the white blood cells are fixed, using a Lyse/Fix buffer. This extra step allows stimulation to occur in the whole blood and enables

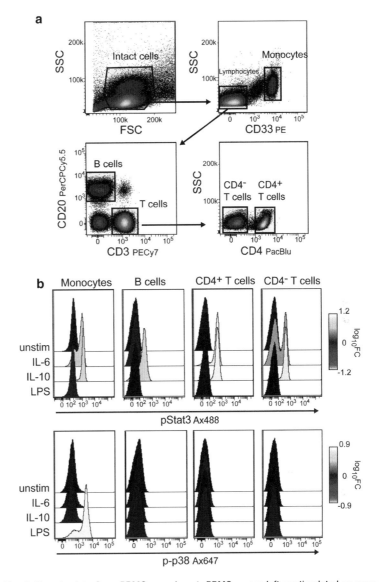

Fig. 2. Sample data from PBMC experiment. PBMCs were left unstimulated or were treated with IL-6, IL-10 (both 50 ng/mL), or LPS (1 μg/mL). Cells were fixed, permeabilized, and stained with phospho-specific antibodies against Stat3 (pY705) and p38 (pT180/Y182) conjugated to Ax488 and Ax647, respectively. (**a**) Identification of cellular subsets by surface gating. Intact cells were identified based on forward and side scatter characteristics. Intact cells were then gated based on CD33 expression, with CD33^{hi} cells representing monocytes, and CD33$^-$ cells representing lymphocytes. Lymphocytes were separated into B- and T-cell populations based on CD20 and CD3 expression, respectively. Finally, the T-cell population was divided into CD4$^+$ and CD4$^-$ (or CD8$^+$) subsets. (**b**) Histogram analysis of phosphorylation levels in gated cellular subsets. For each of the populations identified in (**a**), the levels of Stat3 and p-p38 were analyzed for each of the four treatment conditions. IL-6 induced Stat3 phosphorylation in monocytes, CD4$^+$ and CD4$^-$ T cells, but not B cells. IL-10 induced Stat3 phosphorylation in all the cell types analyzed. LPS-induced p-p38 levels in monocytes only. Histograms are colored (in the electronic version) according to fold change in phosphorylation relative to the unstimulated control (using Cytobank software).

rapid termination of the stimulation reaction at the desired time point. Once the red blood cells are lysed, the samples are treated just as PBMC samples or the U937 cell line.

Note: Prior to starting the experiment, prepare 1× Lyse/Fix buffer by diluting 5× buffer with purified water and warming to 37°C in a water bath.

1. Obtain human whole blood drawn into heparin tube (typically green top) (see Note 13).

2. Place in 37°C water bath for 30 min to ensure blood is warm for stimulation (see Note 14).

3. Aliquot 200 µL of blood into five FACS tubes, labeled 1–5 (see Note 1).

4. Add stimuli as follows:

Tube 1	2	3	4	5
None	None	IL-6 (50 ng/mL)	IL-10 (50 ng/mL)	LPS (1 µg/mL)

5. Incubate for 15 min at 37°C.

6. Add 4 mL of 1× Lyse/Fix buffer (prewarmed to 37°C).

7. Mix thoroughly by inverting the tube with cap ten times or by pipetting up and down ten times (see Note 15).

8. Incubate for 15 min at 37°C in a waterbath or incubator.

9. Centrifuge the samples ($500 \times g$, 4°C, 5 min).

10. Aspirate or decant the supernatant.

11. Vortex the cells for 3–5 s to dislodge the cell pellet in the residual volume.

12. Add 4 mL of ice-cold PBS (see Note 16).

13. Repeat steps 9–11.

14. Add 1 mL of ice-cold methanol.

15. Place the samples on ice for 15–30 min (see Note 5).

16. Add 3 mL of staining medium on top of the methanol.

17. Repeat steps 9–11.

18. Add 4 mL of staining medium to the cell pellet.

19. Repeat steps 9–11.

20. Add 50 µL of staining medium to the pellet. With the residual volume typically left after decanting the supernatant, each tube should have 80–120 µL of cell suspension at this point.

21. Transfer 50 µL of the cell suspension into a fresh set of five FACS tubes, numbered A1–5 (see Note 6).

22. Create the antibody cocktail below and add 80 µL to tubes A2–5 (total volume will be 130 µL). Add 80 µL of staining

medium to tube A1. Sample A1 serves as an unstained control for cytometer setup. Antibody cocktail should be made so as to have enough for at least one extra sample, to ensure having enough for all samples and accommodating small pipetting errors (in this case, enough cocktail should be prepared for five samples).

Antibody	Volume per sample (μL)	Total volume(μL)
CD3 PE-Cy7	10	50
CD4 PacBlu	5	25
CD20 PerCP-Cy5.5	20	100
CD33 PE	5	25
pStat3 Ax488	20	100
p-p38 Ax647	20	100
Total	80	400

23. Incubate for 1 h at room temperature.

24. Add 4 mL of staining medium to the cells.

25. Repeat steps 9–11.

26. Add 100 μL of staining medium to the cells.

27. Acquire 100–200,000 events per sample on the flow cytometer. Expected results are shown in Fig. 3. The results are nearly identical to the PBMC experiment. However, the presence of neutrophils and granulocytes adds many large cells to the forward versus side scatter plots, and makes gating of the monocytes slightly more difficult. Monocytes are high for CD33 expression, while neutrophils and granulocytes are intermediate (with the suggested clone of CD33 antibody). Monocytes have a much more robust response to LPS, so accurate gating is important for observing maximal induction.

3.4. Surface Marker Antibody Validation and Sequential Staining

Perhaps the most important, and most difficult, part of phospho flow is the proper identification of cellular subsets (e.g. T cells, B cells, and monocytes), within heterogeneous samples such as peripheral blood. This is accomplished by staining antigens (e.g. CD3, CD20, and CD33) present on the cell surface of each cell subset. Surface staining can be performed at several points in the phospho flow protocol, each with its advantages and disadvantages, summarized in Table 1.

In our laboratory, most "surface" staining is performed on fixed/permeabilized cells with antibodies that have been carefully validated for use in this protocol. This enables us to stain all markers, both intracellular and surface, simultaneously,

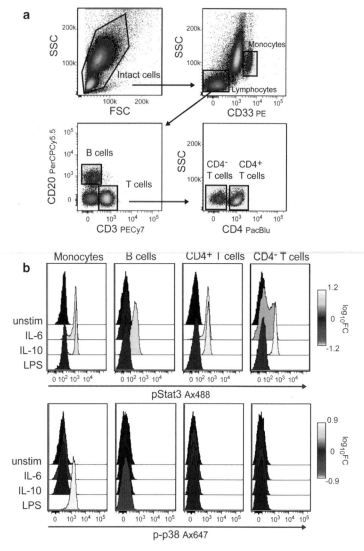

Fig. 3. Sample data from experiment in whole blood. Whole blood was obtained and treated as in the PBMC experiment. (**a**) The gating strategy in whole blood is the same as that for the PBMC experiment shown in Fig. 2. Note, however, that whole blood has a large population of neutrophils, which are high in the side scatter dimension and have intermediate expression of CD33 (with this antibody clone). Typical PBMC isolation procedures remove neutrophils; therefore, this population is unique to whole blood experiments. (**b**) Histogram overlay analysis of signaling in whole blood. The results are very similar to stimulation in PBMC, shown in Fig. 2.

greatly simplifying the protocol. This is also the most flexible way to stain samples, because they can be stored in methanol for months at a time, and then stained with different antibody cocktails as the research project develops and new hypotheses need to be tested.

Table 1
Comparison of different surface staining strategies for phospho flow

Surface stain timing	Advantages	Disadvantages
On live cells, before or during stimulation of cells	Use standard Ab titrations	Limited in fluorophore choices May affect signaling by activating/blocking surface receptors
On fixed cells, before permeabilization	Use standard Ab titrations Most Abs stain similar to live cells	Limited in fluorophore choices Adds extra steps in protocol Some antigens blocked by fixative
On fixed/permeabilized cells	Simplest protocol, all Abs added at same time Works for any fluorophore Quickest protocol Flexibility to stain any surface markers at later time	Some Abs have higher background in permeabilized cells Some Abs do not bind to their antigen once denatured by methanol Requires re-titration of Abs to get optimal titer

Table 2
Fluorophore selection for sequential staining technique

Can be used	Cannot be used
Small molecule dyes (e.g. Pacific Blue, FITC, Alexa dyes, and DyLight dyes)	Protein fluorophores (e.g. PE, PerCP, and APC[a])
Quantum dots	Protein tandem dyes (e.g. PE-TR, PE-Cy5, PerCP-Cy5.5, and APC-Cy7)

[a]APC conjugates are not completely quenched by methanol treatment, but lose over 90% of their signal. Therefore, Ax647 conjugates are recommended for this detection channel for sequential staining

However, some antibodies do not work when staining fixed/permeabilized cells. In these cases, the background staining may increase dramatically, or positive staining may decrease. In either case, one can no longer resolve the positive population. To solve this problem, we adopt a "sequential staining" approach, where stimulated cells are fixed, stained for these difficult surface antigens, then permeabilized, and stained for intracellular epitopes. Note, however, that fluorophore choices are limited in the sequential staining protocol (see Table 2). Protein fluorophores such as PE, PerCP, and APC are denatured by methanol and lose their fluorescence if used in sequential methods. Therefore, antibodies must be conjugated to small molecule fluorophores such as FITC, the Alexa dyes, and DyLight dyes, or to Quantum dots, for use in the sequential methods.

With proper testing and validation, nearly all surface antigens can be stained effectively in the phospho flow protocol.

1. Prepare or thaw 10×10^6 PBMCs.

2. Resuspend at 2×10^6 cells/mL in prewarmed, 37°C RPMI-10.

3. Place 2.5 mL (5×10^6) of cells into two FACS tubes.

4. Tube 1 (live cells): Place on ice. Proceed to step 6.

5. Tube 2 (fixed/permeabilized cells): these steps are the same as Subheading 3.1, steps 5–12.

   (a) Add 250 μL of 16% formaldehyde.

   (b) Incubate for 10 min at room temperature.

   (c) Pellet the cells by centrifugation ($500 \times g$, 4°C, 5 min).

   (d) Decant or aspirate the supernatant.

   (e) Vortex 3–5 s to resuspend the cells in residual volume.

   (f) Add 1 mL of ice-cold methanol.

   (g) Incubate on ice for 15–30 min.

6. Add 2–3 mL of staining medium.

7. Centrifuge the samples ($500 \times g$, 4°C, 5 min).

8. Decant the supernatant by inverting and flicking the tubes, leaving the pellet of cells at the bottom.

9. Resuspend the cell pellet in the residual medium (normally 30–70 μL) by vortexing for 3–5 s or shaking the tubes vigorously by hand.

10. Add 4 mL of staining medium.

11. Repeat steps 7–9.

12. Resuspend the cells by adding 500 μL of staining medium to each tube.

13. Aliquot 100 μL of cells into five tubes each for live and fixed/permeabilized cells.

14. Add varying amounts of the surface marker antibody being tested. Titrate from high to low concentration. Begin with ~1 μg/mL of antibody, or at the manufacturer's recommended dilution. Threefold dilutions work well, and cover a large enough range to obtain a good titer for most antibodies (most monoclonal antibodies have optimal titers between 10 ng/mL and 1 μg/mL).

15. Analyze on the flow cytometer.

16. Compare staining in live cells versus fixed/permeabilized cells. For proper validation, antibody must show (see Note 17):

    (a) Same percentage of cells in both cases.

    (b) Adequate resolution, or separation, of positive cells from negative populations.

17. If antibody does not stain proper percentage of cells, or lacks resolution/separation, sequential staining may be preferred.

3.4.2. Sequential Staining

1. Prepare and stimulate cells as in Subheadings 3.2 and 3.3.

2. Fix cells with 100 μL of 16% formaldehyde per 1 mL of cells (1.5% final concentration).

3. Centrifuge the samples ($500 \times g$, 4°C, 5 min).

4. Decant the supernatant by inverting and flicking the tubes.

5. Resuspend the cell pellet in the residual medium.

6. Wash the cells by adding 4 mL of staining medium.

7. Centrifuge the samples ($500 \times g$, 4°C, 5 min).

8. Resuspend the cells at $\sim 1 \times 10^6$ cells/100 μL in staining medium.

9. Transfer 100 μL of cells into fresh tubes.

10. Stain with the appropriate dilution of surface marker antibodies (typically the same amount as that used for live cell staining). Important note: antibodies must be conjugated to small molecule fluorescent dyes or quantum dots for sequential staining. Antibodies conjugated to protein fluorophores such as PE, PerCP, and APC will lose fluorescence when treated with methanol (see Note 18 and Table 2).

11. Incubate for 30 min on ice.

12. Wash the cells by adding 3 mL of staining medium.

13. Centrifuge the samples ($500 \times g$, 4°C, 5 min).

14. Decant the supernatant by inverting and flicking the tubes.

15. Resuspend the cell pellet in the residual medium.

16. Add 1 mL of ice-cold methanol per mL of starting cell volume.

17. Incubate for 15–30 min on ice.

18. Repeat steps 12–15 twice to wash methanol from cells.

19. Resuspend the cells at $\sim 1 \times 10^6$ cells/100 μL in staining medium.

20. Transfer to fresh FACS tube.

21. Stain the samples with phospho-specific antibodies and any surface marker antibodies conjugated to PE, PerCP, or APC (and their tandems), which have been validated for use in fixed/permeabilized cells.

22. Incubate for 1 h at room temperature.

23. Wash the cells and analyze on flow cytometer as above.

3.5. Data Analysis in Phospho Flow

Phospho flow data analysis is somewhat different than typical flow cytometry experiments, because different samples must be compared quantitatively to a control sample. In phospho flow, one wants to measure the amount of phosphorylation induced by treating a sample with a particular cytokine, or the change in phosphorylation associated with a diseased sample versus that in a normal sample.

Different cell populations within the sample are first gated based on their surface marker staining. The median fluorescence intensity is then calculated for the phospho-specific antibody channel for each population. Medians, as opposed to means, are utilized to avoid the effects of outliers. However, it is important to be cognizant of your data. Medians may not be appropriate for a bimodal distribution, for instance. Comparisons are made within cell types under different conditions, e.g. stimulated versus unstimulated B cells. However, it is difficult to compare between different cell types, e.g. B cells versus monocytes, due to differences in background binding of phospho-specific antibodies, as well as autofluorescence differences due to cell size/shape.

Our laboratory has developed web-based software, called Cytobank, for storing, sharing, analyzing, and visualizing flow cytometry data sets. In particular, Cytobank is well-suited to analyze phospho flow data with its ability to create heatmaps and histogram overlays without requiring third party software. In fact, the steps outlined below can all be performed automatically within Cytobank, eliminating the need for spreadsheet programs. All figures in this chapter were generated using Cytobank.

1. During or after acquiring data on the flow cytometer, compensate data with appropriate compensation controls.

2. Draw gates around cell populations of interest. A standard gating method is to first gate on intact cells based on their forward and side scatter characteristics.

3. Then, use the other fluorescent parameters to identify cell types of interest. For instance, CD3$^+$ cells are T cells, CD20$^+$ cells are B cells, and CD33$^+$ cells are monocytes.

4. For each population that has been gated, calculate the median fluorescence intensity (MFI) of the phospho-protein channel (typically Alexa 488 or Alexa 647).

5. Apply this gating and statistic to all the samples that were acquired.

6. Export or copy/paste the MFI values for each population into a spreadsheet program.

7. Calculate the fold change in phosphorylation induced by each particular stimulation or treatment (e.g. cytokine) with the following equation:

$$\text{Fold change} = \text{MFI}_{\text{stimulated}} / \text{MFI}_{\text{unstimulated}.}$$

(a) This equation simply compares the MFI value of the stimulated samples to the MFI value of the control/unstimulated sample.

(b) If there is no change in phosphorylation upon stimulation, then fold change = 1.

(c) If the phospho-specific antibody staining intensity doubles upon stimulation, then fold change = 2.

8. For visualizing the data (and to accommodate negative changes), it is useful to represent no change in phosphorylation as zero. Therefore, we often calculate the $\log_2$ or $\log_{10}$ fold change (Table 3):

$$\text{Log}_2 \text{fold change} = \log_2 \left(\text{MFI}_{\text{stimulated}} / \text{MFI}_{\text{unstimulated}} \right).$$

(a) Here, if no change in phosphorylation is observed upon treatment, $\log_2 \text{FC} = 0$.

(b) If the staining intensity doubles, $\log_2 \text{FC} = 1$.

(c) If the staining intensity is halved, $\log_2 \text{FC} = -1$.

9. The $\log_2 \text{FC}$ values can now be exported to heatmap analysis software, often used for DNA microarray analysis. A positive change is often represented as yellow, and a negative change as cyan. No change is represented as black (see Fig. 4). This allows rapid visual identification of stimulations or conditions that lead to a change in phosphorylation levels. However,

Table 3
Sample calculations from PBMC experiment (Subheading 3.2)

		pStat3 median	Fold change	Log$_2$FC	Log$_{10}$FC
Monocytes	Unstim	204.6	1.00	0.00	0.00
	IL-6	708.9	3.46	1.79	0.54
	IL-10	1172.2	5.73	2.52	0.76
	LPS	165.5	0.81	−0.31	−0.09
B cells	Unstim	29.2	1.00	0.00	0.00
	IL-6	32.9	1.13	0.17	0.05
	IL-10	190.5	6.53	2.71	0.81
	LPS	28.0	0.96	−0.06	−0.02
CD4+ T cells	Unstim	28.1	1.00	0.00	0.00
	IL-6	393.0	13.99	3.81	1.15
	IL-10	406.4	14.47	3.85	1.16
	LPS	27.1	0.96	−0.05	−0.02
CD4− T cells	Unstim	31.0	1.00	0.00	0.00
	IL-6	83.0	2.68	1.42	0.43
	IL-10	393.5	12.71	3.67	1.10
	LPS	30.3	0.98	−0.03	−0.01

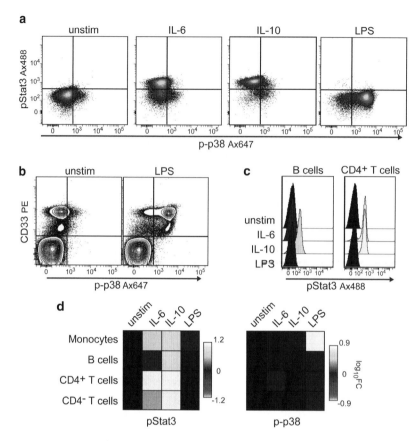

Fig. 4. Data visualization methods for phospho flow. Many methods are available for viewing changes in phosphoryla-tion, and the optimal method should be chosen based on the level of detail required. For instance, heatmaps provide a broad overview of phosphorylation trends when performing large signaling profiling experiments. Histogram overlays provide more detail and enable the identification of heterogeneous, or bimodal, signaling responses. Two-dimensional contour and density plots provide the most detailed analysis level; however, it is difficult to view/analyze many two-dimensional plots. Therefore, a common workflow is to proceed from heatmaps, through histogram overlays, and finally down to the two-dimensional plot level. (**a**) Two-dimensional density plot showing levels of two phospho-proteins against each other. Data are shown for the monocyte subset in PBMC. (**b**) Two-dimensional contour plot showing a surface marker, CD33 versus a phospho-protein, p-p38. This enables the identification of specific CD33-expressing subsets that respond to LPS treatment. (**c**) Histogram overlay analysis of B-cell and CD4[+] T-cell phosphorylation of Stat3 in response to treatments in PBMC. (**d**) Heatmap analysis of the PBMC experiment in Fig. 2, enabling rapid identification of both cell subsets and treatments that lead to phosphorylation of either Stat3 or p38. Data analysis performed with Cytobank software.

since only the median is utilized, it is critical to examine the responding samples to determine whether all of the cells in a population responded (unimodal), or if only a fraction responded (bimodal or multimodal peaks). Multimodal peaks typically indicate that a particular population contains sub-populations of cells that need to be identified with more surface markers.

10. As mentioned above, Cytobank, a software suite developed in our laboratory, is able to perform all of these calculations for the user. In addition, Cytobank enables visualization of the data in heatmap and histogram overlay format (Fig. 4).

4. Notes

1. Because biochemical events inside cells are highly temperature dependent, perform all operations outside the incubator as quickly as possible to avoid drops in temperature. Although cell lines such as U937 cells respond well even at lower temperatures, some primary cell types will be affected more dramatically by slight changes in temperature.

2. In these experiments, the cytokine stimulations are added at doses that are 10–50 times above the EC_{50} concentration, which for most cytokines is in the range of 0.1–1 ng/mL. This ensures complete receptor binding and stimulation of the cells.

3. Because cytokines are proteins, it is best to avoid multiple freeze/thaw cycles or storage of cytokine stocks at 4°C. Therefore, we typically prepare stock solutions at 10 µg/mL (1,000×), and freeze dozens of aliquots at −80°C. Each aliquot is then thawed before an experiment and discarded after that experiment (not re-frozen). This ensures consistent stimulation and dose–response experiments.

4. After adding formaldehyde, typical media such as RPMI will turn yellow temporarily. After about 1 min, the original color will return. This is normal and is an easy way to monitor if formaldehyde was added correctly to all the tubes/wells.

5. If samples need to be stored for longer periods of time, they must be stored at colder temperatures. For storage up to 6 h, placing in a −20°C freezer is acceptable. For overnight storage, or storage up to 6 months, place the samples at −80°C. We have analyzed samples stored at −80°C in methanol for over 6 months and found phospho-protein and surface antigen staining to be nearly identical to that of the same samples analyzed after only a few days of storage.

6. One of the most critical parameters to getting consistent staining between samples is to have the same sample volume. However, after decanting the supernatant, the residual volume left in FACS tubes or plates can vary from 30 to 70 µL depending on how skillfully the researcher decants. Therefore, we recommend transferring samples to a new set of tubes, so that all samples have identical volume. Antibodies should be

titrated at this sample volume to obtain the optimal concentration of antibody. Not having the same sample volume will lead to "wobble" in the median fluorescence staining intensity between samples, and poor statistics. The number of cells in the sample volume does not affect staining intensity significantly when the cell concentration varies less than threefold. If some samples contain ten times more or less cells, then variation in staining intensity may be observed, depending on the antibody being used.

7. Staining can be done in the dark if the researcher prefers, but we have not found any significant difference between staining on the benchtop in typical laboratory lighting versus staining in the dark or by covering the tube rack.

8. Samples can be stored on ice for several hours before acquisition on the flow cytometer. In our laboratory, we have stored stained samples overnight at 4°C and acquired the next day, with results nearly identical to those of samples run immediately. However, this should be tested empirically with different antibodies, since they may have different off-rate kinetics.

9. Freezing the cells slowly is important for viability upon thawing. Several manufacturers produce slow freezing devices for −80°C freezers, including Nalgene-NUNC. These devices are typically plastic containers that insulate the cryovials with isopropanol to produce slower temperature changes.

10. Human primary blood cells should always be treated as potentially biohazardous until the cells have been fixed with formaldehyde. Therefore, the supernatant from live cells should be treated as though it contains infectious agents, and should be poured into bleach before proper disposal.

11. Cell clumping is usually caused by lysis of cells and the release of their DNA into the medium, and this DNA aggregates with cell surface proteins to cause larger cellular clumps. Addition of DNAse to the RPMI-10 used for washing the cells can decrease cell clumping.

12. Cells must be warmed to 37°C to obtain optimal stimulation. We have found that a 1 h rest time allows ample time for warming, but avoids artifacts of increased basal phosphorylation that are often observed with longer resting periods, such as 2–6 h.

13. Phospho flow experiments have been performed successfully on blood drawn into heparin and EDTA as anticoagulants. Heparin tubes have performed most consistently and, therefore, have become our standard tubes. Citrate tubes have shown slightly higher variability in cellular responses, although this has not been tested exhaustively. The blood draw tube that is

chosen should be used throughout the entire study, to avoid any differences the anticoagulant may produce.

14. Human blood should be treated as potentially biohazardous at all times until adequately fixed and/or bleached.

15. After adding the Lyse/Fix buffer to the blood, the tubes must be mixed thoroughly to ensure complete red blood cell lysis. The Lyse/Fix buffer simultaneously lyses red blood cells and fixes the white blood cells. Inadequate mixing may lead to large red blood cell pellets and heterogeneity between samples. The samples should be mixed until the solution is bright red and mostly clear. Cloudiness represents incomplete lysis. Red blood cells appear on the flow cytometer as events that are slightly smaller than the lymphocytes, and can often make lymphocyte identification difficult. Therefore, this step is critical to the whole blood phospho flow method, and should be focused on.

16. The PBS wash is performed to remove as much of the red blood cell debris as possible after the lysis step.

17. In most cases, staining of surface antigens after fixation and permeabilization results in increased backgrounds and/or decreases in positive signals. This leads to an overall decrease in separation between positive cell events and negative cell events. Often, using lower concentrations of antibody is more optimal in permeabilized cells because this leads to lower background staining (13).

18. Protein fluorophores such as PE, PerCP, and APC are denatured by methanol and, therefore, lose fluorescence. APC is not completely quenched, but loses ~90% of its brightness upon methanol treatment. Therefore, with some well-separated signals, APC may be usable in sequential staining methods. In some of our studies, we have observed that using 70% ethanol as the permeabilization reagent may retain protein fluorophore fluorescence and also lead to sufficient permeabilization for good phospho-protein staining. Therefore, with optimization, it may be possible to use these protein fluorophores in sequential staining protocols.

References

1. Krutzik, P. O., Irish, J. M., Nolan, G. P., and Perez, O. D. (2004) Analysis of protein phosphorylation and cellular signaling events by flow cytometry: techniques and clinical applications. *Clin Immunol* **110**, 206–21.

2. Schulz, K. R., Danna, E. A., Krutzik, P. O., and Nolan, G. P. (2007) Single-cell phospho-protein analysis by flow cytometry. *Curr Protoc Immunol* Chapter 8, Unit 8.17.

3. Hotson, A. N., Hardy, J. W., Hale, M. B., Contag, C. H., and Nolan, G. P. (2009) The T cell STAT signaling network is reprogrammed within hours of bacteremia via secondary signals. *J Immunol* **182**, 7558–68.

4. O'Gorman, W. E., Dooms, H., Thorne, S. H., Kuswanto, W. F., Simonds, E. F., Krutzik, P. O., Nolan, G. P., and Abbas, A. K. (2009) The initial phase of an immune response

functions to activate regulatory T cells. *J Immunol* **183**, 332–9.

5. Krutzik, P. O., Hale, M. B., and Nolan, G. P. (2005) Characterization of the murine immunological signaling network with phosphospecific flow cytometry. *J Immunol* **175**, 2366–73.

6. Hale M. B., Krutzik, P. O., Samra, S. S., Crane, J. M., and Nolan, G. P. (2009) Stage dependent aberrant regulation of cytokine-STAT signaling in murine systemic lupus erythematosus. *PLoS One* **4**, e6756.

7. Irish, J. M., Czerwinski, D. K., Nolan, G. P., and Levy, R. (2006) Altered B-cell receptor signaling kinetics distinguish human follicular lymphoma B cells from tumor-infiltrating nonmalignant B cells. *Blood* **108**, 3135–42.

8. Kotecha, N., Flores, N. J., Irish, J. M., Simonds, E. F., Sakai, D. S., Archambeault, S., Diaz-Flores, E., Coram, M., Shannon, K. M., Nolan, G. P., and Loh, M. L. (2008) Single-cell profiling identifies aberrant STAT5 activation in myeloid malignancies with specific clinical and biologic correlates. *Cancer Cell* **14**, 335–43.

9. Irish, J. M., Kotecha, N., and Nolan, G. P. (2006) Mapping normal and cancer cell signalling networks: towards single-cell proteomics. *Nat Rev Cancer* **6**, 146–55.

10. Krutzik, P. O., Crane, J. M., Clutter, M. R., and Nolan, G. P. (2008) High-content single-cell drug screening with phosphospecific flow cytometry. *Nat Chem Biol* **4**, 132–42.

11. Krutzik, P. O. and Nolan, G. P. (2006) Fluorescent cell barcoding in flow cytometry allows high-throughput drug screening and signaling profiling. *Nat Methods* **3**, 361–8.

12. Krutzik, P. O. and Nolan, G. P. (2003) Intracellular phospho-protein staining techniques for flow cytometry: monitoring single cell signaling events. *Cytometry A* **55**, 61–70.

13. Krutzik, P. O., Clutter, M. R., and Nolan, G. P. (2005) Coordinate analysis of murine immune cell surface markers and intracellular phosphoproteins by flow cytometry. *J Immunol* **175**, 2357–65.

Chapter 10

Multiparametric Analysis of Apoptosis by Flow Cytometry

William G. Telford, Akira Komoriya, Beverly Z. Packard, and C. Bruce Bagwell

Abstract

Flow cytometry is the most widely used technology for analyzing apoptosis. The multiparametric nature of flow cytometry allows several apoptotic characteristics to be combined in a single sample, making it a powerful tool for analyzing the complex progression of apoptotic death. This chapter provides guidelines for combining caspase detection, annexin V binding, DNA dye exclusion, and other single apoptotic assays into multiparametric assays.

This approach to analyzing apoptosis provides far more information than single parameter assays that provide only an ambiguous "percent apoptotic" result, given that multiple early, intermediate and late apoptotic stages can be visualized simultaneously. This multiparametric approach is also amenable to a variety of flow cytometric instrumentation, both old and new.

Key words: Apoptosis, Caspase, Flow cytometry, Annexin V, 7-Aminoactinomycin D, Propidium iodide, Pacific Blue, Hoechst 33258

1. Introduction

The importance of apoptosis in the regulation of cellular homeostasis has mandated the development of accurate assays capable of measuring this process. Apoptosis assays based on flow cytometry have proven particularly useful. They are rapid and quantitative; they provide an individual cell-based mode of analysis (rather than a bulk population) (1). The multiparametric nature of flow cytometry also allows the detection of more than one cell-death characteristic to be combined in a single assay. For example, apoptosis assays that utilize DNA dyes as plasma membrane permeability indicators (such as propidium iodide) can be combined with assays that assess different cellular responses

Teresa S. Hawley and Robert G. Hawley (eds.), *Flow Cytometry Protocols*, Methods in Molecular Biology, vol. 699,
DOI 10.1007/978-1-61737-950-5_10, © Springer Science+Business Media, LLC 2011

associated with cell death, including mitochondrial membrane potential and annexin V binding to "flipped" phosphatidylserine (PS) (2–5). Combining measurements for cell death into a single assay has a number of important advantages; it provides simultaneous multiple confirmation of apoptotic activity (important in a process that has proven highly pleiotrophic in phenotype). It also provides a much more comprehensive and multidimensional picture of the entire cell-death process.

Recognition of the pivotal role of caspases in the death process has led to the recent development of assays that can measure these important enzymes in situ. Caspase activation represents one of the earliest easily measurable markers of apoptosis (6). In most cases, caspase activation precedes degradation in cell permeability, DNA fragmentation, cytoskeletal collapse, and PS "flipping"; caspases are in fact both signaling agents and mediators of these downstream manifestations of cell death. Combining fluorogenic assays of caspase activation with fluorescence-based assays for later characteristics of cell death (such as PS "flipping" and loss of membrane integrity) can provide a very information-rich view of cell death. It can be particularly helpful in distinguishing the "early" stages of cell death from later events, allowing better signal transduction studies in cells prior to the complete collapse of the cell structure (7–11).

Several fluorogenic assays for caspase activity have also been described, including the OncoImmunin PhiPhiLux system, the FLICA substrates, and the NucView substrates (12–17). All of these assays have both advantages and drawbacks. In this chapter, we describe the combination of the PhiPhiLux caspase substrate system with two simultaneous assays for later stages of cell death, annexin V binding to "flipped" PS residues, and cell membrane integrity using a DNA binding dye (17). The PhiPhiLux caspase substrates have several characteristics that make them useful for integration with other "live" cell apoptosis assays; they are cell-permeable and possess good caspase specificity. They are also relatively non-fluorescent in the intact state and become fluorescent upon caspase cleavage, with a signal-to-noise ratio of roughly 40 between the two states. They are also based on fluorescent probes with spectral characteristics similar to commonly used probes like fluorescein and rhodamine; this makes them easy to combine with other fluorescent probes (15–17). The ability to observe and measure multiple apoptotic phenotypes in a single assay gives a powerful picture of the overall apoptotic process. It is applicable to both suspension cells by traditional flow cytometry, and adherent cells using laser scanning cytometry (17). This assay can take advantage of newer flow cytometers with multiple lasers, but is also accessible to older cytometers with a single 488 nm laser.

2. Materials

1. PhiPhiLux G1D2 fluorogenic caspase 3/7 substrate (OncoImmunin, Inc., Gaithersburg, MD): OncoImmunin manufactures a series of fluorogenic enzyme substrates that fluoresce upon cleavage of an incorporated consensus domain. The fluorogenic caspase 3/7 substrate (PhiPhiLux G1D2) consists of an 18-amino acid peptide corresponding to the recognition/cleavage sequence from PARP, a target for caspase 3/7 (18). The substrate is homodoubly labeled with one of several fluorophores (in this case, a fluorescein-like molecule) on opposite sides of the molecule; in this conformation, the fluorochrome molecules are in close physical proximity and the fluorescence of the resulting complex is largely quenched (16, 19). After the substrate enters a cell by passive diffusion and is cleaved by caspase 3 or 7, the unquenched fluorescent fragments will be largely retained on the side of the membrane where the cleavage took place (16, 19).

   (a) The PhiPhiLux nomenclature indicates both its substrate specificity and the conjugated fluorochrome. The first letter refers to the substrate specificity: G refers to caspase 3/7, E to caspase 1, L to caspase 8, J to caspase 6, etc. The first number refers to the conjugated fluorochrome: 1 is the fluorescein-like fluorochrome, 2 to the rhodamine-like molecule and 6 to the sulforhodamine-like molecule. So G1D2 is specific for caspase 3 with the fluorescein-like probe, and E2D2 is specific for caspase 1 with the rhodamine-like probe. R2D2 is a special case and refers to the Cy5-like molecule, with the caspase indicated beforehand (3-R2D2 for caspase 3). Excitation and emission spectra for all the fluorochrome conjugates (generically referred to as X1D2, X2D2, etc.) are shown in Fig. 1.

   (b) PhiPhiLux G1D2 spectrally resembles fluorescein and can be excited with the standard 488 nm argon-ion or solid state laser found on most flow cytometers. The excitation and emission spectra for this conjugate and others are shown in Fig. 1. PhiPhiLux G1D2 is spectrally compatible with propidium iodide or 7-aminoactinomycin D (which can be used for measuring apoptotic cell permeability) and either phycoerythrin- or allophycocyanin-conjugated annexin V (for the detection of PS "flipping" during apoptotic death).

   (c) The PhiPhiLux reagents are roughly 40-fold dimmer in the uncleaved state than following caspase activation.

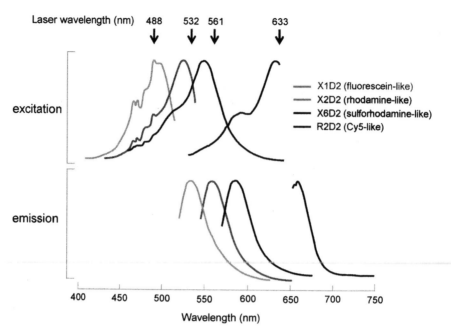

Fig. 1. Spectral characteristics of fluorogenic caspase substrates. Excitation and emission spectra (*top* and *bottom rows*) for four caspase substrate conjugates: X1D2 (fluorescein-like, used in most of the data presented in this chapter), X2D2 (rhodamine-like), X6D2 (sulforhodamine-like), and R2D2 (Cy5-like). The wavelengths of commonly used lasers are indicated.

The expected signal-to-background ratio between unlabeled and substrate-loaded viable and apoptotic cells is shown in Fig. 2, where cycloheximide-treated EL-4 thymoma cells were labeled with PhiPhiLux G1D2 and analyzed by flow cytometry; the apoptotic cells possess one- to three-orders of magnitude higher fluorescence than the viable cells. Primary cell cultures may show somewhat lower levels of caspase activation than cell lines, with subsequent lower levels of substrate fluorescence; however, background fluorescence may be lower with these cells as well (see Note 1).

(d) The PhiPhiLux reagents are also available with other fluorescent tags, including rhodamine- and sulforhodamine-like fluorochromes, and a proprietary Cy5-like fluorochrome that can be excited with a red laser. The excitation and emission spectra for these alternative fluorochrome conjugates are shown in Fig. 1. The rhodamine and sulforhodamine substrates can be readily excited using green or yellow lasers, including 532 and 561 nm sources. These lasers are becoming more widespread on commercial cytometers. See Note 2 for more information on these conjugates.

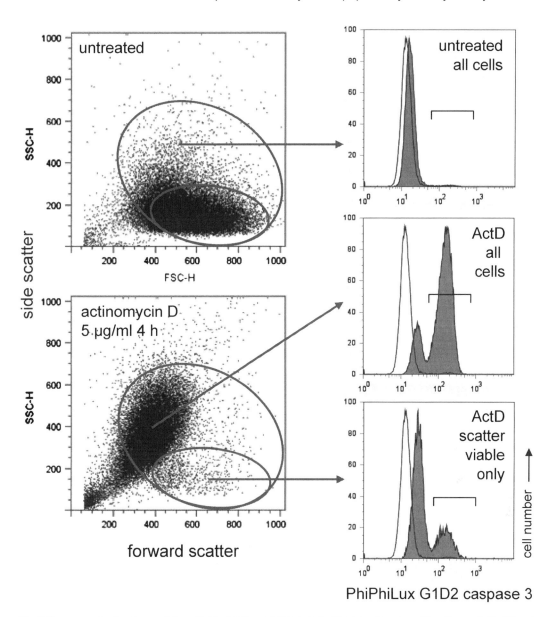

Fig. 2. Caspase activation in apoptotic cells. EL-4 cells were incubated with no treatment or with actinomycin D at 5 μg/ mL for 4 h, followed by labeling with PhiPhiLux G1D2. The *left dot plots* show untreated or treated cells displayed for forward versus side scatter. Gates for the total cell population or the scatter-viable cells only are indicated. The *right histograms* show PhiPhiLux G1D2 caspase 3/7 activation for the entire cell population of the untreated cells (*top*), the entire cell population for the drug-treated cells (*middle*), and the scatter-viable cells for the drug-treated sample (*bottom*). Cells with no PhiPhiLux loading are shown as unshaded peaks.

(e) The PhiPhiLux reagents are commercially provided at concentrations of 5–10 μM in sealed aliquots and can be stored at 4°C prior to opening; once the ampule is opened, any remaining substrate should be stored at –20°C. Avoid repeated freezing and thawing. Shelf life at 4°C is approximately 3–6 months, over 1 year at –20°C.

2. Phycoerythrin (PE)- or allophycocyanin (APC)-conjugated annexin V (available from multiple sources, including Invitrogen Life Technologies, Carlsbad, CA): Annexin V can be conjugated to a variety of fluorochromes, and binds to apoptotic cells with "flipped" PS residues on their extracellular membrane leaflet. Damaged or necrotic cells with a high degree of membrane permeability can also bind annexin V to their intracellular membrane leaflet, despite their uncertain apoptotic nature; therefore, a DNA binding dye as a cell permeability indicator should always be incorporated into annexin V binding assays. Cells that are both annexin V and DNA binding dye positive may therefore be either advanced apoptotic or necrotic.

3. DNA binding dyes: Keep in mind that all of the DNA binding dyes described here have differing cell permeability characteristics. This will affect the ultimate data analysis (see Note 3).

(a) Propidium iodide (PI) is an inexpensive and widely available intercalating DNA binding dye. PI excites at 488 nm and emits in the 570–630 nm range. Dissolve in deionized water at 1 mg/mL and store in the dark at 4°C for up to 3 months.

(b) 7-aminoactinomycin D (7-AAD) (available from Sigma Chemical Co., St. Louis, MO and Invitrogen Life Technologies) is an intercalating/groove binding DNA binding dye that also excites at 488 nm, but emits in the far-red, with an emission peak at approximately 670 nm. 7-AAD is a good alternative to PI where a longer wavelength probe is desired. 7-AAD is also somewhat more cell permeable than PI. Dissolve 7-AAD in EtOH at 1 mg/mL and store at –20°C. Solublized stocks are good for 6 months. Diluted stocks should be used within 24 h.

(c) Red- and violet-excited DNA binding dyes: The proliferation of cytometers with multiple lasers has greatly expanded the fluorochromes available for apoptotic analysis. Several red- or violet-excited DNA binding dyes can be substituted for PI or 7-AAD to increase total fluorochrome capability or to reduce fluorescence compensation requirements. Hoechst 33258 is a widely available minor groove DNA binding dye that is well-excited by ultraviolet or violet lasers; it has cell permeability characteristics similar to PI. Prepare Hoechst 33258 as a 1 mg/mL stock in distilled water, and store at 4°C for up to 3 months. Sytox Red and Sytox Blue (both available from Invitrogen Life Technologies) are red- and violet-excited DNA binding dyes that can be used; they are somewhat more cell-permeable than PI. Both Sytox Red and Sytox Blue are sold as stock solutions at 5 mM in DMSO with storage at –20°C, and should be diluted immediately prior to use. Very cell permeable DNA dyes like Hoechst 33342 (distinct from Hoechst 33258), the DyeCycle dyes (Invitrogen Life Technologies), and DRAQ5 (Biostatus Limited, Shepshed, Leicestershire, UK) should probably be avoided for most cytometric analysis of apoptosis, since they do not discriminate viable from apoptotic cells clearly enough. However, they may allow recognition of apoptosis-associated chromatin damage by microscopy and scanning cytometry.

4. Complete medium: RPMI supplemented with 10% FBS, L-glutamine, and penicillin/streptomycin

5. Wash buffer: Dulbecco's PBS (containing calcium and magnesium) supplemented with 2% fetal bovine serum. This is used for cell washing prior to DNA dye addition. The inclusion of divalent cations is critical for annexin V binding.

6. Flow cytometer equipped with one, two, or three lasers.

7. (Optional) GemStone analysis software (Verity Software House, Topsham, ME).

3. Methods

3.1. Combinations of Fluorochromes

This assay combines fluorescent labels for three characteristics of cell apoptosis, namely caspase activation, PS "flipping", and cell permeability. There is considerable flexibility of fluorochrome selection for the investigator depending on the flow cytometric instrumentation available. Three possible combinations are described below, one for analysis on instruments equipped with a single 488 nm laser, a second for instruments equipped with dual 488 nm/red diode or red HeNe lasers, and a third for instruments equipped with a violet laser diode.

3.1.1. Single 488 nm Laser Instruments

These instruments are limited to a single laser, and tend to have only three or four fluorescent detectors. Examples of these include: flow cytometers from BD Biosciences (San Jose, CA) such as the BD FACScan, single laser FACSort or FACSCalibur; flow cytometers from Beckman Coulter (Fullerton, CA) such as the Coulter Epics XL or Cell Lab Quanta. The following combination should be used when analysis is limited to this instrument type:

1. PhiPhiLux G1D2 (similar to fluorescein): Detect this fluorochrome in the fluorescein or FITC detector on most commercial instruments.

2. PE-conjugated annexin V: Detect this fluorochome in the PE detector on most instruments. Apply fluorescence compensation

to separate the PE signal from PhiPhiLux G1D2 and 7-AAD.

3. 7-AAD: Detect this far-red emitting DNA binding dye in the far-red (or PE-Cy5) detector on most commercial instruments.

3.1.2. Dual 488 nm/Red Laser-Equipped Instruments

Several more recent benchtop flow cytometers are equipped with more than one laser, most commonly a red source (such as a 635 nm red diode or 633 nm red HeNe laser). The BD FACSort, FACSCalibur, LSR, and LSR II (BD Biosciences) fall into this category, as do the FC500 (Beckman Coulter) and Accuri C6 (Accuri Cytometers, Inc., Ann Arbor, MI). A red laser allows several red-excited fluorochromes to be incorporated into flow cytometry assays, including APC. Another group of PhiPhiLux caspase substrates incorporating a proprietary red-excited fluorochrome analogous to Cy5 can also be used on these instruments. The DNA dye Sytox Red can also be incorporated if a red laser is available. The following combination is suggested for dual laser instrumentation:

1. PhiPhiLux G1D2 (similar to fluorescein): Detect this fluorochrome in the fluorescein detector on most commercial instruments.

2. APC-conjugated annexin V: Excite this fluorochome with either a red diode or HeNe laser, and detect in the far-red range. Little fluorescence compensation is required to separate its signal from PhiPhiLux G1D2 or the DNA binding dyes described below, making post-acquisition analysis easier. Annexin V conjugates with Cy5 and Alexa Fluor 647 (which are spectrally similar to APC) can be analyzed in the same way.

3. PI or 7-AAD DNA binding dyes can be incorporated into a cell-death assay with PhiPhiLux G1D2 and APC-annexin V. Detect both in the far-red detector (usually with a mid-600 nm bandpass (BP) or longpass (LP) filter) on most flow cytometers.

4. Further substitutions: If a red-excited DNA dye like Sytox Red is used, move annexin V to another detector (such as the PE detector).

3.1.3. Triple 488 nm/Red Laser/Violet Laser Diode-Equipped Instruments

Many modern cytometers are equipped with more than two lasers; violet laser diodes (~405 nm) are typically included as a third excitation source. Instruments include the BD LSR II, LSR Fortessa (BD Biosciences), Gallios (Beckman Coulter), Stratedigm S1400 (Stratedigm, Inc., Campbell, CA), and Partec CyFlow (Partec GmbH, Münster, Germany). Violet-excited annexin V conjugates and DNA binding dyes can be easily incorporated into apoptotic assay combinations. Violet-excited probes do not significantly overlap into other fluorescence channels, making them very useful for multicolor assays. Two examples are listed below:

1. PhiPhiLux G1D2 (similar to fluorescein): Detect this caspase substrate in the fluorescein detector. Combine it with:

    (a) *PI, 7-AAD or Sytox Red*: Either a 488 nm or red-excited DNA binding dye can be used (multilaser cytometers are typically equipped with both red and violet laser sources).

    (b) *Pacific Blue-annexin V*: Pacific Blue is a relatively bright violet-excited fluorochrome, and is available in an annexin V conjugate. Pacific Blue does not overlap significantly into other fluorescent channels, and other fluorochromes do not overlap significantly into it, making it very applicable for multiparametric assays.

2. Another possible combination still uses the fluorescein detector for PhiPhiLux G1D2, but uses:

    (a) *Hoechst 33258 or Sytox Blue*: These DNA binding dyes use the violet laser for excitation. Sytox Blue is somewhat more cell-permeable than Hoechst 33258, which is roughly equivalent to PI.

    (b) *APC-annexin V*: A red laser can be used to excite APC-annexin V. This combination uses three lasers to excite three fluorochromes; as a result, virtually no spectral overlap occurs, and almost no fluorescence compensation is required.

3.2. Preparation of Cells

EL-4 cells treated with transcriptional or translational inhibitors such as cycloheximide at 50 µg/mL or actinomycin D at 5 µg/mL for 4 h were used to illustrate this assay in Figs. 2–6. This cell line is easily grown and hardy, and can make a useful positive control for more general use (see Note 1).

1. Harvest cell lines grown in suspension or cultured primary cells. Transfer cells to 12×75 mm cell culture tubes, and centrifuge at $400 \times g$ for 5 min.

2. Decant supernatant. Maximum removal of the supernatant is critical; the volume of remaining supernatant should be as low as possible to cause minimal dilution of the caspase substrate. Although cells can be washed prior to labeling, performing the assay in the remaining complete medium supernatant will reduce the amount of incidental cell death occurring during the assay. If cells are obtained from clinical or other in vivo sources, they should be centrifuged and resuspended in a complete tissue culture medium (such as RPMI containing 10% FBS) prior to use, then centrifuged, and decanted as described above.

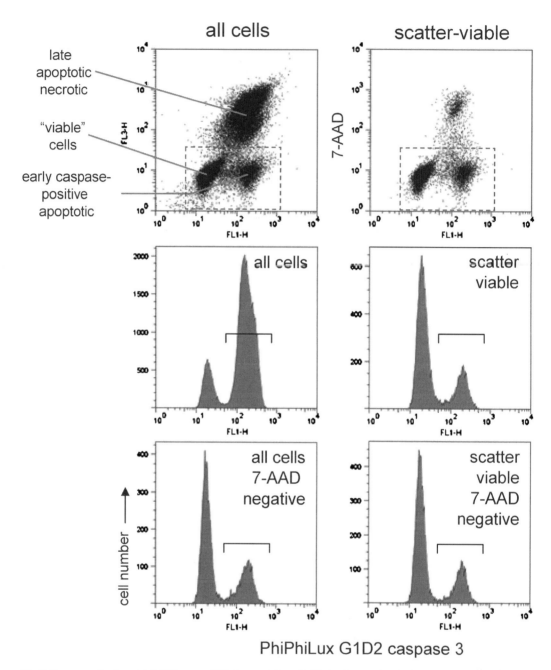

PhiPhiLux G1D2 caspase 3

Fig. 3. Caspase activation and 7-AAD permeability in apoptotic cells. EL-4 cells were incubated with no treatment or with actinomycin D at 5 μg/mL for 4 h, followed by labeling with PhiPhiLux G1D2 and the DNA binding dye 7-AAD. *Top dot plots* show drug-treated cells gated for all cells (*left plot*) or scatter-viable cells (*right plot*) displayed for PhiPhiLux G1D2 caspase 3/7 versus 7-AAD fluorescence. The *middle row* of histograms shows PhiPhiLux G1D2 fluorescence for the above cells (both 7-AAD-negative and -positive). The *bottom row* of histograms shows PhiPhiLux G1D2 fluorescence for the gated 7-AAD-negative portion of the above cells. A significant amount of caspase 3 activity is present in the 7-AAD-negative cell fraction.

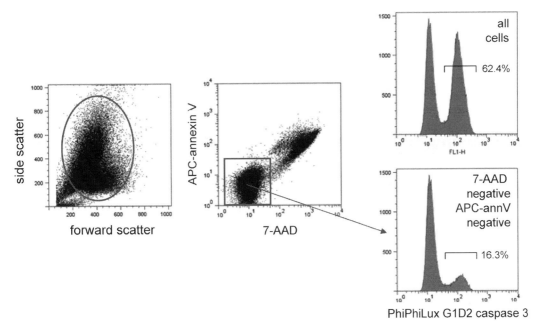

Fig. 4. Caspase activation, annexin V binding and 7-AAD permeability in apoptotic cells. EL-4 cells were incubated with no treatment or with actinomycin D at 5 μg/mL for 4 h, followed by labeling with PhiPhiLux G1D2, APC-conjugated annexin V, and 7-AAD at 5 μg/mL. Drug-treated cells are shown in the forward versus side scatter dot plot at *left*. All cells were then gated into the middle dot plot showing 7-AAD versus APC-annexin V fluorescence. The *right-most histograms* show the PhiPhiLux G1D2 caspase 3/7 fluorescence for the entire cell population (*top histogram*) or for the annexin V-negative 7-AAD-negative gated cell fraction (*bottom histogram*).

3. Label 0.5 to 1×10^6 cells per sample; increasing this number will saturate the detection reagents and reduce caspase and annexin V labeling efficiency. Adherent cells pose special challenges for apoptotic analysis due to the physical trauma and membrane damage that occur with cell dissociation; analysis in the adherent state by laser scanning cytometry is much preferable to flow cytometry under these circumstances (see Note 4).

3.3. Fluorogenic Caspase Substrate Labeling

Cells are initially incubated with the PhiPhiLux caspase substrate. Substrate concentration and incubation time are critical factors in cell-permeable substrate loading.

1. Ensure that as much supernatant is removed, to maximize final substrate volume. Tap each tube to resuspend the cell pellet in the remaining supernatant. The supernatant in the tubes will be approximately 50 μL in volume (but not exceeding 100 μL).

2. Add 50 μL of the PhiPhiLux reagent to each tube and shake gently. The PhiPhiLux reagent should be diluted as little as

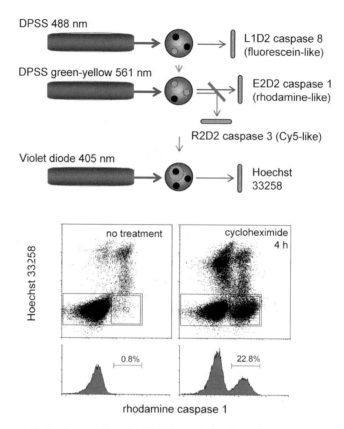

Fig. 5. Multiple caspase activation in apoptotic cells. (*Top*) The analysis scheme for analyzing three caspases simultaneously. A BD LSR II was configured with a 488 nm laser for L1D2 caspase 8 substrate excitation (similar to fluorescein), a yellow 561 nm laser for both E2D2 caspase 1 substrate (similar to rhodamine) and R2D2 caspase 3/7 substrate (similar to Cy5) excitation, and a violet laser diode for Hoechst 33258 excitation. PE-annexin V was also incorporated into the assay and detected using the 488 nm laser. (*Bottom*) EL-4 cells induced with cycloheximide at 50 µg/mL for 4 h and labeled for caspase 1 activation using PhiPhiLux E2D2, with excitation of the substrate using a 561 nm laser. Dot plots display caspase 1 activity versus Hoechst 33258 DNA dye permeability for untreated or drug-treated cells (*left* and *right*, respectively).

possible for maximum detection, hence the need for minimal sample supernatant. PhiPhiLux reagent solutions are typically prepared at 10 µM; this will give a final concentration between 3 and 5 µM (in approximately 100–150 µL of total volume).

3. Incubate the tubes for 45 min at 37°C, in a water bath or an incubator. An incubator may be preferred if CO_2 conditions are desired. For both optimal labeling and reasons of economy, the PhiPhiLux reagent can be titered and tested for use between 0.5 and 5 µM. However, this should be done with caution (see Note 5).

3.4. Annexin V Labeling

Cells are then labeled with fluorochrome-conjugated annexin V. Since centrifuge washings are minimized in this method to reduce

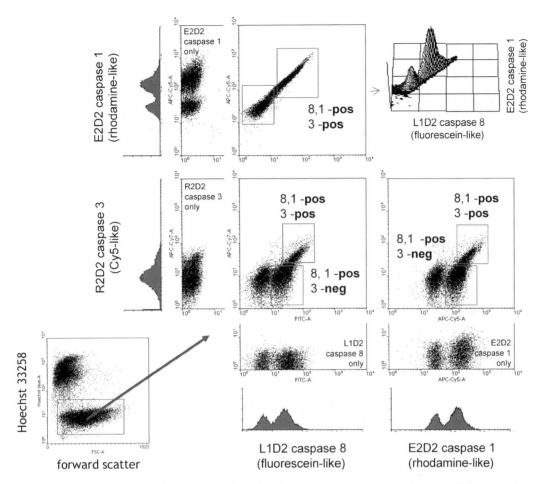

Fig. 6. Multiple caspase activation in apoptotic cells. EL-4 cells were incubated with no treatment or with cycloheximide at 50 μg/mL for 4 h, followed by labeling with PhiPhiLux L1D2 (specific for caspase 8), PhiPhiLux R2D2 (specific for caspase 3/7) and PhiPhiLux E2D2 (specific for caspase 1), followed by PE-conjugated annexin V and Hoechst 33258. The three center dot plots are all gated for Hoechst dye-negative cells, and show all combinations of caspase 1/3/7/8, with labeling control dot plots and single label histograms on the periphery. Based on this analysis, caspase 1 and 8 activities appear first and are simultaneous, followed by caspase 3.

assay-associated cell death, the cells are not washed following caspase substrate loading, but are labeled immediately with fluorochrome-conjugated annexin V. Since most cell culture media and serum supplements contain calcium and magnesium, it is assumed that cation concentrations are sufficient to allow annexin V binding to "flipped" PS residues. However, this should be verified; even brief removal of divalent cations will cause immediate dissociation of the annexin V reagent. Subsequent cell washing is therefore done in PBS containing calcium and magnesium supplemented with FBS (see Note 6).

1. After 45 min of caspase substrate incubation, add the appropriate fluorochrome-conjugated annexin V (in this case, either

PE or APC). Annexin V is generally available in suspension at concentrations ranging from 0.1 to 1 mg/mL. Cell labeling should be carried out at approximately 0.5–5 µg annexin V per sample. Therefore, add 5 µL of a 1 mg/mL annexin V solution to the tubes described in Subheading 3.3, step 3. Again, fluorochrome-conjugated annexin V labeling should be titered in advance of actual use.

2. Incubate at room temperature (or in a 37°C incubator if CO_2 is desired) for 15 min.

3. Add 3 mL of wash buffer to each tube. Centrifuge at $400 \times g$ for 5 min, and decant supernatant.

3.5. DNA Binding Dye Labeling

Depending on the instrumentation available, cells can be subsequently labeled with a DNA binding dye for assessment of cell permeability in the later stages of apoptosis (6). Remember that 7-AAD can be used with single laser instruments. PE-annexin V and PI can be used together on a single laser instrument, but they are spectrally similar, and compensation of fluorescence may be an issue. 7-AAD is therefore preferable when using PhiPhiLux and PE-annexin V. PI is more readily used with dual-laser instruments (blue-green and red), since annexin V can be detected using the APC detector. Sytox Red requires a red laser; Hoechst 33258 and Sytox Blue require violet excitation. (See Note 3.)

1. Prepare a solution of DNA binding dye in complete medium: PI at 2 µg/mL, 7-AAD at 5 µg/mL, Sytox Red or Sytox Blue at 5 µM, or Hoechst 33258 at 2 µg/mL.

2. Add 0.5 mL of the DNA binding solution to each of the tubes described in Subheading 3.4, step 3. Maintain samples at room temperature and analyze within 60 min (see Note 7).

3.6. Flow Cytometric Analysis

Cells should be analyzed as quickly as possible to minimize post-assay apoptotic death. The instrument should be set up and ready for sample acquisition immediately upon completion of the assay. Although the cleaved caspase substrate has a lower membrane permeability than the uncleaved molecule, even the cleaved form will eventually diffuse into the surrounding medium. Samples should be kept at room temperature until analysis; storage at 4°C may reduce dye dissociation, but can itself induce unwanted apoptosis. The choice of fluorescent reagents for both single- and dual-laser flow cytometers was described in Subheading 3.1; fluorescent detector assignments and analysis issues are described here.

1. *PhiPhiLux G1D2*: This fluorescein-like caspase substrate is detected through the fluorescein detector on most flow cytometers (often with the designation "FL1") using a 530/30 nm or similar narrow BP filter. The spectral properties

of PhiPhiLux G1D2 is similar to fluorescein, requiring some spectral compensation when used simultaneously with PE or PI (and to a lesser extent with 7-AAD).

2. *PE-conjugated annexin V*: Like most PE-conjugated reagents, this reagent is detected through the PE detector on most flow cytometers (often with the designation "FL2") using a 575/26 nm or similar BP filter. PE requires some spectral compensation when used with PhiPhiLux G1D2 and 7-AAD.

3. *APC-conjugated annexin V*: APC is excited with a red laser source and detected through the APC detector on many flow cytometers (sometimes with an "FL4" designation) using a 660/20 nm or similar BP filter. An advantage of APC in multicolor assays is its minimal need for color compensation; there is no significant spectral overlap between PhiPhiLux G1D2, PI, or 7-AAD. Cy5 or Alexa Fluor 647 conjugates are spectrally similar to APC, and can be analyzed in the same way.

4. *Pacific Blue-conjugated annexin V*: Pacific Blue is analyzed using a violet laser; most instruments so equipped have at least two detectors aligned to this laser source. A 450/50 nm or similar filter is typically used to detect this fluorescent probe. Cascade Blue and Alexa Fluor 405 are spectrally similar to Pacific Blue, and are analyzed in the same way.

5. *PI*: This DNA binding dye is very bright even at low concentrations, and has a broad emission range, requiring compensation when used with PhiPhiLux G1D2. It can be detected in either the PE (575/26 nm filter) or far-red detection channel (red 650 BP or LP filter). The second choice is preferable to reduce spillover into the fluorescein detector. PE and PI can be analyzed together on older single laser instruments using the traditional PE detector ("FL2" detector, 575/26 nm) for PE detection, and the longer PE-Cy5 detector ("FL3" detector, 650 LP dichroic, or 675/20 nm) for PI. However, the close proximity of their spectra makes this analysis difficult. Substitution of PI with 7-AAD is preferable. PI is highly charged, and will contaminate instrument tubing, causing unwanted "shedding" of the dye into later samples. After PI use, the instrument should be thoroughly cleaned with 10% bleach or similar detergent to remove the dye.

6. *7-AAD*: This DNA binding dye is not as bright as PI and emits in the far-red, allowing its detection in the far-red channel on most single laser flow cytometers (the PE-Cy5, or "FL3" detector) with a 675/20 nm BP, 650 LP dichroic or similar filter. Compensation will be required when used with PhiPhiLux G1D2 and PE.

7. *Hoechst 33258*: Hoechst 33258 is very bright, and can be excited using either an ultraviolet or violet laser source. It is detected through a 450/50 nm or similar filter. It will have minimal spectral overlap into other fluorochromes. Like PI, it is highly charged and will adhere tightly to instrument tubing; the instrument should be cleaned thoroughly with 10% bleach or other detergent after use.

8. *Sytox Red and Sytox Blue*: These dyes can be analyzed using the conditions for APC and Hoechst 33258, respectively. Both are very bright, and are somewhat more cell-permeable than PI or Hoechst 33258.

3.7. Gating for Flow Cytometry

Good gating is critical for meaningful analysis of apoptosis. A typical gating scheme is illustrated in Fig. 4 and is described in more detail in Subheading 3.9. Some guidelines are listed here.

1. *Scatter gating*: Many cell lines and some primary cells show a dramatic alteration in forward and side scatter measurements late in the onset of apoptosis. Forward and side scatter are approximate indicators of cell size and optical density, respectively, and reflect both the cell volume loss and intracellular breakdown occurring during apoptotic death. It therefore seems logical to draw a gate around both the scatter-viable population AND the scatter-shifted apoptotic cells, and look at caspase activation, annexin V binding and DNA dye uptake in this total population.

   However, the scatter-apoptotic population is also usually at very advanced stage of apoptotic death; the cells are already positive for all markers. The advanced physical perturbation of the cells in this group can also produce positive, but highly variable labeling results, interfering with the identification of earlier apoptotic stages. It is therefore also useful to gate only on the scatter-viable cells, and examine early apoptotic markers such as caspase activation only within this group of cells. This dual approach allows an overall picture of both early and late apoptotic stages, as well as examination of the earliest apoptotic cells. It is therefore recommended that both gating approaches be applied to get a clear picture of the apoptotic process (Fig. 2).

2. *Annexin V binding and DNA binding dye exclusion*: Exclusion gating can also be useful for markers other than scatter. Annexin V binding and DNA dye uptake usually occur after caspase activation and are considered "later" markers of apoptosis. Therefore, subpopulations negative or positive for annexin V and DNA dye binding can be gated for discrimination of "early" and "late" apoptotic cells. The annexin V-negative DNA dye-negative cells can be gated as in step 1 to allow detailed examination of the earlier stages of apoptosis such as caspase activation (Figs. 3 and 4).

3. *Differences in DNA dye permeability*: DNA dyes are not completely interchangeable with regard to exclusion by apoptotic cells (see Note 3). For example, 7-AAD is somewhat more cell-permeable than PI and will label an earlier subset of apoptotic cells; the Sytox dyes will also label earlier apoptotic cells than either PI or Hoechst 33258. This will affect the overall analysis. For example, if 7-AAD-positive cells are excluded from the analysis (in an attempt to quantify very early apoptotic events), this dye's greater cell permeability will result in a lower apparent number of caspase-positive cells that are DNA dye-negative than if PI were used instead. These differences should be kept in mind when analyzing these early apoptotic subsets.

4. *Caspase substrate background fluorescence*: Viable cells labeled with a caspase substrate will have somewhat higher background fluorescence levels than completely unlabeled cells. Care should be taken to identify both the viable and apoptotic fraction without using an unlabeled control as a cutoff.

3.8. Simultaneous Immunophenotyping

The protocol described in this chapter is very compatible with simultaneous antibody immunophenotyping of the "viable" subpopulations. For example, PE-conjugated antibodies against a marker of interest could be combined with PhiPhiLux G1D2, 7-AAD, and APC-annexin V labeling as a very stringent "filter" for the removal of dead cells from the phenotyping analysis. This is similar to the common inclusion of PI or another viability probe in cell phenotyping protocols, to exclude dead cells from the analysis; incorporating a multicolor apoptotic assay with immunolabeling for dead cell exclusion is even more powerful. While a natural extension of this method would appear to be the immunophenotyping of early apoptotic cells (such as caspase-positive/7-AAD-negative/annexin V-negative), this should be approached with great caution (see Note 8).

3.9. Sample Results

In all of the illustrated results, apoptosis was induced in EL-4 murine thymoma cells by treatment with either actinomycin D or cycloheximide for 4 h. These drugs rapidly induce apoptosis via the caspase 3 pathway in many rapidly dividing cell lines. The figures both illustrate expected results for the individual components of the multiparametric cell-death assay, and demonstrate how the simultaneous analysis of multiple cell-death characteristics in a single assay gives a multidimensional picture of the total apoptotic process.

1. *Forward and side scatter*: Figure 2 shows a typical shift in forward and side scatter during apoptosis in EL-4 cells treated with actinomycin D. In this case, both the entire population (excluding debris) and the scatter-viable cells are gated, and

subsequently analyzed for caspase activation, annexin V binding, and DNA dye permeability.

2. *Fluorescence distribution of PhiPhiLux G1D2 labeling:* Figure 2 also illustrates the typical signal-to-background ratio between "viable" and apoptotic EL-4 cells labeled with the PhiPhiLux G1D2 substrate (shown here without annexin V and DNA binding dye labeling). The caspase substrate was readily detectable in the fluorescein channel by flow cytometry, in this case on a BD FACSCalibur. The substrate is much less fluorescent in the uncleaved state; signal-to-noise ratios of 1- to 3-log orders of magnitude are normally seen between "viable" and apoptotic cells loaded with PhiPhiLux G1D2. Unlabeled cells are slightly less fluorescent than "viable" labeled cells; this background fluorescence can be more dramatic in some cells types and does not necessarily indicate caspase activity in viable cells.

   It should be noted that the "viable" and apoptotic distribution based on scatter measurements does not strictly correlate with caspase activation. The scatter-viable cells have a large percentage of caspase-positive cells, indicating that cells activate caspases prior to gross changes in scatter morphology. In some cases, the scatter-apoptotic population may also have some caspase-negative cells. While some of these cells may be advanced apoptotic or necrotic cells with diminished or degraded caspase activity, there may also be viable cells in this population. Previous studies have shown that cells may undergo transient volume fluctuations very early in the apoptotic process, well before caspase activation. These results indicate the importance of gating on both the total scatter-viable/apoptotic population, as well as the scatter-viable only cells.

3. *PhiPhiLux G1D2 and 7-AAD labeling:* Figure 3 shows the addition of the DNA dye 7-AAD labeling to the PhiPhiLux G1D2 assay. The dot plots at the top of the figure show 7-AAD labeling versus caspase activation for both drug-treated EL-4 cells gated for either the entire population (left dot plot) or the scatter-viable cells. Even with only two probes for apoptosis, three distinct subpopulations were apparent: a "viable" population at lower left, a caspase-positive population that had not progressed to 7-AAD permeability (lower right), and a caspase-positive population that was permeable to 7-AAD (upper right). Sometimes, a fourth population is also apparent that is also labeled with 7-AAD, but had little caspase activity. If present, this fourth population of cells likely contained necrotic or advanced apoptotic cells, where caspases had leaked out of the cells, or were proteolytically digested. Another important potential source of this population is cells that have undergone apoptosis in the incubation

period following PhiPhiLux labeling but prior to flow analysis. Cells in this region demonstrate the importance of analyzing cells promptly at the completion of the assay, since apoptosis is still occurring. It also illustrates the importance of minimizing cell trauma during the assay; centrifugations and pipet transfers should be kept to a minimum.

At this point, the investigator can either include in the analysis all cells based on scatter (left column), or only the scatter-viable cells (right column). Excluding the advanced apoptotics can allow better resolution of the early-stage apoptotic cells. In addition, DNA dye labeling can now be used to exclude the more advanced apoptotic cells for specific measurement of the earlier dying cells. The bottom row of histograms show caspase 3/7 levels in 7-AAD negative cells. Caspase activation clearly precedes DNA dye permeability in this cell type.

4. *PhiPhiLux G1D2, 7-AAD and APC-annexin V labeling:* Figure 4 shows the final simultaneous analysis of caspase, annexin V, and DNA dye in a single assay. The left dot plot shows the forward and side scatter profile for apoptotic EL-4 cells; the entire cell population is then gated into a dot plot for annexin V binding versus DNA dye permeability (middle dot plot). Either the entire cell population or the annexin V-negative 7-AAD-negative cells can then be displayed for caspase 3/7 activation. A significant population of caspase-positive cells is present even in the annexin V-negative 7-AAD-negative population; caspase activation again precedes both of these characteristics. Layering multiple apoptosis assays into a single multiparameter assay therefore allows a comprehensive assessment of the apoptotic process in a cell population.

5. *Detection of multiple caspases by flow cytometry:* The PhiPhiLux system can incorporate a number of both consensus peptides for different caspase specificities, and fluorochromes for flow cytometric detection. It is therefore possible to load cells with more than one PhiPhiLux reagent, if they possess specificity for different caspases, and if they can be spectrally distinguished from one another by flow cytometry. This is illustrated in Figs. 5 and 6, where three caspase substrates were loaded simultaneously into apoptotic EL-4 cells, along with the DNA binding dye Hoechst 33258. The three substrates used were modifications of the PhiPhiLux reagent described earlier. Cells were simultaneously loaded with PhiPhiLux L1D2 (specific for caspase 8, conjugated to a fluorescein-like fluorochrome), PhiPhiLux R2D2 (specific for caspase 3/7, conjugated to a Cy5-like probe), and PhiPhiLux E2D2 (specific for caspase 1, conjugated to a rhodamine-like probe). The substrate concentrations were increased to allow simultaneous loading with all three substrate conjugates while maintaining

the 3–5 mM concentration specified in Subheading 3.3, step 2. A BD LSR II equipped with 488, 561, and 405 nm lasers was used to excite this combination of fluorochromes (Fig. 5). The 561 nm laser was used to excite the rhodamine-like substrate, and provided adequate excitation for the Cy5-like substrate as well. Cytometers equipped with 532 and 561 nm lasers are now commercially available and becoming more common, giving access to these alternative substrate conjugates (Fig. 1). The rhodamine caspase 1 substrate used in this example was readily excited at this wavelength (Fig. 5).

6. *Multiple caspase results and probability state analysis.* Figure 6 shows the three-caspase activation profile, gated for the apoptotic (Hoechst 33258-negative) cell population. Caspase activation was clearly not simultaneous; caspase 1 and 8 are activated first, followed by caspase 3. This was confirmed using the probability state analysis software GemStone, which plots

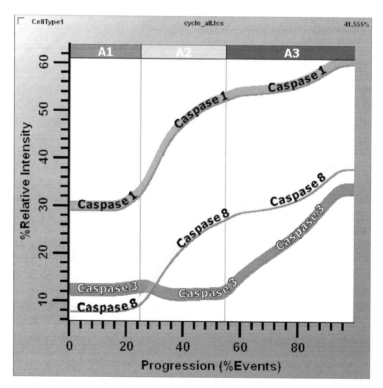

Fig. 7. Probability state analysis of multiple caspase activation. The data from Fig. 6 was analyzed using the probability state analysis software package GemStone, using the known apoptotic events of scatter, annexin V binding, and DNA dye permeability to construct a kinetic model for apoptosis activation with which caspase activation events can be applied. The cells were gated for "viable" events based on normal forward and side scatter and low Hoechst 33258 fluorescence. The X-axis of the graph is time progression; plots for caspase 1, 3/7 and 8 activity are shown. Caspase 1 and 8 activities both increase prior to caspase 3 activation.

changes in flow cytometric parameters as functions of time, relative to a computer model (see Chapter 2 by Bagwell for discussion on Probability State Modeling, *this volume*). This analysis is shown in Fig. 7, where caspase 1 and 8 are upregulated prior to caspase 3. By using multiple lasers and caspase substrates conjugated to multiple fluorochromes, multiparametric assays for apoptosis can become much more informative. This modification to the multiparametric cell-death assay allows an even earlier stage of cell death to be distinguished and identified. Rather than just assaying for cell viability, investigators can collect important information about the signal transduction and effector processes of apoptotic death.

7. These collective results are consistent with many immune cell types and established cell lines; however, wider variations in apoptotic phenotype between different cell types should be expected (see Note 9).

4. Notes

1. *Controls:* Good "viable" and apoptotic controls are important for apoptotic analysis of apoptosis, and should be used, especially when a new cell type or apoptotic stimulus is being investigated. Where possible, an untreated negative control and an independent positive control should be included, the latter being induced by an agent other than that under study (such as a cytotoxic drug). The EL-4 cells used in this study are a good example of a control system that is easy to maintain and is reliable. Samples with both the absence and presence of the PhiPhiLux reagents are also important to include as controls, since the substrate does possess some low, but detectable intrinsic fluorescence in the uncleaved state that can be erroneously interpreted as apoptosis without the appropriate control samples. However, unlabeled cells should not be used as a strict guide for gating on PhiPhiLux labeled cells; they only allow determination of the increase in background from PhiPhiLux labeling.

2. *Fluorogenic caspase substrates with alternative fluorophores:* Fluorogenic caspase substrates coupled to rhodamine-like, sulforhodamine-like, and Cy5-like fluorophores are also available. The spectra of these probes are shown in Fig. 1, as illustrated in Figs. 5 and 6. None of these probes excite well at 488 nm; the rhodamine and sulforhodamine probes require a green or yellow laser source (532 or 561 nm DPSS laser), and the Cy5 probes require yellow or red excitation. The green- and yellow-excited probes were originally designed for

epifluorescence microscopes, which are usually equipped with mercury arc lamp filters that can provide 546 and 577 nm (green and yellow) excitation light. Many flow cytometers are now equipped with lasers in this range, however, making these probes potentially useful for cytometric analysis. Using the rhodamine and sulforhodamine probes limits DNA dye choices; the spectrum of propidium iodide coincides too well with them, and therefore cannot be used simultaneously. All of these longer wavelength probes may give better sensitivity than the fluorescein-like versions; cellular autofluorescence is significantly reduced with green to red excitation (compared to 488 nm blue-green), so overall signal-to-background signal is likely to increase.

3. *DNA binding dyes:* All DNA binding dyes do not have identical cell permeability characteristics. Some DNA dyes will gradually cross the plasma membranes of even viable cells, while others are better excluded. These differences can affect the results obtained from the assay. For example, 7-AAD is somewhat more cell permeable than PI, and will give a greater percentage of apoptotic cells when compared directly to PI. Similarly, Sytox Blue is more cell permeable than Hoechst 33258, and will also identify an earlier set of apoptotic cells. This difference should be kept in mind while designing cell-death assays, and may dictate the use of 7-AAD when this property is desired. Highly permeable DNA binding dyes such as Hoechst 33342, the DyeCycle dye series, and DRAQ5, will enter cells and label their chromatin regardless of their viability state. This may limit their usefulness as apoptotic reagents for flow cytometry. They have however been used to identify morphological changes in chromatin during apoptosis by microscopy and laser scanning cytometry.

4. *Multiparametric analysis of apoptosis in adherent cells:* Flow cytometric analysis of apoptosis in adherent cell lines poses special challenges, since the removal of cells from their growth substrate may itself induce apoptosis. In addition, cell removal methods (such as trypsinization) can trigger false apoptotic indicators, such as aberrant annexin V binding in the absence of true cell death. By far the best solution to this problem is to utilize a laser scanning cytometer (LSC) for the analysis of apoptosis in these cell types; this specialized flow cytometer can perform cytometric analysis of cells on a flat surface, allowing minimal disruption during cell preparation (20). Several apoptosis assays utilizing caspase substrates using laser scanning cytometry have been described (17, 21, 22). The cell-labeling protocol is similar to that for suspension cells as described in Subheading 3.3, using cells cultured on tissue culture microslides as described previously (17).

5. *Caspase substrate specificity and background*: While the PhiPhiLux substrates seem reasonably specific for their target caspases, no synthetic substrate is exclusively specific for any particular enzyme. This should be kept in mind for any assay involving specific proteolytic activity. In general, a considerable excess of substrate will encourage low levels of non-specific cleavage, increasing the non-caspase background of the assay. Titration of the substrate to the lowest concentration able to distinguish activity may be necessary when the specificity of the assay is in doubt.

6. *Annexin V*: Calcium and magnesium are critical for annexin V binding; even brief removal of divalent cations after the binding reaction will result in rapid dissociation from PS residues. The cells must therefore remain in a calcium/magnesium buffer at all stages up to analysis, including all wash buffers.

7. *Incubation periods and sample storage*: All incubation periods and conditions are critical parameters for this assay, as is prompt analysis of samples following the labeling procedure. Insufficient incubation time for the PhiPhiLux substrates will result in poor labeling; prolonged incubation periods will increase the level of non-specific substrate binding and cleavage, resulting in high background fluorescence and decreased signal-to-noise ratios. In addition, prolonged storage of cells following removal of the surrounding PhiPhiLux substrate will eventually result in leakage of the cleaved substrate from the cell, despite its reduced cell permeability in the cleaved state. Overly long annexin V incubation periods will also increase the amount of non-specific binding to cells, making discrimination of "viable" and apoptotic cells more difficult. Although PI (and to a lesser extent 7-AAD) are relatively impermeant to viable cells, prolonged incubation will cause uptake even in healthy cells. If laboratory conditions do not allow prompt analysis of sample, cell-death assays involving fixed cells (such as TUNEL assays or immunolabeling of active caspases) should be considered as alternatives.

8. *Simultaneous immunophenotyping of "viable" and early apoptotic cells*: The protocol described in this chapter is readily amenable to the incorporation of antibody immunophenotyping along with the cell-death markers, resulting in a very sophisticated "screening out" of dead cells for measurement of receptor expression in "viable cells". A potentially exciting extension of this method would appear to be the phenotyping of early apoptotic cells, positive for caspase expression, but negative for later markers. This method should be approached with care; from a cellular standpoint, caspase activation is probably not an "early" event in cell death, and many

alterations in the plasma membrane may have occurred by this timepoint, resulting in aberrant antibody binding to cells as is observed in later cell death. Any cell surface marker expression results obtained by such methodology should be therefore be interpreted with caution.

9. *Pleiotrophy in apoptosis*: Apoptosis is a highly variable process involving multiple biochemical pathways; therefore, there are no universal morphological or physiological characteristics that are common to apoptosis in all cell types. Cell death in different cell types (even in physiologically or morphologically similar ones) may present very different phenotypes, and may not necessarily be detectable by the same assays. Multiparametric assays for apoptosis are very amenable to the nature of apoptosis, since the investigator is not limited to one characteristic of cell death. Investigators should also be willing to try other apoptotic assays to fully characterize their particular system.

Acknowledgments

The authors wish to acknowledge Veena Kapoor and Nga Tu Voong of the National Cancer Institute for excellent technical assistance, and Dr. Z. Darzynkiewicz of the New York Medical College for helpful discussion. Bill Godfrey, Jolene Bradford, and Gayle Buller at Invitrogen Life Technologies (formerly Molecular Probes) provided valuable technical information regarding fluorescent probes. Parts of this work were supported by intramural research fund provided by the Center for Cancer Research, National Cancer Institute.

References

1. Telford, W. G., King, L. E., and Fraker, P. J. (1994) Rapid quantitation of apoptosis in pure and heterogeneous cell populations using flow cytometry. *J. Immunol.* **172**, 1–16.

2. Darzynkiewicz, Z., Juan, G., Li, X., Gorczyca, W., Murakami, T., and Traganos, F. (1997) Cytometry in cell necrobiology: analysis of apoptosis and accidental cell death (necrosis). *Cytometry* **27**, 1–20.

3. Del Bino, G., Darzynkiewicz, Z., Degraef, C., Mosselmans, R., and Galand, P. (1999) Comparison of methods based on annexin V binding, DNA content or TUNEL for evaluating cell death in HL-60 and adherent MCF-7 cells. *Cell Prolif.* **32**, 25–37.

4. Vermes, I., Haanen, C., and Reutelingsperger, C. (2000) Flow cytometry of apoptotic cell death. *J. Immunol. Methods* **243**, 167–190.

5. Pozarowski, P., Grabarek, J., and Darzynkiewicz, Z. (2003) Flow cytometry of apoptosis, in *Current Protocols in Cytometry* (Robinson, J. P. et al., eds.), John Wiley and Sons, New York, NY, pp. 18.8.1–18.8.34.

6. Henkart, P.A. (1996) ICE family proteases: mediators of all cell death? *Immunity* **14**, 195–201.

7. Ormerod, M. G., Sun, X.-M., Snowden, R. T., Davies, R., Fearhead, H., and Cohen, G. M. (1993) Increased membrane permeability in apoptotic thymocytes: a flow cytometric study. *Cytometry* **14**, 595–602.

8. Castedo, M., Hirsch, T., Susin, S. A., Zamzami, N., Marchetti, P., Macho, A., and Kroemer, G. (1996) Sequential acquisition of mitochondrial and plasma membrane alterations during early lymphocyte apoptosis. *J. Immunol.* **157**, 512–521.

9. Green, D. R. and Reed, J. C. (1998) Mitochondria and apoptosis. *Science* **281**, 1309–1312.

10. Overbeek, R., Yildirim, M., Reutelingsperger, C., and Haanen, C. (1998) Early features of apoptosis detected by four different flow cytometry assays. *Apoptosis* **3**, 115–120.

11. Earnshaw, W. C., Martins, L. M., and Kaufmann, S. H. (1999) Mammalian caspases: structure, activation, substrates and functions during apoptosis. *Ann. Rev. Biochem.* **68**, 383–424.

12. Koester, S. K. and Bolton, W. E. (2001) Cytometry of caspases. *Methods Cell Biol.* **63**, 487–504.

13. Gorman, A. M., Hirt, U. A., Zhivotovsky, B., Orrenius, S., and Ceccatelliu, S. (1999) Application of a fluorimetric assay to detect caspase activity in thymus tissue undergoing apoptosis in vivo. *J. Immunol. Methods* **226**, 43–48.

14. Belloc, F., Belaund-Rotureau, M. A., Lavignolle, V., Bascans, E., Braz-Pereira, E., Durrieu, F., and Lacombe, F. (2000) Flow cytometry of caspase-3 activation in preapoptotic leukemic cells. *Cytometry* **40**, 151–160.

15. Bedner, E., Smolewski, P., Amstad, P., and Darzynkiewicz, Z. (2000) Activation of caspases measured in situ by binding of fluorochrome-labeled inhibitors of caspases (FLICA); correlation with DNA fragmentation. *Exp. Cell Res.* **260**, 308–313.

16. Komoriya, A., Packard, B. Z., Brown, M. J., Wu, M. L., and Henkart, P.A. (2000) Assessment of caspase activities in intact apoptotic thymocytes using cell-permable fluorogenic caspase substrates. *J. Exp. Med.* **191**, 1819–1828.

17. Telford, W. G., Komoriya, A., and Packard, B. Z. (2002) Detection of localized caspase activity in early apoptotic cells by laser scanning cytometry. *Cytometry* **47**, 81–88.

18. Lazebnik, Y., Kaufmann, S. H., Desnoyers, S., Poirier, G. G., and Earnshaw, W. C. (1994) Cleavage of poly(ADP-ribose) polymerase by proteinase with properties like ICE. *Nature* **371**, 346–347.

19. Packard, B. Z., Topygin, D. D., Komoriya, A., and Brand L. (1996) Profluorescent protease substrates: intramolecular dimers described by the exciton model. *Proc. Natl. Acad. Sci.U S A* **93**, 11640–11645.

20. Kamentsky, L. A., Burger, D. E., Gershman, R. J., Kametsky, L. D., and Luther, E. (1997) Slide-based laser scanning cytometry. *Acta Cytol.* **41**, 123–143.

21. Packard, B. Z., Komoriya, A., Brotz, T. M., and Henkart, P. A. (2001) Caspase 8 activity in membrane blebs after anti-Fas ligation. *J. Immunol.* **167**, 5061–5066.

22. Smolewski, P., Bedner, E., Du, L., Hsieh, T.-C., Wu, W. M., Phelps, D. J., and Darzynkiewicz, Z. (2001) Detection of caspases activation by fluorochrome-labeled inhibitors: multiparameter analysis by laser scanning cytometry. *Cytometry* **44**, 73–82.

Chapter 11

Multiparameter Cell Cycle Analysis

James W. Jacobberger, R. Michael Sramkoski, and Tammy Stefan

Abstract

Cell cycle-related cytometry and analysis is an essential experimental paradigm for the cell biology of yeast, mammalian, and drosophila cells. Methods have not changed much for many years. The most common is DNA content analysis, which has been well-published and reviewed. Next most common is analysis of 5-bromo-2-deoxyuridine (BrdU) incorporation, detected by specific antibodies – also well-published and reviewed. A new measurement approach to S phase labeling utilizes 5'-ethynyl-2'-deoxyuridine (EdU) incorporation and a chemical reaction to label substituted DNA. The approach is new, but published work indicates that it is equivalent to BrdU incorporation. Finally, multiple antibody labeling to detect epitopes on cell cycle-regulated proteins is the most complex of the cytometric cell cycle assays, requiring knowledge of the chemistry of fixation, the biochemistry of antibody–antigen reactions, and spectral compensation. Because all of this knowledge is relatively well presented, methodologically, in many papers and reviews, this chapter presents a bare-bones Methods section for one mammalian cell type and an extended Notes section, focusing on aspects that are problematic or not well described in the literature.

Key words: Cell division cycle, Mitosis, Antibodies, Intracellular antigens, Cyclins, Cell states

1. Introduction

1.1. Background

For yeast and drosophila cell lines, most published work is simple in cytometric terms, usually using DNA binding or other dyes in single or two color assays, e.g., live/dead assays using Hoechst 33342 and PI (see Subheading 1.2 for abbreviations). See ref. (1) for an interesting application of imaging flow cytometry for cell cycle analysis of budding yeast. This chapter refers only to mammalian cells.

Four common reasons for cytometric cell cycle analysis are to determine the fraction of proliferating cells, the fractions of cells in kinetic compartments (phases, subphases, and states), compartment transit times, or the cell cycle-related expression of a

Teresa S. Hawley and Robert G. Hawley (eds.), *Flow Cytometry Protocols*, Methods in Molecular Biology, vol. 699, DOI 10.1007/978-1-61737-950-5_11, © Springer Science+Business Media, LLC 2011

biochemical activity. In the same order, examples of each modality are: (1) stimulated lymphocyte assays (2, 3), dye labeling dilution (4), and proliferation antigen detection (5, 6); (2) S phase fraction analysis of tumors and hematopoietic malignancies (7, 8); (3) G_1 timing (9–11); and (4) SV50 large T antigen and cyclin B1 expression (12–14).

Simple cell cycle analysis consists of DNA content measurements. By itself, this is insufficient for measuring the fraction of proliferating cells or complete phase fraction analysis. Quiescent or G_0 cells have the same DNA content as G_1 cells, and when assaying cell populations that contain nonproliferating and proliferating cells, the best that can be achieved by DNA content measurements is to obtain a proliferation index by quantifying the S phase fraction. A better assay is to measure the S fraction of labeled cells after continuous labeling over some time period beyond the expected length of the G_1 phase (thus, all cycling cells will be labeled). Equally, the 4C fraction can be composed of G_2, M, and endoreduplicating or binucleate G_1 cells from a subpopulation cycling from 4C 8C. A common mistake in literature describing drug and radiation treatments that perturb G_2 and M transit is either failure to realize or failure to communicate that the increased fraction of "G_2 + M" cells can contain a substantial fraction of 4C G_1 cells.

The limitations of DNA content measurements were overcome over a 20-year period from 1985 through 2005. During this time, the value of monoclonal antibodies as specific, quantitative probes for epitope expression in fixed cells and tissues became clear. The entire approach rests on one caveat that is difficult to validate – that is, that the probe detects an unbiased fraction of the epitope. Antibody assays are always performed as a function of something else. If the fraction of epitope that is available for reaction with the antibody is changing as a function of "that something else," then the assay is subject to misinterpretation, unless that changing availability can be evaluated or measured. In lieu of direct validation, the caveat is supported by the large body of work that leads to the same answer whether the assay is done in tissue, whole cells, fractionated cells, or cell lysates. The level of possible "masking" (a description of unavailable epitope) follows the order tissue, whole cells, etc., therefore, the agreement between, e.g., western blots and cytometry is relatively powerful validation.

After the caveat, the next weakness of the approach is the variability in antibody specificity and affinity. Before cytometric or microscopic assays can be relied on, the antibodies involved need to be validated. There are many published examples wherein this was not done. For a critical analysis of one group of widely used antibodies to the p53 protein see ref. (15).

The methods for validating an antibody are not standard and not established. In 2009, many commercial antibody catalog sites still do not present validating data. For those that do (of which we are familiar), the data are never quantitative, always visual, always anecdotal, never rigorous, and not universally applied – i.e., they sell some with and some without evidence of validation. One assumes that the most common method of screening mono-clonal antibodies is by ELISA using the purified antigen. This may be fine for producing antibodies that work for immunoblots that get around cross-reactions peptide/protein fractionation, but for cell-based assays in which specific and nonspecific reac-tions are either integrated (as in flow cytometry) or only crudely differentiable (i.e., by low resolution localization), it is not suffi-cient for validation. In cytometric assays, the quality of a reaction is defined by the antigen–antibody avidity versus nonspecific binding and cross-reactive binding to other cell constituents. It is avidity that matters because the increased probability that one or the other antigen combining site is bound significantly reduces the antibody off rate, even though most of the antibody is most likely bound monovalently at any one time. For epitopes on infre-quent cells in heterogeneous samples like blood and bone mar-row, the ratio of epitope to nonspecific binding sites is very small, therefore, the antibodies should be tested and staining optimized in these samples. Obviously, this level of assay-specific validation and optimization would be difficult and costly for commercial enterprises to perform, therefore, it is left to the investigator to validate antibodies when using them in previously unvalidated circumstances.

Our approach to validation is the following. First, we prefer-entially choose antibodies from companies that we trust and that provide some evidence for specificity. Generally, at a minimum, that means a western blot of whole cell lysates for positive and negative cells with the full molecular weight range displayed. If the epitope is localized in the cell, and/or modulated by drug treatment, evidence of correct localization and/or modulation will lead us to choose one antibody over another. After choosing an antibody, we perform both immunoblot and flow cytometric titers with negative and positive cell samples (if a negative cell source is unavailable, we default to a negative control with a secondary antibody). We design these with sufficient concentra-tion points to generate a curve so that we can evaluate signal to noise (16, 17). Third, we try to obtain a biological test – siRNA knockdown, gene transfer, virus infection, cytokine stimulation, drug treatment, etc. Fourth, we either perform fluorescence microscopy or laser scanning cytometry (18) to make some check on localization. The working concentrations are defined by the cytometric titration. If our endpoint analysis is laser scanning cytometry, then we retiter by twofold dilutions around the con-centration determined by flow cytometry. This is because we work with higher staining volumes and volume matters to the signal to noise ratio (19).

Unlike immunophenotyping, most of the interesting epitopes for cell cycle analysis are inside the cell. This means cells have to be stabilized (fixed) – proteases, nucleases, transporters, channels, and other active molecules need to be inactivated, and the cell needs to be made permeable to large molecules. This has been reviewed extensively (20–29). Briefly, in our opinion, there are two basic modalities. The first uses denaturation and begins with formaldehyde fixation sufficient to stabilize cells, and then is followed by alcoholic dehydration. For epitopes that are sensitive to formaldehyde, the formaldehyde step can be omitted because the denaturation process inactivates all the activities that formaldehyde inactivates. The second modality is nondenaturing, using formaldehyde followed by nonionic or zwitterionic detergent, or saponin. In this case, the formaldehyde is not dispensable because it also serves to cross-link molecules, creating a matrix through which large molecules diffuse slowly – thus, allowing staining and measurement of even soluble epitopes. Nothing used in this latter process efficiently denatures large molecules, and therefore, this is a native state system. In both processes, some molecules are extracted rather completely, some partially, some displaced. If the target epitope is involved in tight binding to other molecules, denaturation may be required to "unmask" the epitope. If enough formaldehyde is used for a long enough time, it is possible to make penetration of antibodies difficult, and it is possible to promote "masking" relative to cells fixed with less formaldehyde. In all of this, there are many variables that can be played with. For example, different salts can be used to differentially extract molecules during the fixation step (30). Another example is to permeabilize with detergent first then fix with formaldehyde and alcohol (6). This removes loosely bound proteins and other molecules. Use of the latter protocol gives a pattern of PCNA staining coupled with DNA staining that identifies S phase better than DNA staining alone. Two elegant studies using this approach are the exploration of new cell cycle states defined by correlated analysis of Ki-67 and PCNA (31) and correlated analysis of Mcm-6 and PCNA (32).

In our opinion, there have only been two advances in the development of fixation/permeabilization methodology in recent years. The first is described in a paper by Chow et al. (33) that identifies an alcoholic denaturing procedure that works on whole blood or bone marrow, leaving light scatter patterns and surface staining intact enough to identify the major subpopulations by standard immunophenotyping procedures. The second has a similar goal and is described in a patent awarded to Keith Shults and uses heat as the denaturant (34, 35). A third effort, while not development per se, is worth mentioning: Krutzik and Nolan did a careful analysis of several fixation/permeabilization variants and arrived at formaldehyde followed by MeOH as the overall best general approach to phosphoepitopes (36).

To see a recent short review of cell cycle analysis by cytometry, see ref. (37). In the following sections, a single protocol is presented to identify cell cycle states in mitosis, using one cell line. The variations on this same theme can be extended indefinitely by multiple staining for the many epitopes that play an important role in cell cycle regulation, and therefore, create interesting patterns in multiparameter cytometric assays. For other markers of mitosis, see refs. (38, 39). Additionally, for a very large number of very high quality, specific cell cycle assays, one could start in no better place than to consult the work of Zbigniew Darzynkiewicz, Frank Traganos, and their colleagues.

1.2. Abbreviations

Abbreviations used in this chapter are: *A488* Alexa Fluor 488, *A647* Alexa Fluor 647, *APC/C* anaphase promoting complex/cyclosome, *bp* band pass, *C* the genome complement, *DAPI* 4′,6-diamidino-2-phenylindole, *FSC* forward scatter, *PE* phycoerythrin, *PHH3* phospho-S10-histone H3, *PI* propidium diiodide, *FITC* fluorescein isothiocyanate, *SSC* side scatter, *UV* ultraviolet.

2. Materials

2.1. Cell Culture

1. Molt4 cells (ACC 362 from DSMZ, Braunschweig, Germany; or CRL-1582 from ATCC, Manassas, VA). These are an easily grown human T cell leukemia line with a single stem line, low 4C→8C subpopulation. Cells are grown in suspension.

2. RPMI-1640 Medium supplemented with antibiotics and 10% fetal bovine serum (FBS) from any reputable supplier (see Note 1).

3. Tissue culture dishes, flasks, or multiwell plates, any size or type provided they hold more than 2 mL of media.

2.2. Biochemicals and Reagents

1. Formaldehyde 16%, methanol free, Ultra Pure (Polysciences, Inc., Warrington, PA) (see Note 2).

2. Methanol, spectrophotometric grade, >99% (Sigma-Aldrich, St. Louis, MO).

3. Phosphate-buffered saline (PBS): 150 mM NaCl, 10 mM phosphate, Na counter ion, pH 7.4; made in deionized water; 0.2 μm filtered (biochemicals from any reputable supplier).

4. PBS–BSA: Bovine serum albumin – Fraction V (Sigma-Aldrich) at 2% (wt/vol) in PBS.

5. Antibodies reactive with: phospho-S10-histone H3-A488 (#9708; Cell Signaling Technology, Waverly, MA; see Note 3), cyclin A2-PE (Beckman Coulter, Inc., Brea, CA; see Note 4),

cyclin B1-A647 (clone GNS1, #554176; BD Biosciences, San Jose, CA; see Note 4).

6. DAPI stock solution 1 mg/mL in double distilled water (Invitrogen); working solution is diluted to 1 μg/mL in PBS.

2.3. Labware

1. Adjustable pipettors (e.g., Pipetman Gilson, Middleton, WI) and 1–20 and 100–1,000 μL; tips (any reputable supplier); glass or disposable pipettes (1, 5, 10, 25 mL).

2. Humidified CO_2 incubator Biosafety hood Class II.

3. Electrical impedence particle counting instrument (Beckman Coulter) or hemocytometer (Hausser Scientific, Horsham, PA).

4. Phase contrast inverted microscope with 10, 20, and 40× lenses (Olympus, Nikon, Leica or Zeiss).

5. Microfuge tubes: any reputable supplier. Snap-cap tubes for assays that are processed within a few days to weeks. If fixed samples are to be stored for weeks to months, then tubes with rubber o-rings and screw-cap tops are required (otherwise, the samples will dry out). Tubes should be polypropylene.

6. Microfuge: Variable speed, swinging out rotor, set at one-third max (see Note 5).

7. Suction device for removing supernatants. We use house vacuum hooked to a side arm flask with pasteur pipette hooked to rubber tubing hooked to a glass tube through a cork in the top of the side arm flask.

8. Flow cytometer with UV or violet, blue, and red lasers.

3. Methods

3.1. Culture

1. In the example here, cells are growing exponentially as a suspension. Molt4 cells do not adhere to the dish, but many hematopoietic cells lines do adhere loosely. Gentle pipetting is used to remove these cells.

2. To obtain an exponential culture, Molt4 cells should be serially passaged with splits of 1–2 every 2–3 days if cells are approaching their upper density of 2×10^6/mL. Cultures should be split before reaching 2×10^6/mL otherwise cells begin to die as the culture becomes denser.

3. Examine with phase contrast inverted microscope. Cells should be free floating single cells. To obtain an even cell suspension, repeatedly pipette the culture. Pipette cells into a large test tube (15 or 50 mL) for cell counts (necessary if pooling flasks/dishes for large numbers of cells) and/or volume adjustments (see step 4).

4. Count the cells. Adjust cell concentration to 2×10^6 either by adding media or centrifuge to concentrate and then add media for correct density.

5. Dispense aliquots of 2×10^6 cells to 1.5 mL microfuge tubes. This works for antigens that are not labile during processing. If they are (e.g. some phospho epitopes), then formaldehyde is added directly to the tissue culture vessel or the collection test tube prior to counting the cells in a sufficient quantity (generally, 0.125 – 1%) to inhibit changes in the epitope. After addition of formaldehyde, the protocol is the same as what follows.

3.2. Fixation

1. Move to cold room. All the lab procedures after cells are removed from the incubator and counted are in a cold room at 4°C. Pipettors, pipette tips, PBS, PBS-BSA reside at 4°C in the cold room (see Note 6).

2. Pellet (centrifuge for 30 s to1 min) then wash with 1.0 mL PBS. Resuspend in 50 μL PBS.

3. Add 450 μL MeOH (stored at –20°C). At this point, samples can be stored (see Note 7).

3.3. Staining

1. Pellet cells from fixative and aspirate supernate. Wash cells twice with 1.0 mL PBS and then wash once with 0.5 mL PBS-BSA. The second wash is to begin the blocking process.

2. Resuspend pellet in 50 μL PBS-BSA containing 0.125 μg anti-cyclin A2-PE, 0.06 μg anti-cyclin B1-A647, and 0.0125 μg anti-phospho-S10-histone H3-A488 (see Note 8).

3. Incubate at 37°C for 30–90 min (see Note 9).

4. Cool to 4°C then wash two to three times (in cold room) with 0.5 mL PBS-BSA (see Note 10).

5. Resuspend in 0.5 mL of DAPI solution.

3.4. Measurement

1. Use any flow cytometer with UV or violet, blue, and red lasers. We use a BD Biosciences LSR II with a standard filter configuration. UV and Violet laser: 450/40 bp; 488 nm laser: 530/30 bp; red laser: 670/40 bp.

2. Figure 1 shows a standard set up at the BD LSR II in our laboratory.

3. Forward and side scatter are plotted and the voltage adjusted to put each on scale (Fig. 1a). Note that the black dots are mitotic cells; mitotic cells are large (about 2× bigger than the average G_1), and the correlation between forward and right angle light scatter is verified.

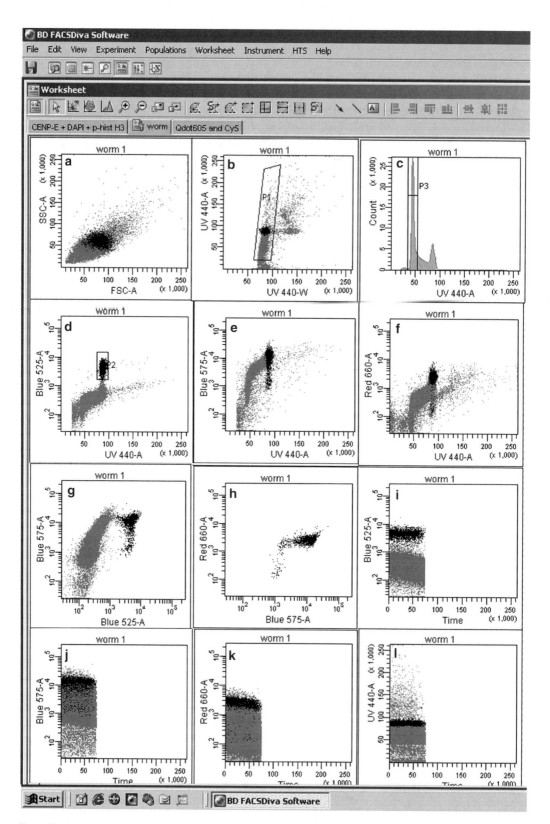

Fig. 1. Display at the instrument during data acquisition. *See* Subheading 3.4 for details. Protocol name = worm1. The histogram acronym codes are: excitation laser – band pass emission filter – signal processing. *SSC* side scatter, *FSC* forward scatter, -*A* area, -*W* width. Therefore, UV 440-W = data derived from the width of the signal from the UV laser through the 440 nm bp filter.

4. The DAPI integrated and peak or width signals at 440 nm from the UV and violet lasers are plotted to form a "doublet discriminator." Figure 1b shows the integrated and width measurements from the UV laser in the example experiment. A gate is set around the singlets (see Note 11). The purpose of this plot is to center the G_1 peak at one-fifth the integrated signal maximum scale and to clean up the subsequent histograms.

5. A single parameter plot of the integrated DAPI signal from the UV and violet lasers is plotted to check the G_1 position more precisely. Figure 1c shows the histogram for the UV laser.

6. From this point on, all histograms are plotted using the integrated signals.

7. In Fig. 1d, phospho-S10-histone H3 (PHH3) is plotted versus DAPI and a mitotic gate is set and color coded. In the example, the mitotic events are black. The PHH3 signal is set to put the G_1 cells in the first or second decade. This may be readjusted later.

8. Next, cyclin A2 versus DAPI is plotted (Fig. 1e). This is done to put the G_1 cells in the first or second decade. The G_2 and M cells should be tenfold or more intense. The pattern of negative G_1 cells and linear expression through S phase with an additional increase in G_2 appears to be universal, and therefore, we check the pattern to ensure that culture, fixation, and staining went well.

9. Next, cyclin B1 versus DAPI is plotted (Fig. 1f). This is done to put the G_1 cells in the first or second decade. The G_2 and M cells should be 10- to 15-fold more intense. The pattern of negative and positive G_1 cells and exponential increase in S at one rate and in G_2 at an increased rate also appears to be universal, and therefore, we also check that pattern (see Note 12).

10. We also plot PHH3 versus cyclin A2 (Fig. 1g). This is done primarily to fine tune the voltage settings for A488 and PE. In the example, compensation was not required, but normally, one can expect the PHH3 signal to be very bright, and for some software to work well the subtraction of A488 from PE should not be more than 100%. Therefore, if one bisects the plot with a diagonal line from the origin to the upper right corner, then the bright mitotic cells should be on the upper side of that diagonal, when the control sample is run (control is stained only with PHH3 and DAPI).

11. We also check the cyclin B1 versus cyclin A2 pattern for the mitotic cells (Fig. 1h). This also appears to be universal and looks like a clockwise rotated "L" (see Note 12).

12. Finally, we plot each fluorescence parameter versus time to determine the stability of the run. For whatever reason

(we think it is temperature change), dim signals often decrease as a function of time during the run. If it appears severe, we restart the run. If it is not severe (here it is not), then we note it. This can be corrected during analysis.

13. Generally, we set a region around a population of interest. If that population is rare, our goal is to collect at least 400 events, or as in this case, as many events as we can. The region of interest here defines the mitotic cells (Fig. 1d), and we have color coded the mitotic events black.

14. Analysis is done offline.

3.5. Analysis

1. Figure 2 shows a typical analysis. There are three things that can be obtained from these data. The first is the percentages of cells in different cell cycle phases or states. The second is the intensity of expression of the epitopes that have been measured in any phase or state, and the third is the entire cell cycle-related expression of those epitopes.

2. In Fig. 2a, the doublet discriminator has been recapitulated – this time using signal peak height rather than width, and a region has been set around the singlets (R1). All signals after this are integrated (area) values. Here, we use the integrated signal from the violet laser and the peak from the UV laser because the violet laser was apparently sharply focused but wide (integrated and peak are equivalent) and the UV laser was apparently not as well focused but narrow (integrated and peak are not equivalent). We do not know the reason, but on any given day, one laser performs better than the other.

3. The histogram in Fig. 2b (PHH3 vs. DAPI) has been gated on R1, and a region has been set around the mitotic population (R2). This is a hyperlog plot, and some of the PHH3 signal has been subtracted as a function of side scatter (not shown).

4. The histograms shown in Fig. 2c, d (cyclin A2 vs. side scatter) shows the "before" (2C) and "after" (2D) for subtraction of background from the cyclin A2 negative G_1 cells. Note that this centers the G_1 cluster around "0" cyclin A2. This is a convenient analytical bottom to the data for cyclin A2 expression. This same procedure was applied to cyclin B1 expression (not shown), and PHH3 expression in G_1. For the cyclins, this procedure is relevant to the biology since cyclins A and B are expressed below the level of detection in most of G_1. Since PHH3 expression is real in G_1, the interpretation of this is that we are setting interphase expression of PHH3 plus background to approximately zero. If the interphase levels of PHH3 were the focus of a study, we would subtract background based on a control antibody. Here, the subtraction is used to obtain cleaner plots of mitotic PHH3.

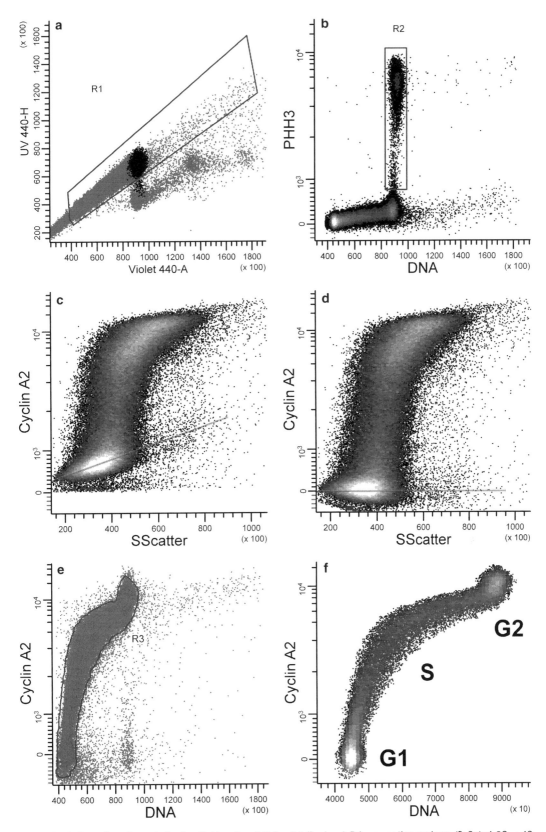

Fig. 2. Logic for cell cycle analysis. See Subheading 3.5 for details. (**a–e**) Primary gating regions. (**f**) Gated 2C→4C interphase cells, phase labeled.

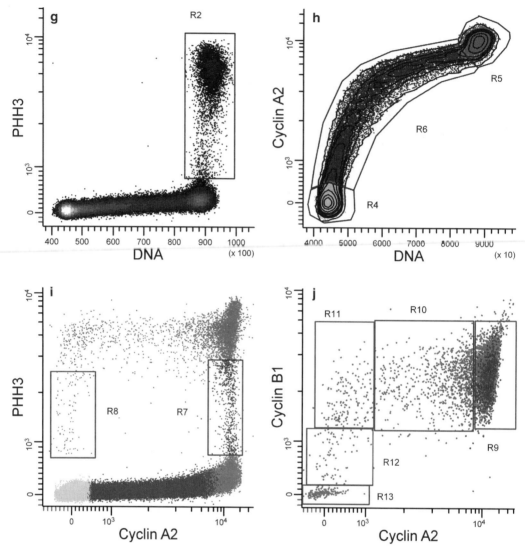

Fig. 2. (continued) (**g**) 2C → 4C complete cycle (logical OR of interphase and mitotic cells). (**h–j**) Analytical regions defining phase and state boundaries. (**k**) Color-coded three-dimensional view of the cell cycle. The parameters plotted are cyclin A2 (*X*-axis), PHH3 (*Y*-axis) and cyclin B1 (*Z*-axis). The axes, which are barely visible, are labeled with *yellow lettering*. The names for each of the cell cycle mitotic states are described in Subheading 3.5. The P1–3 designation for the early mitotic cells in which PHH3 is increasing are designated as such because we have been able to detect three states within that group using other markers. States within those states would be sublabeled with small letter additions, e.g., as has been done for LM2.

5. In Fig. 2e, cyclin A2 versus DAPI has been plotted for the interphase cells only by applying the logical gate R1 NOT R2. A contour region (R3) has been set at the 99% data boundary to exclude the 4C → 8C interphase population and 2C → 4C cells in S phase without cyclin A2. We do not know the origin of these cells, but likely they are either damaged, dying, or dead at the time of fixation.

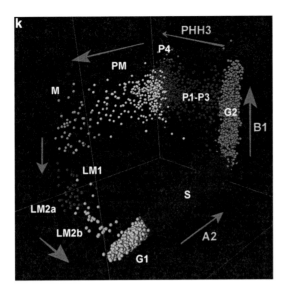

Fig. 2. (continued)

6. In Fig. 2f, the 2C→4C interphase population has been plotted in density mode with labels identifying the G_1, S, and G_2 clusters. An additional gate (R14) has been set around an expanded version of cyclin A2 versus DAPI to clean the distribution further (not shown). It amounts to shrinking R3 to 99% of 99% to eliminate outliers.

7. Figure. 2g shows PHH3 versus DAPI plotted using a logical Gate1 = (R1 AND R3 AND R14) OR R2. This has the effect of adding the 2C→4C interphase cells to the 2C→4C mitotic cells. This process has subtracted the 4C→8C cycling population, apoptotic, cells, and outliers from the analysis.

8. In Fig. 2h (cyclin A2 vs. DAPI), regions have been set around the G_1, S, and G_2 cells of the 2C→4C stemline population (gated as in Fig. 2f)

9. In Fig. 2i, PHH3 versus cyclin A2 has been plotted (using Gate1) and regions set to include the cells in which PHH3 expression was increasing (R7) at the time of fixation and those in which it was decreasing (R8) at the time of fixation.

10. Figure 2j (cyclin A2 vs. cyclin B1) shows the expression of cyclins A2 and B1 for only the 2C→4C mitotic cells (gated on R2). Note that we do not gate on R1 AND R2 because late telophase and cytokinetic cells that have not separated can register as doublets (see the black dots in Fig. 2a). The probability that R2 cells are doublets is relatively nonexistent because PHH3 positive or half-positive doublets would register at 6 and 8C. Also see ref. (40). In this plot, regions

were set on mitotic cells that express cyclins A2 and B1 maximally (R9); cells in which cyclin A2 was undergoing APC/C-mediated degradation and cyclin B1 remained at maximum at the time of fixation (R10); cells in which cyclin A2 has been degraded to background levels – essentially negative, and cyclin B1 was at maximum or beginning to be degraded by APC/C (R11); cells in which cyclin B1 was being degraded by APC/C (R12), and cells in which cyclin B1 has been degraded to background (R13). In Fig. 2i, the cells in R7 and R8 have been color-coded black. In Fig. 2j, R7 cells can be seen to be entering in R9, and R8 cells can be seen to be exiting R13. In the online pdf, color is used to distinguish R7 (blue) and R8 (red).

11. In Fig. 2k, each of the regions, R4–R13 have been color coded and a 3D plot that traces the cell cycle "worm" can be traced through the three-dimensional state space defined by the expression levels of cyclins A2 and B1 and phospho-S10-histone H3.

12. We have named the states that we have defined in mitosis **P1–3** (R7), **P4** (R9 NOT R7), **PM** (R10), **M** (R11), **LM1** (R12), **LM2a** (R13 NOT R8), and **LM2b** (R8). These correlate with prophase (P1–3 and P4), prometaphase (PM), metaphase (M), anaphase > telophase (LM1), and telophase + cytokinesis (LM2). The correlations have been verified by cell sorting and microscopy, inhibitor treatments, and laser scanning cytometry (to be published elsewhere). However, despite the correlation, these biochemical states should not be equated with the stages of mitosis, but thought of as biochemically defined transition states. For a review/commentary piece on mitotic states and stages, see ref. (41). In this example, these measurements were done for three independently processed samples from the same culture. The percentages of cells in each region are presented in Table 1 and Fig. 3. The accuracy is fairly remarkable considering that the frequencies span nearly three orders of magnitude.

13. Although we are not showing it here, the frequency of cells for each region (state) can be plotted on the X-axis, and the corresponding levels of each marker within each state plotted on the Y-axis to obtain the continuous expression of the marker as a function of time in the cell cycle. The frequency data needs to be corrected for the aging of the population (42), but even without the correction, this calculated expression profile is likely to be more accurate than that which can be achieved in the more standard, nonquantitative practice of synchronizing cells, releasing them, then measuring total expression levels as a function of time by western blot.

Table 1
Phase and state fraction analysis

Phase	Mean (%)	SD (%)	CV (%)
G_1	36	5.1	14
S	39	4.8	12
G_2	14	2.3	16
P1–3	0.53	0.11	21
P4	3	0.56	21
PM	0.68	0.15	22
M	0.11	0.023	20
LM1	0.05	0.007	15
LM2a	0.05	0.012	22
LM2b	0.07	0.011	16

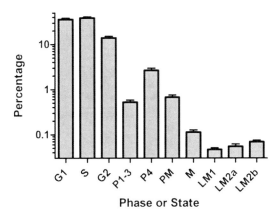

Fig. 3. Bar graph of the phase and state frequencies calculated from staining/measurement/analysis of three independent samples from the same population.

4. Notes

1. DSMZ suggests 20% FBS. These cells are derived from a leukemia and are not very sensitive to serum concentration and grow well in 10% FBS.
2. First, no one fixes cells with paraformaldehyde, which is an insoluble polymer of formaldehyde. For detailed discussions, search the Purdue University Cytometry Laboratories (PUCL) Web site (http://www.cyto.purdue.edu/) on paraformaldehyde. Second, this particular product comes sealed

in glass under nitrogen. Third, the reason to use this high quality product is that it is supplied at a known concentration (using formaldehyde produced from hydrolyzed paraformaldehyde does not provide a product with a known concentration because one isn't sure of the degree of hydrolysis or evaporation during the process); also, the "ultrapurity" helps to ensure that the least "autofluorescence" is induced in fixed cells.

3. We have used most of the phospho-S10-histone H3 antibodies (pH3) from Cell Signaling Technology (CST). The unconjugated rabbit polyclonal (#9701) works very well and provides flexibility for secondary reagents. The same is true for the conjugated versions (#9704, #9708). We haven't tried the newer rabbit monoclonal versions but expect them to perform well. The mouse monoclonal (clone 6G3) works well but does not appear to be available in conjugated form (previously, it was). The rabbit polyclonal was available in biotinylated form, which was very useful when other rabbit antibodies were being used in the assay. The rabbit monoclonal is currently available in biotinylated form (#3642). When using the A488 version with anti-cyclin A2-PE, the spectral overlap of A488 to PE is problematic. Despite compensation, this seriously degrades the cyclin A2 signal by increasing variance after compensation. It is useful to minimize this by titering the conjugated form of pHH3 to a minimum acceptable signal and substituting unconjugated antibody to keep the concentration at optimum.

4. We obtain the GNS1 clone in bulk form and conjugate it ourselves using the Alexa Fluor conjugating kits or the active dye from Invitrogen (Carlsbad, CA). We obtain the cyclin A2-PE antibody as a gift from Vince Shankey at Beckman Coulter. An equivalent cyclin A antibody is available from BD Biosciences (clone BF683). We have done head to head comparison with the Beckman Coulter antibody using unconjugated antibodies in secondary reactions.

5. We use a Fisher Scientific Microcentrifuge (Model 59A). This centrifuge is 24 years old, and as far as we can tell is not available. Cole Parmer (Vernon Hills, IL) lists a swing out rotor (EW-17405-76) that works with Hermle microcentrifuges. Any microfuge centrifuge can be used. The value of this one is that it can be set to spin at very low speed in increments of minutes, or manually turned on and off to spin for seconds. The combination of low speed and swing out rotor is that the cells are pelleted gently and are easily resuspended. This helps for cell recovery and reducing clumping.

6. The value of a cold room is that MeOH fixed cells stick to plastic surfaces and cold inhibits this. Working in the cold

room is ideal (except for human comfort). If one is not available, this can be done working at room temperature or on ice with room temperature equipment.

7. Fixed cells in MeOH can be stored for years. However, there is loss of reactivity over time. We believe that loss of epitope reactivity and loss of masking come to equilibrium at about half the available epitope. For antigens, we have checked, the patterns of expression are the same (43). This is further evidence that the epitopes are exposed in an unbiased manner. They can be stored for days to weeks without concern. Storing at −80°C retards epitope loss.

8. CST has not printed the antibody concentration or amount on their bottles of antibodies in the past. Previously, we phoned to get the concentration from them. The antibodies are pretitered at CST and they sell them as a number of reactions. It is a major improvement in policy that they are now labeling their antibodies with concentrations. The way we optimize is to determine the antibody to target ratio – the target in this case being the amount of antigen in a positive cell line, averaged over two million cells and the antibody defined in nano- or micrograms. The assay is to react the cells and the antibody in the smallest feasible volume. This produces optimal staining in terms of signal to noise ratio (19). When manufacturers sell us antibodies at dilute concentrations, they force upon us reaction volumes that are not optimal.

9. In terms of signal to noise ratio, a longer staining time (90 min) is better. However, for routine purposes, 30 min is sufficient to achieve good results. Depending on the desired output, the time can be shorter or longer. Nonspecific antibody diffuses into MeOH fixed cells to an equilibrated higher concentration relative to the outside of the cell in 5 min. Specific antibody achieves approximate equilibrium staining (reaches an asymptotic cusp) in 15 min (20, 28). Staining can also be done at 4°C or room temperature. The reason we stain at 37°C is that antibodies develop in vivo at body temperature (39°C for mice and rabbits).

10. Three washes are better than two, but two are sufficient for routine purposes.

11. See the chapter by Peter Rabinovitch or books by Alice Givan or Howard Shapiro for descriptions of how doublet discrimination works (44–46). Essentially, identify the G_1 cells, then set a gate with constant slope the width of the G_1 cells, widening it out as a function of intensity to ~2× the width of the G_1 population at 4C and ~4× at 8C, etc.

12. Unscheduled expression of cyclins has been described (47) and is defined as patterns of expression wherein the cyclin is

expressed higher than it would be in normal cells relative to the next earliest expression state. For example, high expression of cyclin B1 in G_1 would be "unscheduled." Villiard et al. went to great lengths to show that cyclin B1 was expressed in G_1 cells of human T cell lymphoma lines but either not expressed or expressed at a lower level in normal human T cells (48). These levels were small but significant. We have measured cyclin B1 in G_1 in all cells that we have examined with GNS1-A647. We have not observed what we would consider large variation in cyclin A2 and B1 expression patterns in human solid tumor cell lines, hTert-immortalized human cell lines, K562, Molt4, and normal human T cells. However, we have not tested the cell lines displaying unscheduled expression presented by Gong et al. (47) for cyclin B1. The features of the Molt4 patterns shown here describe the cyclin A2 pattern relatively universally in our experience. For cyclin B1, we see essentially two patterns, one with higher G_1 expression coupled with larger variance in the early S phase cells in one pattern relative to the other. The first pattern is typified by K562 cells and normal T cells, and the second pattern is typified by Molt4 cells. For some examples of the magnitude of expression differences for cyclin B1 for interphase, see ref. (14). For both of these proteins, there is variation in expression to be sure, but these features – (1) cyclin B1 expression starting before cyclin A2; (2) linear net synthesis of cyclin A2 and nonlinear synthesis of cyclin B1; and (3) degradation of cyclin B1 after cyclin A2, seems to be universal in our experience. In view of our experience, the unscheduled expression of cyclin B1 appears to be largely confined to the differing overall levels of expression and different levels of expression in G_1 (and perhaps differing fractions of cells expressing at the higher levels) rather than different patterns of expression. We have not explored unscheduled expression under conditions of growth imbalance, which has been studied by Gong et al. (49) for Molt4 cells. In this case, each of the cyclins D3, E, A, and B1 displayed unscheduled expression as defined above in cultures synchronized with mimosine or double thymidine block. The abnormalities in cyclin A expression were especially striking after mimosine treatment.

Acknowledgments

This work was supported by grants from NCI, R01CA73413 to JWJ and P30CA43703 to Stan Gerson, which supports the Cytometry and Imaging Microscopy Core facility in which the cytometry is performed. Additional thanks go to Vince Shankey

(Beckman Coulter), Chuck Goolsby (Northwestern University), David Hedley (Ontario Cancer Institute), and Elena Holden (Compucyte) for advice and comments on this work for many years; Sue Chow (Ontario Cancer Institute) on fixation and staining, and Bruce Bagwell and Ben Hunsberger (Verity Software House) for many innovations in WinList that facilitate the analytical part of the work.

References

1. Calvert, M. E., Lannigan, J. A., Pemberton, L. F. (2008) Optimization of yeast cell cycle analysis and morphological characterization by multispectral imaging flow cytometry. *Cytometry A* **73**, 825–33.

2. Crissman, H. A. and Tobey, R. A. (1974) Cell-cycle analysis in 20 minutes. *Science* **184**, 1297–8.

3. Fattorossi, A., Battaglia, A., and Ferlini, C. (2001) Lymphocyte activation associated antigens, in *Cytometry, Part A, Third Edition* (Darzynkiewicz, Z., Crissman, H. A., and Robinson, J. P., eds.), Academic Press, San Diego, CA, *Methods Cell Biol* **63**, 433–63.

4. Lyons, A. B., Hasbold, J., and Hodgkin, P. D. (2001) Flow cytometric analysis of cell division history using dilution of carboxyfluorescein diacetate succinimidyl ester, a stably integrated fluorescent probe, in *Cytometry, Part A, Third Edition* (Darzynkiewicz, Z., Crissman, H. A., and Robinson, J. P., eds.), Academic Press, San Diego, CA, *Methods Cell Biol* **63**, 375–98.

5. Endl, E., Hollmann, C., and Gerdes, J. (2001) Antibodies against the Ki-67 protein: assessment of the growth fraction and tools for cell cycle analysis, in *Cytometry, Part A, Third Edition* (Darzynkiewicz, Z., Crissman, H. A., and Robinson, J. P., eds.), Academic Press, San Diego, CA, *Methods Cell Biol* **63**, 399–418.

6. Larsen, J. K., Landberg, G., and Roos, G. (2001) Detection of proliferating cell nuclear antigen, in *Cytometry, Part A, Third Edition* (Darzynkiewicz, Z., Crissman, H. A., and Robinson, J. P., eds.), Academic Press, San Diego, CA, *Methods Cell Biol* **63**, 419–31.

7. Ault, K. A., Bagwell, C. B., Bartels, P. H., Bauer, K. D., Beckmann, E., Braylan, R. C., Carter, W. O., Clevenger, C. V., Cornelisse, C. J., Crissman, J. D. et al. (1993) *Clinical Flow Cytometry* (Bauer, K. D., Duque, R. E., and Shankey, T. V., eds.), Williams & Wilkins, Baltimore, MD, p. 635.

8. Hedley, D. W., Shankey, T. V., and Wheeless, L. L. (1993) DNA cytometry consensus conference. *Cytometry* **14**, 471.

9. Sladek, T. L. and Jacobberger, J. W. (1992) Simian virus 40 large T-antigen expression decreases the G1 and increases the G2 + M cell cycle phase durations in exponentially growing cells. *J Virol* **66**, 1059–65.

10. DiSalvo, C. V., Zhang, D., and Jacobberger, J. W. (1995) Regulation of NIH-3T3 cell G1 phase transit by serum during exponential growth. *Cell Prolif* **28**, 511–24.

11. Zhang, D. and Jacobberger, J. W. (1996) TGF-beta 1 perturbation of the fibroblast cell cycle during exponential growth: switching between negative and positive regulation. *Cell Prolif* **29**, 289–307.

12. Jacobberger, J. W., Sramkoski, R. M., Wormsley, S. B., and Bolton, W. E. (1999) Estimation of kinetic cell-cycle-related gene expression in G1 and G2 phases from immunofluorescence flow cytometry data. *Cytometry* **35**, 284–9.

13. Frisa, P. S., Lanford, R. E., and Jacobberger, J. W. (2000) Molecular quantification of cell cycle-related gene expression at the protein level. *Cytometry* **39**, 79–89.

14. Frisa, P. S. and Jacobberger, J. W. (2009) Cell cycle-related cyclin B1 quantification. *PLoS One* **4**, e7064.

15. Bonsing, B. A., Corver, W. E., Gorsira, M. C., van Vliet, M., Oud, P. S., Cornelisse, C. J., and Fleuren, G. J. (1997) Specificity of seven monoclonal antibodies against p53 evaluated with Western blotting, immunohistochemistry, confocal laser scanning microscopy, and flow cytometry. *Cytometry* **28**, 11–24.

16. Jacobberger, J. W., Sramkoski, R. M., Zhang, D., Zumstein, L. A., Doerksen, L. D., Merritt, J. A., Wright, S. A., and Shults, K. E. (1999) Bivariate analysis of the p53 pathway to evaluate Ad-p53 gene therapy efficacy. *Cytometry* **38**, 201–13.

17. Jacobberger, J. W., Sramkoski, R. M., Frisa, P. S., Ye, P. P., Gottlieb, M. A., Hedley, D. W.,

Shankey, T. V., Smith, B. L., Paniagua, M., and Goolsby, C. L. (2003) Immunoreactivity of Stat5 phosphorylated on tyrosine as a cell-based measure of Bcr/Abl kinase activity. *Cytometry A* **54**, 75–88.

18. Kamentsky, L. A. and Kamentsky, L. D. (1991) Microscope-based multiparameter laser scanning cytometer yielding data comparable to flow cytometry data. *Cytometry* **12**, 381–87.

19. Srivastava, P., Sladek, T. L., Goodman, M. N., and Jacobberger, J. W. (1992) Streptavidin-based quantitative staining of intracellular antigens for flow cytometric analysis. *Cytometry* **13**, 711–21.

20. Jacobberger, J. W. (1989) Cell cycle expression of nuclear proteins, in *Flow Cytometry: Advanced Research and Applications, Volume 1* (Yen, A., ed.), CRC Press, Boca Raton, FL, pp. 305–326.

21. Jacobberger, J. W. (1991) Intracellular antigen staining: quantitative immunofluorescence. *Methods* **2**, 207–218.

22. Clevenger, C. V. and Shankey, T. V. (1993) Cytochemistry II: Immunofluorescence measurement of intracellular antigens, in *Clinical Flow Cytometry, First edition* (Bauer, K. D., Duque, R. E., and Shankey, T. V., eds.), Williams & Wilkins, Baltimore, MD, pp. 157–75.

23. Bauer, K. D. and Jacobberger, J. W. (1994) Analysis of intracellular proteins. *Methods Cell Biol* **41**, 351–76.

24. Camplejohn, R. S. (1994) The measurement of intracellular antigens and DNA by multiparametric flow cytometry. *J Microsc* **176**, 1–7.

25. Jacobberger, J. W. (2000) Flow cytometric analysis of intracellular protein epitopes, in *Immunophenotyping, Cytometric Cellular Analysis* (Stewart, C. A. and Nicholson, J. K. A., eds), Wiley-Liss, Inc., New York, NY, pp. 361–405.

26. Koester, S. K. and Bolton, W. E. (2000) Intracellular markers. *J Immunol Methods* **243**, 99–106.

27. Koester, S. K. and Bolton, W. E. (2001) Strategies for cell permeabilization and fixation in detecting surface and intracellular antigens. *Methods Cell Biol* **63**, 253–68.

28. Jacobberger, J. W. (2001) Stoichiometry of immunocytochemical staining reactions. *Methods Cell Biol* **63**, 271–98.

29. Jacobberger, J. W. and Hedley, D. W. (2001) Intracellular measures of signalling pathways, in *Cytometric Analysis of Cell Phenotype and Function* (McCarthy, D. A. and Macey, M.

G., eds.), Cambridge University Press, Cambridge, UK.

30. Bruno, S., Gorczyca, W., and Darzynkiewicz, Z. (1992) Effect of ionic strength in immunocytochemical detection of the proliferation associated nuclear antigens p120, PCNA, and the protein reacting with Ki-67 antibody. *Cytometry* **13**, 496–501.

31. Landberg, G., Tan, E. M., and Roos, G. (1990) Flow cytometric multiparameter analysis of proliferating cell nuclear antigen/cyclin and Ki-67 antigen: a new view of the cell cycle. *Exp Cell Res* **187**, 111–8.

32. Frisa, P. S. and Jacobberger J. W. (*in revision*) Cytometry of chromatin bound mcm6 and PCNA identifies two states in G1 that are separated functionally by the restriction point.

33. Chow, S., Hedley, D., Grom, P., Magari, R., Jacobberger, J. W., and Shankey, T. V. (2005) Whole blood fixation and permeabilization protocol with red blood cell lysis for flow cytometry of intracellular phosphorylated epitopes in leukocyte subpopulations. *Cytometry A* **67**, 4–17.

34. Shults, K. E., Flye, L. A. (2008) Esoterix, Inc., assignee. Cell fixation and use in phospho-proteome screening. United States.

35. Shults, K. E., Flye, L. A., Green, L., Daly, T., Manro, J. R., and Lahn, M. (2009) Patient-derived actute myeloid leukemia (AML) bone marrow cells display distinct intracellular kinase phosphorylation patterns. *J Cancer Manag Res* **2009:1**, 1–11.

36. Krutzik, P. O. and Nolan, G. P. (2003) Intracellular phospho-protein staining techniques for flow cytometry: monitoring single cell signaling events. *Cytometry A* **55**, 61–70.

37. Darzynkiewicz, Z., Crissman, H., and Jacobberger, J. W. (2004) Cytometry of the cell cycle: cycling through history. *Cytometry A* **58**, 21–32.

38. Jacobberger, J. W., Frisa, P. S., Sramkoski, R. M., Stefan, T., Shults, K. E., and Soni, D. V. (2008) A new biomarker for mitotic cells. *Cytometry A* **73**, 5–15.

39. Darzynkiewicz, Z. (2008) There's more than one way to skin a cat: yet another way to assess mitotic index by cytometry. *Cytometry A* **73**, 386–7.

40. Gerashchenko, B. I., Hino, M., and Hosoya, H. (2000) Enrichment for late-telophase cell populations using flow cytometry. *Cytometry* **41**, 148–9.

41. Pines, J. and Rieder, C. L. (2001) Re-staging mitosis: a contemporary view of mitotic progression. *Nat Cell Biol* **3**, E3–6.

42. Bagwell, C. B. (1993) Theoretical aspects of flow cytometry data analysis, in *Clinical Flow Cytometry*, 1st edition (Bauer, K. D., Duque, R. E., and Shankey, T. V., eds.), Williams & Wilkins, Baltimore, MD, pp. 41–61.

43. Sramkoski, R. M., Wormsley, S. W., Bolton, W. E., Crumpler, D. C., and Jacobberger, J. W. (1999) Simultaneous detection of cyclin B1, p105, and DNA content provides complete cell cycle phase fraction analysis of cells that endoreduplicate. *Cytometry* **35**, 274–83.

44. Rabinovitch, P. S. (1993) Practical considerations for DNA content and cell cycle analysis, in *Clinical Flow Cytometry*, 1st edition (Bauer, K. D., Duque, R. E., and Shankey, T. V., eds.), Williams & Wilkins, Baltimore, MD, pp. 117–42.

45. Givan, A. G. (2001) *Flow Cytometry First Principles*, 2nd edition. Wiley-Liss, Inc., New York, NY, p. 273.

46. Shapiro, H. M. (2003) *Practical Flow Cytometry*, 4th edition. John Wiley & Sons, Inc., Hoboken, NJ, p. 681.

47. Gong, J., Ardelt, B., Traganos, F., and Darzynkiewicz, Z. (1994) Unscheduled expression of cyclin B1 and cyclin E in several leukemic and solid tumor cell lines. *Cancer Res* **54**, 4285–8.

48. Viallard, J. F., Lacombe, F., Dupouy, M., Ferry, H., Belloc, F., and Reiffers, J. (2000) Different expression profiles of human cyclin B1 in normal PHA-stimulated T lymphocytes and leukemic T cells. *Cytometry* **39**, 117–25.

49. Gong, J., Traganos, F., and Darzynkiewicz, Z. (1995) Growth imbalance and altered expression of cyclins B1, A, E, and D3 in Molt-4 cells synchronized in the cell cycle by inhibitors of DNA replication. *Cell Growth Differ* **6**, 1485–93.

Chapter 12

Rare Event Detection and Analysis in Flow Cytometry: Bone Marrow Mesenchymal Stem Cells, Breast Cancer Stem/Progenitor Cells in Malignant Effusions, and Pericytes in Disaggregated Adipose Tissue

Ludovic Zimmerlin, Vera S. Donnenberg, and Albert D. Donnenberg

Abstract

One of the major strengths of Flow Cytometry is its ability to perform multiple measurements on single cells within a heterogeneous mixture. When the populations of interest are relatively rare, analytical methodology that is adequate for more prevalent populations is often overcome by sources of artifacts that become apparent only when large numbers of cells are acquired. This chapter presents three practical examples of rare event problems and gives detailed instructions for preparation of single cell suspensions from bone marrow, malignant effusions, and solid tissue. These examples include detection of mesenchymal stem cells in bone marrow, characterization of cycling/aneuploid cells in a breast cancer pleural effusion, and detection and subset analysis on adipose-derived pericytes. Standardization of the flow cytometer to decrease measurement variability and the use of integrally stained and immunoglobulin capture beads as spectral compensation standards are detailed. The chapter frames rare event detection as a signal-to-noise problem and provides practical methods to determine the lower limit of detection and the appropriate number of cells to acquire. Detailed staining protocols for implementation of the examples on a three-laser cytometer are provided, including methods for intracellular staining and the use of DAPI to quantify DNA content and identify events with $\geq 2\,N$ DNA. Finally, detailed data analysis is performed for all three examples with emphasis on a three step procedure: (1) Removal of sources of interference; (2) Identification of populations of interest using hierarchical *classifier* parameters; and (3) Measurement of *outcomes* on classifier populations.

Key words: Rare-event analysis, Flow cytometry, Breast cancer, Cancer stem cells, Adult stem cells, Adipose pericytes, Bone marrow-derived mesenchymal stem cells

1. Introduction

This chapter will detail the flow cytometric identification and quantification of three rare populations: one in liquid tissue (mesenchymal stem cells in a bone marrow aspirate), one that

Teresa S. Hawley and Robert G. Hawley (eds.), *Flow Cytometry Protocols*, Methods in Molecular Biology, vol. 699,
DOI 10.1007/978-1-61737-950-5_12, © Springer Science+Business Media, LLC 2011

exists as aggregates in suspension (breast cancer stem/progenitor cells in pleural effusion), and one in solid tissue (pericytes in disaggregated adipose). Detailed protocols will be given for preparation of single cell suspensions from these sources. This will be followed by instructions for efficient staining with multiple antibodies, sample acquisition, and data analysis.

2. Materials

2.1. Supplies and Equipment

1. Polypropylene conical tubes: 15 mL, 50 mL, and 225 mL Falcon tubes.
2. 70 μm nylon and 26 mm diameter Falcon cell strainers.
3. 425 and 180 μm sieves (W.S. Tyler, Mentor, OH).
4. 1,000 mL Nalgene jars.
5. 1.5 mL Eppendorf microcentrifuge tubes.
6. Tubes for flow cytometry: 12×75 mm Falcon tubes with or without 35 μm filter cap.
7. Hemacytometer; Drummond Pipet-Aid pipetting device; 10 mL pipettes.
8. Shaking water bath (such as Bellydancer; Stovall Life Science, Inc., Greensboro, NC).
9. Flow cytometer: Beckman Coulter CyAn, Beckman Coulter Gallios (Beckman Coulter, Inc., Fullerton CA), BD LSR II (BD Biosciences, San Jose, CA), or other flow cytometer equipped with 405 nm or UV laser, blue laser and red laser.
10. Data analysis software such as VenturiOne (Applied Cytometry, Sheffield, UK) or Kaluza (Beckman Coulter).

2.2. Tissues and Cells: Minimum Tissue Requirement

1. Bone Marrow: ten million cells (0.5 mL at 20×10^6/mL) (see Note 1).
2. Solid epithelial tissue: 1 g.
3. Adipose tissue: 20 g.
4. Pleural effusion: ten million cells.

2.3. Reagents and Solutions

1. Ficoll/Hypaque (Histopaque-1077; Sigma-Aldrich, St. Louis, MO).
2. Human Serum Albumin (Albuminar-25, NDC 0053-7680-33).
3. Mouse serum and newborn calf serum.
4. Phosphate Buffered Saline without calcium or magnesium (PBS-A).
5. Staining buffer: PBS-A, 4% newborn calf serum, 0.1% (w/v) sodium azide.

6. Phosphate Buffered Saline with $CaCl_2$ and $MgCl_2$ (PBS).

7. Collagenase type I/DNase solution in PBS: 0.4% Collagenase type I (Sigma-Aldrich), 350 KU/mL DNase (Sigma-Aldrich), and 1% albumin.

8. Collagenase type II solution in PBS: 2.5 g/L Collagenase type II (Worthington Biochemical, Lakewood, NJ), and 1% albumin.

9. Ammonium chloride (NH_4Cl) lysing solution (Beckman Coulter) diluted to 1× (see Note 2).

10. EDTA buffer: PBS-A, 1 mM EDTA, and 0.1% albumin.

11. Permeabilization solution: PBS, 0.1% (w/v) saponin, and 0.5% albumin.

12. DAPI (4′,6-diamidino-2-phenylindole, dihydrochloride) stock solution: 200 μg/mL in PBS.

13. 4% Formaldehyde in hypertonic PBS-A (100 mL): 14 mL 10× PBS-A, 40 mL 10% methanol-free formaldehyde (Polysciences, Warrington, PA), and 46 mL dH2O.

14. Solutions for cell counting: Tuerk's solution and Trypan Blue.

15. Beads for monitoring flow cytometer performance: SpectrAlign beads (DAKO, Glostrup, Denmark); 8-peak Rainbow Calibration Particles (Spherotech, Lake Forest, IL).

16. Beads for compensation: Calibrite beads (unlabeled, FITC, and PE), Calibrite beads (APC); CompBeads Anti-mouse Ig, κ (all from BD Biosciences).

2.4. Antibodies

1. Anti-human antibodies for surface staining of:

   (a) *Bone marrow mesenchymal cells*: CD105-FITC (61R-CD105-DHUFT; Fitzgerald Industries International, Acton, MA), CD73-PE (550257; BD Biosciences), CD34-ECD (IM2709U; Beckman Coulter), CD90-PC5 (IM3703; Beckman Coulter), CD117-PC7 (IM3698; Beckman Coulter), CD31-APC (FAB3567A; R&D Systems, Minneapolis, MN) or CD133-APC (130-090-854; Miltenyi Biotech, Gladbach, Germany), CD45-APC-Cy7 (348805; BD Biosciences).

   (b) *Pleural fluid*: CD44-PE (MCA89PE; Serotec, Raleigh, NC), CD90-biotin (555594; BD Biosciences), Streptavidin-ECD (IM3326; Beckman Coulter), CD14-PC5 (IM2640U; Beckman Coulter), CD33-PC5 (IM2647U; Beckman Coulter), Glycophorin A-PE-Cy5 (559944; BD Biosciences), CD133-APC (Miltenyi Biotech), CD117-PC7 (IM3698; Beckman Coulter), and CD45-APC-Cy7 (BD Biosciences).

   (c) *Adipose pericytes*: CD3-FITC (IM1281U; Beckman Coulter), CD146-PE (A07483; Beckman Coulter),

CD34-ECD (IM2709U; Beckman Coulter), CD90-PC5 (Beckman Coulter), CD117-PC7 (Beckman Coulter), CD31-APC (R&D Systems), and CD45-APC-Cy7 (BD Biosciences).

2. Anti-human antibody for intracellular staining of pleural effusion cells: pan cytokeratin-FITC (IM2356; Beckman Coulter).

3. Methods

3.1. Preparation of Single Cell Suspensions

3.1.1. Bone Marrow

Bone marrow is usually obtained by aspiration from the posterior iliac crests (1). It can also be obtained from discarded surgical specimens (femoral head in hip replacements (1, 2), rib section in lung lobectomy (3), or from cadaveric vertebrae (4)). Bone marrow should be aspirated into heparinized syringes (sodium heparin, 10 U/mL) and held at ambient (not refrigerated) temperature.

1. Bone marrow mononuclear cells, depleted of mature erythrocytes and mature granulocytes, are prepared by Ficoll/Hypaque gradient centrifugation (5).

2. Dilute aspirated bone marrow or bone marrow flushed from bones to 30 mL with PBS–0.1% albumin (see Note 3). The total number of cells per tube should not exceed 50×10^6.

3. Pipette 15 mL of Ficoll/Hypaque into a 50 mL polypropylene conical tube.

4. Using a 10 mL pipette, carefully layer 30 mL of diluted bone marrow over the Ficoll/Hypaque (see Note 4). Set centrifuge to 25°C, and spin at $400 \times g$ for 45 min (brake off).

5. After centrifugation, aspirate and discard the upper layer (diluent) with a 10 mL pipette (see Note 5).

6. Using a 10-mL pipette, carefully collect the Buffy Coat and pipette into a new 50-mL conical tube (see Note 6).

7. Wash the Buffy Coat twice with 50 mL of PBS–0.1% albumin ($400 \times g$, 7 min, 25°C).

8. Resuspend in 2 mL of PBS–0.1% albumin and hold on ice.

9. Count cells and record cell concentration and volume (see Note 7).

3.1.2. Malignant Pleural Effusions

Pleural effusions are collected into plastic containers by suction. Ideally they should be heparinized (sodium heparin, 10 U/mL) because they often contain serous fluid capable of clotting. They contain cells in suspension, but tumor is usually in clumps that range from microscopic to visible. The cell count and volume vary considerably from sample to sample (see Note 8).

1. Record the volume and cell count of the unmanipulated sample (see Note 9).

2. Concentrate the cells and cell clumps by centrifugation ($400 \times g$, 7 min, 4°C).

3. Discard supernatant and resuspend the pellet in 10 mL Collagenase type I/DNase solution (see Note 10).

4. Place sample in a shaking waterbath (e.g. Bellydancer) at 37°C for 30 min, maximum agitation setting.

5. Place a 70 μm cell strainer in the mouth of a new labeled 50 mL conical tube.

6. Using a 10 mL pipette, transfer material from the collagenase digestion tube into the cell strainer.

7. Add 10 mL of PBS–0.1% albumin to the digestion tube and transfer any remaining material to the cell strainer. Discard strainer.

8. Bring volume of strained cells to 50 mL with PBS–0.1% albumin, centrifuge at $400 \times g$ for 7 min, 4°C and discard supernatant.

9. Add 45 mL of NH_4Cl lysing solution and mix (see Note 11).

10. Centrifuge at $400 \times g$, 4°C for 10 min.

11. Pour off the supernatant and loosen cell pellet.

12. Resuspend in 2 mL of PBS–0.1% albumin and hold on ice.

13. Count cells on a hemacytometer (Tuerk's solution to eliminate red blood cells (RBC) and Trypan blue for viability) (6).

3.1.3. Whole Adipose Tissue

Adipose tissue is a byproduct of aesthetic surgery. It may be removed in the form of solid tissue or lipoaspirate. This protocol assumes solid adipose tissue. The expected yield of stromal/vascular cells is approximately 1×10^6 cells/g of tissue.

1. Record the weight of adipose tissue.

2. Cut the tissue in large pieces using the sterile scissors and distribute into 50 mL conicals (approximately 10 g of fat per tube).

3. Thoroughly mince tissue in the 50 mL conical tubes using scissors.

4. Add 30 mL of Collagenase type II solution per tube.

5. Vortex the conical tube and incubate at 37°C in a shaking water bath for 15 min, maximal agitation.

6. Examine the tubes to estimate efficient digestion. The presence of excessive clumps indicates under-digestion. The appearance of a clear yellow lipid layer indicates over-digestion. Reincubate in the water bath if full digestion is not achieved. Repeat every 5 min for a maximum of 30 min total digestion time.

7. Add 10 mL of EDTA buffer to neutralize ongoing collagenase activity.

8. Centrifuge at $400 \times g$ for 10 min at 25°C.

9. Collect all semi-solid top fat layers from all tubes into 50 mL polypropylene conical tubes. Add 30–40 mL of PBS–0.1% albumin and shake thoroughly to homogenize (see Note 12).

10. Vortex the remaining contents of the digestion conical tubes and pass the contents through the 425 μm and 180 μm large sterile sieves into the 1,000 mL Nalgene jar. Use a glass pestle if necessary.

11. Wash the sieves with up to 100 mL of PBS–0.1% albumin.

12. Collect the sieved sample from the basin to 50 mL or 200 mL conical tubes.

13. Add the diluted fat from the top layer of the digest to the sieves and pass through using the glass pestle.

14. Centrifuge the sieved cells at $400 \times g$ for 7 min at 4°C.

15. Discard supernatant and combine all cell pellets in one to two 50-mL conical tubes.

16. Wash the cells with PBS–0.1% albumin (50 mL per tube).

17. Centrifuge at $400 \times g$ for 7 min at 25°C.

18. Discard supernatant and loosen the cell pellet.

19. Add 45 mL of NH_4Cl lysing solution per tube and mix.

20. Centrifuge at $400 \times g$, 4°C for 10 min.

21. Pour off the supernatant and loosen cell pellet.

22. Resuspend in 2 mL of PBS–0.1% albumin and hold on ice.

23. Count cells on a hemacytometer (Tuerk's solution to eliminate RBC and Trypan blue for viability) (6).

3.2. Surface Staining

1. Pellet cells at $400 \times g$ for 10 min at 4°C. Discard supernatant (see Note 13).

2. Resuspend cell pellet in 5 μL of neat decomplemented (56°C, 30 min) mouse serum (see Note 14).

3. Pellet cells ($400 \times g$, 10 min, 25°C) and aspirate the supernatant as thoroughly as possible without disturbing the pellet (see Note 15).

4. Stain the dry pellet for surface markers by the addition of 2 μL of each monoclonal antibody (see Note 16).

5. Add antibodies in the following order (see Note 17):

   (a) Bone marrow mesenchymal cells
   - CD105-FITC
   - CD73-PE

- CD34-ECD
- CD90-PC5
- CD117-PC7
- CD31-APC or CD133-APC
- CD45-APC-Cy7

(b) Pleural fluid

- CD44-PE
- CD90-biotin-streptavidin-ECD
- Lineage cocktail: CD14-PC5, CD33-PC5, and Glycophorin A-PE-Cy5.
- CD133-APC
- CD117-PC7
- CD45-APC-Cy7

(c) Adipose pericytes

- CD3-FITC
- CD146-PE
- CD34-ECD
- CD90-PC5
- CD117-PC7
- CD31-APC
- CD45-APC-Cy7

6. Incubate for 30 min on ice in the dark.

7. Dilute surface stained cell pellets in 1 mL of staining buffer.

8. Centrifuge at $400 \times g$ for 7 min at 25°C.

9. Discard supernatant and loosen the cell pellet.

3.3. Intracellular Staining

Staining for intracellular antigens requires fixation and permeabilization, and is usually performed after surface staining.

1. Following surface staining, pellet cells at $400 \times g$ for 10 min at 4°C. Discard supernatant with care to create a *dry pellet*.

2. Fix stained cells for 20 min at ambient temperature with 200 µL of PBS and 200 µL of 4% methanol-free formaldehyde in hypertonic PBS-A (see Note 18).

3. Centrifuge at $400 \times g$ for 7 min at 25°C.

4. Discard supernatant with care to create a *dry pellet* and flick to loosen.

5. Permeabilize fixed cells by addition of 200 µL of permeabilization solution containing saponin (10 min at ambient temperature).

6. Intracellular cytokeratin staining of pleural effusion cells:

   (a) Centrifuge at $400 \times g$ for 7 min at 25°C.

   (b) Decant supernatant with care to create a *dry pellet* and flick to loosen.

   (c) Add 5 μL of neat mouse serum to cell pellet, incubate at ambient temperature for 5 min, centrifuge, and decant to *dry pellet*.

   (d) Loosen cell pellet and add 2 μL of anti-pan cytokeratin-FITC for 30 min.

   (e) Dilute cytokeratin stained cell pellets in 500 μL of staining medium.

   (f) Centrifuge at $400 \times g$ for 7 min at 25°C and discard supernatant to a *dry pellet*.

   (g) Loosen cell pellets and dilute to a cell concentration of ten million cells/400 μL $(25 \times 10^6/mL)$ of staining buffer.

   (h) Transfer to 12×75 mm tubes with 35 μm filter caps for flow cytometry.

7. DNA content by DAPI staining for bone marrow mesenchymal cells, adipose pericytes, and pleural effusion cells:

   (a) Disaggregate cell pellets and dilute to a cell concentration of ten million cells/400 μL of staining buffer.

   (b) Add 16 μL of DAPI stock solution to a final concentration of 8 μg/mL (7) (see Note 19).

   (c) Transfer to 12×75 mm tubes with 35 μm filter caps for flow cytometry.

3.4. Instrument Setup and Standards

3.4.1. IgG Capture Bead Staining

1. Vortex CompBeads thoroughly before use (see Note 20).

2. Label a separate 1.5 mL Eppendorf tube for each mouse monoclonal antibody conjugated to a tandem dye (e.g. ECD, PE-Cy5, PE-Cy7, and APC-Cy7).

3. Add one full drop (approximately 60 μL) of anti-mouse Ig CompBeads to each Eppendorf tube.

4. Centrifuge for 10 min at $400 \times g$. Carefully aspirate supernatant to ensure a "dry pellet" (see Note 21).

5. Sonicate each tube for 10 s in a water bath sonicator (see Note 22).

6. Add 2 μL of each antibody directly to beads (one antibody per tube) and gently reflux.

7. Incubate for 15 min at room temperature in the dark (see Note 16).

8. Add 1 μL of mouse serum and incubate for 5 min at room temperature (see Note 23).

9. Add 100 µL of staining buffer and reflux.

10. Sonicate each tube for 10 s.

11. Add 1 mL of staining buffer. For manual compensation, add one drop of negative CompBeads to each test tube that contains antibody stained beads (see Note 24).

12. Centrifuge beads for 10 min at $400 \times g$, decant and carefully blot to remove residual supernatant (see Note 25).

13. Resuspend washed beads in 0.5 mL of staining buffer.

14. Transfer to 12×75 mm snap cap tubes for flow cytometry.

15. Sonicate for 10 s prior to acquisition on the flow cytometer.

3.4.2. Instrument Setup and Sample Acquisition

These protocols have been validated on Beckman Coulter CyAn and Gallios cytometers. This protocol is applicable to all cytometers equipped with a UV or violet laser, a blue laser, and a red laser.

1. The cytometer is calibrated to predetermined photomultiplier target channels prior to each use using SpectrAlign beads and 8-peak Rainbow Calibration Particles (see Note 26).

2. The data from 8-peak beads can also be used to monitor instrument sensitivity (resolution of dim peaks) and instrument linearity (8).

3. All fluorescence parameters are collected in the logarithmic mode, with the exception of DAPI emission at 455 nm which is collected in the linear mode (see Note 27).

4. Acquire unstained cells or beads first, and then each single stained sample (bead or cells) from the shortest emission wavelength (FITC) to the longest (e.g. CD45-APC-Cy7).

5. Run a rinse tube. Then run analytical samples with a rinse tube in between if needed (see Note 28). Do not apply spectral compensation. This will be done offline with analytical software.

3.5. How Many Events to Acquire

Not everything that is detected by a flow cytometer is a cell. In fact, in messy samples like disaggregated solid tissues, cells are sometimes in the minority. Rare event problems sometimes necessitate the acquisition of millions of events in order to obtain the required number of cells. The correct number depends on three factors: (1) The proportion of cells to debris; (2) The signal-to-noise ratio of the population of interest compared to all other events; (3) The frequency of the population of interest.

Discriminating between genuine events and potential sources of artifact such as debris will be illustrated in each example. The inclusion of dead or dying cells, autofluorescent events, or subcellular debris distorts both the numerator (population of interest) and denominator, and makes for an unreliable analysis.

Much of the literature dealing with flow cytometry of cultured or disaggregated cells suffers to a greater or lesser extent from inclusion of such sources of artifact.

The signal-to-noise ratio encompasses both the distance (in fluorescence or scatter intensity) between positive and negative populations, and the variability (spread) of those populations in multiparameter space. Isotype controls or other definitive negative populations are often helpful to determine the boundaries defining a population of interest. Even when positive and negative populations appear to be well separated, acquisition of millions of events often reveals a scattering of false positive events within the multiparameter space of the population of interest. Creative gating, including elimination of sources of artifact and use of multiple parameters (both positive and negative) to define the population of interest, can often increase the signal-to-noise ratio. Ultimately, the frequency of false positive events determines the lower limit of detection and thus the maximum number of cells to be acquired. Figure 1 shows an example in which five false

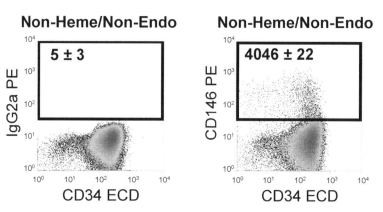

Fig. 1. Estimation of the lower limit of detection from an isotype control. The lower limit of detection is determined by the signal-to-noise ratio and thus is dependent on the mean fluorescence of negative and positive populations and their respective spreads. In this example we show an FMO (fluorescence minus one) control (*left panel*) for the detection of pericytes (*right panel*) in the stromal vascular fraction of adipose tissue. An equal number of events (910,000) are shown in each histogram and both have been gated on viable singlet events that are both non-hematopoietic (CD45−/CD3−) and non-endothelial (CD31−). The gates identifying CD146+ pericytes (*right*) and false positive events (*left*) are identical and, in this case have been placed to exclude all but 5 ± 3 (mean ± SD of triplicate determinations) events in the isotype control tube. The lower limit of detection is calculated as the log mean false positive rate plus $1.6445 \times$ SD. This is an estimate of the 95th percentile of the false positive frequency. The antilog of this value, 0.0085% (1 event in 11,790) is considered the lower limit of detection and defines the lowest proportion of pericytes that can be reliably detected. Combining this information with the methods shown in Fig. 2, the optimal number of total events to acquire for this assay is $100 \times 11,790 = 1,179,000$. The frequency of CD146+ pericytes (*right histogram*) in this example is 2.5% of clean non-heme events or 0.278% of total events. In our example, acquiring 4046 CD146+ events would give a CV of 2% according to Poisson statistics.

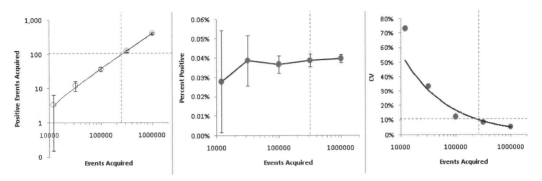

Fig. 2. How many events to acquire? Here we are quantifying CD45−/CD34−/CD105+/CD73+ mesenchymal stem cells in freshly isolated human bone marrow. A single file was acquired for a total of 4,074,618 events. We divided the file into three equal segments of varying number to simulate the acquisition of different numbers of cells in triplicate (13). The *graph on the left* shows the number of positive events (mesenchymal stem cells) as a function of total events acquired (mean of triplicate determinations, bars = 1 SD). The *dashed lines* show the number of total events (*X*-axis) required to detect 100 positive events (*Y*-axis). The *center graph* shows the calculated mean percent of mesenchymal stem cells (bars = 1 SD) as a function of the number of events acquired. The dashed line intersects with the estimated percent positive when 100 positive events are acquired. The *right graph* shows the calculated coefficient of variation (SD/mean) of the percent positive (*red circles*) as a function of the total number of cells acquired. The *solid black line* is a predicted CV based on Poisson counting statistics in which the CV% is equal to the 100/√positive events acquired (14). The dashed lines demonstrate that a CV of 10% is obtained when 100 positive events are detected. In this example, this requires a total of 250,000 events.

positive events were detected in an isotype control sample in which 910,000 gated events were analyzed. By performing replicate determinations, we were able to calculate the 95th percentile of the false positive event frequency is 0.0085% or 1 event in 11,790. Since acquiring 100 events of a population of interest is sufficient to give a CV of 10% (see Fig. 2), acquiring 1,179,000 events will take the assay to its limit of sensitivity, and acquiring more cells will do nothing to improve the result.

Figure 2 shows the effect of acquiring triplicate samples of increasing numbers of cells. As expected, the number of positive events increases linearly with the number of events acquired and the variability between triplicate frequency estimates (Percent positive) decreases. Importantly, the measured coefficient of variation (CV, calculated as the standard deviation/mean) of the triplicate determinations decreases markedly. This empirical demonstration also shows the adequacy of the CV determined by a simpler alternative approach, Poisson statistics. Poisson statistics deal with the probability distribution of rare events. The Poisson CV of the counting error, defined as 100/√positive events counted, is 10% when 100 events are counted. In this example, the frequency of positive events (0.04%) requires that 250,000 events be acquired to achieve a CV of 10%.

There is one more aspect of signal-to-noise that counting statistics do not take into account, and that is the distribution of the population of interest in multiparameter space. Figure 3 shows

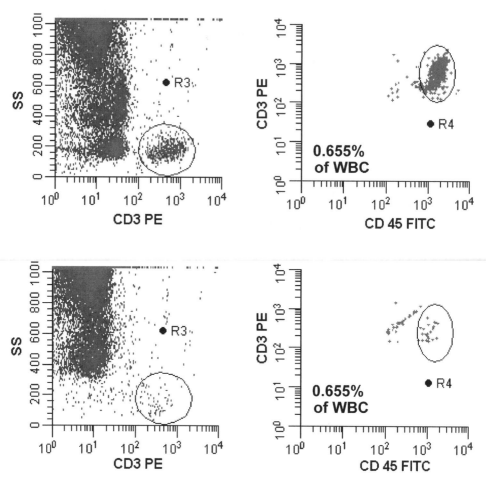

Fig. 3. Knowledge of the spatial distribution of a population of interest can be used to increase the signal-to-noise ratio. T-cell determinations were performed at two time intervals on a patient undergoing immunoablative therapy with cytoxan, fludarabine, and anti-thymocyte globulin. In the upper histograms a plot of CD3 versus side scatter reveals a clear population of CD3+ cells. Further gating on CD45 versus CD3 demonstrates that T-cells have relatively little variability in the expression of either marker. This permits the elimination of several outlying events that in all probability do not represent viable T cells. The bottom row shows the same analysis on a sample that contains far fewer T-cells. Knowledge of the location of *bona fide* T cells in multiparameter space, and careful standardization of the instrument permit the same analytical gates to be applied, even though there is not a clearly discernable population in the plot of CD3 versus side scatter. Projection of these events onto a plot of CD45 versus CD3 reveals a small population of T cells (23 events) and a more numerous diagonal streak of artifactual events. Although the Poisson CV of the T-cell population is an unreliable 20.9%, its distribution in space lends credibility to the analysis. The actual assays were performed in triplicate, using 7-AAD exclusion as a criterion of viability (not shown) to help substantiate the conclusion that the T cells were genuine and viable, even though they represented only 0.028% of CD45+ events.

peripheral T-cell determinations performed on patients undergoing immunoablative therapy (9). Given the knowledge that T-cells are very constrained with respect to the range of CD3 and CD45 expression, very low numbers of positive events (in this case 23 CD3+ events, Poisson CV = 20.9%) can be reliably detected.

3.6. Data Analysis of Three Examples: Bone Marrow Mesenchymal Stem Cells, Cytokeratin + Cells in a Malignant Pleural Effusion, and Pericytes in Human Adipose Tissue

We perform spectral compensation and data analysis offline, using VenturiOne or Kaluza software both of which have been designed to accommodate very large datafiles. After creating compensation matrices using the data from our single-stained beads, we create a playlist of datafiles in which each datafile is associated with an analysis template and a compensation matrix. Arriving at an adequate analysis template is an iterative process that used to be quite painful. The use of playlists to organize datafiles and compensation files from different directories, and the ability to export results to spreadsheets greatly facilitates reanalysis and revisiting data sets with new questions. Our analysis strategy usually proceeds in three steps:

1. First we eliminate sources of interference with logical gates. These may include event bursts (from transient fluidic disturbances), cell–cell doublets and clusters, subcellular debris, dead cells, and autofluorescent events.

2. Next, we decide on our *classifier* parameters. These are used to define the major population(s) of interest.

3. Next *outcome* parameters to be measured on each classifier population are determined. The distinction between classifier and outcome parameters is sometimes clear, but sometimes it is quite fluid, especially when exploring a new combination of analytes. At this stage it is often helpful to color-event cells positive for the outcome parameters and examine their expression on plots of the classifier parameters (10).

3.6.1. Elimination of Sources of Interference

Interference can come from many sources including fluidic disturbances, nonspecific binding of antibodies (dead and dying cells are adept at this), and cells with intrinsic autofluorescence (particularly cultured cells and some populations from fresh disaggregated tissues).

1. Fluidic disturbances alter laminar flow within the flow cell and results in increased variability in measurements, particularly light scatter. It is easy to spot transient fluidic disturbances by plotting a time parameter (or event count) versus log side scatter. Such a disturbance can be seen in the pleural effusion example (Fig. 4). These events can be examined in isolation and removed from the analysis with a logical gate, if they prove to have altered marker expression.

2. Cell doublets and clusters, resulting from physical aggregation or coincidence are also problematic when large numbers of events are acquired. For DNA analysis, a doublet appears as a single cell with 4N DNA (11). For phenotypic analysis a T-cell/B-cell conjugate looks like a single cell with coexpression of T- and B-cell markers. Clusters are easily removed by pulse analysis of the triggering parameter. In our examples,

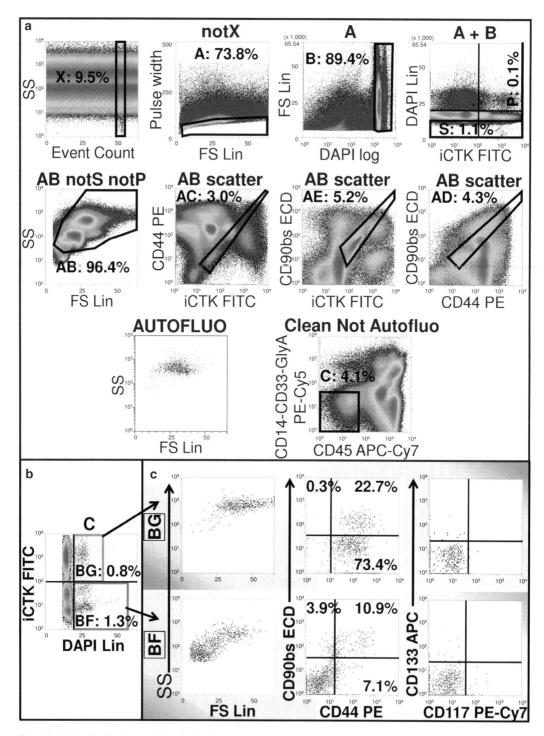

Fig. 4. Analysis of cycling/aneuploid cells in a breast cancer pleural effusion. (**a**) Elimination of sources of artifact prior to analysis. This data is from a cryopreserved breast cancer malignant pleural effusion. Pleural effusions are heterogeneous and often have more inflammatory cells than tumor cells. The goal was to detect aneuploid/cycling cells in subpopulations of cytokeratin+ (epithelial) cells. The first step in data analysis is to remove events which can interfere with detection of the events of interest. We always monitor light scatter as a function of time. Log side scatter is very sensitive to

we compare forward scatter pulse height (labeled FS lin) to forward scatter pulse width (labeled Pulse width) and eliminate event clusters that are too wide (i.e. have too long a time of flight) for their pulse height. In our examples the frequency of clusters ranges from 5% (Fig. 6, bone marrow) to 26% (Fig. 4, malignant pleural effusion).

3. Dead and dying cells may also have altered marker expression, so it is always desirable to eliminate them from the analysis. This can be approached in several ways. In all of the examples given here, cells were permeabilized after staining and fixation in order to facilitate DAPI staining of cellular DNA. This has two benefits: (1) subcellular debris and hypodiploid (apoptotic) cells are easily identified and removed on a plot of DAPI log fluorescence intensity versus FS; (2) display of DAPI fluorescence on a linear scale provides a low-resolution cell-cycle analysis (see Notes 29 and 30). In our adipose example (Fig. 5), 14% of events, most of them subcellular debris resulting from tissue digestion, had $<2\,N$ DNA. Even after limiting the analysis to cells with DNA content $\geq 2\,N$, early apoptotic events with intact DNA may still be present. These can be identified and eliminated by their characteristic light scatter profile (generally low forward scatter with too much side scatter relative to forward scatter). In our adipose example the T-cell marker CD3 was used as a "Dump Gate" and CD3+ T cells were color evented and backgated on the light scatter plot (red colored events in the electronic version,

Fig. 4. (continued) fluidic perturbations which show up as a decrease in FS. Here a transient disturbance was eliminated with "NOT" gate (X). The next step is to use pulse analysis to remove doublets and cell clusters. Pleural effusions are susceptible to clumping, and cell clusters will have marker expression characteristic of the aggregate. In this case 73.8% of events were singlets. The cells in this analysis were permeabilized with saponin after surface staining and fixation. DAPI can be used to measure DNA content in permeabilized cells. DAPI log is useful for eliminating events with no or little DNA (about 10% of singlets). In the linear mode, DAPI fluorescence is used for cell-cycle determinations. Here it is used to remove hypodiploid (apoptotic) cells (1.1% of nucleated singlets). In the same histogram a small proportion of events (0.1%) are so brightly stained with cytokeratin-FITC that they have saturated the PMT and therefore appear in the last channel. These events cannot be spectrally compensated, because their true fluorescence intensity is unknown. Next a conservative gate eliminates events with too little light scatter to be viable cells. Together, the last three histograms in the second row are used to identify and eliminate cells with autofluorescence. Any events falling within the intersection of regions AC, AE and AD (AUTOFLUO) are candidates for elimination. The high light scatter of autofluorescent cells suggests that they are granulocytes. Finally, CD45 negative, hematopoietic lineage negative cells are identified (gate C) as tumor cell candidates. These cells represented only 2.54% of the 3.6 million events acquired and represent the denominator for subsequent rare event analysis. (b) After removing sources of artifact, and excluding cells expressing hematopoietic lineage markers, we defined our classifier populations (populations of interest) as cytokeratin+ (BG) and negative (BF) cells with $>2\,N$ DNA content, as determined by DAPI staining. We detected a total of 767 Cytokeratin+ cells with $>2\,N$ DNA content, representing 0.8% of heme-lineage negative cells and 0.02% of acquired events. (c) Outcomes measured on these two populations included expression of the adhesion molecule CD44 (96% of cycling cytokeratin+ cells) and the stem cell marker CD90 (23%). Credible subpopulations of CD117+ and CD133+ cells were not seen. This analysis was repeated on heme-lineage negative, cytokeratin negative aneuploid/cycling cells, which have much lower light scatter, and smaller proportions of CD44+ and CD90+ cells.

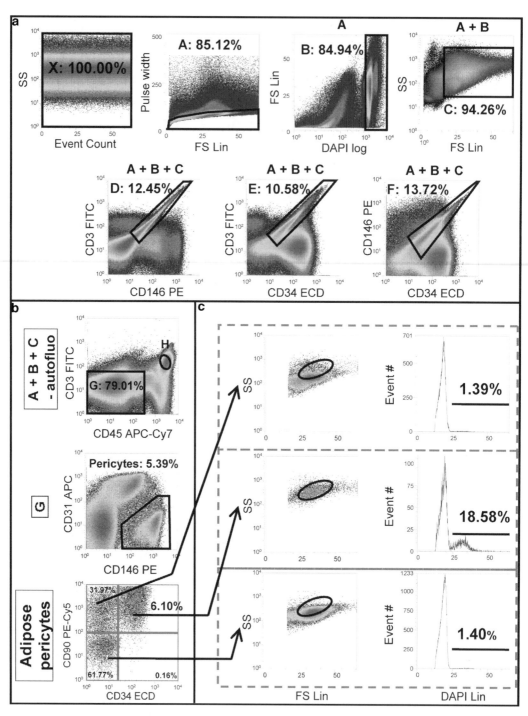

Fig. 5. Flow cytometric analysis of pericytes subsets in adipose tissue. (**a**) Elimination of irrelevant events by selection of cellular nucleated singlets. *Left to right.* We check for disturbances using time versus log side scatter. Cell clusters are eliminated using pulse analysis (*region A*). Nucleated cells were stained with the nuclear dye DAPI after gentle permeabilization (*region B*). Cell-events can be further defined using known internal cellular landmarks. In this example, resting CD3+ CD45+ T-lymphocytes (H) have been color-evented red (within *region C*). The knowledge of the light scatter localization of resting T cells serves as a landmark to exclude smaller subcellullar events and preapoptotic cells. Selection of

Fig. 5). This gives us a point of reference to eliminate events with lower forward light scatter. This method assumes that the cells of interest have at least as high light scatter as small resting T-cells (see Note 31).

4. Elimination of events with saturating fluorescence. It is highly desirable to adjust PMT gains such that all positive events are on scale. In rare cases this is not possible without unbalancing PMT gain or obscuring dim positive events. In Fig. 4, cytokeratin expression occupies such a large dynamic range that 0.1% of all events fall within the last channel (i.e. are saturated with respect to green fluorescence). Although this is a relatively small proportion of events, they must be removed from the analysis. Because their fluorescence is unknown, they cannot be spectrally compensated and will appear positive in the adjacent PE channel.

5. Cellular autofluorescence results from expression of naturally fluorescent biomolecules such as flavinoids (12). Autofluorescence can be distinguished from fluorescence specific to most of the dyes used in cytometry by it's broad emission spectrum. Autofluorescent biomolecules are often excited better with short wavelength light than with long wavelengths. If the cell population of interest is autofluorescent, one is generally limited to long excitation wavelengths (e.g. red diode laser). In our examples, mesenchymal stem cells, adipose pericytes, and cytokeratin + breast cancer cells, the cells of interest are not autofluorescent, so we can use compound logical gates to eliminate events that excite with 488 nm light and emit in the ranges detected by FL1, FL2, and FL3 channels (e.g. using 530/30, 584/42, and 675/20 bandpass filters, respectively). In the pleural effusion example (Fig. 4), relatively few events fall within all three diagonal gates, and these are uniformly cells with high light scatter.

6. Nonspecific fluorescence (autofluorescent or caused by nonspecific antibody binding) can be eliminated using a *dump gate*, that is, a marker which the population of interest is known not to express. If the dump gate uses a dye such as FITC,

Fig. 5. (continued) autofluorescent events in the three first fluorescent channels (FITC, PE, and ECD) was performed as in Fig. 4. (**b**) Selection of adipose pericytes by use of classifiers (CD3, CD45, CD31, and CD146). Remaining autofluorescent events are eliminated using the CD3-FITC signal as a dump gate. Adipose pericytes are non-hematopoietic (*region G*), non endothelial (CD31− cells), and express the cell adhesion molecule CD146 (Region Pericytes). In this example, pericytes account for 5.39% of non hematopoietic events, 4.25% of the adipose stromal vascular fraction, and 2.62% of total events. CD90 and CD34 serve as secondary classifiers, as pericytes can be divided into three distinct subpopulations based on CD90 and CD34 expression. The majority of pericytes do not express CD34 (93.1% of pericytes). One third express CD90 exclusively, and a minor population coexpress CD34 and CD90 (0.16% of total events). (**c**) Outcome: The rarest population, CD34+ CD90+ pericytes, is larger (higher light scatter compared to CD34 negative pericytes) and shows a higher proliferation level (DAPI lin >2*N*, 18.58% of CD34+ CD90+ pericytes). CD34+ CD90+ pericytes may represent a transit amplifying population of cells in transition from pericytes to supra-adventitial adipose stromal cells (15).

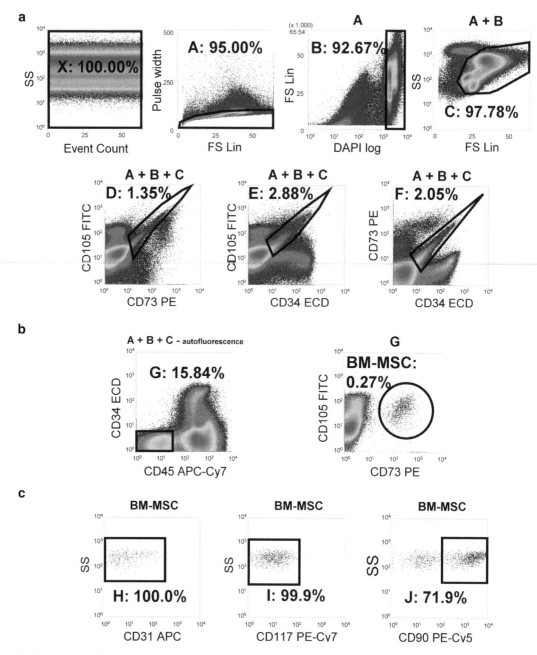

Fig. 6. Detection of bone marrow mesenchymal stem cells in freshly isolated bone marrow. A total of four million events were acquired. (**a**) Elimination of sources of interference. Fluidic disturbances, elimination of cell clusters, selection of nucleated events, and exclusion of debris, subcellular particles and early apoptotic cells were done as in Figs. 4 and 5. The lower row shows the selection of autofluorescent events in the three first fluorescence channels. All events present concomitantly in *regions D, E, and F* were eliminated. (**b**) Identification of a classifier population. Bone marrow mesenchymal stem cells are detected as CD34−/CD45−/CD73+/CD105+ cells. (**c**) Measurement of outcomes on the classifier population. CD31 (an endothelial marker) and CD117 (c-Kit) are all negative on BM-MSC. The majority of BM-MSC are positive for the stem cell marker CD90.

which is within the range of autofluorescent emission, it eliminates cells binding the antibody specifically and nonspecifically, and autofluorescent events as well. This can be seen in Fig. 5, where non-hematopoietic cells are identified as CD3-FITC negative, as well as CD45 negative, prior to the identification of adipose pericytes. Nonspecific antibody binding is also minimized by a preincubation/blocking step with normal mouse serum prior to staining.

3.6.2. Identification of Classifier Parameters

Classifier parameters are those used to identify populations of interest. Thinking of parameters as either primary or secondary markers, or outcomes creates a hierarchical model, focuses the analysis, and eliminates the "all possible combinations" problem encountered in multi-parameter analysis.

1. Primary classifiers can be strung together as a Boolean AND gate. In Fig. 6, the primary classifiers identifying mesenchymal stem cells are absence of CD45 and CD34 expression AND presence of CD73 and CD105.

2. Secondary classifiers branch into multiple populations of interest. In Fig. 5 adipose pericytes, identified by primary classifiers (CD3–/CD45–/CD31–/CD146+) are subsetted into three subpopulations by the secondary classifier markers CD34 and CD90. In Fig. 4, the secondary classifier markers cytokeratin and DAPI identify populations of cycling/aneuploid cells for further analysis.

3.6.3. Measurement of Outcome Parameters on Populations Identified by Classifiers

The division between classifiers and outcomes is not rigid and often depends on how one frames a biologic question. For example, do we wish to start with cells that secrete interferon gamma (classifier) and examine lymphocyte subsets as outcomes, or wish to define lymphocyte subsets (classifiers) and determine (as outcomes) the proportion of cells secreting interferon gamma in each? As such, outcomes are final branch points in our hierarchical mode. In Fig. 4, the outcomes are light scatter, CD44, CD90, CD117, and CD133 expression on cycling/aneuploid cytokeratin + and negative non-hematopoietic cells. In Fig. 5, they are light scatter and DNA content in adipose-derived pericytes. In Fig. 6, the outcome parameters are CD90, CD117, and CD31 expression on mesenchymal stem cells.

4. Notes

1. Bone marrow aspirates are naturally diluted in peripheral blood. When correctly performed, a bone marrow aspirate should contain approximately 20×10^6 mononuclear cells/mL. Lower concentrations indicate excessive hemodilution.

2. If you use a different product make certain that it does not also contain a fixative.

3. Bovine serum albumin or human serum albumin may be used. The recommended concentration (0.1%) is equivalent to 2% serum. Two percent newborn calf serum is more economical, equally effective, and can be used for many applications. Serum albumin protects cells from shear stress encountered during centrifugation.

4. Watch your favorite bartender prepare a "Black and Tan." With the tube slightly tilted, slide the pipette down the wall of the tube until it is close to the Ficoll/Hypaque. Using a pipetting device (Drummond Pipet-Aid), dispense the cell suspension slowly and evenly, carefully sliding the pipette tip up the wall as you go. A successful gradient shows a sharp interface between the bone marrow suspension and the Ficoll/Hypaque layer. Some investigators prefer to add the cell suspension to the tube first and then carefully underlayer the Ficoll/Hypaque using a pipette or a syringe and a bone marrow biopsy needle.

5. The bone marrow mononuclear cells will appear as a whitish-tan layer at the Ficoll/Hypaque interface. This is often referred to as the Buffy Coat. Take care not to disturb the Buffy Coat when removing and discarding the diluent.

6. Prolonged exposure to Ficoll/Hypaque is toxic to the cells. Take care to include as little as possible when recovering the Buffy Coat.

7. We routinely use a Beckman-Coulter Act10 hematology analyzer for cell counting. This instrument detects cells by a change in conductivity that occurs when cells displace electrolyte. It can be fooled by cell-sized debris and fat globules. Train your eyes to estimate cell number from the size of the cell pellet. Hemacytometers are inexpensive and yield reliable counts after a little practice.

8. Tumor cells are usually in the minority in malignant effusions, the majority population being acute or chronic inflammatory cells. Mesothelial cells are also often seen.

9. If the sample collected is more than 200 mL, use 225 mL polypropylene conical tubes.

10. DNase and collagenase require divalent cations and will not work in calcium/magnesium free PBS. The albumin serves as a buffer against cell digestion by nonspecific proteases.

11. Treatment with ammonium chloride lyses red blood cells. This step may be omitted if sample is not visually bloody.

12. This fatty layer often contains trapped cells.

13. The number of cells to be stained in a single pellet can range from 0.5×10^6 to 10×10^6 or more without changing the amount of antibody added. The key issue in antibody staining is the final concentration of the antibody in the reaction mixture and not the number of cells. Best practice is to determine the optimal antibody concentration by titration. Alternatively, read the manufacturer's instructions paying attention to the recommended dilution of stock antibody, but not the number of cells. For example, if the instructions call for suspending the cells in 0.1 mL of buffer and then adding 10 µL of antibody, this means that the antibody can withstand a tenfold dilution. By minimizing the sample volume you will minimize the amount of antibody required per test.

14. Addition of mouse serum, which contains mouse immunoglobulins, blocks nonspecific binding of murine monoclonal antibodies.

15. We refer to this as a *dry pellet* elsewhere in the protocol. Cells pelleted in a 15 mL conical tube and aspirated dry actually contain anywhere from 10 to 50 µL of residual liquid.

16. Tandem dyes are easily degraded by exposure to light. Reagents, stained cells, and stained beads must be carefully protected from ambient light. We perform staining in an unilluminated biological safety cabinet, and cover stained cells and beads with aluminum foil to minimize light exposure.

17. The choice of antibodies and fluorochromes is specific to the question being addressed and the available instrumentation, and can be modified at will. The order of antibody addition may influence staining as binding occurs very rapidly and sequential addition of many antibodies progressively reduces the concentration of the individual antibodies in the mixture.

18. The choice of formaldehyde is critical. We always use EM grade methanol-free material purchased at a low concentration stock solution (10%). It is not necessary or advisable to make up fresh paraformaldehyde solution from powder.

19. Unlike antibody staining, the concentration of cells and the concentration of dye are critical for consistent results. This protocol is specific to cells at $25 \times 10^6/mL$.

20. Single stained compensation standards are essential for correct spectral compensation. We find it best to use internally stained Calibrite beads for FITC, PE and APC, and Ig capture beads (CompBeads) for antibodies conjugated to tandem dyes. DAPI does not require spectral compensation against these dyes so no single stain preparation is required.

21. Beads do not pack as tightly as cells so care must be taken not to aspirate the beads.

22. Sonication is not essential, but is very useful because it disaggregates bead clumps better than vortexing or refluxing.

23. Addition of mouse serum prevents clumping of beads due to antibody crosslinking.

24. Negative beads are not required for some automated compensation algorithms, such as those implemented on Beckman-Coulter instruments or in VenturiOne software.

25. Decanting and blotting must be done in one smooth motion to prevent loss of beads.

26. Determining balanced PMT settings is an art in itself and has been addressed by us elsewhere (9). Once these settings have been established, and bead target channels have been determined, they can be reproduced from experiment to experiment by adjusting PMT voltage to place the brightest peak of the Rainbow particles in the predetermined target channel. The intensity of DAPI staining will vary somewhat from specimen to specimen, because it is measured on a linear scale and is highly dependent on the total amount of DNA in the sample. PMT voltage for the DAPI channel is adjusted for each sample to place the median fluorescence of the $2N$ population at about channel 64 of 1,023.

27. Newer three-laser cytometers make use of high resolution digital pulse analysis. All native data is linear. Logarithmic transformations are performed mathematically. Referring to data collected in the "logarithmic mode" is a throwback to the days when analog pulse measurements were directed to either a log or linear amplifier.

28. Sample carryover can be a killer in rare event analysis. Instruments vary widely in the degree of sample carryover. You can tell if you have a carryover problem if you see excessive events when running a rinse (filtered water or PBS) tube after your sample tube.

29. Cell-cycle analysis can be optimized by using fixation and permeabilization methods less favorable for surface marker staining.

30. If permeabilization is not possible (as is the case when sorting viable cells), uptake of DAPI or another DNA-binding dye such as propidium iodide or 7-aminoactinomycin-D, can be used to identify and exclude dead cells, but this method does not exclude subcellular debris or early apoptotic cells.

31. Some cells (e.g. embryonic stem cells) are very small and have lower light scatter.

Acknowledgments

The authors would like to thank Melanie Pfeifer and E. Michael Meyer for their excellent technical assistance, Peter Nobes and David Roberts (Applied Cytometry Systems) for their valued collaboration in data analysis and Cindy Collins and Brad Calvin (Beckman Coulter) for providing instrumentation, software, and the support of their flow cytometry group. This work was supported by the following granting agencies: Department of Defense (BC032981 and BC044784), NIH NHLBI (Production Assistance for Cellular Therapy (PACT) under contract N01-HB-37165, R01-HL-085819), the Hillman Foundation, and the Glimmer of Hope Foundation.

References

1. Triebel, F., Robinson, W. A., Hayward, A. R., and Goube de Laforest, P. G. (1981) Existence of a pool of T-lymphocyte colony-forming cells (T-CFC) in human bone marrow and their place in the differentiation of the T-lymphocyte lineage. *Blood* **58**, 911–5.

2. Lee, H. S., Huang, G. T., Chiang, H., Chiou, L. L., Chen, M. H., Hsieh, C. H., and Jiang, C. C. (2003) Multipotential mesenchymal stem cells from femoral bone marrow near the site of osteonecrosis. *Stem Cells* **21**, 190–9.

3. Poulin, E. C. and Labbe, R. (1997) Fully thoracoscopic pulmonary lobectomy and specimen extraction through rib segment resection. Preliminary report. *Surg Endosc* **11**, 354–8.

4. Gao, I. K., Noga, S. J., Wagner, J. E., Cremo, C. A., Davis, J., and Donnenberg, A. D. (1987) Implementation of a semiclosed large scale counterflow centrifugal elutriation system. *J Clin Apher* **3**, 154–60.

5. Böyum, A. (1968) Isolation of mononuclear cells and granulocytes from human blood. Isolation of monuclear cells by one centrifugation, and of granulocytes by combining centrifugation and sedimentation at 1 g. *Scand J Clin Lab Invest Suppl* **97**, 77–89.

6. Turgeon, M. L. (2005) *Clinical Hematology: Theory and Procedures.* Lippincott Williams & Wilkins, Philadelphia, PA, p. 570.

7. Park, C. H., Kimler, B. F., and Smith, T. K. (1985) Comparison of the supravital DNA dyes Hoechst 33342 and DAPI for flow cytometry and clonogenicity studies of human leukemic marrow cells. *Exp Hematol* **13**, 1039–43.

8. Hoffman, R. A. and Wood, J. C. (2007) Characterization of flow cytometer instrument sensitivity. *Curr Protoc Cytom* **1.20**, 1–18.

9. Donnenberg, A. D. and Donnenberg, V. S. (2008) Understanding clinical flow cytometry, in *Handbook of Human Immunology, Second Edition* (O'Gorman, M. R. and Donnenberg, A. D., eds.), CRC Press Taylor and Francis, Boca Raton, FL, pp. 181–220.

10. Donnenberg, A. D. and Donnenberg, V. S. (2005) Phenotypic and functional measurements on circulating immune cells and their subsets, in *Measuring Immunity: Basic Biology and Clinical Assessment* (Lotze, M. T. and Thomson, A. W., eds.), Elsevier Academic Press, San Diego, CA, pp. 237–56.

11. Robert, P. W., Francis, J. C., James, F. L., Christa, M., Stetler-Stevenson, M., and Edward, G. (2001) Doublet discrimination in DNA cell-cycle analysis. *Cytometry* **46**, 296–306.

12. Aubin, J. E. (1979) Autofluorescence of viable cultured mammalian cells. *J Histochem Cytochem* **27**, 36–43.

13. Donnenberg, A. D. and Donnenberg, V. S. (2007) Rare-event analysis in flow cytometry. *Clin Lab Med* **27**, 627–52, viii.

14. Shapiro, H. M. (2003) *Practical Flow Cytometry*, 4th edition. Wiley-Liss, Hoboken, NJ, p. 681.

15. Zimmerlin, L., Donnenberg, V. S., Pfeifer, M. E., Meyer, E. M., Peault, B., Rubin, J. P., and Donnenberg, A. D. (2010) Stromal vascular progenitors in adult human adipose tissue. *Cytometry A* **77**, 22–30.

Chapter 13

Flow Cytometry-Based Identification of Immature Myeloerythroid Development

Cornelis J.H. Pronk and David Bryder

Abstract

Precursor cells of the myeloerythroid cell lineages give rise to mature cells of the granulocyte, monocyte, erythroid, and/or thrombocytic lineages. High-resolution profiling of the developmental stages, from hematopoietic stem cells to mature progeny, is important to study and understand the underlying mechanisms that guide various cell fate decisions. In addition, this approach provides greater insights into pathogenic events such as leukemia, diseases that are most often characterized by halted differentiation at defined immature precursor levels. In this chapter, we provide protocols and discuss approaches concerning the analysis and purification of immature myeloerythroid lineages by multiparameter flow cytometry. Although recent data have demonstrated the feasibility of similar approaches also for the human system, we will focus our chapter on C57BL/6 mice, in which immunophenotypic applications have been most widely developed. This should also allow for its application in genetically modified models on this background. For maximal reproducibility, all protocols described have been established using reagents from commercial vendors to be analyzed on a three-laser flow cytometer with factory standard configuration.

Key words: Hematopoiesis, Hematopoietic stem cell, Differentiation, Myeloid, Myeloerythroid, Progenitor, Multiparameter, Purification, FACS, Flow cytometry

1. Introduction

Since most effector cells have a limited life span, appropriate homeostasis in the blood system is contingent upon the constant production of new blood cell elements. This production is the result of multiple differentiation events, where the many different types of mature effector cells are ultimately derived from bone marrow-residing hematopoietic stem cells (HSCs); a cell type capable at the single-cell level to both self-renew and differentiate into separate hematopoietic lineages, thereby allowing for lifelong hematopoiesis (1).

The immediate progeny of HSCs are multipotential progenitor cells that retain full lineage potential but have lost extensive

Teresa S. Hawley and Robert G. Hawley (eds.), *Flow Cytometry Protocols*, Methods in Molecular Biology, vol. 699, DOI 10.1007/978-1-61737-950-5_13, © Springer Science+Business Media, LLC 2011

self-renewal ability. Such multipotent progenitors, in turn, give rise to a set of oligopotent progenitors with more restricted developmental potential. Upon further differentiation, these oligopotent progenitors mature into lineage-restricted progeny, from which all the mature blood cells eventually arise (Fig. 1a). As these subsets constitute only a minor fraction of all cells found in a measurement of a bulk population of unenriched cells, it is obvious that measurement of the bulk population in most cases is not sufficient to deduce information about a defined cellular stage or

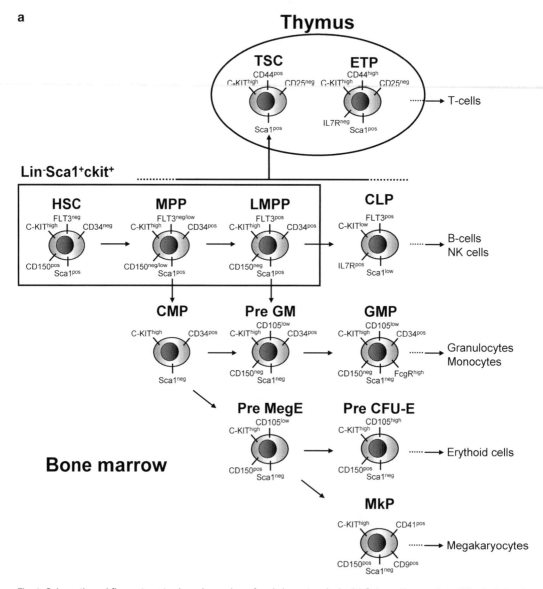

Fig. 1. Schematic and flow cytometry-based overview of early hematopoiesis. (a) Schematic overview of the first developmental stages in the differentiation from the multipotent HSC to progeny possessing oligo- and unilineage potentials. For each cell type, a selection of cell surface protein expression capable of flow cytometry-based detection is depicted.

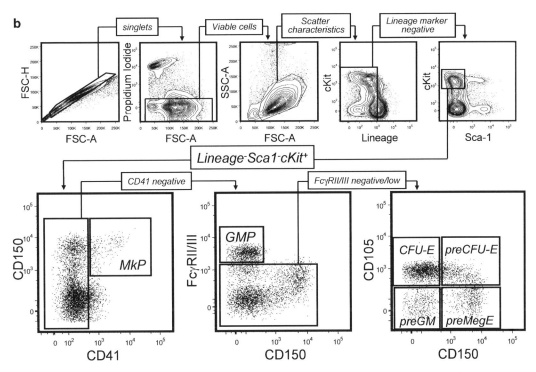

Fig. 1. (continued) (**b**) Flow cytometry-based strategy for high-resolution polychromatic fractionation of murine myeloerythroid precursors according to the protocol described herein. *HSC* hematopoietic stem cell, *MPP* multipotent progenitor, *LMPP* lymphoid primed MPP, *TSC* thymus seeding cell, *ETP* early thymic progenitor, *CLP* common lymphoid progenitor, *CMP* common myeloid progenitor, *Pre-GM* pre-granulocyte/macrophage, *GMP* granulocyte/macrophage progenitor, *Pre-MegE* pre-megakaryocyte/erythrocyte, *Pre-CFU-E* pre-colony forming unit – erythrocytes, *MkP* megakaryocyte progenitor.

developmental pathway. The subset of interest would simply "drown" in the noise contributed by other cell types. Therefore, to detail hematopoietic lineage development, it is necessary to purify cells not only with appropriate lineage affiliation but ultimately also at defined developmental stages.

Due to technological advances in recent years, flow cytometry nowadays routinely allows for the simultaneous assessment of multiple proteins on cells (2). Thereby, it is possible to resolve complex combinatorial expression patterns that associate with functionally distinct cellular properties. Multiparameter analysis is the key here, since no single protein identified to date can be used alone to define HSCs and/or their downstream immature progeny phenotypically (1). Rather, a panel of markers has to be evaluated simultaneously.

Using five-parameter flow cytometry, the prospective identification of oligopotent common lymphoid progenitors (CLPs) (3) and common myeloid progenitors (CMPs) (4) proposed that a first step in hematopoietic lineage restriction is the mutual exclusion of myeloerythroid from lymphoid potential. However, as much of this fundamental concept was challenged in reports detecting myeloid output from CLPs (5–8) and reports demonstrating loss of megakaryocyte/erythroid (Meg/E) potential prior to lymphoid and granulocytic/monocytic divergence (9–11), we screened the myeloid progenitor compartment for additional cell surface markers that could potentially reveal heterogeneity. This effort led to the identification of three markers; CD150, CD105, and CD41 that in combination with several previously described markers can be used to define a hierarchy of myeloid progenitors at high resolution (Fig. 1b) (12, 13).

This chapter provides protocols for flow cytometric detection and purification of immature myeloerythroid progenitor subsets from mouse bone marrow, including early bipotent progenitors for the erythroid/megakaryocyte lineages (preMegE), early monopotent erythroid (preCFU-E and CFU-E) and megakaryocyte progenitors (MkP), and primitive granulocyte/macrophage progenitors (preGM and GMP). We provide detailed protocols for cell isolation and cell surface staining, and describe approaches that can be used to manage instrument settings, data acquisition, and data analysis appropriately.

2. Materials

All reagents mentioned in this section should be stored in dark, at 4°C, unless otherwise indicated.

2.1. Isolation and Preparation of Mouse Bone Marrow Cells

1. C57BL/6 mice (The Jackson Laboratory, Bar Harbor, ME, or other local vendor). All procedures involving experimental animal work must have approval from the local Ethics Committees and performed according to national legislation.

2. Sterile or ethanol-cleaned surgical instruments: Fine scissors, bone-cutting scissors, and two forceps.

3. Mortar and pestle, sterile pipettes, sterile polypropylene tubes, 80-μm Filcon cup type filters (BD Biosciences, San Jose, CA), and a vacuum suction device.

4. Staining and cell preparation buffer: Phosphate-buffered saline (PBS) with 2% (v/v) fetal bovine serum.

5. 70% Ethanol.

6. Refrigerated centrifuge.

7. Cell counting device (Bürker counting chamber or automatic cell counter).

8. Erythrocyte lysing solution. Prepare a 10× stock by dissolving 16.58 g NH_4Cl, 2 g $KHCO_3$, and 0.744 g EDTA to a volume

of 200 mL H$_2$O. Adjust pH to 7.4. Dilute stock to a 1× working solution with distilled H$_2$O immediately prior to use.

2.2. Pre-enrichment of Mouse Bone Marrow Cells

All reagents described herein are used at predetermined optimal concentrations.

1. Materials described under Subheading 2.1.

2. Degassed staining and cell preparation buffer (see Note 1).

3. Sterile or clean pipette tips and sterile or clean Eppendorf tubes.

4. Biotin-conjugated "anti-lineage" antibodies: Ter119 (clone TER119), CD4 (clone GK1.5), CD8 (clone 53-6.7), B220 (clone RA3-6B2), Gr-1 (clone RB6-8C5), and CD11b (or Mac1; clone M1/70).

5. Anti-biotin MicroBeads, autoMACS cell separator including all necessary buffers, or MACS MS/LS cell separation columns with corresponding MACs separator magnet (all Miltenyi Biotec, Bergisch Gladbach, Germany).

2.3. Cell Surface Staining of Mouse Bone Marrow Cells

All reagents described herein are used at predetermined optimal concentrations.

1. Materials described under Subheadings 2.1 and 2.2.

2. Quantum-dot (Qdot) 605-conjugated Streptavidin (Invitrogen, Carlsbad, CA).

3. Fluorochrome-labeled anti-mouse antibodies against indicated cell surface markers (see Note 2), as depicted in Table 1; Sca1 (clone E13-161.7), CD41 (clone MWReg30), FcγRII/III (clone 2.4G2), CD105 (clone MJ7/18), CD150 (clone TCF15-12F12.2), and cKit (or CD117, clone 2B8).

4. Viability dye: a non-cell membrane permeable DNA-binding dye, such as propidium iodide (PI) at 1 mg/mL solution in water (Invitrogen) or 7-amino-actinomycin D (7-AAD; Sigma–Aldrich, St Louis, MO) (see Note 3).

2.4. Compensation Procedures

1. Materials described in Subheadings 2.1–2.3.

2. Machinery: Flow cytometer with lasers and filter setup concordant with the excitation and emission spectra of the used fluorochromes. The protocols in this chapter are based on acquisition on a FACSAria (BD Biosciences) flow cytometer equipped with three lasers (375-nm Violet laser, 488-nm Blue laser, and 635-nm Red laser), with three detectors for the Violet, six detectors for the Blue, and three detectors for the Red laser.

3. Antibody capture beads (BD™ CompBeads; anti-rat/hamster or anti-mouse Igκ; BD Biosciences) and/or splenocytes, used to generate single flurochrome-labeled samples (SS).

Table 1
Staining protocol

	PACIFIC BLUE	QDOT605	FITC	PE	CY7PE	PETXR	APC	APC-ALEXA780
Unstained[a]								
SS[b]	+							
SS[b]		+						
SS[b]			+					
SS[b]				+	⸱			
SS[b]					+			
SS[b]						–[c]		
SS[b]							+[d]	
SS[b]								+
FMO FITC[e]	Sca1	Lineage[f]	–	FcgR	CD105	PI	CD150	cKit
FMO Cy7PE[e]	Sca1	Lineage[f]	CD41	FcgR	–	PI	CD150	cKit
FMO APC[e]	Sca1	Lineage[6]	CD41	FcgR	CD105	PI	–	cKit
Sample	Sca1	Lineage[6]	CD41	FcgR	CD105	PI	CD150	cKit

Experimental staining protocol, including single-stained (SS) controls for instrument setup, as well as fluorescent minus one (FMO) controls for gate setting/analysis purposes
[a]Unstained cells (or negative capture beads) are used to set detector PMTs
[b]SS = single-stained controls, using capture beads or, for instance, splenocytes for automated compensation
[c]PI (Propidium iodide) is detected in most channels of the blue laser and thereby the choice of detector for dead cell exclusion is relatively flexible. No SS sample is required for detecting PI; however, this parameter should be included in the compensation matrix in order to compensate for spectral leakage from other channels into the viability channel
[d]The CD150 antibody does not bind to the capture beads that we are referring to here. Therefore, use another APC-conjugated antibody, preferably from the same provider
[e]FMO = fluorescence minus one. Usage and preparation are outlined in the text
[f]Lineage = cocktail with mature blood cell lineage markers. In this protocol, we use biotinylated anti-Ter119, B220, CD4, CD8, Gr1, and CD11b antibodies that are subsequently visualized with a secondary Streptavidin-Qdot605 reagent

2.5. Acquisition, Gating Strategies, and Sorting of Target Cells

1. Materials described in Subheadings 2.1–2.4.
2. When sorting: Flow cytometer with a cell-sorting device, including a single-cell depositor to perform clone sorting.
3. Appropriate fluorescence minus one (FMO) controls (see Table 1 and Note 4).

2.6. Analysis and Presentation of Data

1. Flow cytometry analysis software (see Note 5).

3. Methods

3.1. Isolation and Preparation of Mouse Bone Marrow Cells

In this section, we describe the dissection of mouse bones and subsequent recovery of bone marrow cells.

1. Euthanize mice according to locally approved procedure.

2. Disinfect the skin of the mouse with 70% ethanol using a spray bottle. Prepare 1 mL of cold buffer per mouse to collect the bones.

3. Using clean instruments, make a transverse cut in the abdominal area and draw the skin laterally to open the abdominal lavage.

4. Inspect liver and spleen for possible signs of disease, abdominal carcinomas, or organomegaly. Signs of disease should be a very rare event in young wild-type mice.

5. Cut off both feet. Hold the knee joint with one forceps and the proximal part of the tibia with the other forceps, and bend the latter anteriorly to break off the tibia. Clear the tibia from tissues (muscles, tendons, etc.) using a scalpel and put bones in cold buffer.

6. Grab distal femur with one pair of forceps and the knee joint with the other, and bend the latter anteriorly to break off the knee joint. Cut the femur loose as proximal as possible, clear from other tissues using a scalpel and transfer to cold buffer.

7. To isolate the Iliac crest (see Note 6), hold with a forceps the site (bone fragment) from which the femur was cut, cut about 2 cm upwards medially from the forceps and pull out the crista. Remove other tissues and transfer to cold buffer.

8. After isolation of all bone fragments, transfer to a mortar and gently crush the bones. Flush crushed bones with isolation buffer and pipette up and down with cold buffer, followed by filtering the suspension (80-μm cup filter) into an appropriately sized collection tube (see Note 7).

9. Centrifuge the cell suspension at $400 \times g$ for 10 min and resuspend cell pellet in 1× lysis buffer if red blood cell lysis is performed (see Note 8); 200 μL/mouse. Incubate for 1 min at room temperature. Add isolation buffer (1 mL per mouse) and filter to rid clumps of cell debris. Centrifuge again at $400 \times g$ for 10 min and resuspend in an appropriate volume of isolation/staining buffer.

This relatively fast method should yield a recovery of bone marrow cells in the range of 1–1.5×10^8 cells per mouse (=one mouse BM equivalent).

3.2. Pre-enrichment of Mouse Bone Marrow Cells

In this section, we describe the enrichment for the target populations using negative selection of cells expressing mature "lineage markers," so-called lineage depletion (see Note 9). These lineage-depleted cells are the cells of interest, since cells within the more immature bone marrow compartments lack expression of these markers (14, 15). For several reasons, we find pre-enrichment of the sample advantageous (see Note 10). Enrichment for lineage-negative cells is performed using either an autoMACS according to instructions from the supplier, or using MACS separation columns and separator magnets as described below. For quantitative analysis of population frequencies in unfractionated bone marrow, the sample or an aliquot of the sample is not enriched (see Note 11) and bone marrow cells are directly targeted for cell surface staining (see Subheading 3.3).

1. Centrifuge cell suspension (at $400 \times g$ for 10 min; this speed and time are used throughout) and resuspend one mouse BM equivalent in 200 μL of buffer containing anti-lineage antibodies at predetermined concentrations.

2. Incubate on ice for 15 min, wash with 1 mL of buffer per mouse equivalent, and centrifuge.

3. Resuspend in 100 μL of buffer per mouse equivalent. Vortex anti-biotin Microbeads stock, add 10 μL of Microbeads stock per mouse equivalent, mix, and incubate on ice for 20 min. Vortex once or twice during incubation.

4. Wash with 1 mL of buffer per mouse equivalent, centrifuge, and resuspend in 250–500 μL of buffer per mouse equivalent (although a minimum of 1 mL).

5. Place the column on the magnet (for up to three mice: use MS columns; for 4–7 mice: use LS columns) and rinse the columns (twice with 1 mL of buffer for MS and twice with 3 mL of buffer for LS columns).

6. Place a 80-μm Filcon cup type filter on top of the column and apply the cells to the column. Apply three times washing volume (1 mL for MS and 3 mL for LS columns) to the filter/column and collect the lineage-depleted fraction.

7. If lineage-positive cells are needed (for instance, to evaluate enrichment efficiency), take the column from the magnet, add one volume of washing volume, and flush out the lineage-positive enriched fraction (see Note 12).

3.3. Cell Surface Staining of Mouse Bone Marrow Cells

Prior to the actual staining of cells, antibody cocktails (containing antibodies at predetermined concentrations in staining buffer) are prepared for the staining steps in which the cells are stained with two or more antibodies simultaneously. This includes antibody cocktails for both the actual samples and the FMO (see Note 4) controls (see Note 13). The staining volumes indicated herein can be adjusted depending on cell numbers.

1. Following enrichment by lineage depletion, aliquot a small fraction of the enriched cells to separate Eppendorf tubes for FMO controls. Centrifuge cells and resuspend all FMOs and sample(s) in FcγRII/III-PE containing buffer; 50 μL per mouse equivalent (see Note 14).

2. Incubate for 2 min and add an equal volume of antibody cocktail containing the remaining antibodies (at double the optimal concentrations to reach a final optimal concentration of antibodies).

3. Incubate for 30 min on ice in the dark.

4. Wash and resuspend at a cell density of 10^8 cells/mL of buffer containing PI at a 1:1,000 diluted concentration.

5. Store in dark on ice until acquisition on the flow cytometer.

3.4. Compensation Procedures

As compensation in multi-color stainings becomes increasingly complicated for each additional parameter, we use automatic software compensation. Ideally, compensation is performed using identical material (cells) and antibodies as in the actual experiment. However, many of the cell surface proteins in the described stainings are expressed on very infrequent population and/or at dim levels, complicating compensation procedures. Capture beads provide a good alternative and are, therefore, recommended in these protocols (see Note 15).

1. Prepare single-stained compensation controls by aliquoting 100 μL of buffer into an Eppendorf tube for each of the fluorochromes used herein and adding 25 μL (or a small drop) of CompBeads to each tube (vortex bead stock first). No compensation control is generated for PI.

2. Add 1 μL of primary antibody to each tube and mix. In the case of markers that are detected with secondary reagents (i.e., biotinylated lineage antibodies), stain capture beads first with the primary antibody (for instance, B220-biotin), then wash and add the secondary reagents. In the case where primary fluorochrome antibodies are not recognized by capture beads (i.e., wrong isotype of species of primary antibody), use an appropriate fluorochrome-conjugated primary antibody recognized by the capture beads (see Note 15).

3. Vortex and incubate for 10 min on ice in the dark. Resuspend each compensation control in 400 μL of buffer and put aside in dark on ice.

4. Aliquot 500 μL to an Eppendorf tube and add two drops of negative control CompBeads.

5. Take the unstained cells to the flow cytometer to set PMT values for all detectors, including the channel dedicated to the PI signal.

6. Acquire all single-stained compensation controls and calculate compensation values across all included detectors according to software instructions.

3.5. Acquisition, Gating Strategies, and Sorting of Target Cells

3.5.1. Acquisition

We recommend to always filter samples directly prior to acquisition to minimize the risk of clogs.

1. Filter an aliquot of the lineage-enriched cell fraction and run on the flow cytometer to set appropriate gains for forward and side scatter.

2. Filter samples and run all FMO controls, followed by the actual sample(s).

3. When sorting, acquire a sufficient amount of events to set sorting gates.

4. For frequency determination, acquire sufficient amounts of events to obtain statistically sound data, as discussed in Notes 10 and 11.

In our experiments, we use the BD FACSAria with a 70-μm nozzle and run the machine at high pressure (70 psi). We follow the manufacturer's recommendations on drop drive frequency at this pressure (typically 88–90 kHz).

3.5.2. Gate Setting Strategies

In this section, we provide a strategy on how to set gates that define the different immature myeloerythroid cellular subsets. We will base this discussion on the plots as depicted in Fig. 1b (see Note 16). Please note that most parameters are presented using a biexponential, or "logicle," display (16) that uses alternative scaling of the lower end of the axis. This allows for presentation also of negative values and avoids events from "sticking to the axes," thereby maximizing visualization of data.

1. Figure 1b, upper plots, most left: Displaying FSC area against FSC height allows for exclusion of most cellular doublets.

2. Figure 1b, upper plots, second from the left: PI-positive cells (i.e., dead cells) are excluded.

3. Figure 1b, upper plots, middle: Plotting FSC area versus SSC area excludes smaller or larger particles such as unwanted cells and debris.

4. Figure 1b, upper plots, second from the right: Lineage-negative cells are defined by plotting cKit versus lineage. Most cKit high cells are lineage negative/low. This gate could be difficult to define on an enriched sample (as in Fig. 1b) and is easier set in an unenriched sample, such as in Fig. 3b, right plot.

5. Figure 1b, upper plots, most right: This gate is set based on other "reference cells" within the same plot. Sca-1 negativity

is based on absent Sca-1 expression in the majority of cKit medium expressing cells. High cKit expression is compared to cKit expression in cKit⁺Lin⁻Sca1⁺ (KLS) cells. A more generous gate for Sca-1 expression will include cells that are more immature, and a more generous gate for cKit expression will primarily include more immature erythroid-restricted progenitors.

6. Figure 1b, lower plots, left: We see two alternatives to set these gates. When performing these staining for the first time, we strongly recommend including FMOs for gate setting controls. The gates in this plot are set based on FMO-APC and FMO-FITC, as presented in Fig. 2a, left panels. However, preparation of FMOs for each parameter could be time consuming and/or difficult. In addition, FMOs are of little use to separate cells that express medium versus high levels of a certain antigen (as could be the case for CD105 in some of the myeloerythroid staining protocols (13)). Therefore, we frequently use internal reference populations (IRP) and find these of great value. IRPs are cell populations within the tested sample that can serve as positive or negative references, as illustrated in Fig. 2b. A requirement for the use of IRPs is

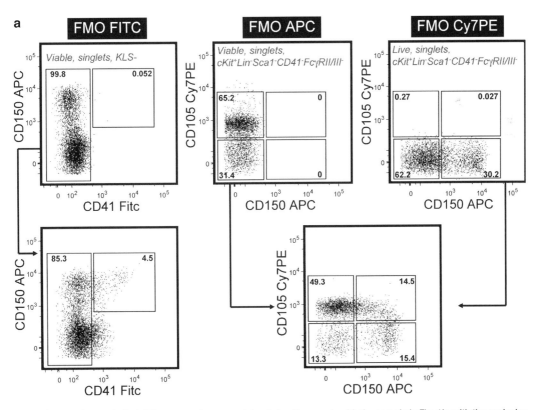

Fig. 2. Gate setting controls. (a) These samples were stained simultaneously with the sample in Fig. 1b, with the exclusion of the indicated fluorochromes, allowing for fluorescence minus one analysis as illustrated.

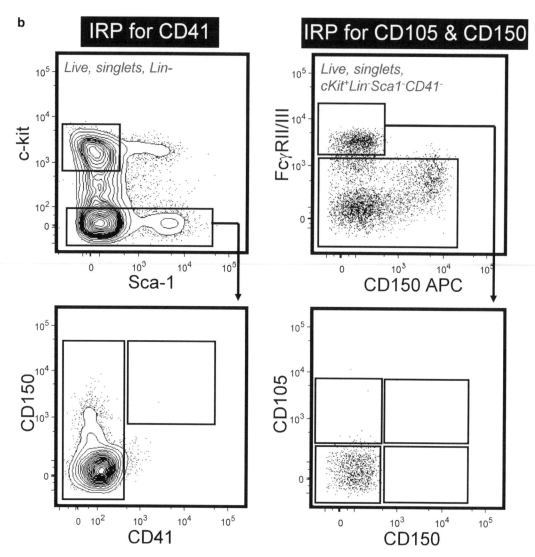

Fig. 2. (continued) (**b**) Within the sample in Fig. 1b, previously defined internal reference populations (IRPs) were used to define negative versus positive expression levels of CD41 (*left panels*) or CD105 and CD150 (*right panels*) within the cKit⁻Lin⁻Sca1± and cKit⁺Lin⁻Sca1⁻ CD41⁻FcγRII/III⁺ populations, respectively.

pre-existing knowledge of cell surface expression of the marker of interest within these IRP, obtained, for instance, through the use of FMO in earlier experiments and/or based on information from the literature.

7. Figure 1b, lower plots, middle: FcγRII/III high versus negative/low-expressing cells are defined by FcγRII/III expression in the CD150-positive cells in this plot.

8. Figure 1b, lower plots, right: The gates in these plots are defined by FMOs for Cy7PE and APC, as illustrated in Fig. 2a, or can alternatively be set based on an IRP, as presented in Fig. 2b.

3.5.3. Single-Cell and Bulk Sorting of Target Cells

1. Set up the machine as describe above. Take the samples (including FMOs) and run enough events (about 20,000–30,000 of lineage-depleted cells) to allow for proper gate setting, as described in Subheading 3.5.2 and Note 16.

2. Optimize the FACS machine for cell sorting (set drop delay, position side stream, etc.) according to the manufacturer's protocol. As the cellular subset in these protocols is relatively infrequent, it is important to sort in the highest purity mode.

3. Adjust speed of sample acquisition depending on the purity of the sample. Too high speed increases the electronic abort rate, while viability can be affected if sorting procedure takes too long. For bulk sorting, we typically run at a higher speed (about 2,000–4,000 events/s) than for single-cell sorting (about 1,000 events/s).

4. Prior to cell sorting, decide the number of desired target cells, and sort the exact number of desired cells directly into the appropriate media used in downstream applications (cell culture media, lysis buffer, etc.) (Note 17).

5. When performing co-cultures with, for instance, stromal cells or transplantation with bone marrow support cells, sort the target cells in a medium that already contains the other cell type.

6. When sorting cells into cell culture plates (96- and 48-well Terasaki plates or other formats), we sort our cells directly into plates containing appropriate media and thereafter transfer plates directly to the incubator (Note 18).

7. When sorting into tubes, vortex the collection tube immediately prior to sorting (for preventing sorted cells from sticking to the sides) and then vortex immediately to mix cells in the media. We recommend using low-retention tubes to prevent the cells from adhering to the tube wall.

3.6. Analysis and Presentation of Data

Some principles for the analysis of these data were already discussed in Subheading 3.5.2. We see some additional considerations that should be taken into account.

1. For frequency determination of cellular subsets in total bone marrow cells, calculate frequencies as percentage of total live cells (Fig. 1b, gate in middle plot, upper row), as presented in Fig. 3b.

2. In plots with many events, use contour plots. However, in cases of very low numbers of events, these plots may become misleading; then use dot plots instead.

3. Of the different commercial analysis software we have tested to date, we find Mac-based FlowJo analysis software to be a good compromise in terms of options, stability, and speed. This latter point may become an issue in cases where multiple, very large data files are to be analyzed.

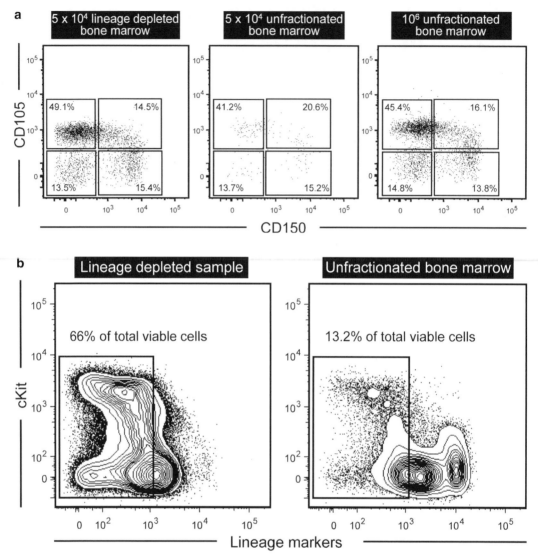

Fig. 3. Advantage of pre-enrichment for analysis and sorting purposes. Parallel to the staining of the enriched sample in Fig. 1b, an aliquot of unenriched (unfractionated) bone marrow cells was stained according to similar procedures. (a) Acquisition of an equal amount of events (50,000) gives a sufficient number of target event and statistical power for the enriched sample (*left plot*), but not for the unenriched sample (*middle plot*). Only by acquisition of many more events (1 × 10⁶) from the unenriched sample can sufficient numbers of targets cells be collected, although at the expense of signal qualities in terms of population discrimination and unspecific staining/events (*right plot*). (b) Pre-enrichment of a sorting sample results in higher frequencies of lineage-negative live cells (*left plot*), compared to an unenriched sample (*right plot*), highlighting its major advantage during cell sorting, as discussed in Note 10.

4. Notes

1. Manufacturer's protocol recommends degassed buffer to obtain optimal MACS-based selections and recovery. However, we have found that using media prepared at least

24 h prior to experimentation, combined with a filter placed on top of the column when loading the cell suspension, also yields acceptable selection and recovery.

2. Antibodies used herein are commercially available at one or several suppliers; we obtain most antibodies from BD Biosciences, e-Bioscience (San Diego, CA), or BioLegend (San Diego, CA).

3. We tend to prefer PI because of its brightness, relative long shelf life, and easier handling (i.e., no need to prepare intermediate stock solutions).

4. FMO. These controls are used as biological or "gate setting" controls and are important when defining a negative versus positive gate for cells expressing dim levels of a certain cell surface marker (i.e., no clear separation between a negative and positive population).

5. Analysis can be performed using the acquisition software of the cytometer, although we prefer more specialized flow cytometry analysis software (i.e., FlowJo for Macintosh, TreeStar Inc, Ashland, OR). Although FlowJo is also available on a PC platform, it is our experience that the PC version does not work as smoothly as the Macintosh version. Because the files to be analyzed are typically very large, this is an important consideration.

6. The recovery from two femurs and two tibiae of a one mouse is around 0.8–1.2×10^8 unfractionated bone marrow cells (variability is dependent on the size of mice and typically therefore of mouse age and gender). By isolating cells also from the two hip bones, although technically somewhat more demanding, this yield can be increased to another 0.2–0.4×10^8 cells.

7. It is our experience that crushing rather than flushing bones using a syringe renders a higher yield of cells. Other investigators have in addition claimed that certain cellular subsets do not release effectively using flushing (17).

8. We usually perform lysis of red blood cells when the cells are only subjected to analysis. When staining cells for sorting and subsequent functional enumerations, we usually do not lyse the sample as we suspect a negative impact of the lysing procedure on the viability of certain cell subsets.

9. Apart from depletion of lineage-negative cells, we have, as an alternative, frequently enriched our samples by MACS-based positive selection of cKit (CD117)-expressing cells. However, we have noticed that cKit enrichment influences expression levels of cKit, presumably because the fluorochrome-conjugated 2B8 clone used to visualize cKit expression is the same antibody coated on cKit magnetic beads, hence partially

blocking efficient staining. This makes careful titration of cKit beads necessary (the recommended amount of beads from supplier does not allow appropriate visualization of cKit with 2B8). Also, we previously performed lineage depletion using the Dynal bead system (Invitrogen) that uses selection for sheep-anti-rat magnetic Dynabeads binding to purified anti-lineage antibodies, but find this method inferior in terms of enrichment efficiency, handling speed, and cost.

In this protocol, we do not use any purified antibodies that are typically visualized by fluorochrome-conjugated goat-anti-rat antibodies. However, if one chooses to include a purified antibody, then both the primary and secondary stains are performed prior to enrichment as the anti-lineage antibodies used herein were generated in rat.

10. Pre-enrichment of the sample, as opposed to staining and sorting from unfractionated bone marrow cells, is advantageous for several reasons. *First*, cell surface staining on a pre-enriched sample that contains higher frequencies of the target cells will give better signals, less unspecific staining, and less noise, as indicated in Fig. 3a. *Second*, obtaining a sufficient amount of events for proper statistical analysis often requires pre-enrichment, as a sufficient number of acquired events from an unenriched sample will lead to very large data files (Fig. 3a). This could cause problems when handling/analyzing these files. *Third*, when sorting for these infrequent cellular subsets, pre-enrichment of the sample increases the frequencies of the cells of interest (Fig. 3b), giving superior purity, yield, and recovery of the cells of interest when compared to sorting from an unenriched sample, in addition to dramatically reducing the sorting time.

11. In several cases, it is desirable to obtain information on frequencies of these subsets as a percentage of total live bone marrow cells, and here a pre-enriched sample does not provide with this information. This issue can be dealt with in different ways: (1) cell surface staining is performed on unfractionated bone marrow cells. Subsequently, for each sample, a file is collected with total events, followed by a file in which only "gated events" (for instance, only cells within a lineage-negative or a cKit-positive gate) are collected on the flow cytometer, allowing for back-calculation of actual frequencies. (2) For each individual, an unenriched and enriched sample is prepared, stained, and acquired. This allows for recalculation of population frequencies as a percentage of total live bone marrow cells.

12. Here, we recommend recovering the lineage-positive fraction and including an aliquot in the staining protocol in order to evaluate enrichment efficiency (i.e., no or hardly any

lineage-negative cells should be present in this sample), especially when inexperienced with the methodology. In addition, a small aliquot of lineage-enriched cells is useful to later set FSC/SSC gains on the flow cytometer when setting up the machine (see Subheading 3.5.1, step 1).

13. In this protocol, there is no real need for the preparation of a separate antibody cocktail for each of the FMO, as well as the sample(s). Instead, make a "base-cocktail" that contains 2× concentrations of Sca1-PacBlue, cKit-APC-Alexa780, and Streptavidin-Qdot605 (NB: FcγRII/III-PE is excluded; see Note 14). Take 3×100 μL from this cocktail for each of the FMOs and add the appropriate (lacking) two antibodies for each of the FMOs at 2× concentration. Thereafter, complete the "base-cocktail" by adding CD41, CD105, and CD150 at 2× concentration (NB: now the volume is altered and requires adjustment of the antibody volume added to obtain correct concentrations).

14. "Traditional Fc-blocking" to prevent unspecific binding cannot be used herein, as FcγRII/III is one of the targeted parameters in this protocol. This issue can, however, be circumvented by incubating the enriched cells in FcγRII/III-PE only, prior to staining with the additional antibodies. The staining volume can be decreased when pooling cells from multiple mice. Lineage-depleted cells from up to 10 mice are stained in approximately 500–600 μL of antibody cocktail, and cells from up to 15 mice can be stained in approximately 1 mL of antibody cocktail. These volumes are, of course, dependent on the efficiency of the pre-enrichment.

15. Be aware of the availability of different types of capture beads, each possessing affinities for different classes of antibodies. The capture beads used herein bind to most antibodies used in protocols established for mouse cells (because historically most anti-mouse antibodies are of rat species). However, in some specific cases, no antibody binding occurs to these beads, as is the case for the anti-CD150 antibodies (a mouse IgG2a lambda isotype). In such cases, to prepare the single-stained compensation control, use another antibody conjugated with a similar fluorochrome instead, preferably from the same supplier. An alternative is the use of splenocytes, instead of capture beads, to prepare compensation controls. However, in such cases, it needs to be established that the antigen of interest is expressed on splenocytes, and preferably at relatively high levels. Washing and centrifugation procedures for capture beads and splenocytes are identical to those for bone marrow cells.

16. Gate setting purely for analytic purposes allows for the gates to be immediately adjacent to one another, in order to not

exclude any events from the analysis. However, when sorting, we use more restrictive gate setting compared to the gate settings in Fig. 1b in order to enhance purity of sorted cells.

17. We prefer not to wash cells following sorting, as this will inevitably mean loss of cells. However, one exception might be in the case of sorting large numbers of cells into a relative small volume since the sheath buffer we use (commercial from BD Biosciences) contains a low amount of detergent. In such cases, one could consider washing and counting the cells afterwards, alternatively to setup and run the FACS machine with PBS instead.

18. Performing cell culture experiments and especially more long-term culturing always involves a risk of contamination during incubation time. We prepare and stain our cells non-sterile on the laboratory bench, and use non-sterile but relatively fresh staining buffers and non-sterile but clean plastics. However, target plates and target media are always (prepared) sterile. Opening a cell culture plate by the FACS sorter increases the risk of obtaining contaminating particles in the culture medium. In addition to running a long clean cycle with ethanol in the FACS machine, we always clean all surfaces on and around the FACS machine with 70% ethanol, use gloves, and try to minimize traffic around the machine. Taking these precautions, we find contamination not to be a problem.

Acknowledgments

Bengt Johansson-Lindbom is gratefully acknowledged for reading and commenting the contents of this chapter. This work was supported in part by grants from the Swedish Cancer Foundation, the Swedish Medical Research Council, the Swedish Pediatric Childhood Cancer Foundation, and the Swedish Strategic Research Foundation to DB, and a public health grant (ALF) to CJHP.

References

1. Bryder, D., Rossi, D. J., and Weissman, I. L. (2006) Hematopoietic stem cells: the paradigmatic tissue-specific stem cell. *Am J Pathol* **169**, 338–346.

2. Perfetto, S. P., Chattopadhyay, P. K., and Roederer, M. (2004) Seventeen-colour flow cytometry: unravelling the immune system. *Nat Rev Immunol* **4**, 648–655.

3. Kondo, M., Weissman, I. L., and Akashi, K. (1997) Identification of clonogenic common lymphoid progenitors in mouse bone marrow. *Cell* **91**, 661–672.

4. Akashi, K., Traver, D., Miyamoto, T., and Weissman, I. L. (2000) A clonogenic common myeloid progenitor that gives rise to all myeloid lineages. *Nature* **404**, 193–197.

5. Allman, D., Sambandam, A., Kim, S., Miller, J. P., Pagan, A., Well, D., Meraz, A., and Bhandoola, A. (2003) Thymopoiesis independent of common lymphoid progenitors. *Nat Immunol* **4**, 168–174.

6. Balciunaite, G., Ceredig, R., Massa, S., and Rolink, A. G. (2005) A B220+ CD117+ CD19− hematopoietic progenitor with potent lymphoid and myeloid developmental potential. *Eur J Immunol* **35**, 2019–2030.

7. Mansson, R., Zandi, S., Anderson, K., Martensson, I. L., Jacobsen, S. E., Bryder, D., and Sigvardsson, M. (2008) B-lineage commitment prior to surface expression of B220 and CD19 on hematopoietic progenitor cells. *Blood* **112**, 1048–1055.

8. Rumfelt, L. L., Zhou, Y., Rowley, B. M., Shinton, S. A., and Hardy, R. R. (2006) Lineage specification and plasticity in CD19-early B cell precursors. *J Exp Med* **203**, 675–687.

9. Adolfsson, J., Mansson, R., Buza-Vidas, N., Hultquist, A., Liuba, K., Jensen, C. T., Bryder, D., Yang, L., Borge, O. J., Thoren, L. A., *et al.* (2005) Identification of Flt3+ lympho-myeloid stem cells lacking erythro-megakaryocytic potential a revised road map for adult blood lineage commitment. *Cell* **121**, 295–306.

10. Lai, A. Y. and Kondo, M. (2006) Asymmetrical lymphoid and myeloid lineage commitment in multipotent hematopoietic progenitors. *J Exp Med* **203**, 1867–1873.

11. Yoshida, T., Ng, S. Y., Zuniga-Pflucker, J. C., and Georgopoulos, K. (2006) Early hematopoietic lineage restrictions directed by Ikaros. *Nat Immunol* **7**, 382–391.

12. Pronk, C. J., Attema, J., Rossi, D. J., Sigvardsson, M., and Bryder, D. (2008) Deciphering developmental stages of adult myelopoiesis. *Cell Cycle* **7**, 706–713.

13. Pronk, C. J., Rossi, D. J., Mansson, R., Attema, J. L., Norddahl, G. L., Chan, C. K., Sigvardsson, M., Weissman, I. L., and Bryder, D. (2007) Elucidation of the phenotypic, functional, and molecular topography of a myeloerythroid progenitor cell hierarchy. *Cell Stem Cell* **1**, 428–442.

14. Spangrude, G. J., Heimfeld, S., and Weissman, I. L. (1988) Purification and characterization of mouse hematopoietic stem cells. *Science* **241**, 58–62.

15. Uchida, N., Aguila, H. L., Fleming, W. H., Jerabek, L., and Weissman, I. L. (1994) Rapid and sustained hematopoietic recovery in lethally irradiated mice transplanted with purified Thy-1.1lo Lin-Sca-1+ hematopoietic stem cells. *Blood* **83**, 3758–3779.

16. Parks, D. R., Roederer, M., and Moore, W. A. (2006) A new "Logicle" display method avoids deceptive effects of logarithmic scaling for low signals and compensated data. *Cytometry A* **69**, 541–551.

17. Haylock, D. N., Williams, B., Johnston, H. M., Liu, M. C., Rutherford, K. E., Whitty, G. A., Simmons, P. J., Bertoncello, I., and Nilsson, S. K. (2007) Hemopoietic stem cells with higher hemopoietic potential reside at the bone marrow endosteum. *Stem Cells* **25**, 1062–1069.

Chapter 14

Flow Cytometry Immunophenotyping of Hematolymphoid Neoplasia

Katherine R. Calvo, Catharine S. McCoy, and Maryalice Stetler-Stevenson

Abstract

Flow cytometry immunophenotyping (FCI) of hematolymphoid specimens is a powerful tool in clinical medicine aiding in the accurate diagnosis and subclassification of acute and chronic leukemias, non-Hodgkin lymphomas, myelodysplasia, and other hematolymphoid neoplastic processes. Multiple flow cytometric strategies are used to evaluate hematolymphoid populations, including identification of neoplastic populations with aberrant immunophenotypes, abnormal maturation patterns, monotypic kappa/lambda light chain expression, restricted V-beta expression, and abnormal light scatter properties. In this chapter, we present a general approach to analyze hematolymphoid populations. Specimen handling, gating strategies, and appropriate antibody panels for flow cytometric detection of abnormal hematologic populations are presented.

Key words: Leukemia, Lymphoma, Myelodysplastic syndrome, Immunophenotype, Prognosis

1. Introduction

Flow cytometry immunophenotyping (FCI) analysis has many uses in the diagnosis and treatment of hematologic neoplasia. Some of these include (1) diagnostic support and subclassification of hematolymphoid neoplasia; (2) detection of antigenic expression on tumor cells that may be important for prognosis or therapy (e.g., ZAP-70 or CD38 on CLL cells, CD52 expression on tumor cells); and (3) clinical management through detection of minimal residual disease or recurrence. Additional medical indications for FCI include clinical symptoms suspicious for malignancy (e.g., lymphadenopathy, organomegaly, and tumor mass) and abnormal clinical findings (e.g., cytopenias, lymphocytosis, and presence of morphologically atypical cells or blasts on

Teresa S. Hawley and Robert G. Hawley (eds.), *Flow Cytometry Protocols*, Methods in Molecular Biology, vol. 699,
DOI 10.1007/978-1-61737-950-5_14, © Springer Science+Business Media, LLC 2011

peripheral smear or biopsy material) (1). FCI is less useful for conditions associated with sampling difficulties, i.e., a sequestered tumor source (non-circulating, localized tumors), a paucity of viable tumor cells (e.g., necrotic tissue), increased cellular fragility, or insufficient tumor yield (e.g., Hodgkin lymphoma or patchy disease distribution).

Proper collection, processing, analysis, and correlation with ancillary testing are crucial to quality FCI. Adverse storage conditions or use of fixatives before staining could damage the cells of interest or alter the specimen immunophenotype, yielding a false-negative interpretation (e.g., no tumor present). Tumor populations could be missed with inadequate sample collection (biopsy or aspiration of non-malignant material), the antibody panel selection could be misdirected with insufficient clinical history, and gating strategies could be too restrictive or too inclusive to detect and isolate the tumor population adequately. Finally, quality FCI in a clinical setting should be correlated with clinical history and other laboratory analyses performed on the same patient material (e.g., morphology, cytology, histology, cytogenetics, and molecular PCR testing) to confirm the FCI results and/or to provide a more definitive overall diagnosis.

Use of normal cells found in most specimens provides an important "internal control" to further ensure quality FCI. Normal cells can be used to establish autofluorescence and fluorescence patterns for each antibody combination. These normal patterns are useful for comparison with aberrant patterns that are potentially representative of tumor populations within the same sample. Detailed guidelines for other important clinical laboratory quality control, quality assurance, and instrument performance considerations are beyond the scope of this chapter. For more information, please refer to additional referenced documents (2, 3).

2. Preanalytic Considerations

2.1. Specimen Source

Clinical samples frequently sent for FCI of suspected hematolymphoid neoplastic involvement include blood; bone marrow; lymph node and tumor mass aspirates; pleural, acetic, synovial, ocular, or cerebrospinal fluids; bronchoalveolar lavages; and tissue from lymph node, spleen, or other organs. All specimens should be labeled with a unique patient identifier, date of collection, and specimen source. A complete patient history including any interfering medications or therapies is required to aid antibody panel selection and correlation of FCI results.

2.2. Blood and Bone Marrow

In general, patients with leukocyte counts between 3 and 10×10^6/mL will require 10 mL of peripheral blood or 1–3 mL of bone marrow aspirate (see Note 1). The anticoagulant choice for blood or bone marrow is sodium heparin or EDTA (see Note 2). If transport is delayed or storage before processing is necessary, store peripheral blood in the original collection tube at 18–25°C for up to 48 h (heparin) or 24 h (EDTA). To store bone marrow, dilute with an equal volume or more of RPMI 1640 and 10% heat-inactivated FBS (caution- see Note 3) and store at 4–8°C for up to 24 h.

2.3. Fluids and Aspirates

Specimen volume for fluids and aspirates will vary by patient. In general, 0.5–5 mL for cerebrospinal fluid (CSF) and fine needle aspirates, and 10–50 mL for other bodily fluids are sufficient. Fluid and fine needle aspirate specimen sources are often pauci-cellular and do not require an anticoagulant. The cells of interest in these specimens can deteriorate quickly and processing time should be minimized, i.e., cell loss in CSF begins within 30 min of collection. RBC lysis should be limited to specimens with significant RBC contamination to decrease the loss of malignant cells in these paucicellular specimens. If transport is delayed or storage before processing exceeds 30 min, dilute CSF or ocular fluid with an equal volume of RPMI 1640 with 10% heat-inactivated FBS and store at 4–8°C for 18 h. Fluids other than CSF or ocular fluids, can be stored at 4–8°C for up to 18 h.

2.4. Solid Tissue Biopsies

Fresh specimens from solid tissue sources such as intact lymph nodes, tumor masses, or organ and tissue biopsies do not require anticoagulant. Generally, a 0.5–2-cm^3 piece of fresh tissue is sufficient for FCI analysis. The cells may deteriorate rapidly and should be transported to the laboratory quickly in an isotonic solution such as saline or RPMI at 4–25°C. Using mechanical dissociation in an isotonic solution, a cell suspension is prepared and filtered through a 50-μm nylon mesh. Prepare a cytocentrifuge slide for morphologic review to confirm the presence of the cells of interest and begin processing immediately. Store the cells in RPMI 1640 with 10% heat-inactivated FBS at 4–8°C for up to 18 h caution- (see Note 3).

2.5. Suboptimal Specimens

Specimens should not be allowed to dry out, freeze, or come into contact with fixative for any amount of time. Specimens not meeting collection recommendations should be carefully evaluated for recollection or alternative testing. If the specimen is irreplaceable and diagnostically useful information can be obtained from the sample, FCI may be attempted with appropriate cautionary notations in the final report (see Note 4).

3. Analytic Considerations

3.1. Sample Preparation, RBC Lysis, and Viability Assessment

1. *Specimen Processing.* To prevent cell loss, manipulate specimens as little as possible. To prevent labile antigen deterioration, begin processing as soon as possible (see Note 4) to determine specimen quality, viability, and the need for RBC lysis (see Note 5). In general, *adjust* cell concentrations to yield $0.2–2.0 \times 10^6$ cells per tube. For minimal residual disease detection, increase the number of cells per tube to yield at least 200 tumor cells.

2. *Viability.* In general, specimens with less than 75% viability should be recollected; efforts should be made to obtain diagnostically useful FCI from irreplaceable specimens as low viability samples consisting predominantly of neoplastic cells may yield valuable diagnostic information. Viability assessment can be determined manually using trypan blue or on the flow cytometer using fluorescent dyes (e.g., 7AAD or propidium iodide). Manual viability has the advantage of keeping all fluorescent channels available for other determinations, while the use of a fluorescent viability dye can be included in the gating strategy for nonviable versus viable analysis (see Subheading 3.2, step 1).

3.2. Antibody Panel Design, Staining, and Internal Controls

1. *Antibody Panel Design.* FCI of leukemia and lymphoma includes assessment of the presence, absence, and level of expression of antigens on the normal and neoplastic cells in a submitted specimen. Because most antigens lack absolute lineage specificity and neoplastic hematolymphoid cells may be missing or expressing antigens from several lineages or stages of maturation at once, a FCI panel for leukemia and lymphoma detection will include a broader range of antibodies than that required for FCI subset analysis. In addition, antibody panels should be designed with consideration for patient history, specimen source, review of morphology and other concurrent laboratory findings, and the medical indication for FCI. The panel should include enough antibodies to allow the recognition of normal and abnormal populations present in the sample while being as efficient and cost effective as possible (1–3). For example, a panel to evaluate nonspecific anemias will require a larger number of antibodies due to the broad range of potential neoplastic conditions that may cause anemia in contrast to that of a staging panel for a known non-Hodgkin B-cell lymphoma specimen. By designing sufficient antibody panels, FCI can identify the representative populations in a specimen.

   Two alternative approaches based on medical indication and review of clinical history, specimen source, morphologic review, etc., are used for designing hematolymphoid neoplasia panels: (1) a single comprehensive panel of fluorescent antibodies to identify every potential differential diagnosis and (2) multiple sequential panels beginning with a primary screening panel and subsequent reflex panel(s) to narrow the differential diagnosis according to lineage. In general, using a comprehensive panel is economical in terms of processing time and provides high sensitivity and specificity for abnormal cell detection and characterization due to the broad range of antibodies employed; however, reagent costs tend to be higher and may not be offset by the savings in personnel or time compared to the second approach. Appropriate antibody panels based on medical indication are presented in Table 1 (see Note 6).

In addition to antibody selection, panel design requires an understanding of antigen expression patterns in normal and neoplastic cells as well as the efficiency of the fluorescent emission (bright vs. dim intensity) of each fluorochrome. Antigens expressed in low concentration relative to other antigens should be tested using antibodies conjugated to bright fluorochromes, while high concentration antigens can be tested using antibodies conjugated to dim fluorochromes.

The number of fluorochrome-conjugated antibodies in each staining tube will depend on several factors. These factors include the number of fluorescent detectors available on the flow cytometer, the complexity of the FCI required, the availability of appropriate antibodies, and the cellular yield in the submitted specimen (e.g., increasing the number of antibodies in each tube may reduce the total number of tubes needed for paucicellular specimens). While it may not be necessary to use the same number of antibodies in every tube, a minimum of four antibodies per tube in the majority of tubes in a leukemia or lymphoma FCI panel is strongly recommended (2).

An efficient panel utilizes specific antibody combinations in each tube to provide additional immunophenotypic information for each antibody beyond positive or negative. One strategy is to design a "lineage" tube to verify that all expected antigens are represented on a specific cell population (e.g., CD2/CD3/CD5/CD7 form a four-color T-cell lineage tube). Another strategy is to design a "subpopulation" tube to identify related populations. This type of tube generally includes an antibody common to all of the subpopulations being tested. Some examples include a T-cell subpopulation tube with CD3 as the common antibody and CD4 and CD8 as the subpopulation antibodies. A B-cell subpopulation tube may be designed to demonstrate light chain subpopulations (CD19 as the common antibody, kappa and lambda as the subpopulation antibodies) or maturation subpopulations (CD19 as the common antibody and CD10, CD20, and CD34 to identify

Table 1
Minimum antibody panels based upon medical indication for FCI

Medical indication	Minimum panel
Anemia, thrombocytopenia, pancytopenia	kappa, lambda, HLA-DR, CD2, CD3, CD4, CD5, CD7, CD8, CD10, CD11b, CD13, CD14, CD15, CD16, CD19, CD20, CD33, CD34, CD38, CD45, CD56, CD64, CD117
Persistent lymphocytosis	kappa, lambda, CD2, CD3, CD4, CD5, CD7, CD8, CD10, CD19, CD20, CD38, CD45, CD56
Monocytosis	HLA-DR, CD3, CD7, CD11b, CD13, CD14, CD15, CD16, CD19, CD33, CD34, CD38, CD45, CD56, CD64, CD117
Eosinophilia	HLA-DR, CD3, CD4, CD5, CD7, CD8, CD11b, CD13, CD14, CD15, CD16, CD19, CD33, CD34, CD45, CD56, CD64, CD117
Thrombocytosis with abnormal platelet forms	HLA-DR, CD3, CD11b, CD13, CD14, CD15, CD16, CD19, CD33, CD34, CD45, CD56, CD117
Blasts in blood or bone marrow	kappa, lambda, HLA-DR, CD2, CD3, CD4, CD5, CD7, CD8, CD10, CD11b, CD13, CD14, CD15, CD16, CD19, CD20, CD33, CD34, CD38, CD45, CD56, CD64, CD117
Lymphadenopathy	kappa, lambda, CD2, CD3, CD4, CD5, CD7, CD8, CD10, CD19, CD20, CD45, CD56
Extranodal masses	kappa, lambda, CD2, CD3, CD4, CD5, CD7, CD8, CD10, CD19, CD20, CD33, CD38, CD45, CD56
Splenomegaly	kappa, lambda, HLA-DR, CD2, CD3, CD4, CD5, CD7, CD8, CD10, CD13, CD19, CD20, CD33, CD34, CD38, CD45, CD56
Staging B-cell non-Hodgkin lymphoma	kappa, lambda, CD3, CD5, CD10, CD19, CD20, CD45
Staging T-cell non-Hodgkin lymphoma	CD2, CD3, CD4, CD5, CD7, CD8, CD19, CD45, CD56
Skin rash	kappa, lambda, CD2, CD3, CD4, CD5, CD7, CD8, CD10, CD13, CD19, CD20, CD33, CD45, CD56
Atypical cells in body fluids	kappa, lambda, CD2, CD3, CD4, CD5, CD7, CD8, CD10, CD13, CD19, CD20, CD33, CD34, CD38, CD45, CD56
Monoclonal gammopathy	kappa (cytoplasmic and surface), lambda (cytoplasmic and surface), CD3, CD5, CD19, CD20, CD38, CD45, CD56, CD138
Unexplained plasmacytosis	kappa (cytoplasmic and surface), lambda (cytoplasmic and surface), CD3, CD5, CD19, CD20, CD38, CD45, CD56, CD138

maturation subpopulations). To these lineage or subpopulation tubes, additional antibodies can be added to detect tumor antigens or medical indication antigens, i.e., adding CD5 to a B-cell subpopulation tube to demonstrate clonality in a specimen with partial involvement with mantle cell lymphoma or adding CD52 for therapeutic monitoring.

Another important use of antibody combinations is to provide a common feature across all tubes for analysis gating. While FSC and SSC parameters provide two common parameters in all tubes, the use of antibodies in one or more fluorescent channels as a common "anchor" parameter allows increased FCI gating strategies. For example, a two-tube panel with no common antibodies must rely on scatter gating for panel analysis, while one common antibody in both tubes could provide multiple anchor gating strategies in addition to the original scatter gates (see Subheading 3.4). Lineage-specific antibodies are often used as gating anchors (e.g., CD45 for leukocytes, CD19 for B cells, and CD3 for T cells), and viability dyes can be used as gating anchors as well. Some laboratories use the common parameter idea as a negative or "dump" gating strategy. Whether it is used to select positive or negative events, keeping the common parameter fluorochrome choices the same in all of the tubes will simplify analysis.

2. *Staining Cells.* While staining for hematolymphoid FCI utilizes good laboratory practice common to other flow cytometry studies (single cell suspensions, protect antibodies from light, and follow manufacturer's recommendations), unique challenges include removal of cytophilic or residual plasma immunoglobulin (Ig), staining cytoplasmic or intranuclear antigens, and reagent validation for analyte-specific reagents (ASRs).

To meet these challenges, use the following preparation methods: Remove plasma or cytophilic immunoglobulin and avoid false-negative results (e.g. ,plasma immunoglobulin will compete with the surface B-cell antigens, leaving little or no anti-immunoglobulin reagent staining on the B cells) by pre-lysing or washing the specimen with an isotonic solution, i.e., phosphate-buffered saline, before antibody incubation (see Note 7). Cell permeabilization prior to staining is required for detection of cytoplasmic antigens (e.g., TdT, cytoplasmic immunoglobulin, cytoplasmic CD3, and cytoplasmic CD22). Select the optimal permeabilization reagent for the antigen being studied. When staining surface and cytoplasmic antigens in the same tube, follow the manufacturer's recommendations to ensure that surface staining is not affected by the additional fixation and permeabilization steps (2). Most manufacturers recommend a volume of antibody to use per number of cells to ensure maximum separation between the positive and negative populations. However, each laboratory must validate its antibody combinations for specificity and reactivity. While this is done on each specimen using internal controls (see Subheading 3.2, step 3), a new antibody or combination should be validated before testing patient specimens.

3. *Internal Controls*:

    (a) *Negative Internal Controls.* In accordance with the US–Canadian Consensus Recommendations on the Immunophenotypic Analysis of Hematologic Neoplasia by Flow Cytometry (4), there is no need to run an iso-type control with each specimen. The negative cell popu-lations present within the sample serve the function of an isotype control by distinguishing aberrant fluorescent patterns from background fluorescence due to autofluo-rescence and nonspecific antibody binding (see Notes 3, 8, 9 and 10). For example, the normal T cells present in a specimen can serve as a negative control for anti-CD19 (a B-cell antigen). Establishing negative cell populations is especially important when analyzing myeloid leukemias due to increased fluorescence in myeloid lineages.

    (b) *Positive Internal Controls.* Positive controls are required to confirm specificity and reactivity. Most of the mono-clonal antibodies included in leukemia and lymphoma typing panels react with subpopulations of normal leuko-cytes. The staining patterns of the residual normal cells function as positive internal controls. These normal pat-terns can also be used to compare and contrast aberrant patterns as potentially representative of tumor popula-tions within the same sample (see Note 10).

3.3. Data Acquisition

It is important to acquire an adequate number of events from each tube to represent normal cells (for internal positive and neg-ative controls) and malignant cells (for disease characterization). A minimum of 10,000–50,000 total events per collection tube should be acquired as ungated listmode data (2). Increase acqui-sition events to yield at least 200 tumor cells when testing for minimal residual disease (see Note 11).

3.4. Data Analysis

3.4.1. Objectives and Strategies for Creating Analysis Gates

FCI analysis objectives for suspected hematolymphoid neoplasia include (1) evaluating the specimen processing quality using nor-mal populations (internal positive and negative controls, see Subheading 3.2, step 3); (2) isolating the abnormal population(s) from the normal populations using one or more parameters to determine the neoplastic immunophenotype; and (3) quantifying the abnormal populations relative to the normal populations (see Note 12).

    Multiple gating strategies can be utilized to define abnormal populations or to quantitate antigen expression depending on the sample characteristics and the antibody combinations used to stain the specimen. Gates are defined using parameters common to all tubes, i.e., scatter and fluorescent anchors. The following gating strategies are helpful for meeting the FCI analysis objectives.

1. Light Scatter Gating (Fig. 1a). Because forward-angle light scatter (FSC) and orthogonal (90°) side light scatter (SSC) serve as gating parameters common to each tube, the FSC versus SSC gate is useful for analyzing specimens with a single homogenous population (e.g., lymph node cell suspension) as well as specimens with multiple homogeneous populations, i.e., peripheral blood. Every antibody in a panel can be analyzed with FSC versus SSC gates as long as the populations of interest are homogeneous and distinct. For example, lymphocytes with moderate FSC and low SSC can be distinct from monocytes with moderate FSC and moderate SSC. If a neoplastic population is distinct from normal lymphocytes, e.g., a large cell lymphoma with increased FSC or hairy cell leukemia (HCL) with increased SSC, no other gates may be necessary to isolate the neoplastic population.

   However, most specimens contain heterogeneous populations and will require additional gating strategies to isolate the populations of interest. For example, HCL and monocytes are both moderate FSC/moderate SSC; therefore, further gating will be necessary to isolate these tumor cells in the presence of normal monocytes.

2. Fluorescence Gating (Fig. 1b and c). When two or more tubes have at least one fluorescent parameter in common, the common parameter can be used to create an anchor gate for each of the tubes. The anchor parameter is often a viability dye or a lineage antibody, i.e., 7AAD, CD45, CD3, and CD19 (see Subheading 3.2, step 1), and it is often used with a scatter parameter or a second anchor antibody to create gates for isolating and analyzing heterogeneous populations (see Note 13).

   (a) *Fluorescence versus Scatter (population gating)*. CD45 versus SSC is particularly useful for analyzing bone marrow aspirates and specimens with immature hematolymphoid or non-hematolymphoid populations because CD45 expression of these populations is more heterogeneous than FSC, resulting in a better isolation of the heterogeneous populations than with a FSC versus SSC gate (Fig. 1b). These population gates can be used singly and in combination. For example, blast and granulocyte populations can be evaluated as separate gates as well as one combined gate.

   (b) *Fluorescence versus Scatter (lineage gating)*. Lineage gates can be used to isolate normal and neoplastic hematolymphoid populations for further analysis of subsets, antigen quantitation, etc. While SSC may be more familiar due to the prevalence of CD45 versus SSC analysis, using lineage versus FSC can be used as well to isolate large neoplastic cells (high FSC) and debris (low FSC) from

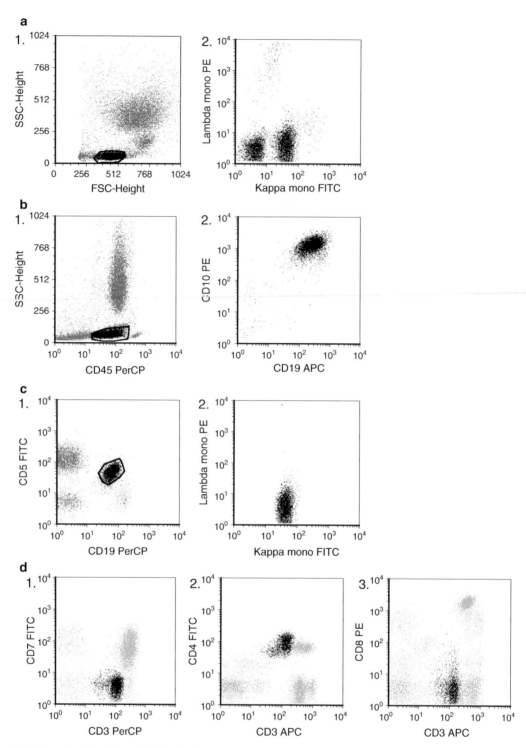

Fig. 1. Flow cytometric data analysis in leukemia and lymphoma. (**a**) Light scatter gating in flow cytometric immunophenotypic analysis. (1) Forward light scatter (FSC) versus side light scatter (SSC) analysis gate in chronic lymphocytic leukemia case. (2) Kappa predominance in FSC versus SSC analysis gate. (**b**) Fluorescence versus scatter gating. (1) CD45 versus SSC analysis gate in acute precursor B-lymphoblastic leukemia bone marrow case. (2) CD19-positive and CD10-positive leukemic cells in CD45 versus SSC gate. (**c**) Antigen gating. (1) Antigen analysis gate based upon CD19 and CD5 co-expression in chronic lymphocytic leukemia case. (2) Kappa-positive and lambda-negative monoclonal B-cell population in CD19 versus CD5 analysis gate. (**d**) Color analysis of abnormal populations: Abnormal CD3 dim T cells are black. (1) Abnormal CD3 dim T cells are CD7 negative. (2) Abnormal CD3 dim T cells are all CD4 positive. (3) Abnormal CD3 dim T cells are all CD8 negative.

other populations. As noted, analysis using lineage versus scatter gating is limited to the tubes containing the specific lineage anchor. Some common lineage antibodies include CD20 for distinguishing B cells; CD3 for T cells; CD38 and CD138 for plasma cells; CD14 for monocytes; CD15 for mature neutrophils and eosinophils; CD71 for erythroid precursors; and CD56 for NK cells and large granular lymphocytes.

Besides isolating populations, the lineage gates can be used to ensure these populations of interest are included in the analysis gate because analysis gating, by definition, limits the data available for evaluation. This "back gating" strategy is especially important in FSC versus SSC gating. For example, gated data containing all CD19-positive events can be evaluated to determine if all fall within the FSC versus SSC scatter gate.

(c) *Fluorescence versus Fluorescence.* Using two fluorescent parameters to set a gate can be useful when characterizing tumor population subsets (Fig. 1c), although this method is limited to tubes containing both antibodies. Once a parameter is used to set a gate, the subsequent histograms should use parameters other than the gating parameter for maximum efficiency. For example, to analyze a CD7/CD26/CD3/CD4 tube for T-cell analysis, set a CD3+ and CD4+ gate to analyze CD7 versus CD26 rather than using the gate to analyze CD7 versus CD3, or CD7 versus CD4, etc.

3.4.2. Using Analysis Gates

Unless the clonal expansion of one tumor has resulted in total replacement of a specimen, gated analysis is often necessary to isolate the cells of interest. In general, the fewer events available or the more similar the neoplastic population is to a normal population, the more a gating strategy may be needed to isolate the population of interest. Each hematopoietic lineage can be recognized by typical staining patterns (e.g., the "checkmark" pattern for CD16 vs. CD13 in normal myeloid populations) in relationship to maturation sequence, antigen expression, and size. Events falling outside these established normal patterns can be further analyzed as potential neoplastic subpopulations.

1. *Ungated Analysis.* Identifying a neoplastic population with a unique immunophenotype can be as simple as displaying the characteristic antibodies in an ungated dual display histogram. For example, in a bone marrow extensively involved with an acute precursor B-lymphoblastic leukemia, a large CD19-positive, CD10-positive, and CD45 dim population would be evident upon simple visual examination of the ungated data. However, small neoplastic subpopulations

with immunophenotypes closer to normal can be more difficult to identify and may require gating strategies.

2. *Color Analysis.* One analysis strategy is to view dual display histograms for each pertinent antibody combination after assigning a color to the population of interest. This can be applied after utilization of a scatter and/or fluorescence gating strategy (Fig. 1d) or on ungated data. Since gating by definition limits the populations that one examines, excluding the population of interest is less likely with color gating than with sequential gating (see Notes 14). Furthermore, since only two parameters are viewed per plot, using color analysis to identify different populations increases the information obtained from each plot (Fig. 1c).

3. *Population Gated Analysis.* Another gating strategy is to view gated dual display plots through a population gate. This strategy is effective when interfering populations make it difficult to isolate a target population. For example, myeloid cells often express higher autofluorescence than lymphocytes or blasts; therefore, excluding granulocytes with an FSC versus SSC gate will help with analysis of low SSC events such as lymphocytes and blasts.

Neoplastic populations frequently show loss of one or more lineage-specific antigens as well as coexpression of antigens not normally expressed for that lineage (see Notes 8–10). Analysis of multiple combinations of lineage and non-lineage antibodies is strongly recommended as an abbreviated analysis may not fully characterize the immunophenotype. For example, T-cell neoplasms typically show aberrant loss of CD7 expression with or without loss of other T-cell antigens, while other T-cell neoplasms may express CD7 with loss of CD3 expression. Targeting T-cell analysis for loss of CD7 alone could miss the neoplastic population altogether if CD3 was not included in the analysis.

3.5. Data Interpretation

Detection, diagnosis, and subclassification of hematolymphoid neoplasms require extensive knowledge of normal light scatter properties and antigen expression on individual hematopoietic lineages throughout the maturation sequences. Detailed guidelines for diagnostic interpretation are beyond the scope of this chapter. For more information, please refer to additional referenced documents. Malignant populations are identified based on multiple features, including aberrant antigen expression, abnormal maturation pattern, monotypic light chain expression on B cells, or restricted V-beta expression on T cells. The flow cytometry report should include a composite immunophenotype of the malignant cells, including aberrant antigen expression, estimate of the percentage of malignant cells in the sample, lineage of the tumor, stage of maturation, and expression of antigens that may be important for prognosis or therapeutic concerns. Ideally, a diagnosis is rendered based on diagnostic criteria outlined in the *WHO Classification of Hematopoietic and Lymphoid Tissues* (5) and should correlate with clinical history as well as morphologic, molecular, or cytogenetic analysis (see Note 15). The general immunophenotypic characteristics of the hematolymphoid malignancies frequently encountered in clinical flow cytometry laboratories are briefly described below.

1. *Acute Leukemias.* FCI is indicated in the diagnostic evaluation of acute leukemia. It is important to establish the lineage of the blast population (i.e., lymphoid vs. myeloid) for appropriate therapy and clinical management. Flow cytometry is sensitive and highly specific for differentiating myeloid from lymphoid lineages. Myeloid leukemias can aberrantly express lymphoid markers and lymphoid leukemias can aberrantly express myeloid markers; hence, a comprehensive panel is necessary for accurate diagnosis. After the patient has been treated with initial therapy, minimal residual disease monitoring by flow cytometry is frequently utilized to guide clinical management and therapeutic decisions.

   (a) *Acute Lymphoblastic Leukemia (ALL).* ALL is a neoplasm of immature B-cell or T-cell precursors arrested in maturation (Fig. 1b). Of the cases of ALL, 75–85% are of B-cell lineage (typically positive for CD19, CD22, CD10, ±CD34, HLA-DR, and TdT with dim to negative expression of CD45) (see Note 16). The remaining percent of cases are of T-cell lineage (variable expression of CD1a, CD2, CD4, CD5, CD7, CD8, and cytoplasmic CD3) (see Note 17).

   (b) *Acute Myeloid Leukemia (AML).* Flow cytometric analysis is sensitive and specific for identifying the immunophenotype and lineage specificity of AMLs in terms of granulocytic, monocytic, erythroid, or megakaryocytic differentiation. CD13 and CD33 are expressed in nearly all cases of AML. Immature AMLs are typically positive for CD117, CD34, HLA-DR, and dim CD45. Monocytic differentiation in AML is evidenced by the expression of CD14, CD4, CD11b, CD11c, CD64, CD36, CD68, and lysozyme. Glycophorin A, bright CD71, and CD36 are often expressed in acute erythroid leukemias. CD36 and the platelet glycophorins CD41, CD61, and CD42 are useful in detecting acute megakaryoblastic leukemias (see Note 18).

2. *Myelodysplastic and Myeloproliferative Disorders.* FCI analysis is increasingly used to detect evidence of myelodysplasia, particularly in cases where the morphologic features are equivocal. Normal myeloid cells mature in a manner that is tightly

controlled and regulated, generating reproducible patterns of antigen expression at different stages of maturation. Although multiple immunophenotypic abnormalities are common in myelodysplastic syndrome (MDS), there is no single MDS-specific immunophenotype, and some abnormalities observed in MDS may be seen in other disorders [such as paroxysmal nocturnal hemoglobinuria (PNH), megaloblastic anemia, and post-growth factor therapy]. The patterns and combinations of multiple abnormalities distinguish MDS from other disease processes. FCI findings characteristic of MDS include abnormal intensity of antigen expression (e.g., increased or decreased levels of CD45 in granulocytes), abnormally low SSC in granulocytes (due to abnormal granularity), absence of normal antigens (e.g., CD10-negative granulocytes), non-myeloid antigens (e.g., lymphoid lineage antigens) on myeloid precursors, and aberrant maturation, such as an asynchronous pattern of maturation antigen expression (i.e., antigen patterns from different stages of differentiation co-expressed) or abnormal blast cell populations. FCI analysis also provides important prognostic information on MDS (6–8). Traditionally, FCI analysis of myeloproliferative disorders has not played a large role in diagnosis unless blast crisis or transformation to AML is clinically suspected.

3. *Mature B-cell Non-Hodgkin Lymphomas and Chronic Leukemias.* In analysis of flow cytometric data for the presence of mature B-cell malignancies, all B cells should be examined for clusters defining cell populations with light scatter or antibody-binding characteristics that fall outside the range observed in normal B cells. Although there are several general abnormalities that are found in the evaluation of B-cell malignancies, the most useful feature of B-cell neoplasia is monotypic kappa or lambda light chain expression or light chain restriction. A B-cell population with restricted light chain expression is, with rare exceptions, considered a B-cell neoplasm (see Note 19). The absence of light chain expression is also an evidence of neoplasia (9) (see Note 20). FCI is able to recognize monoclonal B cells in the presence of normal polyclonal B cells by the simultaneous analysis of other markers that are differentially expressed among benign and malignant cells. In addition to light chain restriction, identification of abnormal patterns of B-cell lineage antigen expression is useful in identifying malignant B cells. Most normal B cells express CD19, CD79b, CD22, and CD20, and failure to express one of these antigens is abnormal. The presence of antigens not normally found on B cells, such as T-cell or myeloid antigens, is also useful. Aberrant expression of CD2, CD4, CD7, and CD8 has been observed in B-cell chronic lymphocytic leukemia/

small lymphocytic lymphoma (B-CLL/SLL), HCL, and B-cell non-Hodgkin lymphomas (10). In addition to the presence or absence of specific antigens on the neoplastic cells, the expression level of various antigens, i.e., abnormally dim or bright expression, is valuable. For example, B-CLL/SLL is characterized by abnormally dim CD20 expression, while HCL is characterized by abnormally bright CD20 expression (11). Demonstration of these abnormally dim or bright staining populations is not only useful in detecting the presence of a malignant B-cell population but may also be instrumental in subclassifying the leukemia or lymphoma into the appropriate diagnostic category. In addition to immunophenotyping, light scatter characteristics provide important data about the cells being studied, and can be useful in detecting malignant B-cell populations. For example, abnormally high FSC can be observed in large cell lymphoma and high SSC is typically seen in hairy cell leukemia. Immunophenotypic profiles of the most common mature B-cell neoplasms are listed in Table 2.

4. *Plasma Cell Neoplasms.* Normal plasma cells have intense expression of CD38; are positive for CD138 and CD19; have some expression of CD45; are negative for CD56, CD117, CD20, CD22, and surface immunoglobulin; and have polyclonal cytoplasmic immunoglobulin. Malignant plasma cells are distinguished from normal cells by the absence of CD19 and CD45, aberrant expression of CD56 and CD117, diminished CD38 and/or CD138, and presence of monoclonal cytoplasmic immunoglobulin light chain.

5. *Mature T-cell Non-Hodgkin Lymphomas and Chronic Leukemias.* Detection of T-cell neoplasia is typically based upon subset restriction; absent, diminished, or abnormally increased expression of T-cell antigens; and presence of aberrant antigens. Expansions of normally rare T-cell populations are indicators of T-cell neoplasia as well. Additionally, T-cell clonality can be directly assessed by FCI analysis of the beta chain variants of the T-cell receptor (TCR). An abnormal expansion of a Vβ population is consistent with a clonal T-cell population, similar to an expansion of light chain restricted B cells in a monoclonal B-cell population. Abnormal T-cell populations can be detected using a panel of antibodies and then anti-Vβ antibodies can be used to determine the clonality of the immunophenotypically defined abnormal T cells. Mature clonal T-cell populations are restricted to CD4+ CD8−, CD8+ CD4−, CD4+ CD8+, or lack of CD4− CD8− (see Notes 21 and 22). The majority of mature T-cell neoplasms fail to express at least one T-cell antigen, so it is important to include multiple T-cell antigens (CD2, CD3,

Table 2
Typical immunophenotype of B-cell disorders

Diagnosis	CD19	CD20	CD22	CD5	CD23	CD10	CD11c	CD103	CD79b	CD25	FMC7	sIg	Comment
BL	+	+	+	-	-	+	-	-	+		+	+	IgM
FL	dim+	+	+	-	±	+	-	-	+	-	+	+	BCL-2+
CLL	+	dim+	dim+	+	+	-	dim±	-	-/dim+	±	-	dim+	CD38, ZAP-70
MCL	+	+	+	+	-	-	-	-	+	+	+	+	cyclinD1+
DLBCL	±	±	±	±	±	±	-	-	±	±	±	±	CD30±
HCL	+	bright+	bright+	-	-	-	bright+	+	+	bright+	+	+	CD123+
HCLv	+	bright+	bright+	-	-	-	bright+ to ±	±	+	-	+	+	CD123±
SMZL	+	+	+	-	-	-	±	-	+	-/+	+	+	CD123-
BALL	+	±	+	-	+	+	-	-	+		-	-	Dim CD45, CD34±

BL Burkitt lymphoma, *FL* follicular lymphoma, *CLL* chronic lymphocytic leukemia, *MCL* Mantle cell lymphoma, *DLBCL* diffuse large B-cell lymphoma, *HCL* hairy cell leukemia, *HCLv* hairy cell leukemia variant, *SMZL* splenic marginal zone lymphoma, *BALL* acute precursor B-lymphoblastic leukemia, *sIg* surface immunoglobulin, + characteristically positive, – characteristically negative, ± variable expression – positive in some cases and negative in others, dim+ expressed at lower than normal levels, bright+ expressed at brighter than normal levels

Table 3

Immunophenotype of T-cell chronic lymphoproliferative disorders

Diagnosis	CD4	CD8	CD2	CD3	CD5	CD7	CD25	CD16	CD56	CD57	Comments
MF	+	−	+	dim+	+	−	±	−	−	−	CD26−
ALCL	±	−	±	±	±	±	±	−	−	−	CD30+, ALK1+
AILT	+	−	+	+	+	±		−	−	−	CD10+
T-PLL	±	±	+	+	+	+		−	−	−	15% CD4+ CD8+
ATL	+	−	+	dim+	+	−	bright+	−	−	−	CD26−
ETTL	−	±	+	+	−	+	−	−	±	−	CD103+, CD30±
HSγδ	−	±	+	+	±	±	−	±	±	−	TCR-γδ+
MCγδ	−	±	+	+	±	−	−	−	±	−	TCR-γδ+
T-LGL	−	+	+	+	dim±	±	−	±	±	+	TCRαβ or TCRγδ+
TLB	±a	±a	+	cyt CD3+; sCD3−/+	+	+					CD1a±

MF mycosis fungoides, *ALCL* Anaplastic large cell lymphoma, *AILT* angioimmunoblastic T-cell lymphoma, *T-PLL* T-cell prolymphocytic leukemia, *ATL* adult T-cell leukemia/lymphoma, *ETTL* enteropathy-type T-cell lymphoma, *HSγδ* hepatosplenic γδ T-cell lymphoma, *MCγδ* mucocutaneous γδ T-cell lymphoma, *T-LGL* T-cell large granular lymphocyte leukemia, *TLB* precursor T-lymphoblastic neoplasm, + positive for antigen, − negative for antigen, ± variable positivity, *cyt CD3* cytoplasmic CD3, *sCD3* surface CD3 aTLB usually CD4 and CD8 double− or CD4 and CD8 double+

CD5, and CD7) in a diagnostic panel (see Note 23). Abnormal intensity of normal T-cell antigens is the most useful method in the detection of neoplastic T-cell populations (see Note 24). Dim CD3 expression is characteristic of cutaneous T-cell lymphoma, and T-cell large granular lymphocytic leukemias typically have abnormally dim levels of CD5 expression. Some clonal T-cell processes are characterized by increased numbers of T-cell subpopulations normally present in low numbers, such as increased CD8+ T-cells coexpressing CD57, CD56, or CD16 in T-cell large granular lymphocytic (LGL) leukemia, and expanded gamma delta T cells in gamma delta T-cell lymphoma or LGL leukemia. Immunophenotypic profiles of the most common mature T-cell neoplasms are listed in Table 3.

6. *Natural Killer (NK) Cell Lymphomas and Leukemias.* NK cells typically express varying levels of CD2, CD7, CD8, CD16, CD56, and CD57 as well as the cytoplasmic cytotoxic proteins perforin and granzyme, but they lack surface CD3, CD5, and TCR molecules. These markers can also be expressed in T-cell neoplasms, hence it is important to demonstrate lack of CD3 and TCR. Flow cytometric analysis of NK receptor (NKR) expression, especially the killer cell immunoglobulin receptors (KIRs) and the CD94/NKG2 complex, can provide evidence of NK cell clonality (12, 13).

4. Notes

1. *Specimen volume (blood)*: Adjust specimen requirements according to patient status to yield adequate normal cells and "cells of interest," i.e., >10 mL for WBC below 3×10^6/mL, <10 mL for WBC with blasts above 10×10^6/mL, and <10 mL for pediatric patients, according to volume restrictions.

2. *Anticoagulant choice*: Sodium heparin is required when splitting samples with the cytogenetics laboratory. Do not store EDTA over 24 h. Lithium heparin is toxic to lymphocytes and should be avoided. Acid citrate dextrose (ACD) is acceptable for blood; do not use it for bone marrow aspirates or other short-draw specimens due to hypotonicity and pH changes.

3. Increased background or autofluorescence noted with myeloid leukemias; use of RPMI or culture media; and poor cell viability. Although addition of RPMI or culture media with 10% FBS to the specimen can increase autofluorescence it can also minimize cell loss during transport and improve viability.

4. *Suboptimal specimen report notes*: "Use caution interpreting results due to possible: (1) selective cell loss in specimens with age >40-h-old, viability <75%, or presence of clots; (2) increased non-specific fluorescence in specimens with age >40-h-old or viability <75%; and (3) labile antigenic sites may be compromised due to chemotherapy treatment/high proliferation tumor/fixation before processing/hemolysis from heating or freezing specimen/ACD sample volume less than 3 ml/storage at 4–8°C."

5. Erythrocyte removal by lysis in blood and bone marrow specimens is preferred as selective loss of potentially important populations may occur with gradient density methods. Unless there is a significant amount of blood contamination in fluids and solid tissue biopsies, do not lyse to minimize cell loss.

6. *Abbreviated panels*: For paucicellular specimens, use an abbreviated primary screening panel from Table 1. For well-characterized minimal residual disease, more cells can be added to yield at least 200 tumor cells in each tube by using an abbreviated primary screening panel.

7. Prewashing specimens or lysing before antibody incubations may remove desirable antigens, i.e., CD23.

8. *Distinguish dim from negative*: Using the same fluorochrome parameter and gating, (1) compare the "dim" population in one tube to the normal internal positive and negative controls in the same tube; (2) compare the "dim" and internal control populations in the first tube to negative events in the second tube for the same fluorochrome and populations; and (3) compare the "dim" and internal normal control populations to another tube with the same antibody conjugated to a different fluorochrome, realizing the relative "positive to negative" shift may be different. In rare cases where the pattern of expression is too dim to determine reliably, report as uncertain dim to negative in the final report.

9. *All populations are negative in a tube*: If none of the events in a clinical sample reacts with any antibodies in one or more tubes, evaluate as a probable process problem rather than a true negative profile. Exception: every antibody represents a rare antigen (e.g., CD30).

10. *Internal versus parallel normal control*: A fresh normal or commercial control can be used to generate normal positive and negative patterns; however, the patient's normal cells are the most effective internal control for determining autofluorescence and fluorescence patterns.

11. *Data acquisition*: Gated data acquisition is not recommended unless there is certainty that analysis of excluded populations is not diagnostically important.

12. Determine "percent of total" for populations rather than "percent reactivity" for individual antibodies when analyzing a heterogeneous mixture of normal and abnormal cell populations.

13. Attempting to use a fluorescence gate to analyze a tube without the common parameter is a frequent novice mistake. For example, a CD19 versus SSC anchor gate cannot be used to characterize B-cell populations in a panel including FMC7/CD79b/CD19/CD23 and CD20/CD22/CD45/CD10 because CD19 is not available in the second tube.

14. *Gating is risky*: Applying an analysis gate without first determining that the population of interest is present in that gate is inherently risky. Neoplastic cells frequently fall outside of normal light scatter regions and fail to express all lineage-associated markers. Use of sequential gating strategies further limits the data being evaluated and increases the probability that an abnormal population will be missed. Initial analysis of ungated data and back gating to validate an analysis gate is highly recommended.

15. *Morphology*: Morphological correlation with touch preps, tissue H&E sections, or cytospin preparations is crucial for diagnosis and may be helpful in assessing for sampling error.

16. *Pitfalls in analysis of normal immature lymphoid cells*: Avoid mistaking non-neoplastic maturing normal lymphoid cells for neoplastic blasts. Maturing normal lymphoid cells demonstrate a FCI pattern of varying intensities of expression of multiple antigens associated with maturation. The expression is synchronous with the maturation sequence and generates a reproducible pattern. In contrast, ALL blasts fall outside of the normal differentiation pattern and typically a FCI distinct population with evidence of expression of an abnormal pattern of maturation associated antigens is observed.

17. *Pitfall in analysis of T-ALL*: T-ALL can on occasion show CD45 expression approaching the expression levels of mature lymphocytes, thus not falling into the CD45/SSC blast region.

18. *Pitfall in analysis of megakaryocytic lineage*: AML blasts or lymphoblasts may be misinterpreted as megakaryocytic due to the adherence of platelets on blasts, giving the appearance of larger blasts with positivity of CD41 and CD61 which are bound to platelets.

19. *Monoclonality in non-neoplastic specimens*: Monoclonal B-cell populations are rarely demonstrated in patients with no definitive evidence of B-cell malignancy (14, 15). The presence of a monoclonal B-cell population must be interpreted within the scope of the clinical history and correlated with morphology.

20. *Mistaking germinal center B cells as surface light chain-negative B cells*: Germinal center cells have dim surface immunoglobulin expression, and may be mistaken as surface light chain-negative B cells, suggestive of malignancy. This error is avoided when one is aware of the normal immunophenotype of germinal center B cells (CD20 bright+, CD10+, and CD19+).

21. *CD4- and CD8-negative T cells*: Increased numbers of CD4- and CD8-negative T cells are present in some immunodeficiency states and are a hallmark of autoimmune lymphoproliferative syndrome (ALPS).

22. Coexpression of both CD4 and CD8 may be due to a technical artifact in staining of unwashed blood. Washing the blood with PBS removes this artifact.

23. *Normal T cells failing to express T-cell antigen*: A small percent of normal peripheral blood CD3-positive T cells are CD7 negative, and a subset of normal γδ T cells do not express CD5. However, large numbers of CD7-negative or CD5-negative non-gamma delta T cells are abnormal.

24. *Levels of expression of T-cell antigens*: When interpreting data, one must remember that CD3 is brighter in gamma delta T cells, CD2 expression is upregulated in reactive T cells, and CD5 is dimmer in normal CD8-positive T cells.

References

1. Davis, B. H., Holden, J. T., Bene, M. C., et al. (2007) 2006 Bethesda International Consensus recommendations on the flow cytometric immunophenotypic analysis of hematolymphoid neoplasia: medical indications. *Cytometry B Clin. Cytom.* **72**, Suppl 1, S5–S13.

2. Stetler-Stevenson, M., Ahmad, E., Barnett, D., et al. (2007) Clinical Flow Cytometric Analysis of Neoplastic Hematolymphoid Cells; Approved Guideline – Second Edition. *Clinical and Laboratory Standards Institute* **27**, Wayne, PA, USA.

3. Wood, B. L., Arroz, M., Barnett, D., DiGiuseppe, J., Greig, B., Kussick, S. J., Oldaker, T., Shenkin, M., Stone, E., and Wallace, P. (2007) 2006 Bethesda International Consensus recommendations on the immunophenotypic analysis of hematolymphoid neoplasia by flow cytometry: optimal reagents and reporting for the flow cytometric diagnosis of hematopoietic neoplasia. *Cytometry B Clin. Cytom.* **72**, Suppl 1, S14–S22.

4. Stelzer, G. T., Marti, G., Hurley, A., McCoy, P., Jr., Lovett, E. J., and Schwartz, A. (1997) U.S.-Canadian Consensus recommendations on the immunophenotypic analysis of hematologic neoplasia by flow cytometry: standardization and validation of laboratory procedures. *Cytometry* **30**, 214–230.

5. Swerdlow, S. H., Campo, E., Harris, N. L., et al. (2008) *WHO Classification of Tumors of Haematopoietic and Lymphoid Tissues*, Forth Edition, International Agency for Research on Cancer (IARC), Lyon, France.

6. Kussick, S. J., Fromm, J. R., Rossini, A., et al. (2005) Four-color flow cytometry shows strong concordance with bone marrow morphology and cytogenetics in the evaluation for myelodysplasia. *Am. J. Clin. Pathol.* **124**, 170–181.

7. Stetler-Stevenson, M., Arthur, D. C., Jabbour, N., et al. (2001) Diagnostic utility of flow cytometric immunophenotyping in myelodysplastic syndrome. *Blood* **98**, 979–987.

8. Stetler-Stevenson, M. and Yuan, C. M. (2009) Myelodysplastic syndromes: the role of flow cytometry in diagnosis and prognosis. *Int. J. Lab. Hematol.* **31**, 476–483.

9. Li, S., Eshleman, J. R., and Borowitz, M. J. (2002) Lack of surface immunoglobulin light chain expression by flow cytometric immunophenotyping can help diagnose peripheral B-cell lymphoma. *Am. J. Clin. Pathol.* **118**, 229–234.

10. Kingma, D. W., Imus, P., Xie, X. Y., et al. (2002) CD2 is expressed by a subpopulation of normal B cells and is frequently present in mature B-cell neoplasms. *Cytometry* **50**, 243–248.

11. Ginaldi, L., De Martinis, M., Matutes, E., Farahat, N., Morilla, R., and Catovsky, D. (1998) Levels of expression of CD19 and CD20 in chronic B cell leukaemias. *J. Clin. Pathol.* **51**, 364–369.

12. Nowakowski, G. S., Morice, W.G., Phyliky, R. L., Li, C. Y., and Tefferi, A. (2005) Human leucocyte antigen class I and killer immunoglobulin-like receptor expression patterns in T-cell large granular lymphocyte leukaemia. *Br. J. Haematol.* **128**, 490–492.

13. Morice, W. G., Kurtin, P. J., Leibson, P. J., Tefferi, A., and Hanson, C. A. (2003) Demonstration of aberrant T-cell and natural killer-cell antigen expression in all cases of granular lymphocytic leukaemia. *Br. J. Haematol.* **120**, 1026–1036.

14. Marti, G. E., Rawstron, A. C., Ghia, P., et al. (2005) Diagnostic criteria for monoclonal B-cell lymphocytosis. *Br. J. Haematol.* **130**, 325–332.

15. Kussick, S. J., Kalnoski, M., Braziel, R. M., and Wood, B. L. (2004) Prominent clonal B-cell populations identified by flow cytometry in histologically reactive lymphoid proliferations. *Am. J. Clin. Pathol.* **121**, 464–472.

Chapter 15

Flow Cytometry Assays in Primary Immunodeficiency Diseases

Maurice R.G. O'Gorman, Joshua Zollett, and Nicolas Bensen

Abstract

The primary immunodeficiency diseases (PIDs) encompass an extremely large and diverse number of clinical disorders caused by mutations in genes that affect virtually every measurable component of our immune systems. Many of the genetic mutations lead to abnormalities that can be detected in circulating peripheral blood cells of suspected patients by flow cytometry and the appropriate combinations of reagents and in vitro manipulations. The flow cytometry procedures that have been developed to detect abnormalities in peripheral blood cells of primary immunodeficiency patients can barely be covered in an entire book, let alone one chapter. Instead of attempting to cover each disease with a specific assay or test, we review three procedures each covering a global aspect of the observed immune abnormality, i.e., detection of lymphocyte subset abnormalities, lymphocyte "marker" abnormalities, and leukocyte function abnormalities.

Key words: Lymphocyte subsets, Primary immunodeficiency disease, Flow cytometry, Oxidative burst, CD40-ligand, Routine immunophenotyping

1. Introduction

The collective development of accurate molecular technologies, an increased understanding of complex molecular signaling pathways and the availability of reagents to assess the latter have allowed scientists to elucidate the molecular etiology of an ever increasing number of primary immunodeficiency diseases (PIDs). At the first WHO sponsored meeting of experts involved in the treatment and investigation of PIDs in 1970 (1), 16 PIDs were identified and classified (2). Prior to the latest meeting of PID experts in Dublin, Ireland in the summer of 2009, over 150 PIDs had been classified and characterized (3). This large group of disorders is most commonly classified into eight individual

Teresa S. Hawley and Robert G. Hawley (eds.), *Flow Cytometry Protocols*, Methods in Molecular Biology, vol. 699, DOI 10.1007/978-1-61737-950-5_15, © Springer Science+Business Media, LLC 2011

Table 1
At the latest meeting of the International Union of Immunological Societies, "Experts Committee on Primary Immunodeficiencies," held in Dublin, Ireland in June 2009, a consensus regarding the classification of newly characterized disorders was published as Tables I–VIII. The category of each of the Tables with a salient example of a disease(s) are listed below (8)

Group	Disease example in each group
Table I. Combined T and B-cell immunodeficiencies	Severe combined immunodeficiency (SCID) and CD40-ligand deficiency
Table II. Predominantly antibody deficiencies	Btk deficiency (X-linked agammaglobulinemia) and common variable immunodeficiency disorders
Table III. Other well-defined immunodeficiency syndromes	Wiscott–Aldrich syndrome and DiGeorge anomaly
Table IV. Diseases of immune dysregulation	Autoimmune lymphoproliferative syndrome (ALPS) and familial hemophagocytic lymphohistiocytosis
Table V. Congenital defects of phagocyte number, function, or both	Congenital neutropenia and chronic granulomatous disease
Table VI. Defects in innate immunity	IL-1 receptor associated kinase 4 (IRAK4) deficiency and MyD88 deficiency
Table VII. Autoinflammatory disorders	Periodic fever syndromes and Blau syndrome
Table VIII. Complement deficiencies	Hereditary angioedema (C1 inhibitor deficiency) and Paroxysmal nocturnal hemoglobinuria

categories (see Table 1). More recently, a comprehensive systematic mathematical classification project based on clinical, pathological, and laboratory parameters identified 11 groups for over 200 clinically defined PIDs of which 167 had known genetic etiologies (4).

Of all of these disorders, the majority have some abnormality that could be detected by a flow cytometry-based application (although it must be acknowledged that not all abnormalities would be "specific" for a particular condition). It would be interesting to attempt to describe each of these applications; however, given that space is limited, we provide only a snapshot of the flow cytometry procedures currently utilized.

To combine an understanding of this inordinately complex group of disorders with an appreciation of how flow cytometry can be applied in their detection, the PIDs are described as one 3 general groups based on their underlying genetic abnormality. Very simply all of the disorders can be grouped according to: (a) mutations in genes that affect the relative representation of a specific subset, i.e., a subset abnormality; (b) mutations in genes that affect the expression of a specific "marker," i.e., marker abnormality, and lastly; (c) mutations in genes that affect a particular cell function, i.e., functional abnormality. Leukocyte subset abnormalities defined by the measurement of the relative and absolute number of specific subsets represent the most common application of clinical flow cytometry. Marker abnormalities are also commonly detected flow cytometrically with the appropriate combination of fluorochromes and monoclonal/polyclonal antibodies. Lastly, there are several physiologic cell functions whose activity can be assessed with the appropriate reagents. In this chapter, we describe one flow cytometry application that has been utilized in a clinical setting to detect PIDs in one of the three categories (i.e., a subset abnormality, a marker abnormality, a functional abnormality). A more comprehensive treatise of the individual PID can be found in (5).

1.1. Primary Immunodeficiency Diseases

Each discovery of a new primary immunodeficiency can be thought of as an "experiment of nature," a term coined originally by Robert A. Good in the mid 1950s (6) in reference to the fact that each case of a new primary immunodeficiency disease has taught us something about the normal functioning of the inordinately complex immune system. From a clinical perspective, it is common to think of patients with a primary immunodeficiency to suffer from an increased susceptibility to both the frequency and severity of infections with both opportunistic and pathogenic organisms. However, we now know that the scope of clinical symptoms encompassed within the category of primary immunodeficiency encompasses more than an increased susceptibility to infections. Autoimmunity, autoinflammatory disorders, atypical hemolytic uremic syndrome, paroxysmal nocturnal hematuria, for example, are now all classified as primary immunodeficiency states. Increased susceptibility to infection is, however, the most common clinical presentation in all of the primary immunodeficiency diseases. A large international effort (the Primary Immunodeficiency Resource Center, http://www.info4pi.org) spearheaded by the Jeffrey Modell Foundation is aimed at increasing the awareness of the public and medical professionals alike and has led to the posting of PID symptoms in most international airports. The ten warning signs (as summarized in these postings) that warrant an evaluation for the possible detection of a primary immunodeficiency disease are: (1) eight or more new ear infections within 1 year; (2) two or more serious sinus infections within 1 year; (3) two or more months on antibiotics with little effect; (4) two or more pneumonias within 1 year; (5) failure of an infant to gain weight or grow normally; (6) recurrent, deep skin or organ abscesses; (7) persistent thrush in mouth or elsewhere on skin, after age 1; (8) need for intravenous antibiotics to clear infections; (9) two or more deep-seated infections; and (10) a family history of primary immunodeficiency.

1.2. Initial Evaluation

A patient suspected of a primary immunodeficiency should first be evaluated with a thorough clinical and family history as well as a physical exam. The first pass laboratory tests would include a CBC with a differential followed by the testing of more specific immune parameters, including quantitative serum immunoglobulin levels and specific antibody determinations. The next stage would be to assess the complement system as well as the cellular components of the immune system. The most appropriate and encompassing screening assessment of the cellular immune system is accomplished by what we refer to as "routine immunophenotyping." Based on the results of the history, physical, and other laboratory tests, more specific flow cytometry procedures (i.e., specific subset, marker, or functional abnormalities) can be assessed. Lastly, once a presumptive diagnosis is obtained, we recommend that a patient sample be obtained to ascertain the molecular etiology and confirm the diagnosis.

2. Materials

2.1. Routine Immunophenotyping Panel for the Screening Diagnosis of Primary Immunodeficiency Disease

1. 1× FACS Lysing solution (BD Biosciences, San Jose, CA): Dilute 10× stock solution 1:10 with deionized H_2O. Store at room temperature. Solution is stable for 3 months.

2. Becton Dickinson MultiTest reagents (BD Biosciences): MultiTest-6 color TBNK (CD3-FITC/CD16+56-PE/CD45-PerCP-Cy5.5/CD4-PE-Cy7/CD19-APC/CD8-APC-Cy7).

3. Becton Dickinson CD3-FITC/HLA-DR-PEs and CD45 PerCP single reagents (BD Biosciences). Each antibody reagent, sufficient for 50 tests, is provided in 1 mL of buffered saline with gelatin and 0.1% sodium azide. When stored at 2–8°C, antibody reagents are stable until the expiration date on label. Antibody reagents should not be frozen or exposed to direct light during storage or during incubation with cells.

4. Single-use BD TruCOUNT tubes, containing a freeze-dried pellet of fluorescent beads (BD Biosciences).

5. Two levels of cellular controls: Streck CD Chex and Low CD4 CD Chex (Streck, Omaha, NE).

6. Isoton II Sheath fluid (Beckman Coulter Inc., Fullerton, CA).

7. BD FACSCanto II® 8 color flow cytometer and computer workstation (BD Biosciences).

8. Vortex mixer; BD electronic reverse pipettor; micropipettes that deliver 10 and 20 μL and pipette tips.

2.2. Oxidative Burst Assay for the Screening Diagnosis of Chronic Granulomatous Disease

1. Dihydrorhodamine 123 (DHR 123) (Invitrogen, Carlsbad, CA) stock solution: Add 2 mL of dimethylformamide (DMF) (Sigma–Aldrich, St. Louis, MO) to a 10 mg vial of DHR 123. Note: DMF dissolves some plastics. Make 100 µL aliquots of the stock solution (5 mg/mL) and store at –70°C for 1 year.

2. Phorbol 12-myristate 13-acetate (PMA) (Sigma–Aldrich) stock: Add 1 mL of dimethyl sulfoxide (DMSO) to a 5 mg vial of PMA, mix. Add 4 mL of DMSO for a stock concentration of 1 mg/mL. Make 100 µL aliquots and store at –70°C. At this temperature, PMA is stable for 6 months.

3. Erythrocyte lysing solution: 10× concentration of ammonium chloride (NH_4Cl) lysing solution (Pharmlyse BD Pharmingen, San Diego, CA).

4. Ca^{2+} and Mg^{2+} free phosphate buffered saline (PBS). Store at room temperature. PBS is stable until the expiration date on the bottle.

5. Washing solution: 2.5 g sodium azide, 10 mL fetal calf serum 1,000 mL PBS. Filter and sterilize. Store at 4°C. Washing solution is stable for 6 months when stored at 4°C.

6. 16% paraformaldehyde (PFA) (Sigma–Aldrich): Prepare a 1% fixative solution using normal saline. Store working solution at 4°C. Working solution is stable for 6 months when stored at 4°C.

7. Isoton II sheath fluid (see Subheading 2.1).

8. Flow cytometer such as a BD FACSCalibur flow cytometer equipped with CellQuest software (BD Biosciences).

9. Vortex; centrifuge; shaking water bath at 37°C; transfer pipettes; 12×75 mm test tubes; micropippets and pippet tips.

2.3. In Vitro Induced CD40-Ligand (CD154) Upregulation for the Screening Diagnosis of X-Linked Hyper IgM Syndrome (CD40-Ligand Deficiency)

2.3.1. Detection of CD154 Upregulation Using a Monoclonal Antibody Specific for CD154

1. PMA (see Subheading 2.2).

2. Calcium Ionophore (Sigma–Aldrich) stock: Add 1 mL of DMSO to a 1 mg vial of calcium ionophore for a stock concentration of 1 mg/mL. Make 100 µL aliquots and store at –70°C.

3. Monoclonal antibodies: CD40L-PE, MsIgG1-PE, CD8-FITC, CD3-FITC, CD69-PE, CD3-PerCP (BD Biosciences).

4. RPMI culture medium: 20 mL RPMI, 0.2 mL penicillin–streptomycin 0.2 mL l-Glutamine.

5. 1× FACS Lysing solution (see Subheading 2.1).

6. Ca^{2+} and Mg^{2+} free PBS (see Subheading 2.2).

7. Flow cytometry wash solution: PBS, 1% FCS, 0.1% sodium azide.

8. 1% PFA (see Subheading 2.2).

9. Flow cytometer such as BD FACSCalibur equipped with appropriate analysis software such as CellQuest.

10. 37°C CO_2 incubator; centrifuge; 12×75 mm test tubes; pipettes.

2.3.2. Detection of CD154 Upregulation Using a CD40-Receptor-Human IgG Chimeric Recombinant Protein

1. PMA (see Subheading 2.2).

2. Calcium Ionophore (see Subheading 2.3.1).

3. Monoclonal antibodies: CD8-FITC, CD3-FITC, CD69-PE, CD3-PerCP (BD Biosciences).

4. Recombinant Human CD40/Fc Chimera (R&D Systems, Minneapolis, MN).

5. Goat F(ab)$_2$ antihuman IgG-PE (Southern Biotech, Birmingham, AL).

6. RPMI culture medium (see Subheading 2.3.1).

7. 1× FACS Lysing solution (see Subheading 2.1).

8. Ca^{2+} and Mg^{2+} free PBS (see Subheading 2.2).

9. Flow cytometry wash solution (see Subheading 2.3.1).

10. 1% PFA (see Subheading 2.2).

11. Flow cytometer such as BD FACSCalibur with appropriate analysis software such as CellQuest.

12. 37°C CO_2 incubator; centrifuge; 12×75 mm test tubes; pipettes.

3. Methods

3.1. Routine Immuno-phenotyping Panel for the Screening Diagnosis of Primary Immunodeficiency Disease

3.1.1. Sample Preparation

1. Label an appropriate number of TruCOUNT tubes for each patient and control. *Note*: Before use, verify that the TruCOUNT bead pellet is intact and within the metal retainer at the bottom of the tube. If not, discard the tube and replace it with another.

2. Add 20 µL of the 6 color MultiTest antibody reagent to the first tube and 10 µL each of the CD3-FITC/HLA-DR-PE Simultest and CD45 PerCP (or CD45-PerCP-Cy5.5 can be substituted) antibodies into the second tube.

3. Add 50 µL of well-mixed EDTA whole blood to each patient and each level of Streck whole blood control tube (we recommend running one set of controls per day) using the BD electronic reverse pipette (see Note 1).

4. Incubate the samples at room temperature in the dark for 15 min.

5. Add 450 µL of 1× FACS Lysing solution to each tube and vortex immediately.

6. Incubate the samples at room temperature in the dark for 15 min.

7. Samples should be stored at 2–8°C and acquired on the FACSCanto within 1 hour of lysing using the FACSCanto software for Tube 1 and FACSDiva software for Tube 2 as described below (see Note 2).

8. Use FACSCanto software to check the bead lot numbers. Choose "*Tools> Lot IDs.*" Choose "*Absolute Count Beads*" and enter or verify the lot number and bead information on the bead-tubes foil pouch is correct. Do not mix lot numbers.

3.1.2. FACSCanto Setup

The FACSCanto cytometer must be calibrated before each use and optimized assay settings must be implemented prior to sample collection. It is highly recommended that manufacturer's instructions be followed. In our laboratory, 7-Color Setup Beads are run before acquisition to automatically set the voltages and compensation for the parameters used in acquiring the 6 color monoclonal antibody combination. These settings are maintained and used for the acquisition of the second tube using the FACSDiva software. List mode data files acquired using both FACSCanto (first tube, 6 color combination) and FACSDiva software (for the second tube, 3 color) are then analyzed in each of the respective software programs. For the 6 color (first tube), the lymphocyte gate (CD45 vs. right angle light scatter) is reviewed and optimized (to include the entire cluster of lymphocytes) and the FACSCanto software automatically calculates the proportions and absolute numbers of the major lymphocyte subsets, i.e., T, B, NK, T helper, and T cytotoxic lymphocyte subset percentages and absolute counts. In summary, we recommend that the manufacturer's instructions for the FACSCanto Setup, acquisition, and analysis of the 6 color-panel first tube be followed. Our second tube, CD45, CD3, and HLA-DR, is a custom combination and is not currently amenable to automated acquisition and analysis of the FACSCanto. FACSDiva software is used for both the acquisition and analysis (i.e., measurement) of the CD3+ HLADR+ lymphocytes. In the final assessment of the routine immunophenotyping panel for each patient, the CD3 lymphocyte percentage and absolute count in each tube are compared to each other and valuable quality control parameters (see Subheading 3.1.5).

3.1.3. FACSCanto Acquisition and Analysis

1. Acquire the 6-color TBNK tubes from both patients and cellular controls using automated FACSCanto software. Verify the accuracy of the "expert gating." All leukocyte populations should be clearly defined, i.e., lymphocytes, monocytes, and granulocytes. The lymphocyte population is gated using low right angle light scatter (also known as side scatter) and bright CD45 fluorescence (see Fig. 1). Lymphocyte subsets

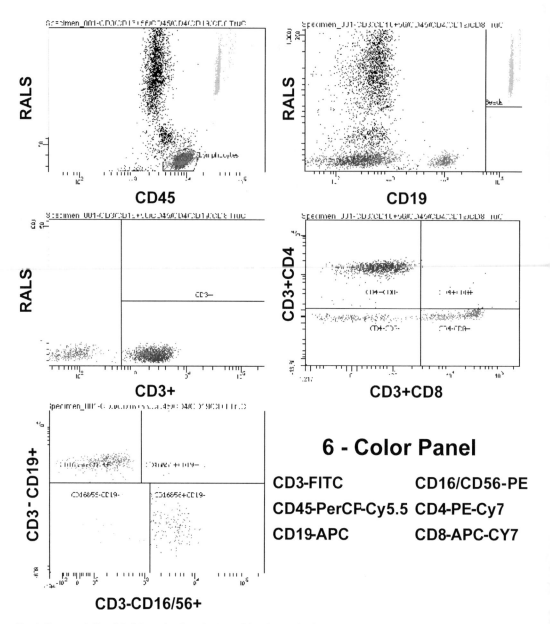

Fig. 1. Representative dot plots and gating strategy of 6 color routine immunophenotyping performed on a FACSCount II flow cytometer using automated FACSCount software.

are displayed as distinct populations (see Note 3). Adjust gates by holding down the left mouse button and dragging them to the appropriate location.

2. When all gates are adjusted, print the screen.

3. After printing, go to the next specimen by clicking the ➤ icon.

4. Repeat steps 1–3 for all specimens in the run.

5. Close FACSCanto software and login to FACSDiva software.

3.1.4. Acquire and Analyze the List Mode Data for the CD3/HLA-DR/45 Tube from Patients and the Cellular Controls Using FACSDiva Software

1. Link the instrument settings to the 7-Color Setup Beads run prior to acquisition using the "Lyse No Wash settings." Collect 2,500 lymphocyte events. Create an acquisition template to include three dot plots (see Fig. 2). On the first dot plot, display all events in a dot plot of right angle light scatter versus CD45-PerCP, with one analysis gate for the total lymphocyte subset and another for the beads (use this dot plot to calculate the total number of bead events acquired). On the second dot plot, display all events in a dot plot of right angle light scatter versus CD3 and draw an analysis gate around the CD3+ T cell cluster (use this dot plot for total number of T cell events). Draw a third dot plot which displays only those

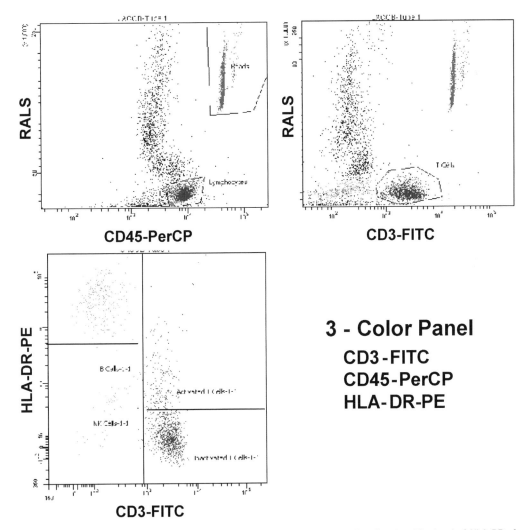

3 - Color Panel

CD3-FITC
CD45-PerCP
HLA-DR-PE

Fig. 2. Flow cytometry evaluation of the coexpression of CD3 and HLA-DR on lymphocytes. The level of HLA-DR of lymphocytes is a measure of the level of in vivo immune activation whereas the absence of HLA-DR expression on lymphocytes has been associated with a primary immunodeficiency.

events which fall into the lymphocyte gate. The CD3 versus HLA-DR plot is used to calculate the percentage of lymphocytes that are both CD3+ and HLA-DR+. Save this setup as your experiment template for future use.

2. During the analysis, adjust the lymphocyte gate on the dot plot of right angle light scatter (Y-axis) versus CD45-PerCP (X-axis). Ensure clean separation between debris (low CD45 and low right angle light scatter) and monocytes (similar CD45 intensity but higher right angle light scatter) by adjusting the lymphocyte gate as required. Set the second gate to encompass the entire bead population (see upper left plot in Fig. 2).

3. In the third graph, adjust the quadrants so that the activated T-cells (CD3 + HLA-DR+) are in the upper right quadrant (see Note 4).

4. Record the following statistics from the analysis: total number of bead events (first dot plot), total number of CD3+ lymphocyte events (second dot plot), as well as percentage of CD3+ T lymphocytes (third dot plot: upper right plus lower right quadrants) and CD3 + HLA-DR+ cells (third dot plot: upper right quadrant only) (see Note 5).

5. Calculate absolute CD3+ T cell counts (see Note 6) and record the percentage of CD3+ events and the percentage of CD3 + HLA-DR+ events.

3.1.5. Quality Control

3.1.5.1. Cellular Controls

The percentages and absolute counts of each of the subsets measured on both the low and the regular cellular control samples must fall within the limits established in-house (highly recommended) or printed on the package insert. If there are gross differences between the results obtained on the cellular controls and the defined ranges for each lymphocyte subsets, do not report patient results until/unless the cause of the problem is identified. Significant abnormalities observed with the cellular control material usually signify a problem. This must be addressed before the results of patient samples can be properly interpreted.

3.1.5.2. Internal Quality Control on Patient Samples

1. The CD3 percentages of Tube 1 (6 color TBNK) and Tube 2 (CD3/HLA-DR) must differ by <5%.

2. The absolute count difference between tubes 1 and 2 must be <15%.

3. The lymphosum must be 100 ± 5.

4. The sum of the CD4 and CD8 subsets must be within 10% of the mean of the CD3 population. If the CD3 mean is 10% or greater, the gamma/delta panel must be run (unless the patient has been tested previously). Acquire and analyze in FACSDiva. CD3 + gamma/delta+ > 15% is one explanation of why the T cell subsets do not add up to the CD3 total.

5. If any of the other criteria above are not within the specified limits, the following actions should be taken:

   (a) All values are reviewed for clerical or transcription errors.

   (b) FCS files are re-analyzed.

   (c) If these actions do not correct the criteria into acceptable ranges whether or not the cellular control results are in range, the assay is repeated.

3.2. Oxidative Burst Assay for the Screening Diagnosis of Chronic Granulomatous Disease

3.2.1. Sample Preparation

1. Prepare working dilution of DHR 123 (45 μg/mL). Add 30 μL of DHR 123 stock solution (5 mg/mL) to 3.33 mL of PBS and vortex. DHR 123 is extremely light sensitive. Keep stock and working dilutions of DHR 123 in the dark at all times.

2. Prepare working dilution of PMA (10 μg/mL). Add 10 μL of PMA stock (1 mg/mL) to 1 mL of $Ca^{2+}Mg^{2+}$ free PBS and vortex. PMA is extremely light sensitive. Keep stock and working dilutions of PMA in the dark at all times.

3. Prepare 1× NH_4Cl lysing solution by adding 2.0 mL of 10× NH_4Cl lysing solution and 18.0 mL of distilled, deionized water to a clean beaker or flask.

4. Label three tubes for each patient and control to be assayed. Tube #1: No dye/unstimulated; Tube #2: Plus dye/unstimulated; Tube #3: Plus dye/stimulated. Add 900 μL of PBS and 100 μL of well mixed whole blood to each tube.

5. Add 25 μL of the DHR 123 solution to tubes #2 & 3 (final concentration = 1.125 μg/mL). Incubate for 15 min at 37°C in the shaking waterbath.

6. Add 10 μL of PMA to Tube #3 (final concentration = 100 ng/mL). Incubate all three tubes for 15 min at 37°C in a shaking waterbath. After this incubation period, centrifuge the tubes at $400 \times g$ and remove supernatant with a transfer pipette.

7. Lyse the pellet by adding 2.5 mL of the NH_4Cl lysing solution and incubate for 15 min in the dark at room temperature. It is imperative to vortex both at the beginning and at the end of the lysing procedure. If after the first lyse there remains many RBC, repeat the lysing procedure.

8. Wash twice with 2 mL of washing solution.

9. Vortex, add 0.5 mL of 1% PFA and vortex again.

10. Specimen is ready to be acquired on the flow cytometer (see Note 7).

3.2.2. Acquisition and Analysis Using CellQuest Software

1. Standard instrument quality control must be followed prior to acquisition.

2. Open the CellQuest software and load appropriately established instrument "settings." Create an acquisition template

of forward versus right angle light scatter and gate on the granulocyte cluster. While in the "Setup," adjust forward and right angle light scatter parameters of Tube #1 such that the lymphocyte, monocyte, and granulocyte clusters are clearly discernable.

3. Set an electronic analysis gate around the granulocyte cluster and adjust the fluorescence (FL1 detector fitted with a 530/30 bandpass filter) to be in the first decade on a 4-decade log scale (<channel 10) (see Note 8).

4. Deselect "setup" option and acquire 10,000 events for each of the three tubes after entering the appropriate patient and tube identifying information.

5. Create an analysis document with forward versus right angle light scatter and set an analysis gate around the granulocyte cluster. Create a single parameter histogram to display the FL1 fluorescence of the events in the granulocyte analysis gate and measure the median fluorescence in each of the tubes (see Fig. 3).

6. Calculate the normal oxidative index (NOI) by dividing the mean fluorescence of Tube #3 by the mean fluorescence obtained in Tube #2 (see Note 9).

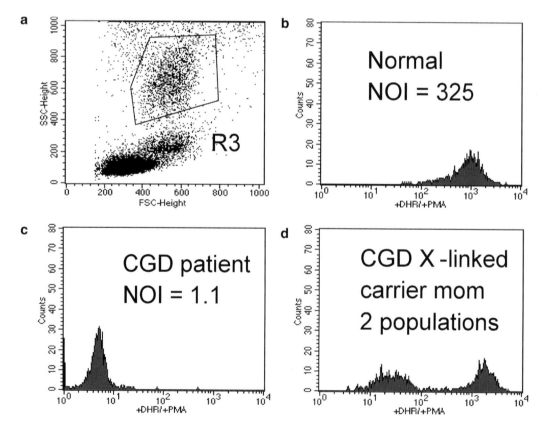

Fig. 3. Flow cytometric assay used as screening diagnostic assay for Chronic Granulomatous Disease.

3.3. CD40-Ligand (CD154) Upregulation Induced In Vitro for the Screening Diagnosis of X-Linked Hyper IgM Syndrome (CD40-Ligand Deficiency)

3.3.1. Detection of CD154 Upregulation Using a Monoclonal Antibody Specific for CD154

1. Prepare working dilution of PMA (final concentration in step 5 = 15 ng/mL):
   (a) Add 3 µL of 1 mg/mL stock to 5 mL of culture medium.
   (b) Add 1 mL of (a) to 1 mL of culture medium, mix.
   (c) Add 1 mL of (b) to 1 mL of culture medium for final dilution.

2. Prepare working dilution of Calcium Ionophore (final concentration in step 5 = 400 ng/mL):
   (a) Add 18 µL of 1 mg/mL stock to 3 mL of culture medium, mix.
   (b) Add 1 mL of (a) to 0.5 mL of culture medium for final dilution.

3. For each sample tested, label two tubes: one for the "unstimulated" control and one for the "stimulated" test sample.

4. To the "unstimulated" tube add 800 µL of culture medium and 200 µL of well-mixed whole blood.

5. To the "stimulated" tube add 600 µL of culture medium, 200 µL of whole blood, 100 µL of PMA working dilution and 100 µL of Calcium Ionophore working dilution.

6. Vortex tubes gently, cap loosely or put parafilm loosely over the tops, and place in a dark CO_2 (5%) incubator at 37°C for 4 h.

7. After incubation, vortex gently. Add 2 mL of $Ca^{2+}Mg^{2+}$ free PBS and spin at an RFC of $700 \times g$ for 5 min.

8. Aspirate supernatant completely. Resuspend pellet in PBS such that final volume is 300 µL (three samples of 100 µL each are stained from each tube).

9. Samples are to be stained with the following monoclonal antibody panel (100 µL of sample and 10 µL of each antibody):
   (a) Tube #1: CD8-FITC/MsIgG1-PE/CD3-PerCP
   (b) Tube #2: CD8-FITC/CD40L-PE/CD3-PerCP
   (c) Tube #3: CD3-FITC/CD69-PE.

10. Label three tubes containing the mAb combinations above for the unstimulated cells and three for the stimulated cells.

11. Add 100 µL of cell suspension to each tube, vortex, and incubate at room temperature for 20 min.

12. At the end of the incubation period, add 2 mL of 1× FACS Lysing solution to each tube, vortex, and incubate for 10 min at room temperature. Vortex extensively for 5 min.

13. Spin tubes at an RFC of $700 \times g$ for 5 min. Decant and wash two times with flow cytometry wash solution (i.e., add 1 mL of wash, vortex, spin, decant, and repeat).

14. Add 0.5 mL of 1% PFA to each tube and vortex.

15. Acquire and analyze 10,000 events on the FACSCalibur flow cytometer using CellQuest software (see Note 10).

3.3.2. Detection of CD154 Upregulation Using a Chimeric CD40-Receptor-Human IgG Recombinant Protein

1. Prepare working dilution of PMA (final concentration in step 5 = 15 ng/mL):

   (a) Add 3 μL of 1 mg/mL stock to 5 mL of culture medium.

   (b) Add 1 mL of (a) to 1 mL of culture medium, mix.

   (c) Add 1 mL of (b) to 1 mL of culture medium for final dilution.

2. Prepare working dilution of Calcium Ionophore (final concentration in step 5 = 400 ng/mL):

   (a) Add 18 μL of 1 mg/mL stock to 3 mL of culture medium, mix.

   (b) Add 1 mL of (a) to 0.5 mL of culture medium for final dilution.

3. For each sample tested, label two tubes: one for the "unstimulated" control and one for the "stimulated" test sample.

4. To the "unstimulated" tube add 800 μL of culture medium and 200 μL of well-mixed whole blood.

5. To the "stimulated" tube add 600 μL of culture medium, 200 μL of whole blood, 100 μL of PMA working dilution and 100 μL of Calcium Ionophore working dilution.

6. Vortex tubes gently, cap loosely or put parafilm loosely over the tops, and place in CO_2 incubator at 37°C for 4 h.

7. After incubation, vortex gently. Add 2 mL of $Ca^{2+}Mg^{2+}$ free PBS and spin at an RCF of $700 \times g$ for 5 min.

8. Aspirate supernatant completely. Resuspend pellet in PBS such that final volume is 300 μL.

9. Samples are to be stained with the following monoclonal antibody panel:

   (a) Tube #1: CD8-FITC/Human IgG-PE/CD3-PerCP

   (b) Tube #2: CD8-FITC/Human CD40/Fc Chimera – antihuman IgG-PE/CD3-PerCP

   (c) Tube #3: CD3-FITC/CD69-PE

10. Label three tubes containing the mAb combinations above for the unstimulated cells and three for the stimulated cells. The staining process is a three step process with complete washing steps after the addition and incubation of each antibody or antibody combination.

11. Add 20 μL of the Recombinant Human CD40/Fc Chimera antibody to Tube #2 for both the stimulated and unstimulated cells. (Tubes #1 and #3 are not stained with any antibody in this step, but undergo the same incubation and washing steps.)

12. Add 100 μL of cell suspension to each tube, vortex, and incubate at room temperature in the dark for 20 min.

13. Add 1 mL of flow cytometry wash solution. Vortex and centrifuge at an RFC of $700 \times g$ for 5 min.

14. Pippet and dispose of supernatant using a transfer pippet.

15. Repeat wash procedure (i.e., add 1 mL of wash solution, vortex, centrifuge, and pippet and dispose of supernatant).

16. Add 10 μL of antihuman IgG-PE to tubes #1 and #2. Incubate at room temperature in the dark for 20 min.

17. Add 1 mL of flow wash solution. Vortex and centrifuge at an RCF of $700 \times g$ for 5 min.

18. Pippet and dispose of supernatant using a transfer pippet.

19. Repeat wash procedure (i.e., add 1 mL of wash solution, vortex, centrifuge, and pippet and dispose of supernatant).

20. Add 10 μL of CD8-FITC and CD3-PerCP to tubes #1 and #2 for both the stimulated and unstimulated tubes. Add 10 μL of CD3-FITC and CD69-PE to Tube #3 for both the stimulated and unstimulated tubes.

21. At the end of the incubation period, add 2 mL of 1× FACS Lysing solution to each tube, vortex, and incubate for 10 min at room temperature. Vortex extensively for 5 min.

22. Spin tubes at an RCF of $700 \times g$ for 5 min. Decant and wash two times with flow cytometry wash solution (i.e., add 1 mL of wash solution, vortex, spin, decant, and repeat).

23. Add 0.5 mL of 1% PFA to each tube and vortex.

24. Acquire and analyze 2,500 CD3+ CD8− events on the FACSCalibur flow cytometer using CellQuest software.

3.3.3. Flow Cytometric Analysis

The lymphocyte activation protocol used in both procedures results in the downregulation of the CD4 molecule on T cells. The antigen of interest, CD40L is preferentially expressed on CD4 + T cells. Since it is not possible to gate on the CD4+ T cells, a negative gating strategy is employed. In both procedures (see Subheadings 3.3.1 and 3.3.2), cells in Tube #1 are labeled with CD8-FITC/IgG-PE and CD3-PerCP. A gate is drawn around the cells which express CD3 but do not express CD8, the majority of which are CD4 positive T cells. Therefore, an initial gate is drawn around the lymphocytes, then using Boolean logic draw a second gate around the population of cells which express CD3

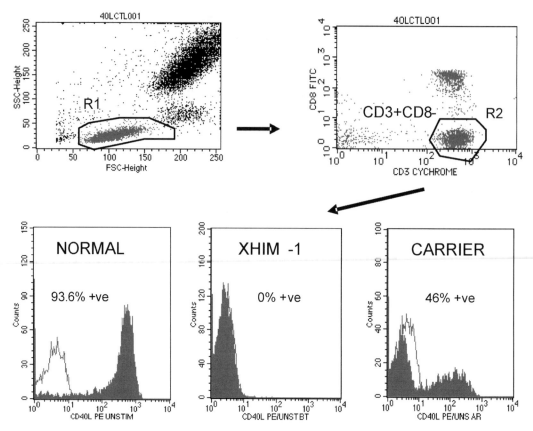

Fig. 4. CD40-ligand (CD154) upregulation induced in vitro for the screening diagnosis of X-linked Hyper IgM Syndrome (CD40-ligand Deficiency).

but do not express the CD8 molecule. Alternatively, if it is not possible to draw a gate around the lymphocyte cluster based on light scatter parameters only, the first gate can be drawn around the CD3 positive cells on a histogram of right angle light scatter versus CD3, and then follow the same steps as above (see Fig. 4).

1. Using Tube #1, set the positive negative/discriminator such that <2% of the cells are positive, i.e., >98% of the cells are negative.

2. The level of CD40L is then determined on the cells in Tube #2 as the percentage of CD3+CD8− cells that express either of the following:

    (a) CD40L fluorescence greater than the isotype control, in the procedure using a monoclonal antibody specific for CD154 (see Subheading 3.3.1).

    (b) Human CD40/Fc Chimera – antihuman IgG PE fluorescence greater than the isotype control, in the procedure using a chimeric CD40-receptor-human IgG recombinant protein (see Subheading 3.3.2).

3. Tube #3 contains a panel of mAb which is used as the in vitro stimulation control. This control consists of measuring the level of CD69 expressed on resting cells versus the expression after 4 h using the same stimulation protocol. Lymphocytes are gated based on light scatter properties and then again based on the positive expression of CD3. The level of CD69 on both resting and activated CD3-positive lymphocytes is then determined.

3.3.4. Quality Control and Normal Reference Range

A normal healthy control is always run in parallel with the patient's specimen. Failure of the control to fall within established ranges for Human CD40/Fc Chimera expression and CD69 expression necessitates a test repeat. The level of expression of CD69 must be >85% and the level of Human CD40/Fc Chimera expression must be >80% (see Note 11).

4. Notes

1. Accuracy of the blood volume is critical. The pippet tip should be wiped on the top of the inside of the primary specimen container to remove any residual blood located on the outside of the pippet tip (i.e., ensure that it does not go into the staining tubes). Deliver the blood volume into the Trucount tube just above the metal retainer. If a droplet of blood has been delivered from the pippet but is still clinging to the pippet tip, lightly touch the droplet to the side of the Trucount tube, this last drop must be in the tube. BD antibodies are titered for 2×10^7 cells/mL so the amount of whole blood used may need to be adjusted for specimens that are leukopenic or leukocytotic. In this case, absolute counts should be derived from CBC and differential results.

2. Proper mixing is important to accurate absolute counts so each specimen should be well mixed prior to acquisition.

3. Patients that do not show distinct populations may have "run-up," which requires a prewash and restaining. We do not report single-platform absolute counts on sample requiring a pre-wash. A CBD and differential on the original sample is obtained and used to calculate absolute counts.

4. The CD3+ T cells expressing HLA-DR is a continuum from negative to positive. As such we have run several samples stained for CD3 and CD45 without HLA-DR in order to determine where to set the cutoff between positive and negative HLA-DR expression on the CD3+ T cells.

5. Lymphocytes that are CD3 negative but strong HLADR+ represent the B lymphocyte population. The latter population

is used as an additional quality control parameter (as B cell percentages between the two tubes must differ by <5%) and the absolute counts must be within 15%.

6. The absolute T cell count (Absolute number of T cells/μL) is calculated by the following formula:

   # of CD3+ events/# bead events X # of beads/tube/test volume.

   Both the percentage of T cells as well as the absolute number of T cells must be agreement between the two different tubes in the panel. These quality control parameters are measured in each assay and if the agreement fails, the analysis is repeated. If the reanalysis fails the QC parameters (<5% difference in percentage of CD3 and <15% difference is absolute counts) the sample is restained and the assay repeated.

7. Samples must be acquired immediately after the addition of PFA. Note: PFA fixation leads to the slow leakage of the oxidized (fluorescent) dye from the granulocytes leading to a significant reduction in fluorescence measured and the geometric mean in Tube #3. The latter can lead to incorrect calculation of the NOI and an inaccurate result.

8. Tube 1, has no dye and has not been stimulated. This is a true negative tube in terms of fluorescence generated specifically from the DHR123 dye (excited at 488 nm with peak emission approx. 529 nm) and is used to evaluate the fluorescence generated in tubes #2 and #3.

9. This assay is used to measure the ability of granulocytes to elaborate an oxidative burst as a screening test for chronic granulomatous disease (CGD). The normal range, measured as the NOI is >30. It is known that the light scatter of the granulocytes from CGD patients is not affected by the genetic mutations. Therefore, the light scatter pattern of the granulocytes from patients and the light scatter pattern of the granulocytes from carriers that express the X-chromosome with the CGD mutation do not differ from healthy controls. Light scatter patterns generated from older samples may generate variable granulocyte light scatter patterns. In such cases, it is recommended that the fluorescence generated from Tube #3 be used ungated. The cluster of granulocytes generating a positive fluorescence result (brightest signal) can then be identified by backgating.

10. Samples can be kept overnight in the dark at 4°C.

11. The CD40-ligand is encoded by a gene located on the X-chromosome. As with other X-linked disorders maternal carriers of the mutated genes have some cells that express the normal gene whereas other cells express the X-chromosomes that

carry the mutated gene. In such carriers, the expression of the CD40-ligand (or the CD40 chimera) shows a bimodal distribution with the normal cells expressing normal levels while those cells expressing the X-chromosome that carry the mutation express abnormal levels of the CD40-ligand. It must also be noted that mutations have been detected that the result in abnormal CD40-ligand signaling but that result in apparently normal CD40-ligand levels when measured by flow cytometry (7). In such cases, it is recommended that the patient's DNA be sequenced for CD40-ligand gene mutations.

References

1. Fudenberg, H. H., Good, R. A., Hitzig, W., Kunkel, H. G., Roitt, I. M., Rosen F. S., Rowe, D. S., Seligmann, M., and Soothill, J. R. (1970) Classification of the primary immune deficiency deficiencies: WHO recommendation, *N. Engl. J. Med.* **283**, 656.

2. Fudenberg H., Good, R. A., Goodman, H. C., Hitzig, W., Kunkel, H. G., Roitt, I. M., Rosen, F. S., Rowe, D. S., Seligmann M., and Soothill, R. J. (1971) Primary immunodeficiencies: report of a World Health Organization Committee. *Pediatrics* **47**, 927–946.

3. Geha, R. S., Notarangelo, L. D., Casanova, J. L., Chapel, H., Conley, M. E., Fischer, A., Hammarström, L., Nonoyama, S., Ochs, H. D., Puck, J. M., Roifman, C., Seger, R., and Wedgwood, J. (2007) Primary immunodeficiency diseases: an update from the International Union of Immunological Societies primary immunodeficiency diseases classification committee. *J. Allergy Clin. Immunol.* **120**, 776–794.

4. Samarghitean, D., Ortutay, C., and Vihinen, M. (2009) Systematic classification of primary immunodeficiencies based on clinical, pathological

   and laboratory paramters. *J. Immunol.* **183**, 7569–7575.

5. O'Gorman, M. R. G. (2008) Role of Flow Cytometry in the Diagnosis and Monitoring of Primary Immunodeficiency Disease, in *Handbook of Human Immunology, Second Edition* (O'Gorman, M. R. G. and Donnenberg, A. D., eds.), CRC Press, Boca Raton, FL, pp. 267–311.

6. Good, R. A. and Lolomon, J. Z. (1956) Disturbances in gamma globulin synthesis as "experiments of nature". *Pediatrics* **18**, 109–146.

7. Seyama, K., Nonoyama, S., Gangsaas, I., Hollenbaugh, D., Pabst, H. F., Aruffo, A., and Ochs, H. D. (1998) Mutations of the CD40 ligand gene and its effect on CD40 ligand expression in patients with X-linked hyper IgM syndrome. *Blood* **92**, 2421–2434.

8. Notorangelo, L. D., Fischer, A., Geha, R. S., et al. (2009) International Union of Immunological Societies Expert Committee on Primary Immunodeficiencies: primary immunodeficiencies: 2009 update. *J. Allergy Clin. Immunol.* **124**, 1161–1178.

Chapter 16

Flow Cytometric Analysis of Microparticles

Henri C. van der Heyde, Irene Gramaglia, Valéry Combes, Thaddeus C. George, and Georges E. Grau

Abstract

Cell-derived microparticles (MPs) are increasingly recognized as important cell-to-cell signaling mechanisms and may exhibit important functions in homeostasis but also in pathogenesis. Indeed, MPs are associated with a number of diseases inhibiting their production that protects against pathogenesis. MPs are distinct from exosomes and apoptotic bodies, often exhibiting the membrane proteins of the activated or apoptotic cell from which they are derived. Electron microscopic analyses have shown that MPs are produced by all cell types tested to date, and ELISA-based assays have established that increased numbers of MPs are produced following cell activation. These approaches do not, however, determine the exact number of MPs and distribution of functional proteins on their surface. Flow cytometry represents an obvious approach to analyze MPs, and we present here a method to assess the number and phenotype of MPs by using a conventional flow cytometer. We also present the caveats with this method and describe a new imaging flow cytometry approach that overcomes these limitations.

Key words: Flow cytometry, Microparticles, Imaging

1. Introduction

Extracellular vesicles are small buds of the cell membrane about 40–1,000 nm in diameter. Currently, three types of extracellular vesicles have been defined (see Table 1 for a summary of differences): microparticles (MPs) generally range from 100 to 200 nm in diameter (1–3) but can be up to 1,000 nm in size; exosomes (EXs) range from 40 to 100 nm; and apoptotic bodies (AB) vary considerably in size and are usually large (>1.5 μm). There are differences in the mechanisms of formation between the vesicle types, which likely affect their function (4, 5). Although it is clear that MP formation is an active process which is continually being performed by cells and enhanced by agonists, the precise molecular mechanisms

Teresa S. Hawley and Robert G. Hawley (eds.), *Flow Cytometry Protocols*, Methods in Molecular Biology, vol. 699,
DOI 10.1007/978-1-61737-950-5_16, © Springer Science+Business Media, LLC 2011

Table 1
Summary of the differences between microparticles, exosomes, and apoptotic bodies

	Apoptotic bodies	Microparticles	Exosomes
Size	1–4 μm	100 nm to 1 μm	40–100 nm
Origin	Shedding from plasma membrane during final stage of apoptosis	Shedding from plasma membrane (psychological activation and early apoptosis)	Secreted from α-granules and multivesicular bodies
Release after activation	–	+++	+++
Isolation	800×g	25,000×g	$2 \times 10^5 \times g$ sucrose fraction
Detection	Flow cytometry	Flow cytometry	Western blot
Annexin V	+++	+++	±
Surface marker	+++	+++	±
Other markers	DNA	Factor X	CD63
Function(s)	???	Activation of coagulation/ antigen transfer	Transmission of activation signal/ antigen presentation

? not yet defined

remain to be determined. The formation of MPs does depend on the reorganization of cytoskeleton. MPs contain a variety of intracellular components, including cytoplasmic proteins, RNA, and DNA, and generally express the proteins from the parent cells. EXs are derived from multivesicular bodies and consequently contain only cytoplasmic proteins. Apoptotic body formation is a terminal step in apoptosis where the cell has released MPs beforehand, activated caspases, fragmented DNA, and dissociated into smaller particles, presumably to facilitate phagocytosis.

Initially, extracellular vesicles were considered cellular debris but accumulating evidence indicates that they play important functions in homeostasis (4). Dysregulation of MP production, in particular, is a biomarker for disease and these MPs may be a required pathway leading to pathogenesis (6). MPs are ubiquitous being formed from every cell type, including nonnucleated platelets and erythrocytes. Activation of platelets by thrombin, ADP, and other agonists leads to the translocation of phosphatidylserine from the inner leaflet to the outer leaflet and the adoption of a high affinity conformation by integrins, such as CD41/ CD61 (also known as αIIb/β3). The negatively charged surface caused by increased levels of phosphatidylserine in the outer leaflet is important for the binding of the prothrombinase and tenase complexes. The release of MPs from activated platelets markedly increases the surface area for the binding of the tenase and pro-thrombinase and facilitates the activation of the coagulation cascade (5). The importance of MPs in the activation of the coagulation system is underscored by the ability of MPs to markedly decrease coagulation times (7). In nucleated cells, agonists that elicit MP formation (e.g., lipopolysaccharide) often also elicit apoptosis (8). We have reported that MPs can carry functional molecules from the membrane of one cell to another (9), providing an important new signaling mechanism that falls outside of the conventional autocrine or paracrine signaling by secreted molecules. Increased numbers of circulating MPs have been found as biomarkers in a number of diseases (cerebral malaria, sepsis, and several autoimmune diseases), and our data indicates that they may play a required role in the pathogenesis of experimental cerebral malaria (6). Antigen-presenting cells of the immune system are an important functional source of EXs (4). EXs exhibit marked adjuvant effects, improving the immune response to vaccine candidates and contain functional MHC II and peptide along with costimulatory molecules. Apoptotic bodies are considered inert and not to elicit a response by immune cells (4). Cell-derived vesicles are a new mode of cell–cell signaling that is now under intense investigation. MPs in infectious disease may also exhibit other important functions, such as carrying antigens that are important for the activation of the immune system (10).

To determine the molecular mechanisms of MP formation, their function and production from selected cell types in response to agonists, it is necessary to separate out the individual cell populations; this separation is described in Subheadings 3.1 and 3.2. The process of separating vesicles from cells and soluble factors by centrifugation is described in Subheading 3.1. In Subheading 3.2, we present agonists and procedures that are commonly used to elicit MP formation in a number of different cell types. The resultant MPs or MPs produced in blood must be first separated from their parent cells and soluble factors so that these variables do not affect the functional analysis of MPs. This step is critical especially if the MPs are frozen for subsequent analysis. In Subheading 3.3, we present protocols for fluorescence labeling of MPs and in Subheading 3.4 their analysis by conventional flow cytometry. There are, however, limitations to conventional flow cytometry; these limitations can be overcome by using the protocols described in Subheading 3.5 using imaging flow cytometry. In Subheading 3.6, we briefly outline alternate bulk assays for MP analysis.

2. Materials

2.1. Collection of Blood and Processing of Cells

1. Human blood:

   (a) Vacutainer™ blood collection tubes (13 × 100 mm) with acid citrate dextrose. Anticoagulant (0.248 M, ratio blood:anticoagulant is 9:1) (Becton Dickinson, Franklin Lakes, NJ).

   (b) Vacutainer™ holder, 21G needles, gauze, alcohol wipes (Becton Dickinson).

2. Mouse venous blood: retro-orbital venipuncture under general anesthesia

   (a) Glass capillaries.

   (b) Acid citrate buffer (3.5%, ratio blood:anticoagulant is 4:1).

3. Rat blood:

   (a) 3 mL syringe with 21G needle (Becton Dickinson).

   (b) Acid citrate buffer (3.5%, ratio blood:anticoagulant is 4:1).

4. Common items for processing blood:

   (a) (Optional) Apyrase, PGE1, and PGI2 (Sigma-Aldrich, St. Louis, MO): may be added to the anticoagulant to provide additional inhibition of platelet activation.

   (b) 15 mL conical tubes (Falcon brand; Becton Dickinson); 1.5 mL microcentrifuge tubes.

2.2. Generation of Cell-Derived Microparticles

1. Tyrode's buffer lacking Ca and Mg; Histopaque-1077 (Sigma-Aldrich).

2. Agonists for generating increased MP numbers from erythrocytes, leukocytes, and endothelial cells and platelets (see Table 2).

3. (Optional) Cell tracking dyes: carboxyfluorescein diacetate succinimidyl ester (CFSE; Invitrogen), PKH67 or PKH26 (Sigma-Aldrich).

2.3. Fluorescence Labeling of MPs

1. Antibodies specific for proteins detected in human MP analysis (see Table 3).

2. (Optional) Vesicle labeling dye: Calcein AM.

3. Hank's Balanced Salt Solution (HBSS).

4. 0.8 and 1.1 μm particles latex beads (Sigma-Aldrich).

5. Equipment:

   (a) Flow cytometer (e.g., Cytomics FC500; Beckman Coulter, Fullerton, CA).

   (b) ImageStream (Amnis Corporation, Seattle, WA).

Table 2
Summary of agonists and working concentrations for generating increased mp numbers from erythrocytes, leukocytes, and endothelial cells and platelets

Agonist	[Agonist]	Cell number	Timing	Vendor
Platelets (markers: CD41 [αiib], CD61 [β3], CD42a [gpIX])				
A23187	10 μM	1×10^7/mL	15–45 min	Sigma-Aldrich
Collagen	10 μg/mL	1×10^7/mL	15–45 min	Sigma-Aldrich
Thrombin	10 U/mL	1×10^7/mL	15–45 min	Sigma-Aldrich
ADP	10 μM	1×10^7/mL	15–45 min	Sigma-Aldrich
Epinephrine	10 μM	1×10^7/mL	15–45 min	Sigma-Aldrich
Platelet storage	4°C	1×10^7/mL	7 days	n/a
Erythrocytes (marker: CD235a [glycophorin A])				
A23187	10 μM	1×10^7/mL	15–45 min	Sigma-Aldrich
Storage	4°C	1×10^7/mL	7 days	n/a
Leukocytes (monocytes: CD14/CD11b; PMN: CD66b; T cells: Thy1, CD4 or CD8)				
Lipopolysaccharide	100 μg/mL	2×10^6/mL	6–18 h	Sigma-Aldrich
Endothelial cells (markers: CD31 [PECAM1], CD51 [αv], CD105 [endoglin])				
A23187	10 μM	2×10^6/mL	45–60 min	Sigma-Aldrich
Lipopolysaccharide	100 μg/mL	2×10^6/mL	18 h	Sigma-Aldrich
IL-1	25–50 μg/mL	2×10^6/mL	18 h	Peprotech
TNF	50–100 μg/mL	2×10^6/mL	18 h	Peprotech

Except for adherent cells, 100 μL of cells are used in each assay

3. Methods

3.1. Collection and Processing of Blood for MP Analysis

1. *Human blood.* The assessment of the number and phenotype of blood MPs during disease requires the standardization of collection and processing, as well as the minimization of steps that may actually elicit MP formation from blood cells. Steps that should be standardized for the analysis of patient samples include: sampling site (cubital vein versus central venous catheter), the size the needle (21G) used during venipuncture, the absence of a tourniquet, the anticoagulant (sodium citrate is optimal), and the use of a 10 mL vacutainer rather than a syringe (constant suction). Apyrase (1 U/mL) and either PGE1 (100 ng/mL) or PGI2 (0.2 μM) may be added to further inhibit platelet activation, but we have found that this is not necessary. Donors should not have taken medication

Table 3
The following table lists proteins detected in human mp analysis and a supplier

Marker	Alternate name(s)	Cell-type	Clone	Vendor
Phosphatidyl serine		Most activated/ apoptotic cells	Annexin V	BD, Beckman Coulter
CD41	αIIb	Platelet	HIP2	BD
CD41 activated			PAC1	BD
CD42a	gpIX		ALMA.16/ SZ1	BD
CD235a	Glycophorin A	Erythrocyte	11E4B	Beckman Coulter
CD45	Common leukocyte antigen	Leukocytes	J.33	Beckman Coulter
CD14		Monocyte	RMO52	Beckman Coulter
CD11b	MAC1	Monocyte	Bear1	Beckman Coulter
CD66b		Granulocyte	80H3	Beckman Coulter
CD4		T-helper	13B8.2	Beckman Coulter
CD8		T-cytotoxic	B9.11	Beckman Coulter
CD19/CD20		B cell	J3-119/B9E9	Beckman Coulter
CD31	PECAM1	Endothelial cell/ monocyte/platelet	WM59	BD
CD54	ICAM-1	Activated endothelial cells/leucocytes	YN1	Beckman Coulter
CD106	VCAM-1	Activated endothelial cells	51-10C9	BD
CD105	Endoglin	Endothelial cell	1G2	Beckman Coulter
CD144	VE-cadherin	Endothelial cell	TEA1/31	Beckman Coulter

BD: Becton Dickinson

(especially one interfering with platelet functions, such as aspirin) for 10 days prior to blood sampling. The samples should not be shaken, chilled, or overheated because shear and temperature stress induces increased MP formation. Once collected, the blood should be rapidly processed within 1 h to prevent further release of MPs from cells. If platelet MPs are a component of the analysis, 1 μL of blood is removed for flow cytometry to provide a positive control for platelet labeling and analysis. To directly analyze MPs in blood, it is necessary to generate cell-free enriched MPs via a two-step centrifugation procedure.

(a) Centrifuge the blood at $1,500 \times g$ for 15 min to generate about 4.5 mL of platelet-poor plasma.

(b) Carefully remove the supernatant and pipette 1.5 mL into microcentrifuge tubes. Centrifuge at $13,000 \times g$ for 2 min to remove all residual platelets and cell fragments larger than 1 µm. The supernatant is MP enriched, platelet-deficient plasma or "MP plasma."

(c) Platelets are the smallest cell type in the blood, so the quality of cell-free, MP-enriched plasma ("MP plasma") is assessed by counting the number of CD41+ platelets present in 1 µL of the blood versus 1 µL of MP plasma. If the MP plasma cannot be analyzed immediately on site, the MP plasma can be frozen at −80°C and analyzed at a later time (see Note 1). Because the cellular sources of MPs have been depleted, this procedure does not, in our experience, change the MP counts significantly (11).

(d) This MP plasma does contain EXs. If it is important to distinguish between the two vesicle populations, then centrifuge the MP plasma at $10,000 \times g$ for 1 h to pellet the MP; the majority of the EXs remain in the supernatant. Purification of the EXs requires centrifugation on a sucrose gradient. Note, some investigators centrifuge at $20,000 \times g$, which pellets vesicles of >200 nm size (12); we have not, however, tested this possibility. The pelleting of MPs is also used to separate soluble factors (e.g., cytokines) in the blood from the MPs; this is necessary to ascertain the function of MPs. The MPs primarily pellet and EXs remain in solution; this separation is, of course, not absolute but does provide a mechanism to analyze each vesicle population.

2. *Mouse blood.* Collect blood samples via bleeding from the retro-orbital plexus or intracardiac puncture, and collect the blood (1:4) into citrate. In our hands, bleeding from the retro-orbital plexus yields few if any activated platelets, and no evidence of coagulation (assessed by coagulation factors in the blood) (6, 13). Process the blood into MP plasma as described in step 1 (see Notes 1 and 2).

3. *Rat blood.* Obtain blood by cardiac puncture or from an indwelling catheter.

3.2. Generation of Microparticles from Cells

MP can be generated from a mixed population of cells, e.g., whole blood or from purified cells (cell culture, isolated subpopulation). If the purpose is to increase the number of MP present in the solution, an agonist that will induce the release of MP without modifying the cell surface phenotype, i.e., no up- or downregulation of adhesion molecules on endothelial cells or glycoprotein on platelets, can be chosen, e.g., calcium ionophores: calcimycin and ionomycin. Because all cell types release a basal level of MP, the cell supernatant always contains some MPs and their number is increase in a time dependently manner regardless of the presence of an agonist. These MPs are used as "control" MP. All other agonists listed in Table 2 not only modify the cell surface protein expression, but also likely increase the number of released MP. The biological questions being addressed determine which cell-specific and activation markers are included in the flow cytometry analysis. Below, we briefly summarize protocols for generating in vitro MPs from these selected cell types (see Note 3).

1. *Platelets.* Obtain blood as described in Subheading 3.1.

   (a) Centrifuge the blood at $100 \times g$ for 10 min to pellet erythrocytes and leukocytes and consequently generate Platelet Rich Plasma (PRP).

   (b) Generate purified platelets by washing PRP with Tyrode's buffer lacking Ca and Mg or by Sepharose CL-4B gel filtration. Wash platelets by centrifuging at $12,000 \times g$ for 1 min.

   (c) Enumerate the platelets in a hemocytometer, e.g., an Improved Neubauer chamber, or by flow cytometry using counting beads or by using a cytometer that determines the number of particles per unit volume (e.g., Beckman-Coulter and Accuri flow cytometers). Resuspend the platelets at 1×10^7/mL in 0.1 mL aliquots and then stimulate with the agonists of Table 2 for the indicated times (see Note 4).

2. *Erythrocytes.* Erythrocytes are the predominant cell type in blood and are easily purified from other cell populations.

   (a) Centrifuge the blood at $120 \times g$ for 10 min; remove both the supernatant and buffy coat (thin layer of white cells on the surface of the erythrocyte pellet) in order to generate purified erythrocytes.

   (b) Centrifuge the blood twice at $120 \times g$ for 10 min, each time removing the supernatant and any residual buffy coat.

   (c) Ascertain the purity of the erythrocyte population by flow cytometry using anti-CD41 to label platelets, anti-CD45 to label leukocytes, and anti-CD235a (glycophorin A) to label the erythrocytes. Addition of counting beads determines the numbers of RBCs/µL.

   (d) Count the RBCs, resuspend at 1×10^7/mL in 0.1 mL aliquots, and then stimulate with the selected agonists.

3. *Leukocytes.* Purify primary mononuclear cells from blood by slowly layering the blood on top of Histopaque-1077, centrifuging ($400 \times g$ for 30 min at room temperature), and collecting

the mononuclear cells at the interface. Further separate the individual cell populations either by Fluorescence Activated Cell Sorting (FACS) or Magnetic Activated Cell Sorting (MACS) using the cell population specific antibodies of Table 3. Stimulate the purified cell populations with the selected agonists in vitro. Centrifuge the cells to separate the cells from MPs for flow cytometry. A variety of cell lines are available for the analysis of MP formation in vitro, which does not require cell purification as described above. However, their transformation into immortal cells may alter their responses to agonists and MP generation. These cells are cultured until in the log phase of replication and then used in the assays.

4. *Endothelial cells.* Culture primary or cell lines on a matrix, such as collagen or gelatin. When reaching subconfluence, cells are stimulated during various periods of time (from 10 min with histamine to 18 h with TNF). Because endothelial cells are adherent, the MP are detected in the culture supernatant without purification step. However, an enrichment/concentration by centrifugation may be wanted to identify poorly represented subsets of MP, i.e., MP carrying low levels of antigens or antigens localized at certain sites on the cell surface and therefore not present on all MP.

For in vitro analyses of MP formation, the membranes of the cells can be fluorescence labeled, confirming the origin of the MPs. Cells are labeled green with CFSE or PKH67 and red with PKH26. CFSE is a colorless "vital" dye passively diffuses into cells until it is cleaved by intracellular esterases removing the acetate group, thereby becoming brightly fluorescent. The succinimidyl ester group reacts with intracellular amines, thereby forming fluorescence that is retained within the cells. The cells are incubated in serum-free medium for 15 min at 37°C (platelets can be labeled at room temperature) in 10 µM CFSE (range: 0.5–10 µM) or using the 2× dye solutions and diluents for the PKH dyes (14). The cells are washed to remove excess dye and then used in the assay. Pelleted MP can be labeled directly with the same dyes (14), and Wagner's group has also labeled MPs with Calcein AM (5 µg/mL) for 20 min at room temperature (15).

3.3. Fluorescence Labeling of Microparticles

If the MP plasma was frozen, thaw the samples rapidly using a water bath and then place it on ice. As described above, activated cells exhibit increased intracellular Ca^{2+} that in turn induces translocation of phosphatidylserine from the inner leaflet to the outer leaflet; thus, most investigators use annexin V labeling to identify MPs. However, others indicate that the phosphatidylserine may not be accessible or expressed on all the MPs (particularly of endothelial cell origin) (16), resulting in an undercounting of

MPs. Alternatively, cell-specific markers may be used, and selected markers for the cell types of Subheading 3.2.

1. *Platelets*: all platelets express CD41/CD61 (gpIIb/IIIa) on their surface and CD41 is only detected on the cell surface of platelets. Accordingly, anti-CD41 is used to identify platelet-derived MPs (PMPs). The activated conformation of CD41/CD61 is recognized by PAC1 (humans) and JonA-PE (mouse). The von Willebrand Factor receptor CD42b-CD42c/CD42a/CD42d (gpIb-αβ/gpV/gpIX) is another important platelet protein complex, which also binds thrombin and P-selectin. CD42b is an activation marker that is proteolytically cleaved upon platelet activation, whereas CD42c is constitutive and represents an alternate platelet marker to CD41. P-selectin is detected on the surface of activated platelets. CD41+ and CD42+ MPs may be elicited differentially during pathogenesis, so enumeration of both populations may be indicated (17).

2. *Erythrocytes*: CD235a (glycophorin A) is used to identify red blood cell-derived MPs (RMPs).

3. *Leukocytes*: CD45 is present on all leukocytes and is used to identify leukocyte-derived MPs (LMPs). For leukocyte subsets, the following mAbs are used: CD14/CD11b (monocytes), CD66b (granulocytes), CD4 (T_H), CD8 (T_C), and CD20 (B cells) (see Note 5). Because CD14 is a raft-associated protein, Freyssinet and colleagues (18) suggest that CD14 may not be present on MPs or at lower levels; others, however, have used this marker (8, 16).

4. *Endothelial cells*: CD105 (endoglin) is not only expressed on ECs, but also weakly expressed on leukocytes. Endothelial-derived MPs (EMPs) are therefore: CD105+, CD45−. CD144 (VE-cadherin) is also used as marker for EMPs. CD54 (ICAM1), CD62E (E-selectin), CD62P (P-selectin), and CD106 (VCAM1) can be markedly upregulated on endothelial cells by various agonists and therefore may provide a measure of whether subsets of EMPs are derived from activated endothelial cells.

3.3.1. Fluorescence Labeling of the MPs for Flow Cytometric Analysis (no annexin V Labeling)

1. Pipette 10 μL aliquots of MP plasma into flow cytometry tubes.

2. Some investigators add 1 μL of EDTA (25 mM) to MP plasma (final concentration 2.5 mM EDTA). EDTA reduces nonspecific labeling but in our opinion is not necessary. It may also affect the binding to some molecules, such as CD41.

3. Add 2 μL of selected cell-specific mAbs (final concentration: 15 μg/mL) and incubate for additional 25 min.

4. Add 500 μL of HBSS and subject to flow cytometry (see Note 6).

3.3.2. Fluorescence
Labeling of the MPs for
Flow Cytometric Analysis
(with annexin V Labeling)

1. 10 µL aliquots of MP plasma are pipetted into flow cytometry tubes.

2. Add 1 µL of ready-to-use annexin V-FITC + 1 µL of binding buffer 10× (to optimize the calcium concentration necessary for annexin V binding in the plasma and counteract the binding inhibitory effect of the anticoagulant); incubate for 10 min. However, one must be aware that the presence of binding buffer containing 2.5 mM Ca^{2+} can trigger plasma coagulation, therefore the total incubation time must be constant and samples must be checked prior to flow analysis for the presence of microclots.

3. Add 2 µL of selected cell-specific mAbs (final concentration: 15 µg/mL) and incubate for additional 20 min.

4. Dilute samples in 0.5 mL of annexin V binding buffer and analyze by flow cytometry.

The choice of fluorescence label(s) depends on the cell population being analyzed in vitro or if determining the cellular origin of the MPs is important. If the cell membrane has been labeled (Subheading 3.2), vesicles derived from the cell membrane will be labeled. Alternatively, the small vesicles can be labeled with Calcein AM (15). If secondary reagents are used, then the MPs are pelleted by high speed centrifugation (25,000×*g*, 30 min, 15°C) before the fluorescence labeled secondary mAb or streptavidin is added. It is important to include negative isotype controls for the selected specific Abs because there is greater nonspecific binding of Abs to MPs than intact cells and the AB may not be removed if washing is omitted.

Polystyrene beads of defined size are often added to the sample to provide rudimentary quality control. In most cases, 1 µm polystyrene beads are added and events exhibiting lower forward scatter are collected and defined as MPs. Polystyrene beads with defined sizes (e.g., 190, 530, 780, and 990 nm) have also been used to provide a guide for size. However, it is important to emphasize that the forward scatter signal derived from these beads cannot be related to the size of the particle (see Subheading 3.5). If a known number of beads are added, these can then be used as counting beads if the flow cytometer does not assess the volume of sample analyzed (Fig. 1).

3.4. Conventional Flow Cytometric Analysis of Microparticles

The analysis of MPs on a Beckman-Coulter's Cytomics FC 500 flow cytometer is summarized in Fig. 1. All parameters, including FSC and SSC, are set in log-mode. The parent cell may be used to set the compensation if several fluorophores are used in the analysis. It is important to ensure that saturating levels of antibody are used for fluorescence labeling of the MPs; we recommend performing titration curves using MPs or the parent cells. The MP sample is then acquired on the flow cytometer.

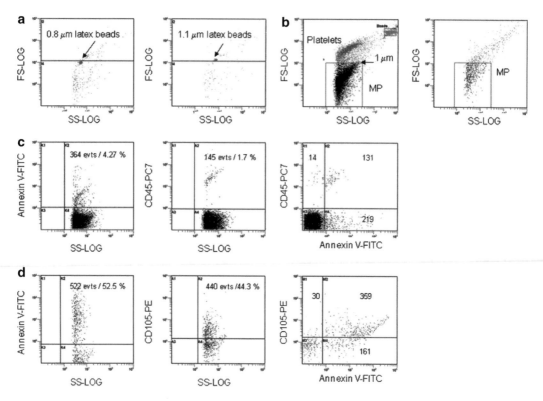

Fig. 1. Conventional flow cytometric analysis of MPs. MPs were extracted from mouse whole blood (**c**) or from cultured brain endothelial cells (**d**) as described in Subheadings 3.1 and 3.2. They were fluorescence labeled with annexin V-FITC and CD45-PC7 for the mouse plasma, and annexin V-FITC and CD105-PE for endothelial cells as described in Subheading 3.3, and then analyzed by flow cytometry on a Beckman-Coulter's Cytomics FC 500. (**a**) Forward and side scatter gating showing the 0.8 and 1.1 μm beads (*left* and *right* cytograms, respectively) and the upper limit of the MP gate. (**b**) *Left* cytogram represents a forward and side scatter of a mouse plasma that has not been completely depleted in platelets. The *top right* population represents the counting beads. MP and platelet populations can be easily distinguished using the 1 μm limit. *Right* cytogram represents a forward and side scatter of a culture supernatant. Few events are found outside the MP gate compared to plasma. They correspond to larger vesicles and cell debris. (**c**) Example of single and double staining of mouse plasma. From *left* to *right*: side scatter-annexin V-FITC, side scatter-CD45-PC7, and annexin V-FITC-CD45-PC7 dot plots. *Numbers* inside cytograms represent the number of positive elements for the marker analyzed and in the first two dot plots, corresponding percentage of the gate population is mentioned. (**d**) Example of single and double staining of endothelial cell culture supernatant. From *left* to *right*: side scatter-annexin V-FITC, side scatter-CD105-PE, and annexin V-FITC-CD105-PE dot plots. *Numbers* inside cytograms represent the number of positive elements for the marker analyzed and in the first two dot plots, corresponding percentage of the gate population is mentioned. Note that for an equivalent number of MP, the percentages of positive MP in a culture supernatant or purified suspension of MP is always higher than that in a plasma or whole as lots of events passing in front of the laser are detected but counted as positive. Therefore, this decreases artifactually the percentages of positive events. Thus, the number of positive MP analyzed per unit of time is very important.

1. Run 0.8 and 1.1 μm particles latex beads alone, and use to set the upper limit of the population to analyze with FSC <1 μm (Fig. 1a) and have these events shown in the fluorescence dot plots.

2. Run positive samples (PRP) if applicable to define the platelet region (Fig. 1b). Note, platelets are the smallest cell type and

are about 2–5 µm in size and must be excluded from the MP analysis (Fig. 1b). Other cell populations are much larger, so they are not problematic.

3. To set the positive gate limit, acquire MP events using isotype controls for the specific antibodies.

4. To verify compensation, acquire MP events with single specific antibodies.

5. Acquire multiple-labeled MP events (Fig. 1c). The slowest flow rate of the instrument provides the greatest sensitivity because the fluorescence signal is integrated the longest but may take too long to sample. Increasing the flow rate above 300/s will increase the number of coincident events and lower the count assessed by the flow cytometer (see Note 7). When possible, more than 1,000 positive MPs should be collected in each sample, and we save all events for the subsequent analysis.

6. Determine the percent positive events and the number of labeled MPs. If the flow cytometer is syringe-driven (e.g., Beckman-Coulter), then the number of MPs/µL sample is calculated by the instrument. Alternatively counting beads are spiked into the sample. The number of MPs per µL = [number assessed by flow cytometer × (beads counted/beads added)]/sample volume (Fig. 1b).

A gate is set to include only events exhibiting FSC <1 µm beads; if the MPs are fluorescence labeled, then a second gate is set to include only events with a fluorescence intensity above that observed with unlabeled MPs. Because MPs from many cell populations express PS, fluorescence labeling with annexin V can also be used to positively identify MPs in the sample. The fluorescence intensity is then measured for each MP, which then indicates the cellular origin of the MP and extended analyses may determine functional proteins on the surface of MPs. For example, we have reported that MPs contain the P-glycoprotein (that confers multidrug resistance to cancer cells), bind to drug sensitive cancer cells, and functionally transfer drug resistance (9). The numbers of positive MPs for a specific protein is also recorded.

3.5. Imaging Flow Cytometric Analysis of Microparticles

Amnis has developed an imaging flow cytometer, which overcomes many of the problems associated with flow cytometric assessment of MPs. First, the instrument provides an actual measurement of size rather than forward scatter. Accordingly, it is not necessary to extrapolate the arbitrary forward scatter measurement of polystyrene beads to the size of biological vesicles, which is not valid anyway. Second, the signal to noise ratio for small particles is markedly better by using imaging cytometry. A conventional flow cytometer measures accurately the fluorescence intensity in a selected range of wavelengths of light (bandpass-filtered), but the small particle occupies only a small portion of that volume. Thus, the small signal is present in a large background, which compromises sensitivity. In contrast, the ImageStream analysis software (IDEAS) measures signal specifically with the pixels that the particle's image occupies. Because signal arising from small objects occupy a minimal number of pixels, there is very little pixel-associated noise, and thus the ImageStream is very sensitive to MP fluorescence. This is especially important for the analysis of MPs, which exhibit low levels of markers on their surface and hence low fluorescence intensity when compared with the parent cell. Furthermore, obvious coincident events can be flagged and potentially included in the analysis. If two particles are captured and are separated spatially in the image, then these can be assessed as distinct MPs in the analysis. The MP plasma is fluorescence labeled and then subjected to Imaging Analysis as described below.

1. Set up the instrument with an upper limit for the raw maximal pixel intensity (background subtracted intensity of the brightest pixel in the image) set to 100 to eliminate the collection of bead calibration reagent.

2. Acquire the unlabeled samples and single color isotype control samples to set up software compensation for the MP analysis. The samples may need to be diluted if there are large numbers of coincident events. We dilute the samples tenfold into a sample volume of 50 µL.

3. Acquire the PRP control samples if included.

4. Acquire all events except bead calibration reagent in the test samples with at least 5,000 events. Use compensation settings from the single color samples.

5. In the analysis phase, set the MP gate to analyze events <1 µm and platelet gate for events >1 µm (Fig. 2, histogram). Depending on the selected fluorescence labels, display the dot plots and place regions to define subpopulations of MPs (Fig. 2, dot plots). The analysis even indicates the number of unlabeled MPs, which exhibit an SSC above background (not shown). Because our sample contained both platelets and MPs, we can compare both populations. The analysis of the regions reports the fluorescence intensity and number of particles. This analysis shows that thrombin activation increased the number of platelet MPs, which decreased their expression of CD42b, a proteolytically cleaved activation marker.

The limitation of the ImageStream is the accurate measurement of size for small particles. Particle cross-section area is measured on the ImageStream by placing a mask or limit around the brightfield object identified, counting the number of pixels in that

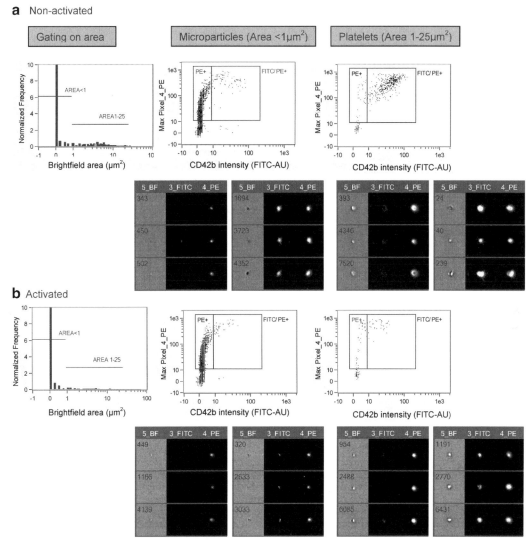

Fig. 2. Imaging (ImageStream) flow cytometric analysis of MPs. Unactivated PRP (**a**) and thrombin-activated (10 U/mL thrombin for 15 min) (**b**) PRP were fixed for 15 min in 1% paraformaldehyde and then fluorescence labeled with anti-CD42b-FITC and anti-CD41-PE as described in Subheading 3.3. ImageStream analysis of platelet MPs in unactivated- and activated-platelets was performed. Histogram: size gating separating MPs (<1 μm^2) from intact cells (platelets) that are larger (1–25 μm^2). These populations are then analyzed in dot plots and representative images (brightfield, SSC, FITC, and PE) are shown for each subpopulation identified in the dot plot.

region, and then converting the number of pixels to actual size based on physical pixel size of the camera and objective magnification. In the experiment of Fig. 2, the objective was 40×, resulting in 0.5 μm^2 pixels. The accuracy of this measurement depends on the ability of the software to place the mask and the focal position of the object. Objects that have low contrast or are smaller than 0.5 μm^2 are under masked and appear smaller. If the object drifts from the optimal focal plan, its image blurs and the software will mask this event as larger than its actual size. The ImageStream therefore has limitations on the assessment of MP size, but we contend that this size discrimination is not possible by conventional flow cytometry either.

3.6. Alternative Bulk Assays for the Analysis of Microparticles

1. An alternate approach to assess MP levels is ELISA MP capture assays; in this ELISA, the MPs are captured by annexin V or antibodies specific to antigens on surface of MPs (e.g., CD41). The levels of MP antigen are then assessed by using a second MP-specific antibody with a colorimetric, fluorescence, or luminescence detection system. This approach while quantitative does not determine the number or phenotype of particles and is sensitive to cell carryover, which gives a large signal compared with MPs.

2. Scanning electron microscopy is used to detect the presence of MPs in samples and is the most accurate approach to generate size ranges for the MPs. Confocal microscopy is useful for detecting the binding and potential transfer of functional proteins from MPs to cells. The flow cytometric assays should be extended by performing functional assays to link increases in the selected phenotype of MPs with a function.

4. Conclusions

In conclusion, conventional flow cytometry is widely accepted approach to measuring the number and phenotype of MPs in blood and after stimulation of cells with agonists. These MPs seem to play an important role in homeostasis and in pathogenesis. There are marked differences in results between laboratories although the overall conclusions are generally similar. One source of differences in results is how the samples are processed. Stimulation of cells during processing may generate MPs and lead to a high background in the control unstimulated group. There are important limitations on conventional flow cytometry, such as the sensitivity of the instrument in measuring the low intensity of fluorescence on MPs due to their small size and low numbers of surface proteins compared with intact cells. Coincident events are often not checked for and may be problematic. Presence of lipids in the plasma, especially in human samples or some gene-deficient mice may also alter the flow cytometric reading. We posit that many of the problems with conventional flow cytometry may be overcome by using imaging flow cytometry because the signal to noise ratios are better than conventional flow cytometers, coincident events can be more readily detected, and the image provides a direct measurement of size.

5. Notes

1. If the plasma still contains platelets detectable by flow cytometry, platelets will not affect the flow cytometry analysis of MPs in a fresh sample as shown in Fig. 1. However, to be able to freeze plasma and analyze it later after thawing, one has to make sure that all platelets are removed, or MP created during freezing/thawing by the breakdown of platelets may affect the analysis. We recommend snap freezing of MP plasma in liquid nitrogen followed by storage for up to several weeks at –80°C or in liquid nitrogen.

2. Because mouse blood contains four to five times more platelets than human, a higher concentration of anticoagulant is used. In addition, generating platelet-free plasma is more difficult and may require an additional high speed centrifugation.

3. These times are guidelines and time course analyses are required to confirm the generation of MPs.

4. There are large numbers of platelets/µL of PRP. Accordingly, 1 µL of PRP may be diluted in 99 µL of Tyrode's buffer; this limits platelet activation during washing and decreases the experimental time. The disadvantage of this approach is that identical numbers of platelets are not added because there is some variation in platelet number between individuals and this can be pronounced during disease (thrombocytopenia).

5. These are not absolutely specific because, for example, CD4 is detected on monocytes and CD11b on granulocytes.

6. Some investigators use 30 µL of MP plasma for flow cytometric analysis (15). Others fix the labeled MPs in freshly made 1% paraformaldehyde solution to prevent additional production and/or alteration.

7. To rule out that coincident events are affecting the MP analysis, it is useful to perform twofold dilutions of the sample and assess the number of MPs following dilution. If coincidence is not problematic, then the number of MPs determined should decline in direct proportion to the dilution.

References

1. Reininger, A. J., Heijnen, H. F., Schumann, H., Specht, H. M., Schramm, W., and Ruggeri, Z. M. (2006) Mechanism of platelet adhesion to von Willebrand factor and microparticle formation under high shear stress. *Blood* **107**, 3537–3545.

2. Kushak, R. I., Nestoridi, E., Lambert, J., Selig, M. K., Ingelfinger, J. R., and Grabowski, E. F. (2005) Detached endothelial cells and microparticles as sources of tissue factor activity. *Thromb Res* **116**, 409–419.

3. Aras, O., Shet, A., Bach, R. R., Hysjulien, J. L., Slungaard, A., Hebbel, R. P., Escolar, G., Jilma, B., and Key, N. S. (2004) Induction of microparticle- and cell-associated intravascular tissue factor in human endotoxemia. *Blood* **103**, 4545–4553.

4. Beyer, C. and Pisetsky, D. S. (2010) The role of microparticles in the pathogenesis of rheumatic diseases. *Nat Rev Rheumatol* **6**, 21–29.

5. Combes, V., Coltel, N., Faille, D., Wassmer, S. C., and Grau, G. E. (2006) Cerebral malaria: role of microparticles and platelets in alterations of the blood–brain barrier. *Int J Parasitol* **36**, 541–546.

6. Combes, V., Coltel, N., Alibert, M., van Eck, M., Raymond, C., Juhan-Vague, I., Grau, G.E., and Chimini, G. (2005) ABCA1 gene deletion protects against cerebral malaria: potential pathogenic role of microparticles in neuropathology. *Am J Pathol* **166**, 295–302.

7. Combes, V., Simon, A. C., Grau, G. E., Arnoux, D., Camoin, L., Sabatier, F., Mutin, M., Sanmarco, M., Sampol, J., and Dignat-George, F. (1999) In vitro generation of endothelial microparticles and possible prothrombotic activity in patients with lupus anticoagulant. *J Clin Invest* **104**, 93–102.

8. Shet, A.S. (2008) Characterizing blood microparticles: technical aspects and challenges. *Vasc Health Risk Manag* **4**, 769–774.

9. Bebawy, M., Combes, V., Lee, E., Jaiswal, R., Gong, J., Bonhoure, A., and Grau, G. E. (2009) Membrane microparticles mediate transfer of P-glycoprotein to drug sensitive cancer cells. *Leukemia* **23**, 1643–1649.

10. Couper, K. N., Barnes, T., Hafalla, J. C., Combes, V., Ryffel, B., Secher, T, Grau, G. E., Riley, E. M., and de Souza, J. B. G. (2010) Parasite-derived plasma microparticles contribute significantly to malaria infection-induced inflammation through potent macrophage stimulation. *PLoS Pathog* **6**, e1000744.

11. Combes, V., Taylor, T. E., Juhan-Vague, I., Mege, J. L., Mwenechanya, J., Tembo, M., Grau, G. E., and Molyneux, M. E. (2004) Circulating endothelial microparticles in malawian children with severe falciparum malaria complicated with coma. *JAMA* **291**, 2542–2544.

12. Simak, J., Holada, K., Risitano, A. M., Zivny, J. H., Young, N. S., and Vostal, J. G. (2004) Elevated circulating endothelial membrane microparticles in paroxysmal nocturnal haemoglobinuria. *Br J Haematol* **125**, 804–813.

13. van der Heyde, H. C., Gramaglia, I., Sun, G., and Woods, C. (2005) Platelet depletion by anti-CD41 (αIIb) mAb injection early but not late in the course of disease protects against Plasmodium berghei pathogenesis by altering the levels of pathogenic cytokines. *Blood* **105**, 1956–1963.

14. Faille, D., Combes, V., Mitchell, A. J., Fontaine, A., Juhan-Vague, I., Alessi, M-C., Chimini, G., Fusai, T., and Grau, G. E. (2009) Platelet microparticles: a new player in malaria parasite cytoadherence to human brain endothelium. *FASEB J* **23**, 3449–3458.

15. Bernimoulin, M., Waters, E. K., Foy, M., Steele, B. M., Sullivan, M., Falet, H., Walsh, M. T., Barteneva, N., Geng, J. G., Hartwig, J. H., et al. (2009) Differential stimulation of monocytic cells results in distinct populations of microparticles. *J Thromb Haemost* **7**, 1019–1028.

16. Gelderman, M. P. and Simak, J. (2008) Flow cytometric analysis of cell membrane microparticles. *Methods Mol Biol* **484**, 79–93.

17. Horstman, L. L., Jy, W., Jimenez, J. J., Bidot, C., and Ahn, Y. S. (2004) New horizons in the analysis of circulating cell-derived microparticles. *Keio J Med* **53**, 210–230.

18. Satta, N., Toti, F., Feugeas, O., Bohbot, A., Dachary-Prigent, J., Eschwege, V., Hedman, H., and Freyssinet, J. M. (1994) Monocyte vesiculation is a possible mechanism for dissemination of membrane-associated procoagulant activities and adhesion molecules after stimulation by lipopolysaccharide. *J Immunol* **153**, 3245–3255.

Chapter 17

Noncytotoxic DsRed Derivatives for Whole-Cell Labeling

Rita L. Strack, Robert J. Keenan, and Benjamin S. Glick

Abstract

Fluorescent proteins (FPs) are invaluable tools for biomedical research. Useful FPs have desirable fluorescence properties such as brightness and photostability, but a limitation is that many orange, red, and far-red FPs are cytotoxic when expressed in the cytosol. This cytotoxicity stems from aggregation. To reduce aggregation, we engineered the surface of DsRed-Express to generate DsRed-Express2, a highly soluble tetrameric FP that is noncytotoxic in bacterial and mammalian cells. Directed evolution of DsRed-Express2 yielded the color variants E2-Orange, E2-Red/Green, and E2-Crimson. These variants can be used to label whole cells for single- and multi-color experiments employing microscopy or flow cytometry. Methods are described for reducing the higher-order aggregation of oligomeric FPs and for analyzing FP cytotoxicity in *Escherichia coli* and HeLa cells.

Key words: Fluorescent protein, DsRed, Whole-cell labeling, Protein engineering, Cytotoxicity

1. Introduction

Since the introduction of the green fluorescent protein (GFP) from *Aequorea victoria* (1), fluorescent proteins (FPs) have become invaluable research tools for microscopy and flow cytometry. Dozens of FPs have been discovered or engineered to span the visible spectrum (2). In particular, a number of red FPs have been engineered for use alone or in multi-color studies with GFP. A major goal has been to engineer monomeric red FPs from proteins that are dimeric or tetrameric in their wild-type forms. These efforts have produced the widely-used mFruits (3, 4) as well as DsRed-Monomer (5), TagRFP (6), TagRFP-T (7), mKate (8), mKate2 (9), mRuby (10), mKO (11), and mKO2 (12).

Although monomeric red FPs are theoretically useful for any application, they tend to be dimmer and more photolabile than the corresponding oligomers (Table 1) (13). Therefore, researchers

Teresa S. Hawley and Robert G. Hawley (eds.), *Flow Cytometry Protocols*, Methods in Molecular Biology, vol. 699,
DOI 10.1007/978-1-61737-950-5_17, © Springer Science+Business Media, LLC 2011

Table 1
Fluorescence properties of orange, red, and far-red FPs[a]

Fluorescent protein	Excitation/ emission maxima	Extinction coefficient	Quantum yield	Relative brightness[b]	Maturation half-time (h)	Photobleaching half-time (s)[c]	pK_a
E2-Orange	540/561	36,500	0.54	0.54	1.3	81 ± 3	4.5
mOrange2	549/563	56,300	0.49	0.75	4.5	40 ± 3	7.5
KO	548/560	72,800	0.55	1.1	3.8	21 ± 2	5.0
mKO2	549/563	54,300	0.82	1.2	1.8	5 ± 1	5.0
WT DsRed	558/583	51,500	0.71	1.0	11	–	–
DsRed-Express2	554/591	35,600	0.42	0.41	0.7	64 ± 4	4.5
DsRed-Max	560/589	48,000	0.41	0.54	1.2	9 ± 1	–
E2-Red/Green (green)	484/498	100,200	0.06	0.17	0.4	236 ± 8[d]	4.0
E2-Red/Green (red)	560/585	53,800	0.67	0.98	1.2	93 ± 3	4.5
DsRed-Express	554/586	33,800	0.44	0.41	0.6	71 ± 3	–
DsRed-Monomer	557/592	27,300	0.14	0.10	1.3	15 ± 1	–
mCherry	585/609	66,400	0.23	0.42	0.6	18 ± 1	–
tdTomato	553/581	85,700	0.69	1.6	2.0	5 ± 1	–
TagRFP	554/582	77,000	0.47	0.98	1.5	8 ± 4	–
TagRFP-T	554/584	67,800	0.40	0.73	1.6	20 ± 2	–
TurboRFP[e]	550/573	–	–	–	1.5	32 ± 1	–
RFP611	555/606	109,700	0.60	1.8	2.7	7 ± 2	–
E2-Crimson	611/646	126,000	0.23	0.79	0.4	26 ± 3	4.5
mPlum	587/649	29,300	0.10	0.08	1.6	31 ± 2	5.5
mRaspberry	594/627	42,000	0.14	0.16	2.1	21 ± 2	5
mKate	584/632	45,500	0.33	0.41	1.3	15 ± 2	–
mKate2	586/630	56,400	0.39	0.60	0.8	23 ± 2	6.5
Katushka	584/631	76,300	0.32	0.67	0.6	15 ± 1	7.5
RFP637	585/637	53,800	0.22	0.32	2.4	60 ± 4	4.5
RFP639	587/639	74,700	0.21	0.43	1.7	29 ± 3	4.5

[a]pK_a values were determined for only a subset of the FPs. Photobleaching was not measured for WT DsRed because maturation of this protein is too slow for our in vivo assay. For details, see refs. (5, 17–19)

[b]Brightness was calculated as the product of extinction coefficient and quantum yield, and was normalized to a value of 1 for wild-type DsRed

[c]Photobleaching half-times during widefield illumination are listed as mean ± s.e.m. for three independent replicates

[d]The photobleaching half-time reported for the green chromophore of E2-Red/Green cannot be compared directly to the photobleaching half-times of red chromophores because a different filter set was used

[e]Because TurboRFP showed very poor solubility during extraction from bacteria, we were unable to perform brightness measurements for this protein

have also created improved oligomeric red FPs such as DsRed-Express (14), Kusabira-Orange (KO), TurboRFP (6), Katushka (8), and RFP611, RFP637, and RFP639 (15). These oligomeric FPs cannot be used as fusion tags, but they are suitable for labeling organelles, whole cells, tissues, and entire organisms.

A serious problem with both monomeric and oligomeric FPs is cytotoxicity (16). Most of the available red FPs show pronounced cytotoxicity that presumably stems from aggregation (17–19). For this reason, we modified the surface of DsRed-Express to create a tetrameric derivative that shows minimal higher-order aggregation. The resulting FP, DsRed-Express2, combines low cytotoxicity with favorable photophysical properties such as brightness, fast maturation, photostability, and pH stability (17). As a result, the tetrameric DsRed-Express2 is an ideal red FP for whole-cell labeling.

Noncytotoxic color variants were then engineered by modifying the interior of DsRed-Express2. This approach yielded three tetrameric derivatives termed E2-Orange (18), E2-Red/Green (18), and E2-Crimson (19). Like the parental DsRed-Express2, these new variants substantially outperform other FPs with regard to cytotoxicity in bacterial and mammalian systems (17–19). E2-Orange is ideal for multi-color experiments involving green and far-red FPs (Fig. 1a), while E2-Red/Green is useful as a "third color" for flow cytometry in combination with red and green FPs (Fig. 1b). E2-Crimson is particularly noteworthy because it is the fastest-maturing red or far-red FP, the brightest far-red FP, and the only known member of the GFP family that is efficiently excited with standard 633-nm lasers. As a result, E2-Crimson is useful for multi-color microscopy and flow cytometry (Fig. 1c). The methods used to engineer and evaluate DsRed-Express2 and its derivatives are detailed below, and can be used as a guide for the development of new noncytotoxic FPs.

2. Materials

2.1. Screening E. coli Colonies for Brightness

1. Luria Broth (LB) plates supplemented with 100 µg/mL ampicillin.

2. Chemically competent *Escherichia coli* (*E. coli*) DH5α cells.

3. Mutagenic library encoded in pQE-60NA (17), which is a modified version of the pQE-60 vector (Qiagen, Valencia, CA) lacking the hexahistidine tag. pQE-60NA is a low-copy vector that constitutively expresses genes from a medium-strength T5 promoter flanked by two *lac* operator sequences.

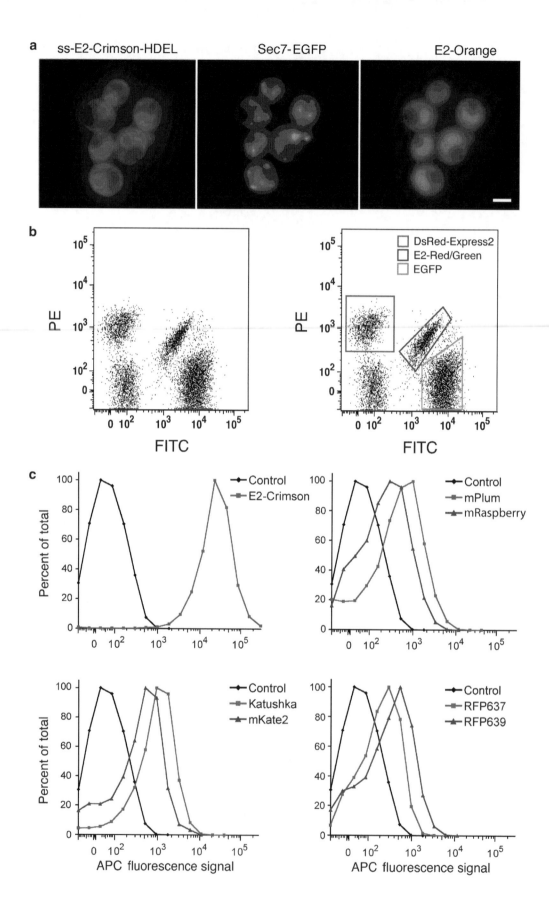

4. Carousel 4200 slide projector (Eastman Kodak Co., Rochester, NY) with a 300-W bulb (General Electric, Fairfield, CT).

5. Glass bandpass filters for excitation (Chroma, Rockingham, VT). The nm ranges are 540/20, 560/20, and 620/20 for orange, red, and far-red FPs, respectively. Excitation filters are hung in front of the slide projector light source using a small hook.

6. Emission filter goggles. Red FP emission was observed using 580-nm longpass goggles (CE-EN207 with Krypton/Copper Vapor Filter; NoIR Laser Co., South Lyon, MI). Goggles for viewing orange and far-red FPs were made in-house by covering standard laboratory goggles with Kodak Wratten 560- or 650-nm longpass filters, respectively.

7. Sterile toothpicks.

2.2. Screening for Aggregation Using the Bacterial Lysis Assay

1. Chemically competent *E. coli* DH5α cells (see Note 1).

2. Liquid LB medium supplemented with 100 μg/mL ampicillin.

3. Sterile, transparent, round-bottom 96-well plates with 300-μL well volume (Costar).

4. Black round-bottom 96-well plates with 300-μL well volume (Costar).

5. Sterile toothpicks.

6. BPER II lysis reagent (ThermoFisher Scientific, Waltham, MA).

7. Microplate fluorometer, such as a Tecan Safire^2 plate reader.

2.3. Assaying Cytotoxicity in Bacteria

1. Chemically competent *E. coli* DH10B cells harboring the pREP4 repressor plasmid (Qiagen).

2. LB plates supplemented with 100 μg/mL ampicillin and 30 μg/mL kanamycin, either with or without 1 mM isopropyl β-D-1-thiogalactopyranoside (IPTG).

3. 8–16% precast SDS–polyacrylamide gels (Pierce).

Fig. 1. Novel applications of DsRed-Express2 derivatives. (**a**) E2-Orange and E2-Crimson are useful for three-color imaging with GFP. *Saccharomyces cerevisiae* cells with the endoplasmic reticulum labeled with E2-Crimson, Golgi cisternae labeled with enhanced GFP (EGFP), and the cytosol labeled with E2-Orange were imaged using widefield microscopy with standard filter sets. (**b**) E2-Red/Green is useful for three-color flow cytometry together with red and green FPs. *S. cerevisiae* cells expressing either EGFP (*lowest box, green* in electronic version), DsRed-Express2 (*highest box, red* in electronic version), E2-Red/Green (*middle box,* or *blue* in electronic version), or no FP were pooled and analyzed by flow cytometry with a 488-nm excitation laser. These four populations were cleanly resolvable. (**c**) E2-Crimson is uniquely suited to flow cytometry with 633-nm excitation lasers. *E. coli* cells expressing various FPs or no FP ("Control") were analyzed by flow cytometry with 633-nm excitation. Only the population expressing E2-Crimson was fully resolvable from nonfluorescent control cells.

4. BupH Tris–HEPES–SDS Running Buffer (Pierce) dissolved in deionized H_2O.

5. PageRuler prestained protein ladder (Fermentas Inc., Glen Burnie, MD).

6. 5× SDS–PAGE sample loading buffer: 280 mM Tris–HCl, pH 6.8; 10% (w/v) sodium dodecyl sulfate (SDS); 22.5% (v/v) glycerol; 0.04% (w/v) bromophenol blue; and 10% (v/v) β-mercaptoethanol (see Note 2).

2.4. HeLa Cell Transient Transfection Assay for Cytotoxicity

1. HeLa cells (CCL-2; ATCC, Manassas, VA).

2. Culture medium consisting of Dulbecco's Modified Eagle's Medium (DMEM) with high glucose and L-glutamine (DMEM High Glucose; HyClone) supplemented with 10% fetal bovine serum (FBS; HyClone) and 1× MEM non-essential amino acids (Invitrogen, Carlsbad, CA).

3. OptiMEM (Invitrogen).

4. 0.05% Trypsin/EDTA for cell dissociation.

5. 1× PBS.

6. Lipofectamine 2000 (Invitrogen) for transient transfections.

7. FPs encoded by a mammalian transient transfection vector with the SV40 origin of replication. The FPs are expressed under control of the CMV promoter with a Kozak sequence immediately upstream of the start codon.

8. Polybrene (Sigma-Aldrich Corp., St. Louis, MO) diluted to 10 mg/mL in sterile deionized H_2O for lentiviral transduction.

9. Hemocytometer.

10. Flow cytometer.

11. Analysis software such as FlowJo (Tree Star Inc., Ashland, OR).

2.5. Generation of Lentiviral Vectors and HeLa Cell Lentiviral Transduction Assay for Cytotoxicity

1. 293-T/17 cells (CRL-11268; ATCC).

2. Culture medium (see Subheading 2.4).

3. Lenti-X Expression Kit (Clontech, Mountain View, CA), including the pLVX-Puro plasmid.

4. QuickTiter Lentivirus Quantitation Kit (Cell Biolabs, Inc., San Diego, CA).

3. Methods

The engineering of useful FPs for whole-cell labeling typically involves multiple rounds of mutagenesis. To generate DsRed-Express2, we employed targeted or random mutagenesis followed by screening for improved variants. Each screen was conducted in two stages, in which a preliminary selection for bright clones was followed by a bacterial-lysis assay to measure aggregation. This general approach can be used to create noncytotoxic derivatives of FPs that have desirable photophysical properties. Oligomeric FPs are best suited to optimization by this strategy, because the multivalent nature of an oligomeric FP enhances the functional affinity that can be detected using the bacterial lysis assay.

In our experience, high solubility of an oligomeric FP in the bacterial-lysis assay correlates strongly with low cytotoxicity in bacterial and mammalian cells. However, cytotoxicity should be measured for each FP. To test for cytotoxicity in bacteria, *E. coli* colonies expressing a given FP are compared to colonies grown under repressing conditions. Smaller colonies indicate greater cytotoxicity. To test for cytotoxicity in mammalian cells, FPs are expressed in HeLa cells either transiently or by lentiviral transduction, and the fluorescence signal is monitored for several days using flow cytometry. With a cytotoxic FP, the average brightness of the cell population decreases over time, whereas with a noncytotoxic FP, such as enhanced GFP (EGFP) or DsRed-Express2, the average brightness remains steady over time.

3.1. Screening E. coli Colonies for Brightness

1. Transform pQE60-NA-based libraries into DH5α cells using standard protocols. Spread the transformation mixture on LB + ampicillin plates at a density that yields approximately 1,000–1,500 colonies per 100-mm Petri plate.

2. After 12–15 h growth at 37°C (see Note 3), examine colony fluorescence using the slide projector, as follows. In a dark room, place an appropriate bandpass filter over the slide projector light source. Wearing an appropriate emission filter goggles, illuminate the Petri plate with the lid off and look for the brightest colonies. Rotate the Petri plate to assure that all regions are examined.

3. Using sterile toothpicks, pick the brightest colonies (approximately 10% of the total) into 96-well plates for screening in the aggregation assay, or restreak promising colonies for maintenance (see Note 4).

3.2. Screening for Aggregation Using the Bacterial Lysis Assay

1. Fill each well of a 96-well round-bottom plate with 175 μL of LB + ampicillin. As described in Subheading 3.1, pick individual fluorescent *E. coli* colonies into the wells. In each plate, set aside four control wells for clones expressing the FP that was the template for mutagenesis (see Note 5). Grow 10–12 h at 37°C (see Note 6), shaking at an angle in an orbital shaker.

2. To store clones for later recovery, use a multichannel pipette to spot 2 μL each from 48 wells onto an LB + ampicillin plate in a grid pattern. Label the Petri plates according to well number.

Incubate the Petri plates at 37°C for 12–15 h, and store them at 4°C.

3. Meanwhile, for each 96-well plate, centrifuge the remaining culture samples for 5 min at $3,000 \times g$. Carefully remove the supernatants using a multichannel pipette.

4. Resuspend each cell pellet in 100 μL BPER II using a multichannel pipette. Allow the lysis to proceed for 20 min at 37°C in an orbital shaker.

5. Centrifuge each 96-well plate for 5 min at $3,000 \times g$. Using a multichannel pipette transfer the supernatants to a pair of black 96-well plates, skipping every other column to leave room for the corresponding pellet fractions. Be careful not to create bubbles during the transfer.

6. Using a multichannel pipette, thoroughly resuspend each pellet of lysed cells in 100 μL BPER II. Transfer the resuspended pellet samples to black 96-well plates, placing each pellet sample next to the corresponding supernatant sample.

7. Quantify the fluorescence signals using a 96-well plate fluorometer. It is important to use a fixed gain that is high enough to produce strong signals without overloading the detector. For red FPs, we typically use 560 ± 20 nm excitation and 600 ± 20 nm emission.

8. Determine the aggregation value for each clone by calculating the percentage of the signal found in the pellet. For a given well, divide the pellet signal by the sum of the pellet and supernatant signals. As a positive control, EGFP should have an aggregation value of 2–4%. Aggregation values for various oligomeric red and far-red FPs are shown in Fig. 2.

9. For each well, compare the aggregation value to the average aggregation value for the four template wells. Calculate the standard error of the mean (s.e.m.) for the four template wells (see Note 7). Multiply the s.e.m. by two and subtract it from the mean aggregation value for the remaining 92 wells, thereby obtaining a threshold for significance. Any well with an aggregation value less than or equal to the threshold should be considered promising for further analysis.

10. To evaluate promising candidates, restreak from the replica spots (step 2) on fresh LB + ampicillin plates. For each candidate, pick four to eight restreaked colonies into 96-well plates, and repeat the aggregation assay with a template control. Candidates that are reproducibly more soluble than the template should be characterized further by preparing miniprep DNA for sequencing.

11. After obtaining the sequences of the improved variants, build a combinatorial library that includes the solubility-enhancing

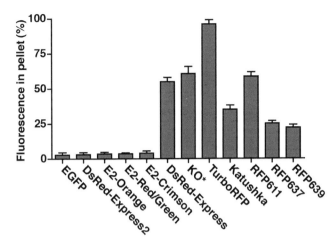

Fig. 2. DsRed-Express2 and its derivatives are as soluble as EGFP and much more soluble than other oligomeric FPs. Oligomeric FPs were expressed in *E. coli*. For each FP, a bacterial-lysis assay was used to determine the aggregation value, which is defined as the percentage of the total fluorescence signal in the pellet fraction. A high aggregation value indicates the formation of higher-order aggregates. As a reference, EGFP has an aggregation value of only 3%. In these assays, green fluorescence was measured with 470 ± 10 nm excitation and 510 ± 10 nm emission, orange fluorescence with 540 ± 10 nm excitation and 560 ± 10 nm emission, red fluorescence with 560 ± 20 nm excitation and 600 ± 20 nm emission, and far-red fluorescence with 595 ± 20 nm excitation and 640 ± 20 nm emission. KO* is a modified version of KO in which synonymous alternative codons were used near the translation start site to enhance bacterial expression (18).

mutations as well as the original codons. In some cases, it may be relevant to test additional mutations at key positions (see Note 8).

12. Screen this library as described above (see Note 9) to identify the optimal combination of mutations.

3.3. Assaying Cytotoxicity in Bacteria

1. Adjust the concentrations of pQE-60NA-based FP expression plasmids to 1 ng/μL in sterile H$_2$O. Transform these plasmids into competent DH10B cells harboring the pREP4 repressor plasmid.

2. Prepare LB + ampicillin + kanamycin plates, one without IPTG (repressing conditions) and one with 1 mM IPTG (derepressing conditions). For each plate, use a marker to demarcate sectors for the FPs that are being tested.

3. Spread equal volumes of a given transformation mixture on the appropriate sector of each plate, aiming to achieve a density of about 10 colonies/cm^2 (see Note 10).

4. Incubate for 12–15 h at 37°C. Compare colony sizes for repressing versus derepressing conditions. Large colonies

under derepressing conditions indicate low cytotoxicity (see Note 11). An example of the colony size assay is shown in Fig. 3a.

5. Colony size is a meaningful measure only if different FPs are expressed at similar levels (see Note 12). Expression is analyzed by SDS–PAGE of cell lysates. For this purpose, grow transformed clones overnight at 37°C in 5-mL cultures of LB + ampicillin + kanamycin to generate saturated precultures. Dilute each preculture to an OD_{600} of 0.05 in 5 mL of LB + ampicillin + kanamycin in a 50-mL baffled flask. After growth for 2 h at 37°C in an orbital shaker to yield an OD_{600} of ~0.5, add 1 mM IPTG to induce FP expression, and incubate with shaking for 4 h at 37°C. Then centrifuge 1 OD_{600} unit of each culture in a microfuge tube, resuspend each cell pellet in 200 µL of 1× SDS–PAGE sample loading buffer, and

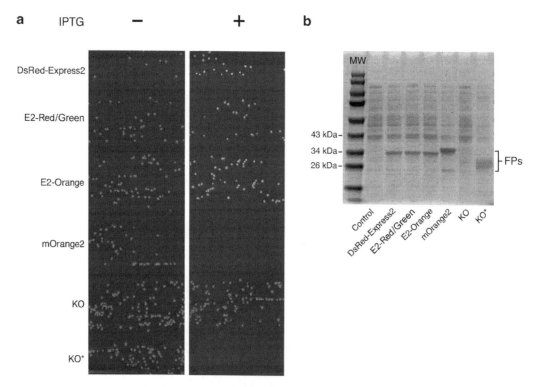

Fig. 3. DsRed-Express2 and its derivatives are noncytotoxic when expressed at high levels in bacteria. (a) Representative colony size assay showing the bacterial cytotoxicities of several red and orange FPs. DH10B cells harboring the pREP4 repressor plasmid were transformed with pQE-60NA encoding either DsRed-Express2, E2-Red/Green, E2-Orange, mOrange2, KO, or KO* (see Fig. 2 legend). In this assay, cells expressing DsRed-Express2, E2-Red/Green, and E2-Orange produced large colonies, indicating low cytotoxicity. Cells expressing mOrange2 and KO* produced very small or pinprick colonies, respectively, indicating cytotoxicity. Cells expressing KO produced large colonies, but these colonies were nearly colorless due to poor expression. (b) Quantitation of FP expression under derepressing conditions. Cells were grown to an OD_{600} of ~0.6 and then treated with 1 mM IPTG for 4 h. Whole-cell lysates were separated using SDS–PAGE followed by staining with Coomassie Blue. Control cells were transformed with the empty pQE-60NA vector. The image shows that KO expression was very low, while KO* expression was comparable to that of other FPs.

boil for 10 min. Centrifuge this mixture at $16,000 \times g$ for 10 min. Without disturbing the pellet, dilute a portion of each sample supernatant 1:4 in SDS–PAGE sample buffer, and load 20 µL in a lane of an SDS–polyacrylamide gel. Load 10 µL of PageRuler prestained protein ladder as molecular weight standards. Run the gel and stain with Coomassie Blue or a comparable dye. FP bands can be identified based on size and intensity, with reference to a control sample from cells carrying an empty expression vector. A typical gel image is shown in Fig. 3b.

3.4. HeLa Cell Transient Transfection Assay for Cytotoxicity

1. Culture HeLa cells in a 100-mm dish to 50–70% confluence.

2. Remove the culture medium by aspiration. Wash the cells with 10 mL sterile PBS. Remove the PBS, and dissociate the HeLa cells by incubating for 4 min at 37°C with 2 mL of Trypsin–EDTA solution. Resuspend the cells in 5 mL of culture medium at 37°C (see Note 13).

3. Add the cell suspension to a 15-mL conical tube and centrifuge for 5 min at $2,000 \times g$.

4. Discard the supernatant, and resuspend the cell pellet in 5 mL of fresh culture medium.

5. Determine the cell density using a hemocytometer.

6. Add $0.5–1 \times 10^5$ cells to each well of several 24-well plates. To test each FP in triplicate, use 15 culture wells per FP plus 5 culture wells for the "No DNA" control. Allow the cells to grow for 16–24 h to attain ~30% confluence.

7. Dilute the FP expression plasmids to 400 ng/µL.

8. For each FP, dilute 30 µL of plasmid in 750 µL of OptiMEM. Mix gently. For the "No DNA" control, use 30 µL of sterile H_2O.

9. In separate tubes, add 15 µL of Lipofectamine 2000 to 750 µL of OptiMEM for each FP. Mix gently and incubate at room temperature for 5 min.

10. For each FP, combine the plasmid mixture with the Lipofectamine mixture. Mix gently and incubate at room temperature for 20 min.

11. For a given FP, add 100 µL of the plasmid/Lipofectamine mixture to each of 15 culture wells. For the "No DNA" control, add 100 µL of the H_2O/Lipofectamine mixture to each of five culture wells.

12. Incubate the cells for 24 h. Remove the Lipofectamine-containing medium and replace with fresh culture medium.

13. Analyze the cells by flow cytometry at daily intervals for 5 days, beginning at 24 h post-transfection. To prepare cells

for analysis on a given day, select three wells per FP and one well for the "No DNA" control. For each well, remove the medium, wash the cells with 1 mL of PBS, then replace the PBS with 200 µL of Trypsin–EDTA and incubate at 37°C for 4 min. Remove the Trypsin–EDTA, and suspend the cells in 2 mL of culture medium.

14. Perform flow cytometry with the suspended cells using suitable laser lines and emission filters. For red FPs, 561-nm lasers and PE filter sets (585/15 nm) are commonly used. Based on side scatter area (SSC-A) and forward scatter area (FSC-A) profiles, gate on living cells. Based on side scatter width (SSC-W) and forward scatter width (FSC-W), gate on single, non-aggregated cells. Gate fluorescent cells based on the "No DNA" control sample.

15. Analyze the fluorescence data with FlowJo software. For each sample, including the "No DNA" control, record the average population brightness of living fluorescent cells. To determine the signal/background ratio, divide the average brightness of the fluorescent cells by the average brightness of the "No DNA" cells. Data for representative orange, red, and far-red FPs are shown in Fig. 4.

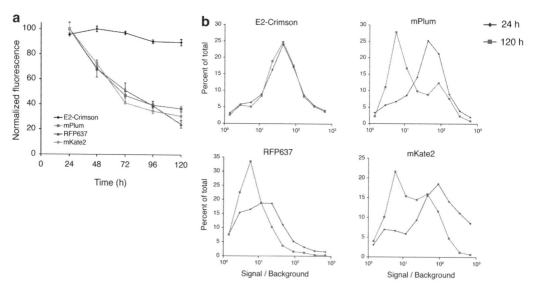

Fig. 4. E2-Crimson is noncytotoxic to HeLa cells under conditions of standard high-level expression. (a) HeLa cells were transiently transfected in 24-well plates for constitutive high-level expression of the indicated FPs. For each FP, three wells per day were analyzed by flow cytometry, and the average brightness of the viable fluorescent cells was measured relative to untransfected cells. Maintenance of a high percentage of the original fluorescence indicates low cytotoxicity. (b) Fluorescence intensity distributions for the same data were analyzed for 24 h (*triangle line*) and 120 h (*square line*) post-transfection. Each data point is a binned value that represents the percentage of cells with fluorescence in a range centered about the data point. Cells expressing E2-Crimson maintained a nearly identical population distribution over the course of the experiment, while cells expressing the other FPs showed dramatic population shifts in favor of dimmer cells.

3.5. Generation of Lentiviral Vectors and HeLa Cell Lentiviral Transduction Assay for Cytotoxicity

1. Generate lentiviral particles for FP-encoding vectors according to the Lenti-X Expression Kit protocol. In parallel, generate control viral particles by transfecting 293-T cells with pLVX-Puro.

2. Aliquot the viral particles and store them at −80°C (see Note 14).

3. Determine the viral titer for each sample using the QuickTiter Lentivirus Quantitation Kit (see Note 15).

4. Culture HeLa cells to 50–70% confluence. Split the cells as described in Subheading 3.4 into 6-well dishes. Add 2.5×10^5 cells per well, with three wells per FP and one well for the control virus. Incubate for 16–24 h at 37°C until the cells are ~50% confluent.

5. Remove the culture medium and replace it with exactly 2 mL of fresh culture medium per well.

6. Add 1.6 µL of 10 mg/mL polybrene to each well for a final concentration of 8 µg/mL (see Note 16).

7. Add equivalent titers of viral particles to each well, resulting in three replicates per FP. Incubate for 24 h at 37°C.

8. Remove the culture medium and excess viral particles, wash each well with 2 mL of culture medium, and add 2 mL of fresh culture medium. This time point is defined as day 0.

9. Incubate for 72 h at 37°C to allow expression of the FPs.

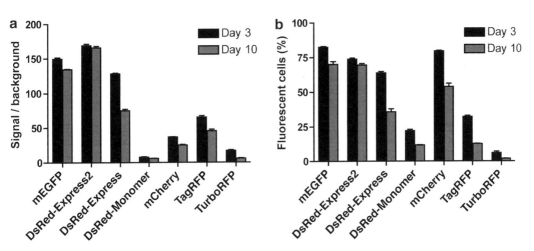

Fig. 5. Fluorescent protein cytotoxicities after lentiviral transduction. (a) HeLa cells were transduced with lentiviral particles encoding the indicated FPs or with a control lentiviral particles lacking an FP gene. At 3 and 10 days after transduction, cells were analyzed by flow cytometry. Plotted are the average fluorescence signals from viable fluorescent cells relative to the control. (b) The percentage of viable cells that were fluorescent was also recorded at 3 and 10 days after transduction. In these experiments, DsRed-Express2 and monomeric EGFP (mEGFP, which is EGFP with the A206K mutation) maintained almost all of the original fluorescence signal and showed only a small decline in the percentage of fluorescent cells. By contrast, all of the other red FPs showed dramatic reductions in average fluorescence and in the percentage of fluorescent cells, indicating that those other FPs were cytotoxic.

10. At the end of this incubation (on day 3), remove the culture medium and wash the cells in each well with 2 mL of PBS. Add 0.5 mL of Trypsin–EDTA to each well, and incubate for 4 min at 37°C.

11. Remove the Trypsin–EDTA, and resuspend the cells in 2 mL of fresh culture medium per well. To allow continued growth, add 0.5 mL of cell suspension to 2 mL of culture medium per well in fresh 6-well plates.

12. Analyze the remaining cells by flow cytometry as described in Subheading 3.4. Record the average brightness of viable fluorescent cells and control cells. Also record the percentage of viable cells that are fluorescent.

13. Repeat the passage and flow cytometry analysis on days 5 and 7. Do a final analysis on day 10. Results for a monomeric green FP and several representative red FPs are shown in Fig. 5.

4. Notes

1. *E. coli* strain DH5α is optimal for use in the bacterial lysis assay. In our experience, other strains do not lyse as efficiently in BPER II.

2. Dissolve all ingredients except β-mercaptoethanol in 90% of the total volume. Add fresh β-mercaptoethanol immediately before use.

3. Screening colonies as early as possible for brightness allows for selection of mutants that are both bright and rapidly maturing.

4. Restreak promising candidates on adjacent sectors of the same plate to examine relative brightness. As a control, streak bacteria expressing the template FP.

5. Due to slight variability in the procedure, each plate should include clones carrying the template FP for direct comparison with mutants.

6. When expressing extremely cytotoxic FPs, grow cells containing the pREP4 plasmid under repressing conditions and then derepress expression with IPTG. In this case, allow the cells to grow for 3–4 h, add 1 mM IPTG, and continue the incubation for 6–12 h.

7. Calculate s.e.m. by dividing the standard deviation by the square root of the number of samples.

8. In our experience, mutation to charged residues such as lysine can help to increase solubility. These mutations may not be

discovered through random mutagenesis by error-prone PCR due to the limited number of codon substitutions that can be obtained from single point mutations.

9. To determine the number of colonies that should be screened in this assay, determine the size of your library by calculating the total number of possible mutant combinations. For example, if two mutations are screened at each of three positions, the library size is 2^3 or eight potential clones. A tenfold over-screening is generally sufficient to identify all of the relevant clones. In the example given above, 80 colonies would be chosen for the bacterial lysis assay.

10. A high density of colonies can bias the results toward small colonies. For this reason, low density is critical for proper evaluation. Plating the right number of cells may require trial and error.

11. If colonies are not visible to the naked eye, they can sometimes be seen with the slide projector assay as fluorescent "pinprick" colonies.

12. For certain FPs, bacterial expression from the endogenous start codon may be very low, yielding large, nearly colorless colonies under derepressing conditions. This effect is likely due to formation of secondary structure near the 5′ end of the mRNA (20, 21). In such a case, the FP cannot be tested using this assay. However, we have successfully introduced silent mutations within the first few codons of FP genes, thereby increasing bacterial expression and allowing for analysis of cytotoxicity. As an example, the original KO and the expression-enhanced KO* (18) are shown in Fig. 3.

13. Trypsin–EDTA and all media should be pre-warmed to 37°C before use. This pre-warming is assumed for all steps that describe Trypsin–EDTA treatments, washes in culture medium, and changes of culture medium.

14. Freezing and thawing lentiviral particles may reduce their efficacy. For this reason, aliquots are recommended for storage.

15. This method measures the total titer of the viral particles, and is not a measure of viral particles that are competent for transduction.

16. 4–10 μg/mL of polybrene is typically used for lentiviral transduction. Therefore, it is usually unnecessary to add more polybrene to adjust for the change in volume caused by addition of viral particles. However, if the viral particles are very dilute and more than 1 mL of viral particle solution is added, adjust the polybrene concentration to 8 μg/mL.

References

1. Chalfie, M., Tu, Y., Euskirchen, G., Ward, W. W., and Prasher, D. C. (1994) Green fluorescent protein as a marker for gene expression. *Science* **263**, 802–5.

2. Shaner, N. C., Steinbach, P. A., and Tsien, R. Y. (2005) A guide to choosing fluorescent proteins. *Nat. Methods* **2**, 905–9.

3. Shaner, N. C., Campbell, R. E., Steinbach, P. A., Giepmans, B. N. G., Palmer, A. E., and Tsien, R. Y. (2005) Improved monomeric red, orange and yellow fluorescent proteins derived from Discosoma sp. red fluorescent protein. *Nat. Biotechnol.* **22**, 1567–72.

4. Wang, L., Jackson, W. C., Steinbach, P. A., and Tsien, R. Y. (2004) Evolution of new nonantibody proteins via iterative somatic hypermutation. *Proc. Natl. Acad. Sci. U. S. A.* **101**, 16745–9.

5. Strongin, D. E., Bevis, B., Khuong, N., Downing, M. E., Strack, R. L., Sundaram, K., Glick, B. S., and Keenan, R. J. (2007) Structural rearrangements near the chromophore influence the maturation speed and brightness of DsRed variants. *Protein Eng. Des. Sel.* **20**, 525–34.

6. Merzlyak, E. M., Goedhart, J., Shcherbo, D., Bulina, M. E., Shcheglov, A. S., Fradkov, A. F., Gaintzeva, A., Lukyanov, K. A., Lukyanov, S., Gadella, T. W., and Chudakov, D. M. (2007) Bright monomeric red fluorescent protein with an extended fluorescence lifetime. *Nat. Methods* **4**, 555–7.

7. Shaner, N. C., Lin, M. Z., McKeown, M. R., Steinbach, P. A., Hazelwood, K. L., Davidson, M. W., and Tsien, R. Y. (2008) Improving the photostability of bright monomeric orange and red fluorescent proteins. *Nat. Methods* **5**, 545–51.

8. Shcherbo, D., Merzlyak, E. M., Chepurnykh, T. V., Fradkov, A. F., Ermakova, G. V., Solovieva, E. A., Lukyanov, K. A., Bogdanova, E. A., Zaraisky, A. G., Lukyanov, S., and Chudakov, D. M. (2007) Bright far-red fluorescent protein for whole-body imaging. *Nat. Methods* **4**, 741–6.

9. Shcherbo, D., Murphy, C. S., Ermakova, G. V., Solovieva, E. A., Chepurnykh, T. V., Shcheglov, A. S., Verkhusha, V. V., Pletnev, V. Z., Hazelwood, K. L., Roche, P. M., Lukyanov, S., Zaraisky, A. G., Davidson, M. W., and Chudakov, D. M. (2009) Far-red fluorescent tags for protein imaging in living tissues. *Biochem. J.* **418**, 567–74.

10. Kredel, S., Oswald, F., Nienhaus, K., Deuschle, K., Rocker, C., Wolff, M., Heilker, R., Nienhaus, G. U., and Wiedenmann, J. (2009) mRuby, a bright monomeric red fluorescent protein for labeling of subcellular structures. *PLoS One* **4**, e4391.

11. Karasawa, S., Araki, T., Nagai, T., Mizuno, H., and Miyawaki, A. (2004) Cyan-emitting and orange-emitting fluorescent proteins as a donor/acceptor pair for fluorescence resonance energy transfer. *Biochem. J.* **381**, 307–12.

12. Sakaue-Sawano, A., Kurokawa, H., Morimura, T., Hanyu, A., Hama, H., Osawa, H., Kashiwagi, S., Fukami, K., Miyata, T., Miyoshi, H., Imamura, T., Ogawa, M., Masai, H., and Miyawaki, A. (2008) Visualizing spatiotemporal dynamics of multicellular cell-cycle progression. *Cell* **132**, 487–98.

13. Shaner, N. C., Patterson, G. H., and Davidson, M. W. (2007) Advances in fluorescent protein technology. *J. Cell Sci.* **120**, 4247–60.

14. Bevis, B. J. and Glick, B. S. (2002) Rapidly maturing variants of the *Discosoma* red fluorescent protein (DsRed). *Nat. Biotechnol.* **20**, 83–7.

15. Kredel, S., Nienhaus, K., Oswald, F., Wolff, M., Ivanchenko, S., Cymer, F., Jeromin, A., Michel, F. J., Spindler, K. D., Heilker, R., Nienhaus, G. U., and Wiedenmann, J. (2008) Optimized and far-red-emitting variants of fluorescent protein eqFP611. *Chem. Biol.* **15**, 224–33.

16. Tao, W., Evans, B. G., Yao, J., Cooper, S., Cornetta, K., Ballas, C. B., Hangoc, G., and Broxmeyer, H. E. (2007) Enhanced green fluorescent protein is a nearly ideal long-term expression tracer for hematopoietic stem cells, whereas DsRed-Express fluorescent protein is not. *Stem Cells* **25**, 670–8.

17. Strack, R. L., Strongin, D. E., Bhattacharyya, D., Tao, W., Berman, A., Broxmeyer, H. E., Keenan, R. J., and Glick, B. S. (2008) A noncytotoxic DsRed variant for whole-cell labeling. *Nat. Methods* **5**, 955–7.

18. Strack, R. L., Bhattacharyya, D., Glick, B. S., and Keenan, R. J. (2009) Noncytotoxic orange and red/green derivatives of DsRed-Express2 for whole-cell labeling. *BMC Biotechnol.* **9**, 32.

19. Strack, R. L., Hein, B., Bhattacharyya, D., Hell, S. W., Keenan, R. J., and Glick, B. S. (2009) A rapidly maturing far-red derivative of DsRed-Express2 for whole-cell labeling. *Biochemistry* **8**, 8279–81.

20. Pfleger, B. F., Fawzi, N. J., and Keasling, J. D. (2005) Optimization of DsRed production in *Escherichia coli*: effect of ribosome binding site sequestration on translation efficiency. *Biotechnol. Bioeng.* **92**, 553–8.

21. Sörensen, M., Lippuner, C., Kaiser, T., Mißlitz, A., Aebischer, T., and Bumann, D. (2003) Rapidly maturing red fluorescent protein variants with strongly enhanced brightness in bacteria. *FEBS Lett.* **552**, 110–4.

Chapter 18

Flow Cytometric FRET Analysis of Protein Interaction

György Vereb, Péter Nagy, and János Szöllősi

Abstract

Investigation of protein–protein interactions in situ in living or intact cells gains expanding importance as structure/function relationships proposed from bulk biochemistry and molecular modeling experiments require demonstration at the cellular level. Fluorescence resonance energy transfer (FRET)-based methods are excellent tools for determining proximity and supramolecular organization of biomolecules at the cell surface or inside the cell. This could well be the basis for the increasing popularity of FRET; in fact, the number of publications exploiting FRET has doubled in the past 5 years. In this chapter, we intend to provide a generally useable protocol for measuring FRET in flow cytometry. After a concise theoretical introduction, recipes are provided for successful labeling techniques and measurement approaches. The simple, quenching-based population-level measurement; the classic ratiometric, intensity-based technique providing cell-by-cell actual FRET efficiencies, and a more advanced version of the latter, allowing for cell-by-cell autofluorescence correction, are described. Finally, points of caution are given to help design proper experiments and critically interpret the results.

Key words: Fluorescence resonance energy transfer, Förster resonance energy transfer, Flow cytometry, Protein interactions, Molecular proximity

1. Introduction

Modern biochemistry and molecular biology have served to identify hundreds of pairs of cellular proteins that are in vitro capable of interacting with each other. These protein–protein interactions are crucial in both maintaining the stable "resting" state of cells and in driving activation processes that allow cells to respond to external stimuli. Hence it becomes more and more important to be able to detect such interactions in situ inside or on the surface of cells. Fluorescence techniques are widely used to quantify molecular parameters of various biochemical and biological processes in vivo because of their inherent sensitivity, specificity, and temporal resolution. Combination of fluorescence spectroscopy

Teresa S. Hawley and Robert G. Hawley (eds.), *Flow Cytometry Protocols*, Methods in Molecular Biology, vol. 699, DOI 10.1007/978-1-61737-950-5_18, © Springer Science+Business Media, LLC 2011

with flow cytometry provided a solid basis of rapid and continuous improvements of these technologies. A major asset in studying molecular level interactions was the application of fluorescence resonance energy transfer (FRET) to cellular systems. Applying fluorescent-labeled antibodies, proteins, lipids, or other biomolecules either using organic dyes or fluorescent fusion proteins, the FRET technique can be used to determine inter- and intramolecular distances of cell surface components in biological membranes, as well as molecular associations inside live or intact cells. Excellent reviews are available on the applicability of FRET to biological systems as well as descriptions and comparison of various approaches. Only a few are quoted here (1–5).

FRET is a special phenomenon in fluorescence spectroscopy during which energy is nonradiatively transferred from an excited *donor* molecule to an *acceptor* molecule (Fig. 1) (6). For the process to occur, a set of conditions have to be fulfilled:

1. The emission spectrum of the donor has to overlap with the excitation spectrum of the acceptor. The larger the overlap, the higher the rate of FRET is.

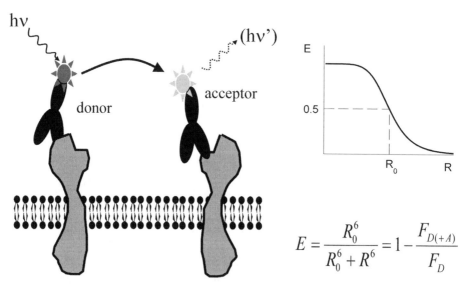

$$E = \frac{R_0^6}{R_0^6 + R^6} = 1 - \frac{F_{D(+A)}}{F_D}$$

Fig. 1. FRET basics. In FRET, an excited fluorescent molecule, called *donor*, donates energy to an *acceptor* molecule by a dipole–dipole resonance energy transfer mechanism. The acceptor then may emit this energy as a photon, provided it is fluorescent. The usual term for characterizing the efficiency of FRET is E, which is the ratio of excited-state donor molecules relaxing by FRET to the total number of excited donors. A simple way to assess this ratio is to measure donor fluorescene in the absence (F_D) and in the presence ($F_{D(+A)}$) of an acceptor, to obtain $E = 1 - F_{D(+A)}/F_D$. The rate of energy transfer is dependent on the negative sixth power of the distance R between the donor and the acceptor, resulting in a sharply dropping curve when plotting E against R, centered around R_0. R_0 is the distance where $E = 0.5$, that is, where there is a 50% chance that the energy of the excited donor will be transferred to the acceptor. As separation between the donor and acceptor increases, E decreases, and at $R = 2R_0$, E is already getting negligible.

2. The emission dipole vector of the donor and the absorption dipole vector of the acceptor need to be close to parallel. The rate of FRET decreases as the angle between the two vectors increases. In biological situations where molecules are free to move (rotate), we generally assume that dynamic averaging takes place, i.e., the donor and the acceptor assume many possible steric positions during the excited-state lifetime of the donor, among them positions that can yield an effective transfer of energy. There are cases when the aforementioned assumption of dynamic averaging is most likely correct (e.g., antibodies labeled by fluorescent dyes bound to the antibody via flexible linkers); however, in some cases, it is certainly incorrect (e.g., fluorescent dyes intercalated into DNA). FRET taking place between proteins tagged by GFP (green fluorescent protein) or its derivatives represents an intermediate case in which the assumption of dynamic averaging is a relatively good approximation.

3. The distance between the donor and acceptor should be between 1 and 10 nm.

This latter phenomenon is the basis of the popularity of FRET in biology: The distance over which FRET occurs is small enough to characterize the proximity of possibly interacting molecules; under special circumstances, it even provides quantitative data on exact distances and, additionally, information on the spatial orientation of molecules or their domains. Hence the very apropos term from Stryer, who equaled FRET to a "spectroscopic ruler" (7). The usual term for characterizing the efficiency of FRET is E, which is the ratio of excited-state donor molecules relaxing by FRET to the total number of excited donors. The rate of the energy transfer process is dependent on the negative sixth power of the distance between the donor and the acceptor, resulting in a sharply dropping curve, practically eliminating FRET above a separation of 10 nm for the usual fluorophores (Fig. 1). Conversely, as the distance reaches values below 1 nm, strong ground-state interactions or transfer by exchange interactions become dominant at the expense of FRET (8).

The occurrence of FRET has profound consequences on the fluorescence properties of both the donor and the acceptor. An additional de-excitation process is introduced in the donor, which decreases the fluorescence lifetime and the quantum efficiency of the donor, rendering it less fluorescent. The decrease in donor fluorescence (often termed *donor quenching*) can be one of the most facilely measured spectroscopic characteristics that indicates the occurrence of FRET. Additionally, since the acceptor is excited as a result of FRET, those acceptors that are fluorescent will emit photons (proportional to their quantum efficiency) also when FRET occurs. This is called *sensitized emission* and can also be a good measure of FRET.

Flow cytometric FRET (FCET) methods that can be implemented on commercial flow cytometers exploit one or both of these phenomena. The main advantage of the flow cytometric approach is the ability to examine large cell populations in a short time, and still provide FRET efficiency values up to single-cell resolution (albeit averaged over each cell). In a simplified scenario, population averages from the flow cytometer can be used to estimate an overall FRET efficiency for the whole cell population measured. While in theory there are at least 22 different possible approaches to quantifying FRET (9), some of these have never been tested, and many require a (usually microscope-based) system where cells or their compartments can be revisited once or several times for completing the measurement. Among the eight approaches applicable to flow cytometry, most require extensive modification of the equipment. Most of these methods are based on the measurement of conventional hetero-FRET in which spectroscopically different donor and acceptor molecules interact with each other. However, there is another modality of FRET interactions which takes place between identical fluorophores, i.e., the donor and acceptor molecules are of the same kind. The only manifestation of the phenomenon called homo-FRET is the decreased fluorescence anisotropy of the fluorophore population. Although instruments capable of recording polarized fluorescence intensities do not abound, the measurement of homo-FRET makes quantitative analysis of large-scale protein clusters possible, since the fluorescence anisotropy changes inversely with the average number of proteins in a cluster (10). The approaches discussed in this chapter are based on hetero-FRET; the simple donor quenching-based FCET method and the more complex, but fully quantitative ratiometric, intensity-based FCET have already been extensively and successfully applied to biological systems. Finally, we shall describe a modification of the latter approach that allows a cell-by-cell correction for autofluorescence and thus can reduce the uncertainty of the quantitative determination of FRET when fluorescence signals are low. The necessary mathematical details will be introduced along with the measurement procedure, so that the importance and utility of the various control samples can be fully appreciated.

2. Materials

2.1. Cellular Systems

1. The easiest targets of FCET measurements are cells growing in suspension, such as those of lymphoid origin that can directly be labeled in suspension.

2. Cells that grow attached to substratum need to be trypsinized. Use a trypsin–EDTA solution that is appropriate for the cell

line examined. Usually, 0.05% w/v trypsin and 2 mM EDTA work well, but cells that have spent a longer time in the same culture dish or express a massive extracellular matrix may need a more vigorous treatment. In our experience, a 5-min trypsin–EDTA treatment does not decrease cell surface receptor levels by more than 5%.

2.2. Fluorophores

1. The choice of fluorophore must be appropriate for the lasers of the flow cytometer to be used (11); the availability of labeled antibodies, compatible with other fluorophores measured co-temporarily; or the native fluorescence of the sample. Since this latter is usually more prominent in the blue-green emission region, it is advisable to use fluorophores that emit in the red and far red spectral range (12).

2. The most straightforward method of labeling is using fluorophore-conjugated antibodies. If possible, direct labeling with conjugated primary antibodies should be the choice. Indirect labeling usually diminishes FRET owing to the greater separation of fluorophores in the sandwich of antibodies, but it might also be possible that FRET will occur between otherwise more distant targets if a large molecular labeling complex is built (12). Also, the polyclonal nature of most secondary antibodies can cause unwanted difficulties in interpreting the results.

3. Labeling the proteins of interest with fluorescent proteins is another option. While the molecular biology behind these approaches exceeds the scope of this chapter, the most useful fluorescent protein pairs are also listed in Table 1 (see Notes 1 and 2).

4. In Subheadings 2.3 and 2.4, we shall describe the use of fluorescent-labeled antibodies; however, the FCET protocols described in the chapter are equally applicable to fluorescent fusion proteins, labeled substrates or toxins, and directly labeled proteins that are microinjected into cells.

Table 1
The most frequently used fluorophore pairs adequate for FRET measurements

Donor	Acceptor
Fluorescein, AlexaFluor488, Atto488, Cy2	Rhodamine, AlexaFluor546, AlexaFluor555, Cy3, Atto550
Rhodamine, AlexaFluor546, AlexaFluor555, Cy3, Atto550, Phycoerythrin	Cy5, AlexaFluor633, AlexaFluor647, Atto647N, Allophycocyanin
CFP and variants, such as Cerulean	YFP and variants, such as Venus
eGFP	tagRFP, mCherry, and other red FPs (see Note 2)

2.3. Labeling Target Epitopes in the Membrane Using Fluorescent Antibodies

1. PBS (for washing): 137 mM NaCl, 3 mM KCl, 10 mM Na$_2$HPO$_4$, and 1.8 mM KH$_2$PO$_4$, pH 7.4 (see Note 3).

2. PBS + 0.1% BSA (for labeling).

3. PBS + 1% formaldehyde (for fixation).

4. Fluorescently labeled antibodies, with donor and acceptor dyes.

2.4. Labeling Intracellular Targets Using Fluorescent Antibodies

1. PBS (for washing).

2. PBS + 0.1% BSA + 0.1% (v/v) Triton X-100 (for permeabilization and labeling, see also Note 4).

3. PBS + 1% formaldehyde (for fixation).

4. Fluorescently labeled antibodies, with donor and acceptor dyes.

2.5. Evaluation Software

For simple donor-quenching measurements, median or trimmed mean data exported from the flow cytometry program of choice can be used. To obtain FRET distribution on a cell-by-cell basis, samples can be analyzed with flow cytometric data analysis programs that can derive further parameters from list mode data using equations. Alternatively, these calculations can be done in a spreadsheet program after obtaining some parameters needed for correction factors and exporting the list mode data from any flow cytometry program of choice. Finally, there is a custom-made program (13) (available to the public from http://www.freewebs.com/cytoflex) that can be used to perform the complete analysis, including the calculation of necessary factors and the cell-by-cell distribution of FRET.

3. Methods

3.1. Sample Preparation and Labeling

3.1.1. Sample Preparation

1. Measurement of donor quenching caused by FRET is simple, but cannot be used for cell-by-cell data analysis of FRET efficiency. In the classical approach introduced for the cell-by-cell measurement of FRET (14–16), the donor and acceptor fluorophores are excited separately using the appropriate laser line. In a perfected approach, autofluorescence at the individual cell's level can be taken into correction during calculations, rather than using the autofluorescence histogram means (12). This improves the dispersion of FRET histograms, but one should always remember that inherent (Poissonian and additive) measurement noise that tends to dominate at low signals (low protein expression levels) cannot be eliminated even with cell-by-cell autofluorescence correction.

2. In general, the following samples are necessary:

   (a) Sample 1. Unlabeled cells.

   (b) Sample 2. Cells with epitope A labeled with the donor.

   (c) Sample 3. Cells with epitope A labeled with the acceptor.

   (d) Sample 4. Cells with epitope B labeled with the donor.

   (e) Sample 5. Cells with epitope B labeled with the acceptor.

   (f) Sample 6. Cells with epitope A labeled with the donor and epitope B labeled with the acceptor.

   (g) Sample 7. Cells with epitope B labeled with the donor and epitope A labeled with the acceptor.

   Sample 3 or 5 may be omitted: The α factor (see later in Eqs. 4, 25, and 26) can be determined either from samples 2 and 3, or from samples 4 and 5, depending on whether epitope A or B has a higher expression rate. The spillage factors (S, see Eq. 2 and onwards) should also be determined from the epitope that gives the better signal. Sample 2 or 4 may be omitted along the same principles (see Note 5).

   Samples 6 and 7 are complementary in the sense that they provide FRET efficiencies in "the two opposite directions," i.e., from epitope A to B and from epitope B to A. If one epitope is expressed in great molar excess of the other, labeling that one with the acceptor yields a more efficient FRET, but interpretation can be somewhat intricate. However, if we can also swap the labels, a more complete picture can be obtained (see also Subheading 3.6, step 6).

3. If a positive control with known molecular interactions is known, it is advisable to have another set of samples labeling these molecules on the same cell type along the principles in step 2. Since major histocompatibility complex (MHC) class I molecules comprising the heavy chain and the light chain (ß2 microglobulin) are expressed on many cells, and the two chains of MHC I are in close proximity, antibodies against these proteins with similar fluorophores as those against the proteins of our interest provide a useful positive control. It is best to label the light chain with donor, as not all heavy chains may be complexed with light chains, but there are no free light chains on the cell surface.

4. Additional information needed for the calculations:

   (a) The dye/protein molar ratio of all antibodies used.

   (b) The molar absorption coefficients of all dyes used.

   (c) The quantum efficiencies of the dyes used, if any labeling yields a fluorescence intensity below three to five times the background.

5. Harvesting cells (for adherent cells only). Adherent cells grown in 75-cm^2 flasks are detached using trypsin–EDTA solution. After rinsing the flask with trypsin–EDTA twice, leave only a thin layer of it over the cells for the optimized short time. Then add 10 mL of medium with FCS to stop the trypsin and restitute the calcium concentration, and homogenize the suspension by pipetting. Let the cells recover for 20 min in the flask. It has been determined that after gentle trypsinization, most cell surface proteins are either unchanged or totally recovered within 20–30 min (see Note 6).

3.1.2. Labeling Extracellular Epitopes

1. Wash the cells with ice-cold PBS and centrifuge the suspension.

2. Repeat washing.

3. Add one million cells per sample tube and store on ice.

4. Label the cells with (usually) 5–50 µg/mL of final concentration of antibodies (should be above saturating concentration that was determined previously) in 50 µL total volume of PBS–0.1% BSA mixture for 10–30 min on ice (see Note 7).

5. Wash cells twice with ice-cold PBS and centrifuge the suspension.

6. Fix the cells with PBS + 1% formaldehyde in 500–1,000 µL.

7. Store the samples in refrigerator or cold room until measurement. The samples can be stored for up to a week and may be kept for backup even after the first measurement.

3.1.3. Labeling Intracellular Epitopes

1. Wash the cells with ice-cold PBS and centrifuge the suspension.

2. Repeat washing.

3. Add one million cells per sample tube and store on ice.

4. Label the cells with (usually) 5–50 µg/mL final concentration of antibodies (should be above saturating concentration that was determined previously) in 50 µL total volume of PBS–0.1% BSA–0.1% Triton X-100 for 30–120 min on ice (see Note 4).

5. Wash the cells twice with ice-cold PBS–0.1% BSA–0.1% Triton X-100: fill the tube with the wash buffer, centrifuge at $600 \times g$ for 6 min, remove the supernatant, fill the tube again with the wash buffer, let unbound label diffuse out for ~5 min, and centrifuge again.

6. Fix the cells with PBS + 1% formaldehyde in 500–1,000 µL.

7. Store the samples in refrigerator or cold room until measurement. The samples can be stored for up to a week and may be kept for backup even after the first measurement.

3.2. Flow Cytometric Measurements

1. Before measurement, resuspend the cells with gentle shaking and if clumps are detected upon examination in the microscope, run the suspension through a fine sieve.

2. Always examine labeled cells dropped on a microscopic slide using the fluorescence microscope to verify proper cellular position (e.g., membrane) of the label.

3. Start with Sample 1 (Background) as a negative control.

4. Set FSC and SSC in linear mode so as to see your population on the scatter plot (SSC/FSC dot plot).

5. Set fluorescence channels in logarithmic mode if wide-range linear digital acquisition is not available. Use the following fluorescence channels:

   (a) Donor excitation – Donor emission (donor channel, I_1).

   (b) Donor excitation – Acceptor emission (FRET channel, I_2).

   (c) Acceptor excitation – Acceptor emission (acceptor channel, I_3).

   (d) Donor excitation – independent emission (optional autofluorescence channel, I_4, only for Subheading 3.5) (see Note 8).

   The choice of excitation wavelengths and emission filters should be driven, in order of preference, by the optimal sensitivity of detection, including minimization of background, followed by the availability of labels that best exploit the lasers and optics at our disposal.

6. Set I_1, I_2, I_3, and I_4 voltages so that mean fluorescence intensities are about 10^1 for the unlabeled sample 1.

7. Run samples 2 and 4 (donor only) and adjust I_1, I_2, and I_4 voltages so that mean fluorescence intensities are in the acquisition range.

8. Run samples 3 and 5 (acceptor only) and adjust I_2 and I_3 voltages so that mean fluorescence intensities are in the acquisition range.

9. Save instrument settings.

10. Create the following plots:

    (a) FSC/SSC dot plot.

    (b) I_1, I_2, I_3, and I_4 histograms for monitoring intensities.

    (c) I_1/I_2, I_1/I_3, and I_2/I_3 dot plots for monitoring the correlation of signals.

11. Run sample 1 and create gate 1 on the FSC/SSC dot plot around intact cells.

12. Format plots so that only events in gate 1 are displayed.

13. Define a statistics window to show the mean fluorescence intensities of all histograms from the gated events.

14. Set the cytometer to acquire 20,000 events.

15. Run all samples.

3.3. Quick Estimation: Donor-Quenching FRET in Flow Cytometry

1. Measurement of donor quenching is probably the easiest way to perform a FRET experiment, but simplicity comes at a price. In these kinds of measurements, the *average* fluorescence intensities of two different samples (donor-labeled and double-labeled, i.e., samples 2 and 6 or samples 4 and 7 in Subheading 3.1.1, step 2) are compared. Therefore, FRET cannot be calculated on a cell-by-cell basis; instead, a population average is measured. It is assumed that the difference between the donor only and the double-labeled samples is due to the presence of the acceptor. Since it is practically impossible to meet this requirement for cells transfected with fluorescent protein variants, FRET measurements based on donor quenching can be done only on antibody-labeled cells.

2. Use the fluorescence from a donor-only labeled sample (e.g., for epitope A, sample 2: $I_{1(2)}$) corrected for the background fluorescence in that channel $(I_{1(1)})$, as well as the fluorescence of the sample double labeled with donor and acceptor $(I_{1(6)})$.

3. FRET efficiency is calculated as

$$E = 1 - \frac{I_{\text{Donor+Acceptor}} - I_{\text{background}}}{I_{\text{Donor}} - I_{\text{background}}} = 1 - \frac{I_{1(6)} - I_{1(1)}}{I_{1(2)} - I_{1(1)}} \qquad (1)$$

Since $I_{1(2)}$ and $I_{1(6)}$ are measured on distinct samples, only the histogram means from the two populations can be considered here. This is one of the main disadvantages of the method and hence it is only suggested as a quick and rough estimate of whether FRET and thus molecular proximity occur. However, it is a quite useful approach when signals are low compared to background/autofluorescence (see Note 9).

3.4. Evaluation of Flow Cytometric FRET Based on Donor and Acceptor Fluorescence (Intensity Based or Ratiometric FCET)

1. Based on Monte-Carlo simulations, Berney and Danuser have suggested that in order to obtain stable FRET measurements, energy transfer is best observed in the FRET channel, i.e., by excitation of the donor and a measurement of the acceptor emission. Methods estimating FRET from the donor signal in the presence and absence of acceptor are less robust (5). The evaluation scheme presented below relies on measuring donor quenching and sensitized acceptor emission at the same time and provides an analytical solution to calculate not only the FRET efficiency from them, but also the corrected fluorescence intensities that could be measured if FRET did not take place.

2. In a sample labeled with both the donor and the acceptor, the intensities measured in each channel (I_1 through I_4, see Subheading 3.2, step 5) are composed of the contribution of donor, FRET, acceptor, and autofluorescence signals to that channel, and the crosstalk signals from all or some of the other channels. In this classical approach, we shall assume that the autofluorescence in each channel does not vary much from cell to cell, and consequently we will reduce our analysis to the donor, FRET, and acceptor channels, taking their respective signals as I_1, I_2, and I_3 after subtracting the average autofluorescence (mean fluorescence of unlabeled cells) from each channel. In analytic terms, this means that $I_{1(1)}$, $I_{2(1)}$, and $I_{3(1)}$ histogram means need to be determined and subtracted as a constant from the respective list mode data for samples 2–7. *All intensities in the following steps of Subheading 3.4 are assumed to be background corrected!*

3. On the donor-only sample that has the greater signal, determine the spectral correction factors S_1 and S_3 characterizing the spillover of donor fluorescence from the donor channel to the FRET and acceptor channels, respectively, according to the following equations (see Note 10):

$$S_1 = \frac{I_2}{I_1}, S_3 = \frac{I_3}{I_1} \tag{2}$$

4. On the acceptor-only sample that has the greater signal, determine the spectral correction factors S2 and S4 characterizing the spillover of acceptor fluorescence from the acceptor channel to the FRET and donor channels, respectively, according to the following equations (see Note 10):

$$S_2 = \frac{I_2}{I_3}, S_4 = \frac{I_1}{I_3} \tag{3}$$

5. Determine factor α. Use a donor and an acceptor-labeled sample which are labeled by the same antibodies, but conjugated to the two different fluorophores. Calculate the mean background-corrected I_1 fluorescence intensity of the donor-only sample, and the mean background-corrected I_2 fluorescence intensity of the acceptor-labeled sample. Determine α according to the following equation (see Note 11):

$$\alpha = \frac{I_{2,a}}{I_{1,d}} \frac{\varepsilon_d L_d}{\varepsilon_a L_a}. \tag{4}$$

Here, ε_d and ε_a are the molar absorption coefficients of the donor and the acceptor, respectively, at the donor excitation wavelength (i.e., the excitation wavelength used for I_1 and I_2), and L_d and L_a are the labeling ratios (i.e., number of fluorophores/antibody) of the donor and acceptor-labeled antibodies, respectively. The use of robust estimators of central tendency (trimmed mean, median) instead of the mean is preferable if the distribution is wide or if there are outlier events significantly distorting the mean. If cells transfected with FP variants are used, a different approach has to be used for the determination of α, which is described in detail elsewhere (17).

6. For the double-labeled samples, the I_1, I_2, and I_3 intensities can be expressed according to the following equations:

$$I_1 = I_D(1 - E) + I_A \cdot S_4 + I_D \cdot E \cdot \alpha \cdot \frac{S_4}{S_2} \tag{5}$$

$$I_2 = I_D(1 - E) \cdot S_1 + I_A \cdot S_2 + I_D \cdot E \cdot \alpha \cdot \tag{6}$$

$$I_3 = I_D(1 - E) \cdot S_3 + I_A + I_D \cdot E \cdot \alpha \cdot \frac{1}{S_2} \cdot \frac{\varepsilon_{\lambda A}^D \cdot \varepsilon_{\lambda D}^A}{\varepsilon_{\lambda D}^D \cdot \varepsilon_{\lambda A}^A} \tag{7}$$

Here, E is the FRET efficiency, I_D and I_A are the unquenched donor and direct acceptor fluorescence intensities, and ε denotes the molar absorption coefficient of the donor (D) or acceptor (A) labeled in the upper index, at the donor (λD) or acceptor (λA) excitation wavelengths.

7. From the above system of equations, E can be calculated as follows:

$$E = \frac{S_2 \left(I_2 - I_1 S_1 - I_3 S_2 + I_1 S_2 S_3 + I_3 S_1 S_4 - I_2 S_3 S_4 \right)}{\alpha \left(\frac{\varepsilon_{\lambda A}^D \cdot \varepsilon_{\lambda D}^A}{\varepsilon_{\lambda D}^D \cdot \varepsilon_{\lambda A}^A} - 1 \right) \left(I_2 S_4 - I_1 S_2 \right) + S_2 \left(I_2 - I_1 S_1 - I_3 S_2 + I_1 S_2 S_3 + I_3 S_1 S_4 - I_2 S_3 S_4 \right)} \tag{8}$$

8. In most cases, the above equation can be simplified by neglecting some of the constants. For example, S3, S4, and the $\frac{\varepsilon_{635}^D \cdot \varepsilon_{488}^A}{\varepsilon_{488}^D \cdot \varepsilon_{635}^A}$ absorption ratio are negligible for the Cy3–Cy5 donor–acceptor pair measured on a FACSCalibur (BD Biosciences, San Jose, CA). In this case, the equation can be rewritten in the following form:

$$E = \frac{S_2 \left(I_2 - I_1 S_1 - I_3 S_2 \right)}{\alpha I_1 S_2 + S_2 \left(I_2 - I_1 S_1 - I_3 S_2 \right)} \tag{9}$$

9. During data analysis, it is advised to follow a general scheme. First, one needs to determine the mean background intensities and the autofluorescence correction factors. Then calculate the alpha factor and spectral overspill parameters (Sfactors) from the acceptor- and donor-labeled samples. With these parameters in hand, now the energy transfer efficiency can be determined on a cell-by-cell basis.

If the fluorescence intensity of the samples is comparable to auto-fluorescence, subtraction of a constant autofluorescence value can result in serious errors in the calculation. In a modified version of the approach described in Subheading 3.4, the fourth fluorescence intensity I_4 is used to correct for the autofluorescence of each cell (12) (see also Note 12). In addition to analysis steps described under Subheading 3.4, perform the following steps:

1. From unlabeled sample 1, determine factors B_1, B_2, and B_3 characterizing the spillover of autofluorescence from the autofluorescence channel to the donor, FRET, and acceptor channels, respectively.

$$B_1 = \frac{I_1}{I_4}, \; B_2 = \frac{I_2}{I_4}, \; B_3 = \frac{I_3}{I_4} \tag{10}$$

2. For the donor-labeled sample used for calculating S_1 and S_3, also determine factor S_5 characterizing the spectral spillover of donor fluorescence from the donor channel to the auto-fluorescence channel:

$$S_5 = \frac{I_4}{I_1} \tag{11}$$

3. For the acceptor-labeled sample used for assessing S_2 and S_4, also determine factor S_6 characterizing the spectral spillover of acceptor fluorescence from the acceptor channel to the autofluorescence channel:

$$S_6 = \frac{I_4}{I_3} \tag{12}$$

4. The fluorescence intensities of the double-labeled samples can be expressed by the following set of equations:

$$I_1 = AF \cdot B_2 + I_D(1 - E) + I_A \cdot S_4 + I_D \cdot E \cdot \alpha \cdot \frac{S_4}{S_2} \tag{13}$$

$$I_2 = AF \cdot B_3 + I_D(1 - E) \cdot S_1 + I_A \cdot S_2 + I_D \cdot E \cdot \alpha \tag{14}$$

$$I_3 = AF \cdot B_4 + I_D(1 - E) \cdot S_3 + I_A + I_D \cdot E \cdot \alpha \cdot \frac{1}{S_2} \cdot \frac{\varepsilon_{\lambda A}^D \cdot \varepsilon_{\lambda D}^A}{\varepsilon_{\lambda D}^D \cdot \varepsilon_{\lambda A}^A} \tag{15}$$

$$I_4 = AF + I_D(1 - E) \cdot S_5 + I_A \cdot S_6 + I_D \cdot E \cdot \alpha \cdot \frac{S_6}{S_2} \tag{16}$$

where AF denotes the autofluorescence intensity of single cells.

5. The above set can be converted to a system of linear equations by substituting the variable $I_D \cdot E$ by X.

$$I_1 = AF \cdot B_1 + I_D + I_A \cdot S_4 + X \cdot \left(\alpha \cdot \frac{S_4}{S_2} - 1 \right) \tag{17}$$

$$I_2 = AF \cdot B_2 + I_D \cdot S_1 + I_A \cdot S_2 + X \cdot (\alpha - S_1) \tag{18}$$

$$I_3 = AF \cdot B_3 + I_D \cdot S_3 + I_A + X \cdot \left(\alpha \cdot \frac{1}{S_2} \cdot \frac{\varepsilon_{\lambda A}^{D} \cdot \varepsilon_{\lambda D}^{A}}{\varepsilon_{\lambda D}^{D} \cdot \varepsilon_{\lambda A}^{A}} - S_3 \right) \tag{19}$$

$$I_4 = AF + I_D S_5 + I_A \cdot S_6 + X \cdot \left(\alpha \cdot \frac{S_6}{S_2} - S_5 \right) \tag{20}$$

6. The solution of this equation system in general can be expressed as

$$X = \frac{\begin{vmatrix} 1 & S_5 & S_6 & I_4 \\ B_1 & 1 & S_4 & I_1 \\ B_2 & S_1 & S_2 & I_2 \\ B_3 & S_3 & 1 & I_3 \end{vmatrix}}{\begin{vmatrix} 1 & S_5 & S_6 & \alpha\frac{S_6}{S_2}-S_5 \\ B_1 & 1 & S_4 & \alpha\frac{S_4}{S_2}-1 \\ B_2 & S_1 & S_2 & \alpha-S_1 \\ B_3 & S_3 & 1 & \alpha\cdot\frac{1}{S_2}\cdot\frac{\varepsilon_{\lambda A}^{D}\cdot\varepsilon_{\lambda D}^{A}}{\varepsilon_{\lambda D}^{D}\cdot\varepsilon_{\lambda A}^{A}}-S_3 \end{vmatrix}}, \quad I_D = \frac{\begin{vmatrix} 1 & I_4 & S_6 & \alpha\frac{S_6}{S_2}-S_5 \\ B_1 & I_1 & S_4 & \alpha\frac{S_4}{S_2}-1 \\ B_2 & I_2 & S_2 & \alpha-S_1 \\ B_3 & I_3 & 1 & \alpha\cdot\frac{1}{S_2}\cdot\frac{\varepsilon_{\lambda A}^{D}\cdot\varepsilon_{\lambda D}^{A}}{\varepsilon_{\lambda D}^{D}\cdot\varepsilon_{\lambda A}^{A}}-S_3 \end{vmatrix}}{\begin{vmatrix} 1 & S_5 & S_6 & \alpha\frac{S_6}{S_2}-S_5 \\ B_1 & 1 & S_4 & \alpha\frac{S_4}{S_2}-1 \\ B_2 & S_1 & S_2 & \alpha-S_1 \\ B_3 & S_3 & 1 & \alpha\cdot\frac{1}{S_2}\cdot\frac{\varepsilon_{\lambda A}^{D}\cdot\varepsilon_{\lambda D}^{A}}{\varepsilon_{\lambda D}^{D}\cdot\varepsilon_{\lambda A}^{A}}-S_3 \end{vmatrix}} \tag{21}$$

7. From this, the FRET efficiency E can be calculated according to the following equation:

$$E = \frac{X}{I_D} \tag{22}$$

When using the Cy3–Cy5, and AlexaFluor546–AlexaFluor 647 FRET pairs, and assigning the FL1 channel of a FACSCalibur to measure the cellular autofluorescence, the S_3, S_4, and S_6 factors and the $\dfrac{\varepsilon_{635}^{D} \cdot \varepsilon_{488}^{A}}{\varepsilon_{488}^{D} \cdot \varepsilon_{635}^{A}}$ absorption ratio become negligible and the equation takes a much simpler form that can be used for calculating E as a new cellular parameter in the list mode file (see Note 13):

$$E = 1 + \frac{\alpha \left(I_4 B_1 - I_1 \right)}{\begin{aligned} & I_2 - B_2 I_4 + \alpha \left(I_1 - B_1 I_3 \right) - I_1 S_1 + B_1 I_4 S_1 - I_3 S_2 + B_3 I_4 S_2 \\ & + S_5 \left(B_2 I_1 - B_1 I_2 - B_3 I_1 S_2 + B_1 I_3 S_2 \right) \end{aligned}} \tag{23}$$

3.6. Interpretation of FCET Data

1. Although FRET can provide very useful information about molecular proximity and associations, it has its own limitations. The most serious drawback of FRET is that it has restricted capacity in determining absolute distances because FRET efficiency depends not only on the actual distance between the donor and acceptor, but also on their relative orientation. It is still quite good at determining relative distances, namely, whether the two labels are getting closer or farther upon a certain treatment/condition. Even when measuring relative distances, care must be taken to ensure that the orientation does not change between the two systems to be compared. Fluorophores attached to proteins via flexible linkers (such as a 6–12 carbon linker in the case of dyes or a few glycines in the case of fluorescent proteins) are relatively free to rotate, which allows for the donor and the acceptor to assume several spatial orientations during the excited-state lifetime of the donor (called dynamic averaging) (18), which allows for optimal FRET measurements in complex biological systems. Conversely, rigid systems or changes of donor–acceptor orientation to positions less favoring FRET can lead to an underestimation of actual proximity. Thus, it is advisable to use alternative approaches in order to prove the reason for observed changes in FRET efficiency convincingly. For example, if fluorescent antibodies are used and if antibodies against different epitopes are available, it is advisable to test them as well.

2. Another problem is that FRET has very sharp distance dependence. Owing to this, it is difficult to measure relatively long distances because the rate of FRET gets very low. At the same time, energy transfer tends to occur on an all-or-none basis: if the donor and acceptor are within $1.63 \times R_0$ distance, energy transfer is detectable; if they are farther apart, energy is transferred with very little efficiency. Due to this sharp decrease, the absence of FRET is not a direct proof of the absence of molecular proximity between the epitopes investigated. Also, absence of FRET can be caused by steric hindrance even for neighboring molecules or protein domains. On the contrary, the presence of FRET to any appreciable extent above the experimental error of measurement is a strong evidence of molecular interactions.

3. Indirect immunofluorescent labeling strategies may be applied to FRET measurements if suitable fluorophore-conjugated mAbs are not available, or as an approach to enhance the specific fluorescence signal (see Note 14). In such cases, special attention should be paid to the fact that the size of the antibody complexes used affects the measured FRET efficiency values. Application of a larger antibody complex causes a decrease in FRET efficiency due to the geometry of the

antibody complexes, since when antibody or $F(ab')$ complexes become larger, the actual distance between the donor and acceptor fluorophores increases (12). FRET values cannot be compared directly to each other if they are obtained using different labeling strategies.

4. Another possibility for increasing the fluorescence signal is to use phycoerythrin (PE)- or allophycocyanine (APC)-labeled antibodies, since PE and APC have exceptional brightness. However, the size of these molecules is comparable to, or even greater than, that of whole antibodies. Due to steric limitations, the measurable FRET efficiency values can be low, at the border of detection limit (19). Nonetheless, it should be noted that even these low FRET efficiency values might have a biological meaning, since the accuracy of measurements is greatly improved owing to the high level of specific signals. Appropriate positive and negative controls can help make the decision whether the given molecules are associated or not on the basis of the measured FRET efficiency values.

5. When studying cells labeled with donor- and acceptor-conjugated monoclonal antibodies, averaging is performed at different levels. The first averaging follows from the random conjugation of the fluorescent label. An additional averaging is brought about by the actual distribution of separation distances between the epitopes labeled with monoclonal antibodies. This multiple averaging, an inevitable consequence of the non-uniform stoichiometry, explains why the goals of FRET measurements are so uniquely different in the case of purified molecular systems on the one hand, and in the case of in situ labeled membrane or cytoplasmic molecules on the other hand. In the former case, FRET efficiency values can be converted into absolute distances, while in the latter, relative distances and their changes are investigated.

6. Calculation of distance relationships from energy transfer efficiencies is easy in the case of a single-donor/single-acceptor system if the localization and relative orientation of the fluorophores are known. If the FRET measurements are performed on the cell surface or inside the cell, many molecules might not be labeled at all; many could be single without a FRET pair, while others may be in smaller groups of heterooligomers, creating higher rates of FRET than expected from stand-alone pairs. A large number of epitopes binding the acceptor increases E simply by increasing the rate of FRET rather than actually meaning that the two epitopes investigated are closer to each other. If, in these cases, reversing the labeling still results in large E values, the proximity can be considered verified. However, if reversing the ratio of donor:acceptor to >1 makes FRET disappear, chances are

that we have previously seen random colocalizations due to the high number of acceptors. One can use model calculations to predict the average distance of donors and acceptors, assuming random distribution, and compare the predictions of the calculations to the observed FRET efficiencies (16, 20–23). Alternatively, the examination of the dependence of FRET on acceptor density and on the donor–acceptor ratio can help decide whether the examined proteins form clusters or are randomly distributed (24). However, the donor/acceptor ratios should be in the range of 0.1–10 (5) if we want to obtain reliable FRET measurements. Outside this range, noise and data irreproducibility propagate unfavorably, rendering accurate FRET efficiency calculations impossible. It also needs to be noted that even random associations resulting from high expression levels of certain oncoproteins must not be regarded as biologically irrelevant.

4. Notes

1. Owing to the very large number of available fluorophores, it is impossible to present a thorough list of even the most widely used combinations. Excitation and emission spectra of molecules for donor–acceptor pair selection can be checked at one of the following web sites: Becton-Dickinson Fluorescence Spectrum Viewer (http://www.bdbiosciences.com/spectra), and Invitrogen-Molecular Probes Spectrum Collection(http://www.invitrogen.com/site/us/en/home/References/Molecular-Probes-The-Handbook/tables/Spectral-characteristics-and-recommended-bandpass-filter-sets-for-Molecular-Probes-dyes.html). The number of available GFP variants has exploded in recent years. A paper to aid the selection of the optimal GFP variant has been published by the Tsien Laboratory (25). A detailed characterization of classical GFP variants for FRET experiments is also available (26). Recent developments are explored in refs. (27–29).

2. Red-emitting fluorescent proteins are known to mature incompletely and, therefore, in tandem chimeric constructs where both donor and acceptor are in the same molecule, one cannot expect a 1:1 ratio for them; in fact, the average ratio will change from cell to cell. Furthermore, due to the incomplete maturation, exact proximity measurements cannot be performed reliably.

3. HEPES buffer (125 mM NaCl, 5 mM KCl, 20 mM HEPES, 1 mM $CaCl_2$, and 1.5 mM $MgSO_4$, pH 7.2) can also be used throughout instead of PBS, both for extra- and intracellular labeling.

4. Instead of 0.1% Triton X-100, 0.1% (w/v) saponin can be used. The saponin stock solution (20% w/v in dH$_2$O) should be stored frozen. Optimizing the signal and minimizing background for intracellular labeling are a must, but are beyond the scope of this chapter.

5. At least one donor-only and one acceptor-only sample are needed for spillage factors, and the same epitope labeled once with donor and once with acceptor is needed for calculating the α factor. It is a good approach, however, to prepare all the samples for obtaining independent determinations.

6. One needs to determine labeling intensity as a function of time after trypsinization for the particular proteins examined.

7. Time of labeling should also be optimized previously. Usually, incubation beyond 10 min does not increase labeling intensity by more than 10%.

8. The excitation and emission wavelengths for this channel I_4 are to be chosen such that the donor and acceptor fluorophores will not contribute considerably to the fluorescence intensity measured here. The main point is to obtain a fluorescence signature which is measured independent of the donor and acceptor channels (i.e., at least the excitation or the emission wavelength should differ from those of other detection channels), but can be used for calculating the contribution of cellular autofluorescence to these channels. There are alternatives to the arrangement proposed here, but this example is the one applicable to most commercial flow cytometers (see Subheading 3.5).

9. There is always one additional control to make, and that is to check for competition of the antibody carrying the acceptor with donor labeling. This should be done with the unlabeled antibody against the "acceptor" epitope, and any decrease of donor fluorescence caused by adding the unlabeled antibody should be attributed to competition rather than donor quenching due to FRET. Needless to say, competition between labeling antibodies is likely also a sign of molecular proximity, albeit not as readily quantitated as FRET.

   In some rare cases, an antibody can increase the binding of another antibody. This enhancement can also lead to misinterpretation of FRET data. For example, if the acceptor-labeled antibody increases the binding of the donor-labeled antibody, the unquenched donor intensity of the donor + acceptor double-labeled sample is larger than that of the donor-only sample, so the FRET calculated by comparing the donor intensity of the double-labeled sample and the donor-only sample will be underestimated. In some cases, the acceptor fluorescence may spill over to the donor channel, and the assumption that the background (i.e., non-donor) fluorescence intensity of the double-labeled sample is equal to the fluorescence intensity of the unlabeled sample does not hold. In such a case, a sample labeled by the acceptor-conjugated antibody and the unlabeled antibody against the donor epitope (to correct for the competition between the two antibodies) is to be used for background subtraction. An equation taking acceptor spillover and competition effects into account can be written in the following form:

$$E = 1 - \frac{I_{\text{Donor+Acceptor}} - I_{\text{Blank\_donor+Acceptor}}}{I_{\text{Donor+Blank\_acceptor}} - I_{\text{background}}}. \tag{24}$$

Here, $I_{\text{Blank\_donor+Acceptor}}$ and $I_{\text{Donor+Blank\_acceptor}}$ denote, respectively, the fluorescence intensities (measured in the donor channel) of the sample tagged with the unlabeled antibody against the donor epitope and the acceptor-conjugated antibody, and that of the sample tagged with the donor-conjugated antibody and the unlabeled antibody against the acceptor epitope.

10. Although the spectroscopic spillover factors $S1$ through $S6$ are not expected to show any cell-by-cell heterogeneity, their cell-by-cell determination also has certain advantages. Performing mathematical calculations with cells having low fluorescence intensity introduces a large error into the calculations. Omitting these cells from the determination of the S factors greatly increases the reliability of these calculations.

11. Factor α is necessary to relate the actual fluorescence emission by the sensitized acceptor measured in the acceptor channel to the fluorescence one could measure in the donor channel from the donor if the quanta that are transferred in FRET had been emitted by the donor. This proportionality factor relates to the Q quantum efficiencies of the acceptor (index A) and donor (index D), and the detection efficiencies η in the acceptor and donor channels for photons emitted by the acceptor and the donor, respectively:

$$\alpha = \frac{Q_A \eta_A}{Q_D \eta_D} \tag{25}$$

However, for practical purposes, α is more easily determined by measuring the same number of excited donor and acceptor molecules, respectively, in the donor and acceptor channels and normalizing them to their molar absorption coefficients (see Eq. 4). In Eq. 4 for α, there is a contribution from the sample labeled only with acceptor excited at the donor wavelength. Usually, this fluorescence intensity is rather small, thus giving the main error source in the calculations. To decrease this error, α should be determined using a protein that is abundant in our cells and recalculated for the actual antibodies used in the experiment.

The fluorescence quantum yields of the dyes may depend on the type of antibody they are attached to and even on the labeling ratio L, thereby affecting the value of α. The α factor determined for a given donor–acceptor antibody pair can be used for other antibody pairs labeled with the same dyes, provided its value is corrected for the differences in the quantum yields:

$$\alpha_2 = \alpha_1 \frac{Q_{A,2}}{Q_{A,1}} \frac{Q_{D,1}}{Q_{D,2}} \qquad (26)$$

Here, subscript "1" refers to the antibody pair for which α has been determined previously and subscript "2" refers to the new antibody pair. Of course, such a correction requires the determination of quantum yields for the other antibody pair.

12. As already pointed out, autofluorescence values that are high compared to the donor and acceptor intensities can seriously disperse the calculated FRET efficiency distributions. Therefore, it is advisable to decrease the autofluorescence level as much as possible. A straightforward way to achieve this is to use yellow or red fluorescent dyes, since cellular autofluorescence becomes progressively weaker in the red region of the visible spectrum (12).

13. The simplified Eq. (23) is tested for the particular filter setup listed in Table 2.

14. Optimizing the sensitivity of FRET measurements is a formidable challenge. To make the measurable range of molecular interactions as large as possible, donor–acceptor dye pairs should be chosen with the maximal spectral overlap. However, this will increase the cross talk between detection channels. At the same time, the higher the amount of spectral spillover

Table 2
Filter setup in a FACSCalibur for cell-by-cell autofluorescence correction with red-shifted fluorophores

Channel	Excitation (nm)	Emission filter (nm)
I_1	488	585 BP
I_2	488	670 LP
I_3	635	661 BP
I_4	488	530 BP

compared to the FRET signal proper, the lower the reliability of the experiment. Therefore, efforts have to be made to minimize spectral spillover. A prudent approach to optimize the choice of fluorophores and filters to reach a balance between these contradictory requirements results in the recognition that the normalized fluorescence of applied dyes is an even more important asset ameliorating the detection of FRET through improved signal-to-noise ratio, and, collectively, AlexaFluor546 with AlexaFluor647 appears to be the most favorable FCET pair in a typical biological system (11).

Acknowledgments

The authors were supported by the following grants: EU FP6 LSHBCT-2004-503467, LSHC-CT-2005-018914, MRTN-CT-2005-019481, and MCRTB-CT-035946; Hungarian National Research Fund K62648, K75752, K68763, K72677; Hungarian National Development Agency TAMOP-4.2.2-08/1-2008-0019 and TAMOP-4.2.11B-09/11KONV-2010-0007; and Hungarian Ministry of Health ETT 362/2009.

References

1. Szöllősi, J., Damjanovich, S., and Mátyus, L. (1998) Application of fluorescence resonance energy transfer in the clinical laboratory: routine and research. *Cytometry* **34**, 159–79.

2. Bastiaens, P. I. H. and Squire, A. (1999) Fluorescence lifetime imaging microscopy: spatial resolution of biochemical processes in the cell. *Trends Cell Biol* **9**, 48–52.

3. Clegg, R. M. (2002) FRET tells us about proximities, distances, orientations and dynamic properties. *J Biotechnol* **82**, 177–9.

4. Vereb, G., Szöllősi, J., Matkó, J., et al. (2003) Dynamic, yet structured: the cell membrane three decades after the Singer-Nicolson model. *Proc Natl Acad Sci U S A* **100**, 8053–8.

5. Berney, C. and Danuser, G. (2003) FRET or No FRET: a quantitative comparison. *Biophys J* **84**, 3992–4010.

6. Förster, T. (1946) Energiewanderung und Fluoreszenz. *Naturwissenschaften* **6**, 166–75.

7. Stryer, L. and Haugland, R. P. (1967) Energy transfer: a spectroscopic ruler. *Proc Nat Acad Sci U S A* **58**, 719–26.

8. Dexter, D. L. (1953) A theory of sensitized luminescence in solids. *J Chem Phys* **21**, 836–50.

9. Jares-Erijman, E. A. and Jovin, T. M. (2003) FRET imaging. *Nat Biotechnol* **21**, 1387–95.

10. Szabó, A., Horváth, G., Szöllősi, J., and Nagy, P. (2008) Quantitative characterization of the large-scale association of ErbB1 and ErbB2 by flow cytometric homo-FRET measurements. *Biophys J* **95**, 2086–96.

11. Horváth, G., Petrás, M., Szentesi, G., et al. (2005) Selecting the right fluorophores and flow cytometer for fluorescence resonance energy transfer measurements. *Cytometry A* **65**, 148–57.

12. Sebestyén, Z., Nagy, P., Horváth, G., et al. (2002) Long wavelength fluorophores and cell-by-cell correction for autofluorescence significantly improves the accuracy of flow cytometric energy transfer measurements on a dual-laser benchtop flow cytometer. *Cytometry* **48**, 124–35.

13. Szentesi, G., Horváth, G., Bori, I., et al. (2004) Computer program for determining fluorescence resonance energy transfer efficiency from flow cytometric data on a cell-by-cell basis. *Comput Methods Programs Biomed* **75**, 201–11.

14. Szöllősi, J., Trón, L., Damjanovich, S., Helliwell, S. H., Arndt-Jovin, D., and Jovin,

T. M. (1984) Fluorescence energy transfer measurements on cell surfaces: a critical comparison of steady-state fluorimetric and flow cytometric methods. *Cytometry* **5**, 210–6.

15. Damjanovich, S., Trón, L., Szöllősi, J., et al. (1983) Distribution and mobility of murine histocompatibility H-2Kk antigen in the cytoplasmic membrane. *Proc Natl Acad Sci U S A* **80**, 5985–9.

16. Trón, L., Szöllősi, J., Damjanovich, S., Helliwell, S. H., Arndt-Jovin, D. J., and Jovin, T. M. (1984) Flow cytometric measurement of fluorescence resonance energy transfer on cell surfaces. Quantitative evaluation of the transfer efficiency on a cell-by-cell basis. *Biophys J* **45**, 939–46.

17. Nagy, P., Bene, L., Hyun, W. C., et al. (2005) Novel calibration method for flow cytometric fluorescence resonance energy transfer measurements between visible fluorescent proteins. *Cytometry A* **67**, 86–96.

18. Dale, R. E., Eisinger, J., and Blumberg, W. E. (1979) The orientational freedom of molecular probes. The orientation factor in intramolecular energy transfer. *Biophys J* **26**, 161–93.

19. Batard, P., Szöllősi, J., Luescher, I., Cerottini, J. C., MacDonald, R., and Romero, P. (2002) Use of phycoerythrin and allophycocyanin for fluorescence resonance energy transfer analyzed by flow cytometry: advantages and limitations. *Cytometry* **48**, 97–105.

20. Wolber, P. K. and Hudson, B. S. (1979) An analytic solution to the Forster energy transfer problem in two dimensions. *Biophys J* **28**, 197–210.

21. Dewey, T. G. and Hammes, G. G. (1980) Calculation on fluorescence resonance energy transfer on surfaces. *Biophys J* **32**, 1023–35.

22. Snyder, B. and Freire, E. (1982) Fluorescence energy transfer in two dimensions. A numeric solution for random and nonrandom distributions. *Biophys J* **40**, 137–48.

23. Szöllősi, J., Damjanovich, S., Balázs, M., et al. (1989) Physical association between MHC class I and class II molecules detected on the cell surface by flow cytometric energy transfer. *J Immunol* **143**, 208–13.

24. Kenworthy, A. K. and Edidin, M. (1998) Distribution of a glycosylphosphatidylinositol-anchored protein at the apical surface of MDCK cells examined at a resolution of <100 A using imaging fluorescence resonance energy transfer. *J Cell Biol* **142**, 69–84.

25. Shaner, N. C., Steinbach, P. A., and Tsien, R. Y. (2005) A guide to choosing fluorescent proteins. *Nat Methods* **2**, 905–9.

26. Patterson, G. H., Piston, D. W., and Barisas, B. G. (2000) Forster distances between green fluorescent protein pairs. *Anal Biochem* **284**, 438–40.

27. Ai, H. W., Hazelwood, K. L., Davidson, M. W., and Campbell, R. E. (2008) Fluorescent protein FRET pairs for ratiometric imaging of dual biosensors. *Nat Methods* **5**, 401–3.

28. Shcherbo, D., Souslova, E. A., Goedhart, J., et al. (2009) Practical and reliable FRET/FLIM pair of fluorescent proteins. *BMC Biotechnol* **9**, 24.

29. Sun, Y., Booker, C. F., Kumari, S., Day, R. N., Davidson, M., and Periasamy, A. (2009) Characterization of an orange acceptor fluorescent protein for sensitized spectral fluorescence resonance energy transfer microscopy using a white-light laser. *J Biomed Opt* **14**, 054009.

Chapter 19

Fluorescent Protein-Assisted Purification for Gene Expression Profiling

M. Raza Zaidi, Chi-Ping Day, and Glenn Merlino

Abstract

Cell type-specific expression of fluorescent proteins allows the purification of rare cells from complex tissues by flow cytometry. This strategy is especially useful for molecular analysis of cancer cells because these cells can be effectively purified away from the noncancerous tumor stroma. Coexpression of biolumi-nescence with fluorescence makes further allows in vivo tracking of cancer cells, which can then be purified at specific tumorigenic stages. Here, we describe protocols for purifying rare skin stem cells, and for in vivo monitoring and purification of cancer cells from lung metastases. Also described is a protocol for the isolation of total RNA from the purified cells for the purpose of performing gene expression profiling.

Key words: Tetracycline-induced system, Green fluorescent protein, Luciferase, Skin stem cells, Label-retaining cells, Pol2 promoter, Syngeneic mice, Lung metastases

1. Introduction

In vivo real-time tracking of cell movement and measurement of cellular growth in animal models by targeted expression of bio-luminescence markers, combined with the ability to purify specific cell types of interest from a complex tissue at a particular point in time via expression of fluorescent proteins are invaluable tools available to biomedical scientists (1, 2). These objectives are achieved by targeting "molecular beacons" to specific cell types, either endogenously by expressing beacon-encoding genes under the control of cell type-specific gene promoters in transgenic mice (3), or exogenously by expressing these genes via vectors in ex vivo or cultured cells followed by transplantation in recipient animals (4, 5). These techniques provide an excellent opportunity to study specific cell types not only at the cellular level, but also at the

Teresa S. Hawley and Robert G. Hawley (eds.), *Flow Cytometry Protocols*, Methods in Molecular Biology, vol. 699, DOI 10.1007/978-1-61737-950-5_19, © Springer Science+Business Media, LLC 2011

molecular level. The two most commonly used molecular beacons are firefly luciferase, encoded by the *luc* gene, and the green fluorescent protein (GFP) from jellyfish. GFP, and other fluorescent proteins (6), provide a unique advantage in that they allow the purification of live-targeted cell types residing in complex heterogeneous tissues, which can then be subjected to molecular analyses, including gene expression profiling. For the purpose of fluorescence-activated cell sorter (FACS) purification, targeted expression of the fluorescent proteins in the cells of interest is superior to labeling cells with fluorescent antibodies because: (1) The cell type of interest may not have a unique cell surface antigen that could be used for antibody labeling, and the use of antibodies against antigens that may be present on other cell types in the tissue risks cross-contamination; (2) Even if a unique cell type-specific cell surface antigen is known, a good antibody against that antigen may not be available; (3) Intracellular antigens cannot be used because those would require fixation and permeabilization of cells, and thus preclude the purification of live cells, which is required for obtaining high-quality RNA for gene expression profiling. Here, we describe two examples of purification of specific cell types from mouse tissues. The following protocols for the preparation of single-cell suspensions of mouse tissues for FACS-purification of fluorescently labeled cells can easily be applied to a wide variety of tissue types (7). The protocol for the isolation of total RNA from a relatively small number of FACS-purified mammalian cells can be applied universally.

1.1. Fluorescence Labeling of Rare Cells in Transgenic Mouse Skin

One of the most exciting applications of flow sorting is the ability to purify rare cell types within complex animal tissues. Tumbar et al. (8) have used a tetracycline inducible system (9, 10) to target GFP to the epidermis and hair follicles in the skin, and identified rare skin stem cells. They used two transgenic mice: (1) K5-tetVP16 transgenic mice expressing the tetracycline-regulated transactivator under the control of the epidermal keratinocyte-specific keratin-5 promoter, and (2) Mice harboring a transgene-containing histone H2B-GFP fusion transcript under the control of tetracycline response element (TRE) promoter. These two transgenic mice were crossed to obtain double transgenic mice that expressed GFP in K5-expressing cells in the skin. This tet responsive system (Tet-off) allowed GFP expression to be shut off in the presence of tetracycline (or its analog doxycycline). Feeding double transgenic mice doxycycline-fortified chow shuts off any new expression of GFP. While the rapidly cycling skin cells completely lose GFP signal within 4–8 weeks after the start of doxycycline diet, the slow-cycling stem cells retain GFP signal even 4 months later. This small population of "label retaining cells" can be selectively FACS-purified from the skin of these mice following enzymatic digestion and mechanical dissociation of the skin tissue into a single-cell suspension.

1.2. Bifunctional Reporter-Aided Monitoring and Purification of Cancer Cells

Tumors are complex heterogeneous mixtures of cancer cells and a variety of noncancerous cell types. For gene expression profiling studies of cancer cells with respect to their growth properties and/or metastatic potential, it is crucial to purify the tumor cells away from the noncancerous tumor-associated cells (stromal cells). Coexpression of luciferase and GFP can be used to noninvasively image and track the growth, trafficking, and metastatic spread as well as FACS-purification of transplanted cancer cells from primary tumors as well as metastatic sites. We have developed lentiviral (LV) reporter (11) that express both luciferase and GFP from a fused transcript under the control of the Pol2 and FerH promoters (4). We linked these two reporter genes (Luc/GFP) to ensure colinear signals from both reporters. This fusion reporter strategy can also overcome the failure of LV infection to reach 100% efficiency because the Luc/GFP-labeled cells can be FACS-purified. We have determined that the Pol2 promoter provides sustainable long-term expression of the Luc/GFP reporter even in immunocompetent mice (4). Luciferase can be used to monitor the primary tumor growth, as well as metastatic spread to distant sites (1). Metastatic cells from distant metastatic tumors, and even dormant micrometastases, can then be FACS-purified using the GFP label (4). Here, we describe a method for tumor cell infection with LV-Luc/GFP, transplantation in syngeneic mice, and FACS purification of lung metastases.

2. Materials

All experiments with mice must be performed under protocols approved by the Institutional Animal Care and Use Committees and the strict NIH guidelines for humane use of research animals.

2.1. Reagents and Cells

1. Liberase TL (Roche Applied Science, Indianapolis, IN). Reconstitute at 38.5 mg/mL (200 U/mL) in dH_2O. Keep in small aliquots at $-20°C$ for 6 months. Avoid repeated freeze/thaw cycles.

2. DNAse I (Sigma-Aldrich Corp., St. Louis, MO). Reconstitute 1 g in 20 mL dH_2O for a final stock concentration of 50 mg/mL. Invert gently to mix. Do not vortex as DNAse I is sensitive to physical force. Store at $-20°C$ in 1 mL aliquots for 1 year.

3. Digestion medium: 5 mL DMEM, 50 μL Liberase TL stock solution (0.385 mg/mL final concentration).

4. DFD solution: 24 mL DMEM, 6 mL fetal bovine serum (FBS), 300 µL DNAse I stock solution. Mix by inverting and chill on ice. Do not vortex, as DNAse I is sensitive to physical force. Make fresh as needed.

5. PBSA solution: 50 mL 1× phosphate-buffered saline (PBS), 0.5 g bovine serum albumin (BSA powder, store at 4°C). Shake well to dissolve. To dissolve completely, incubate at 37°C for 30 min. Shake well and then filter through a 0.45-µm syringe filter to eliminate any undissolved BSA particles. Chill on ice.

6. Cell culture medium: DMEM supplemented with 10% FBS.

7. Transfection medium: Opti-MEM I (Invitrogen, Carlsbad, CA).

8. Inoculation medium: RPMI 1640.

9. ACK lysis buffer.

10. HEK-293 T cells (CRL-11268; American Type Culture Collection, Manassas, VA).

11. ViraPower LV Expression System (Invitrogen).

12. Lipofectamine 2000 (Invitrogen).

13. Trizol reagent.

14. 70% ethanol: dilute absolute ethanol with DEPC-treated water.

15. 80% ethanol: dilute absolute ethanol with DEPC-treated water.

16. RNeasy RNA isolation kit, Mini or Micro (Qiagen, Valencia, CA).

2.2. Equipment and Supplies

1. Dissection instruments; fine-tooth shaver or hair removal cream.

2. Medimachine system, 50-µm Medicons (BD Biosciences, San Jose, CA).

3. Filters: Filcon 70-µm purple filters, syringe-type; Filcon 30-µm green filters, cup-type; 70-µm white cell strainers (all from BD Biosciences).

4. 20-mL Luer Lock syringes, with 18-gauge 1–0.5 in. needles.

5. Fine balance.

6. Refrigerated benchtop centrifuge with bucket rotors for 15-mL conical Falcon tubes.

7. Xenogen bioluminescence imaging system (Caliper LifeSciences, Hopkinton, MA).

8. FACS.

3. Methods

3.1. Isolation of GFP+ Mouse Skin Cells by FACS

3.1.1. Obtaining Mouse Skin Tissue

1. Euthanasia: mice older than 1-week should be euthanized by CO_2 asphyxiation. For certain experimental protocols, with IACUC approval, cervical dislocation can be used for euthanizing mice (see Note 1).

2. Shave the back skin as closely as possible. Alternatively, hair removal cream can be used under some circumstances (see Note 2).

3. Take the entire back skin, from shoulders to just above the tail, and put in a dish with 1× PBS on ice.

4. Immerse the skin in a second dish with cold 1× PBS and using two curved fine forceps remove all the fat and blood vessels from the underside of the skin. Scrape off fat layer until translucent skin is exposed from underneath. Put the cleaned skin in another dish in 1× PBS on ice (see Note 3).

5. Cut the skin tissue in small pieces, ~2–4 mm long, and weigh on a fine balance.

3.1.2. Preparation of Single-Cell Suspensions from Mouse Skin

1. Put 0.6–1 g of skin per 5 mL of digestion medium in a 15-mL Falcon tube. Invert the tube a few times to mix and incubate in 37°C water bath for 1 h. Mix once by inverting a few times at the mid-point of incubation (see Note 4).

2. Put the digested tissue in a dish on ice. Keep the cells cold as much as possible from here on.

3. Put a few skin pieces inside the Medicon chamber while rotating the blades, enough to cover the bottom of the chamber. Put the lid on, insert the Medicon into the Medimachine and run for 2–2.5 min (see Note 5).

4. Put two or three 15-mL conical Falcon tubes on ice and insert 70-μm purple filters in each. Using a 20-mL syringe, dispense 2 mL of DFD solution through the filters into each tube.

5. When the Medicon run is complete, take it out of the Medimachine and take the lid off. Using an 18-gauge needle on the 20-mL syringe, dispense DFD into the Medicon chamber to fill it. Then, insert the needle all the way into the little hole on the side and draw the DFD from the bottom of Medicon into the syringe. After removing the needle from the syringe, put this single-cell suspension in the Falcon tube through the purple filter (see Note 6).

6. Remove the remaining chunks of tissue from the Medicon chamber and repeat the cycle with new tissue pieces. When done with the Medicon, rinse it with one extra volume of DFD to completely draw any remaining cells.

7. When all the tissue has been processed into a single cell suspension, centrifuge the Falcon tubes at ~300×*g* for 5 min at 4°C.

8. Aspirate out as much of the supernatant as possible without disturbing the cell pellet. Resuspend the pelleted cells in one tube in 8 mL of PBSA. Transfer this suspension to other tubes sequentially to combine all of the cells.

9. Filter the pooled cell suspension through a 70-μm white strainer into a 50-mL Falcon tube. Rinse all the tubes with another 5 mL of PBSA and then rinse through the white strainer to combine.

10. Transfer the pooled cell suspension (13 mL) to a 15-mL Falcon tube and centrifuge at ~300×*g* for 5 min at 4°C. Aspirate the supernatant.

11. Resuspend the cell pellet in 8 mL of PBSA and transfer to a new 15-mL Falcon tube. Rinse the old tube with 5 mL of PBSA and combine into the new tube. Repeat centrifugation. Aspirate as much of the supernatant as possible without disturbing the pellet (see Note 7).

12. Suspend the final pellet in 2–4 mL of PBSA. This cell suspension can be directly used for the isolation of GFP+ cells on a FACS equipped with a 488-nm laser and a 530/30 bandpass filter. Just before FACS sorting, filter the cell suspension through a 30-μm green filter to remove clumped cells and debris (see Note 8).

3.2. In Vivo Tracking of Luc/GFP-Expressing Tumor Cells and Lung Metastases

The construction of pSico-Luc/GFP fusion reporter lentivirus vector has been described (4).

3.2.1. Production of Lentivirus

1. The day before transfection (day 1), plate 5×10^6 HEK-293 T cells in T-75 flask in 15 mL of DMEM culture medium containing 10% FBS. Do not include antibiotics in culture medium.

2. On the day of transfection (day 2), remove the culture medium from the HEK-293 T cells and replace with 7.5 mL of Opti-MEM I medium.

3. In a sterile 5-mL tube, mix 9 μg of the ViraPower™ Packaging Mix and 3 μg of LV expression plasmid DNA (12 μg total) in 1.5 mL of Opti-MEM I medium without serum. In a separate sterile 5-mL tube, dilute 36 μL of Lipofectamine™ 2000 in 1.5 mL of Opti-MEM® I Medium without serum. Mix gently and incubate for 5 min at room temperature.

4. Combine the diluted DNA with the diluted Lipofectamine™ 2000 and mix gently (no pipetting). Incubate for 20 min at room temperature and then add the DNA-Lipofectamine™

2000 complexes dropwise to each plate of cells. Mix gently by rocking the plate back and forth. Incubate the cells overnight. The next day (day 3), remove the medium and replace with 15 mL of complete culture medium without antibiotics.

5. 24 h after changing medium (day 4), examine the cells with fluorescence microscope. HEK-293 T cells should express substantial level of GFP if transfection is successful.

6. Harvest virus-containing supernatants twice at 48 and 72 h posttransfection (days 4 and 5, respectively) by transferring medium into a 50-mL sterile, capped, conical tube, and then centrifuge supernatants at $2,000 \times g$ for 15 min at 4°C to pellet debris. Pipet viral supernatants into cryovials in 1 mL aliquots. Store viral stocks at −80°C, which can maintain the titer of the virus stably up to one year. Repeated freezing and thawing is not recommended.

7. If it is necessary to concentrate the virus supernatant, load the supernatant in the outer chamber of an Amicon centriprep-10 unit (Millipore Corp., Bedford, MA) and centrifuge at $3,000 \times g$ for 45–60 min at room temperature or 4°C. Discard the ultrafiltrate from the inner chamber and repeat the centrifugation step. It usually takes 2.5–3 h to concentrate supernatant 10- to 20-fold. Recovery of virus can be quantitated by titering material pre- and postconcentration. Losses are usually about 20%.

8. Titer the viral supernatant on a standard human adherent cell line, such as HOS cells (human osteosarcoma), HeLa, or HEK-293 cells. Plate the target cells 1 day prior to transduction in 24-well format, at 50,000 cells/well (day 1). In the next day, count the cell number from two wells (day 2). This would be the cell number used in the calculation of virus titer later. Incubate increasing amounts of virus supernatant (e.g., 2, 5, 10, 20, 50, 100, and 200 μL) with the target cells. Change the medium the next day. Two days later, observe the transduced cells with fluorescence microscope to confirm the expression of GFP. Bring the cells for flow cytometric analysis to measure GFP-positive percentage (% GFP+), and then identify the virus supernatant amount to reach 1% < %GFP+ < 10%. In this range, cells are assumed to be transduced with a single copy of virus. The titer is calculated as following: Infectious Units (IU)/mL = ([cell number counted at day 2] × %GFP+)/(virus supernatant amount in $\mu L \times 10^{-3}$).

3.2.2. Labeling of Tumor Tissue and In Vivo Monitoring of Growth and Metastasis

1. Put 1 g of fragmented B16BL6 melanoma tumor tissue freshly harvested from subcutaneous sites of mice per 5 mL of digestion medium in a 15-mL Falcon tube. Invert the tube a few times to mix and incubate in 37°C water bath for 30 min. Mix once by inverting a few times at the mid-point of incubation (see Note 4).

2. To make single-cell suspension, follow Subheading 3.1.2, steps 2–7.

3. Aspirate out as much of the supernatant as possible without disturbing the cell pellet. Resuspend the pelleted cells in one tube in 8 mL of PBSA. Count cell number and transfer 10^6 cells into 15-mL Falcon tubes. Centrifuge at ~$300 \times g$ for 5 min at 4°C.

4. Aspirate the supernatant and resuspend the cell pellet with 500 μL of virus supernatant with IU higher than 10^7. Adjust the final volume to 1–1.5 mL with RPMI 1640 serum-free medium.

5. Centrifuge at 800–1,200 ×g for 0.5–1 h (depending on the fragility of tumor cells). Aspirate the supernatant. Resuspend the cells with 100 μL of serum-free RPMI 1640 medium and inoculate subcutaneously into mice immediately. *Caution*: use rotors with biosafety cover for centrifugation.

6. Monitor the tumor growth by measuring the physical size and bioluminescence imaging on the Xenogen biolumines-cence imaging system (Fig. 1). When the tumors reach 1 cm³ in mice, harvest tumor and prepare cell suspension following steps 1–3.

7. Resuspend the pelleted cells in 5 mL of ACK lysis buffer and incubate at room temperature for 10 min. Centrifuge at $300 \times g$, 4°C, for 10 min. Resuspend the cell pellet with 8 mL of PBSA. Transfer this suspension to other tubes sequentially to combine all of the cells.

8. Filter the pooled cell suspension through a 70-μm white cell strainer into a 50-mL Falcon tube. Rinse all the tubes with another 5 mL of PBSA and then rinse through the white strainer to combine. Transfer the pooled cell suspension (13 mL) to a 15-mL Falcon tube and centrifuge at ~$300 \times g$ for 5 min at 4°C. Aspirate the supernatant.

9. Resuspend the final pellet in 2–4 mL of PBSA. This cell sus-pension can be directly used for FACS (Fig. 2).

10. Following sorting, pool the sorted GFP+ cells, centrifuge at $300 \times g$, 4°C, for 10 min. Aspirate the supernatant and resus-pend the cells with serum-free RPMI 1640 medium to reach 10^6 cells/100 μL.

11. Inoculate 10^6 tumor cells subcutaneously into each mouse. The tumors are uniformly labeled with luciferase-GFP fusion gene, allowing in vivo bioluminescence monitoring and isola-tion of GFP+ cells from the harvested tumors.

3.3. Preparation of Total RNA from Sorted Cells for Gene Expression Profiling

1. Centrifuge sorted GFP+ cells for 10 min at ~$300 \times g$ at 4°C.

2. Remove as much of the supernatant as possible without disturbing the pellet (see Note 9).

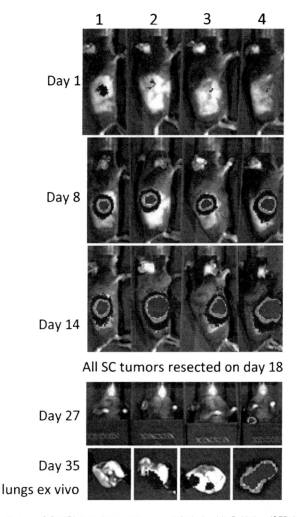

Fig. 1. Monitoring of B16BL6 melanoma tumors labeled with Pol2-Luc/GFP lentiviral vector by bioluminescence (BL) imaging. Tumors derived from B16BL6 cells were infected with Pol2-Luc/GFP lentiviral vector and subcutaneously inoculated into C57BL/6 mice. The tumor growth was monitored by both measurement of physical size and bioluminescence imaging (signal range from 1.2×10^4 to 5.0×10^5, marked as *purple* and *red* in the heat map, in the electronic version). All the tumors were resected at day 18 after inoculation and subjected to FACS analysis (see Fig. 2). The BL monitoring was continued for lung metastasis. At day 27, lung metastases were detected in mouse #4. At day 35, all the mice were sacrificed and the lungs were resected for ex vivo BL imaging. Lung of mouse #4 had diffuse disease while relatively small lesions appeared in the lungs of mice #2 and #3.

3. Lyse the cells in 1 mL of Trizol reagent by vortexing vigorously. Transfer to an Eppendorf tube. This lysate can be stored indefinitely at −80°C.

4. Add 200 µL of chloroform and shake vigorously (or vortex at low setting) for 15 s. Let stand at room temperature for 5 min.

5. Centrifuge at $20,000 \times g$ for 15 min at 4°C.

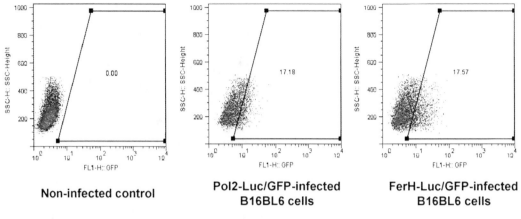

Non-infected control

**Pol2-Luc/GFP-infected
B16BL6 cells**

**FerH-Luc/GFP-infected
B16BL6 cells**

Fig. 2. Flow cytometric analysis of B16BL6 melanoma tumors labeled with Pol2-Luc/GFP and FerH-Luc/GFP. Primary B16BL6 tumors were resected from mice (including those in Fig. 1) to make cell suspensions and analyzed as described in the text. The GFP+ percentages were similar in both labeled tumors, and the difference in median fluorescence intensity reflected the relative promoter activities of the Pol2 and FerH promoters (4).

6. Carefully take the aqueous (top) phase and transfer to a new Eppendorf tube. Measure the volume transferred.

7. Add an equal volume of 70% ethanol and vortex to mix.

8. Transfer the mixture to RNeasy MinElute spin column (see Note 10). Centrifuge at $10,000 \times g$ for 15 s at room temperature. If the volume is more than 700 µL, reload the rest to the column until all the volume has been run through the column. Discard all flow through.

9. Apply 700 µL of RW1 buffer (included in RNeasy kit) and centrifuge at $10,000 \times g$ for 15 s. Discard flow through and collection tube.

10. Transfer spin column to a new collection tube (provided in the kit). Apply 500 µL of RPE buffer (ethanol added, see Note 11) to the column and centrifuge at $10,000 \times g$ for 15 s. Discard the flow through.

11. Add 500 µL of 80% ethanol to the spin column and centrifuge at $10,000 \times g$ for 15 s. Discard the flow through.

12. Repeat wash with 500 µL of 80% ethanol, and centrifuge for 2 min at $10,000 \times g$ (see Note 12).

13. Transfer column to a new collection tube. Open the cap of the spin column and centrifuge at maximum speed for 5 min to dry the column completely (see Note 13). Discard the flow through and collection tube.

14. Transfer column to an Eppendorf tube. 10-14 uL (for Micro kit) or 30-50 uL (for Mini kit) of RNAse-free water directly in the center of the column membrane and centrifuge for 1 min at maximum speed to elute RNA.

15. Store RNA at –20°C, or at –80°C for long-term storage.

4. Notes

1. CO_2 asphyxiation is the method of choice for euthanasia of mice. However, if exposure to CO_2 is suspected to adversely affect the gene expression profile of the cells of interest, then cervical dislocation can be used. Cervical dislocation must be performed with appropriate IACUC approval and by experienced personnel.

2. Hair removal cream can be used, but this can adversely affect the gene expression profiles in the epidermal cells.

3. Removing all the blood vessels and especially fat from the underside of the skin is extremely important because the presence of fat in the final cell suspension causes cells to clump and interfere in the FACS sorting. Do not use too much 1× PBS while scraping because it causes the skin to float up from the sides and makes it difficult to scrape. Scrape the underside of the skin until you expose the translucent skin. Remove as much of the fat as possible.

4. Liberase TL is a cocktail of collagenases and neutral proteases (NP). Keep in mind that given enough time, NP damage the surface proteins and increase cell death. Incubating for longer periods and/or using more Liberase are thus counterproductive. Liberase TL has an NP:Collagenase ratio that is much lower than the majority of other available Liberase formulations. This relatively lower NP level in the TL formulation is important to keep the cell viability relatively high. However, at least some level of cell death is expected with 1 h incubation. On the other hand, 1 h incubation is important for optimal dissociation of tougher (fibrous) tissues, e.g., skin, and thus higher yield in FACS. For less fibrous tissues, 30–45 min incubation is sufficient.

5. Do not overfill the Medicon chamber with tissue, as it might jam. The time of medimachine run can be extended up to 4 min for tougher and fibrous tissues to increase tissue dissociation. Do not exceed 4 min of running time, as longer runs only increase fine debris (e.g., connective tissue) that would clog the Medicon membrane. You can use two Medicons at a time so that you can process/refill one Medicon while the other is running. Change to a new pair midway. This is because the Medicon membrane clogs with debris and cells are trapped.

6. Change the 70-μm purple filters when you change the Medicons midway.

7. The cell pellet is mostly translucent. However, depending on how well the blood vessels and/or blood have been removed,

there may be a reddish hue to the pellet. Small amounts of red blood cells would not affect the FACS yield.

8. Before filtering through the 30 μm filter, it is very important to dilute the final cell suspension appropriately so that it is not too cell dense. Too much cell density causes large clumps, which are filtered out and reduce yield. Large cell pellets should be diluted to 3.5–4 mL before filtration. Smaller pellets should be diluted to 2–3 mL. Volumes larger than 4 mL take too long to process through the FACS.

9. Depending on how many cells have been collected by sorting, a pellet may or may not be visible. Generally, cell pellets containing less than 2×10^5 cells are not visible. In such cases, be careful while aspirating the supernatant. You can leave a little bit of the supernatant in the bottom, as long as it is less than 10% of the volume of Trizol used.

10. Depending on the number of cells being processed, either RNeasy Mini or Micro kit can be used. Use the Mini kit for larger number of cells (millions), and Micro for hundreds to less than a million cells.

11. Buffer RPE is provided in the kit as a concentrate and absolute ethanol must be added before use. See the bottle label for how much ethanol to add.

12. The second 80% wash ensures that no salt is carried over to the eluate. Even small amounts of carryover salts can inhibit subsequent labeling enzymatic reactions.

13. It is important to dry the column completely before RNA elution to make sure that no ethanol is carried over, which can inhibit subsequent enzymatic reactions.

References

1. Contag, C. H. and Bachmann, M. H. (2002) Advances in in vivo bioluminescence imaging of gene expression. *Annu Rev Biomed Eng* **4**, 235–60.

2. Gross, S. and Piwnica-Worms, D. (2005) Spying on cancer: molecular imaging in vivo with genetically encoded reporters. *Cancer Cell* **7**, 5–15.

3. Spergel, D. J., Kruth, U., Shimshek, D. R., Sprengel, R., and Seeburg, P. H. (2001) Using reporter genes to label selected neuronal populations in transgenic mice for gene promoter, anatomical, and physiological studies. *Prog Neurobiol* **63**, 673–86.

4. Day, C. P., Carter, J., Bonomi, C., *et al.* (2009) Lentivirus-mediated bifunctional cell labeling for in vivo melanoma study. *Pigment Cell Melanoma Res* **22**, 283–95.

5. Haas, D. L., Case, S. S., Crooks, G. M., and Kohn, D. B. (2000) Critical factors influencing stable transduction of human CD34(+) cells with HIV-1-derived lentiviral vectors. *Mol Ther* **2**, 71–80.

6. Shaner, N. C., Patterson, G. H., and Davidson, M. W. (2007) Advances in fluorescent protein technology. *J Cell Sci* **120**, 4247–60.

7. Wolnicka-Glubisz, A., King, W., and Noonan, F. P. (2005) SCA-1+ cells with an adipocyte phenotype in neonatal mouse skin. *J Invest Dermatol* **125**, 383–5.

8. Tumbar, T., Guasch, G., Greco, V., *et al.* (2004) Defining the epithelial stem cell niche in skin. *Science* **303**, 359–63.

9. Zhu, Z., Zheng, T., Lee, C. G., Homer, R. J., and Elias, J. A. (2002) Tetracycline-controlled transcriptional regulation systems: advances and application in transgenic animal modeling. *Semin Cell Dev Biol* **13**, 121–8.

10. Romano, R. A. and Sinha, S. (2010) Tetracycline-regulated gene expression in transgenic mouse epidermis. *Methods Mol Biol* **585**, 287–302.

11. Ventura, A., Meissner, A., Dillon, C. P., *et al.* (2004) Cre-lox-regulated conditional RNA interference from transgenes. *Proc Natl Acad Sci USA* **101**, 10380–5.

Chapter 20

Multiparametric Analysis, Sorting, and Transcriptional Profiling of Plant Protoplasts and Nuclei According to Cell Type

David W. Galbraith, Jaroslav Janda, and Georgina M. Lambert

Abstract

Flow cytometry has been employed for the analysis of higher plants for approximately the last 30 years. For the angiosperms, ~500,000 species, itself a daunting number, parametric measurements enabled through the use of flow cytometers started with basic descriptors of the individual cells and their contents, and have both inspired the development of novel cytometric methods that subsequently have been applied to organisms within other kingdoms of life, and adopted cytometric methods devised for other species, particularly mammals. Higher plants offer unique challenges in terms of flow cytometric analysis, notably the facts that their organs and tissues are complex three-dimensional assemblies of different cell types, and that their individual cells are, in general, larger than those of mammals.

This chapter provides an overview of the general types of parametric measurement that have been applied to plants, and provides detailed methods for selected examples based on the plant model *Arabidopsis thaliana*. These illustrate the use of flow cytometry for the analysis of protoplasts and nuclear DNA contents (genome size and the cell cycle). These are further integrated with measurements focusing on specific cell types, based on transgenic expression of Fluorescent Proteins (FPs), and on analysis of the spectrum of transcripts found within protoplasts and nuclei. These measurements were chosen in particular to illustrate, respectively, the issues encountered in the flow analysis and sorting of large biological cells, typified by protoplasts; how to handle flow analyses under conditions that require processing of large numbers of samples in which the individual samples contain only a very small minority of objects of interest; and how to deal with exceptionally small amounts of RNA within the sorted samples.

Key words: Plants, Protoplasts, Nucleus, Gene expression, Sequencing, Microarrays

1. Introduction

1.1. General Constraints Regarding the Flow Cytometric Analysis of Plants

Flow cytometry is a technology that presupposes the availability of samples comprising single-cell suspensions, since it interrogates the fluorescence and light scatter properties of these cells as they stream individually in liquid through foci of intense light sources.

Teresa S. Hawley and Robert G. Hawley (eds.), *Flow Cytometry Protocols*, Methods in Molecular Biology, vol. 699,
DOI 10.1007/978-1-61737-950-5_20, © Springer Science+Business Media, LLC 2011

The flowering plants, at most growth stages, do not exist in the form of single-cell suspensions. Instead, the predominant form is the multicellular sporophyte, with characteristic organs (leaves, roots, flowers, etc.) within which the tissues comprise complex interspersions of different cell types. Multicellularity arises through the controlled division of cells within meristematic zones (1). In contrast to animal species, sporophytic plant cells are non-motile and move away from the meristems solely as a consequence of cell division and subsequent expansion of the daughter cells. These cells, which in general have different fates, remain physically associated as a consequence of the mechanism of cell division, which subdivides the mother cell asymmetrically within the context of a shared, cellulosic cell wall. Therefore, prior to using flow cytometry to analyze plant cells, methods to produce single-cell suspensions are required. This can be done by dissolving the cellulosic cell wall under hyperosmotic conditions, which releases wall-less cells, termed protoplasts. Protoplasts, being bound by a plasma membrane devoid of much of the mechanical resilience found in animal cells, are inherently fragile; furthermore, they are characteristically larger, and sometimes much larger, than the mammalian cells that were used for establishing the original design specifications of flow cytometers. This introduces a number of additional complications in the successful flow analysis and, particularly, sorting of plant protoplasts. Special cases in which natural single-cell suspensions of plants are encountered include pollen and the developing microspores.

As an alternative approach for the flow analysis of plant systems, one can focus not on the examination of single cells, but on the examination of their subcellular contents. For this purpose, plant tissues and organs are converted into cell-free homogenates, and the constituents of the homogenates are subjected to flow analysis. In general, flow cytometers were not designed to accommodate this type of measurement, which is complicated by the fact that the objects of interest comprise an extreme minority of the total objects within the sample, and that the samples contain very high concentrations of objects detectable by the flow cytometer.

In employing flow cytometry and sorting with plant species, whether utilizing protoplasts or homogenates, a primary consideration is the purpose of the analysis. Considerable current interest exists in the use of flow-based methods for providing purified materials, protoplasts, or nuclei of specific cell types, as sources of transcripts for global gene expression profiling. This recognizes the need for single-cell resolution in deriving a genomic understanding of complex tissues and organs, in which different cell types are intricately interspersed.

This chapter, therefore, provides detailed information as to the best ways, using flow cytometry and sorting, to handle the complications presented by large particles and by crude homogenates, and to integrate these with genomic technologies aiming at characterization of global gene expression. I present specific details and helpful tips for successful operation of the cytometric instrumentation. My laboratory has experience working with the Coulter EPICS and Elite, the Cytomation MoFlo, the Becton Dickinson FACScan and LSR II, and the Accuri C6. The described methods are generally applicable to all of these instruments.

1.2. Applying Flow Cytometry and Cell Sorting to Plant Protoplasts

We first described flow analysis, sorting, and culture of plant protoplasts in 1984 (2). In this work, we employed tobacco leaves as the source of protoplasts, since they are readily converted into protoplasts, and these protoplasts are particularly easy to take through tissue culture to regenerate cell walls and to induce organogenesis and thereby produce plants. Dissolution of the cell wall was achieved using mixtures of cellulases, hemicellulases, and pectinases (3, 4), in the presence of an osmoticum, which serves to stabilize the plant plasma membrane under slightly hypertonic conditions. Since that time, we have described flow analysis and sorting of protoplasts from maize leaves (5–7) and *Arabidopsis* roots (8, 9), and these methods are applicable to *Arabidopsis* aerial tissues (10) and leaves (11). Reports of the diameters of *Arabidopsis* protoplasts vary according to tissue and cell type: 10–20 μm for root protoplasts (12), around 30–50 μm for protoplasts prepared from well-expanded leaf tissues (11), and 10–20 μm for protoplasts prepared from aerial portions of plantlets grown on vertical agar plates (Galbraith laboratory, unpublished observations).

1.3. Specifying Parameters for Protoplast Analysis

Parameters available for flow analysis of protoplasts include the standard forward angle and 90° light scatter signals, as well as any fluorescence emission signals produced either from endogenous or from introduced fluorochromes such as the Green Fluorescent Protein (GFP). The shapes of the pulse waveforms over time can also be employed, for example, pulse width time-of-flight as a measure of cell size (13), but this has not found widespread use.

A prerequisite to the use of flow sorting for purification of specific cell types is a method to tag these cells which provides a fluorescent signal that can be detected by the cytometer. Particularly productive has been transgenic expression of members of the family of Fluorescent Proteins, of which the Green Fluorescent Protein (GFP) of *Aequorea victoria* is the archetype (14). The FP family now constitutes a large number of different proteins isolated from marine organisms, complemented by a large and ever-increasing number of sequence variants of these proteins ((15–17), and citations therein).

As observed with other organisms, FPs can be readily expressed in plants (16, 18) and, with caveats, transgenic FP expression appears to be generally non-toxic within plants. FPs can be targeted to essentially all subcellular locations, using translational fusions to topogenic motifs or even entire proteins for this purpose (15, 16, 18). Caveats concerning the effects of excessive levels of expression and associated with mis-targeting should be noted (19). Targeting to subcellular organelles such as the nucleus (20–24) provides increased signal-to-noise ratio, since background autofluorescence is generally dispersed throughout the cytoplasm. This advantage in sensitivity does not translate to flow cytometry when the fluorescence of the entire cell is quantified, but is a factor when analysis of isolated nuclei within cell-free homogenates is done (23, 24).

For use in flow cytometry and sorting, the type of light sources that are available, including spectral quality and quantity, must be considered. Most forms of GFP and YFP are efficiently excited using the commonly available argon laser line at 488 nm, with DsRed and other red FPs being reasonably well excited using the krypton laser line at 568 nm. CFP excitation is optimal at around 425 nm, but it can also be excited at 457 nm (using a tunable argon laser), at 407/413 nm (using a krypton laser), or at 405 nm (using solid state laser diodes). BFP and its derivatives, which are of low brightness, are not routinely employed as fluorescent transgenic reporters.

For higher plants, mesophyll and epidermal cells within leaves can be defined based on the presence or absence of chlorophyll, and protoplasts can be analyzed, flow sorted, and characterized, based on this parameter (13, 25). Plant protoplasts can also be analyzed and sorted based on transient or transgenic expression of GFP (5–10, 12, 26–28), as can nuclei (23, 24). Combined flow analysis and sorting of GFP and RFP has also been recently reported (29). Cell type-specific FP expression involves the production of transgenic plants containing constructions in which the FP coding sequence is placed under the transcriptional control of specific promoters (8–10, 23, 24, 26–28). Promoter/ enhancer trap methodologies can also provide transgenic plant lines within which selected subsets of cells are highlighted by FP expression (see http://www.arabidopsis.org/abrc/haseloff.htm for details). Work toward flow sorting of protoplasts and nuclei based on accumulation of FPs within specific cell types is ongoing at an increasing number of laboratories. Table 1 provides a listing of recent publications employing combinations of FP-labeling and flow cytometry for analysis and sorting of protoplasts and nuclei from specific cell types in plants.

1.4. Dealing with Large Cells

Plant protoplasts, in general, are heterogeneous in size and their diameters can be large in comparison to that of the flow tips with which flow cytometric instrumentation are conventionally equipped. Diameters, for example, frequently exceed 20–30 μm and are sometimes much larger (30), and flow analysis and sorting

Table 1
Reports of the flow sorting and/or analysis of FP-labeled protoplasts and nuclei

Species	Fluorescent protein	Tissue or cell type(s)	Objects sorted	Purpose	References
Maize	GFP	Transfected leaf protoplasts	Protoplasts	Transient expression	(5, 6)
Arabidopsis	GFP	Epidermal atrichoblasts/ lateral root cap/ endodermis/endodermis plus cortex/stele	Protoplasts	Transcript profiling	(8, 9)
Arabidopsis	GFP, DsRed	Four cell types of the shoot apical meristem	Protoplasts	Transcript profiling	(10)
Arabidopsis	GFP	Fourteen cell types within the root	Protoplasts	Metabolite profiling	(12)
Tobacco	GFP	Transgenic plant leaves	Nuclei	Characterization of GFP targeting	(21)
Arabidopsis	GFP, YFP	Five cell types within the root	Nuclei; analysis only	Ploidy status	(23)
Arabidopsis	GFP	Root phloem companion cells	Nuclei	Transcript profiling	(24)
Arabidopsis	GFP	Root quiescent center cells	Protoplasts	Transcript profiling	(26)
Arabidopsis	GFP	19 cell types within the root	Protoplasts	Transcript profiling	(27)
Arabidopsis	GFP	Six cell types within the root	Protoplasts	Transcript profiling	(28)
Arabidopsis	GFP, RFP	Transfected root protoplasts	Protoplasts	Transient expression	(29)

using standard flow tips (50–70 μm diameter) becomes impossible; for successful sorting, flow tips having diameters in the range of 100–200 μm may be needed. The applicable laws of physics impose constraints on the higher limits of actuation frequencies of the piezo drive responsible for synchronizing droplet production (see ref. (31) for a full discussion), which in turn limits the absolute rate of sorting. For this reason, sorting of nuclei, always having smaller diameters than protoplasts and hence allowing the use of smaller flow tips and resulting in higher sorting rates, may be advantageous in specific situations.

1.5. Dealing with Minority Populations in Complex Mixtures

Flow cytometers conventionally employ light scatter signals to trigger the cycle of detection initiated when a particle enters the region of illumination. This is done for the simple reason that all particles, whether cells or of subcellular origin, scatter light; typically only a subset of these particles is fluorescent. Flow cytometric measurements involving mammalian cell suspensions benefit from the observation that most of the objects within the suspensions are those of interest (i.e., are cells). When employing plant homogenates for flow cytometric analysis, a very different situation pertains: the objects of interest comprise a small, and sometimes very small, minority of the light-scattering objects (i.e., debris) in the sample. A popular, important, and widely employed example is the measurement of nuclear DNA contents within plant tissue homogenates using DNA-specific fluorescent staining methods (32). The avalanche of light-scattering signals resulting from subcellular debris can obscure the detection of the objects of interest. Adjustment of discriminator settings to eliminate the contribution of this debris presupposes a knowledge of the light-scattering properties of the nuclei which, for novel samples, may not be a valid assumption. Further complications are introduced by the fact that the subcellular organelles themselves can exhibit autofluorescence, chloroplasts being a prime example. Employing protoplasts, rather than homogenates, may not completely forestall the problem of debris; for example, leaf mesophyll protoplasts contain large numbers of chloroplasts (13) and the breakage of even a minor proportion of these protoplasts results in a considerable excess of chloroplasts over protoplasts within the cellular suspension. For this reason, use of isopycnic gradient flotation for protoplast purification is strongly recommended prior to flow analysis and sorting.

1.6. Dealing with Low Amounts of RNA for Expression Profiling

Although modern flow sorters can usefully operate at sort rates of around 40,000/s, the rate of recovery of specific cells is limited by the proportion of cells within the target tissue, the degree of enrichment/purification specified for the sort operation, and the size of the cells (which serves to define the upper sort rate). The total RNA content of eukaryotic cells is a function of nuclear DNA content, and varies also according to organism, species, tissue type, and developmental stage; typical values are ~1–100 pg of total RNA. Global transcriptional profiling generally employs one of two analytical platforms: DNA microarrays (33–35) or, of increasing popularity, Next Generation DNA sequencing ("RNA-seq"; see, for example, ref. (36)). For both platforms, microgram amounts of input target (targets being defined as the uncharacterized transcript-derived sample) are required. Target amplification techniques are, therefore, essential for studies integrating flow sorting with transcript analysis. Although amplification by a factor of 10^7 to 10^8 may appear daunting, commercial kits have been developed that are both robust and reproducible. In this chapter, we provide a description of methods that provide sufficient target for transcriptional profiling, down to the level of single sorted cells.

2. Materials

General Information and Precautions. Always wear a laboratory coat, disposable gloves, and a protective eyewear. Clean the working areas with 70% ethanol, before and after use. All chemicals are reagent grade, unless otherwise indicated. The following section lists specialty chemicals, kits, and equipment for the described applications and methods.

2.1. Preparation of Protoplasts

1. Polysaccharide hydrolases: Cellulysin (EMD Chemicals, Gibbstown, NJ), pectolyase (Karlan Research Products, Cottonwood, AZ), cellulase RS (Karlan Research Products), and macerozyme R10 (Karlan Research Products).

2. Solution A : For 200 mL of solution, mix 2 mL of 1 M KCl, 400 µL of 1 M $MgCl_2$, 400 µL of 1 M $CaCl_2$, 0.2 g of BSA, 78 mg of 2-(*N*-morpholino) ethanesulfonic acid hydrate (MES), and 21.86 g of mannitol (giving the following concentrations: 10 mM KCl, 2 mM $MgCl_2$, 2 mM $CaCl_2$, 1 mg/mL BSA, 2 mM MES, and 0.6 M mannitol). Adjust pH to 5.5 with 1 M Tris base; typically, less than 50 µL is required.

3. Solution BL (for leaves; 50 mL): 50 mL solution A, 750 mg cellulysin (1.5% w/v), and 50 mg pectolyase (0.1% w/v). Prepare immediately prior to tissue harvest.

4. Solution BR (for roots; 50 mL): 50 mL solution A, 750 mg cellulase RS (1.5% w/v), and 100 mg macerozyme R10 (0.2% w/v). Prepare immediately prior to tissue harvest.

5. 20.5% (w/v) sucrose dissolved in solution A.

6. Supplies: BD Falcon™ 40-, 70-, and 100-µm pore size cell strainers; disposable plastic transfer pipettes; No. 22 scalpel; and single-edged razor blades honed to a thickness of ≤0.2 mm (#55411-050; VWR, West Chester, PA).

2.2. Isolation of Nuclei

1. Chopping buffer: 45 mM $MgCl_2$, 30 mM sodium citrate, 20 mM MES, pH 7.0 adjusted with NaOH. Filter sterilize and store as 50 mL aliquots at –20°C. Once thawed, the buffer may be kept at 4°C for up to 1 week.

2. Propidium iodide (PI) stock solution: 1 mg/mL prepared in deionized water (diH_2O). Store as 1 mL aliquots at –20°C until the day of use.

3. 4′,6-diamidino-2-phenylindole (DAPI) stock solution: 0.1 mg/mL prepared in diH$_2$O. Store as 1 mL aliquots at –20°C until the day of use.

4. 10 mg/mL solution of DNAse-free RNAse A.

5. 10 μm Flow-Check™ Fluorospheres (Beckman Coulter, Brea, CA).

6. Supplies: single-edged razor blades (VWR) and CellTrics® disposable filters, 30-μm porosity (Partec, Münster, Germany).

2.3. Extraction and Amplification of RNA from Protoplasts and Nuolci

1. RNAqueous®-Micro Kit (Applied Biosystems/Ambion, Austin, TX).

2. GeneAmp® 10× PCR Buffer II, supplied with 25 mM MgCl$_2$ solution (Applied Biosystems).

3. Oligonucleotides (180 nmol) (Eurofins MWG Operon, Huntsville, AL):

   (a) V1 primer: 5′-NH$_2$-ATATGGATCCGGCGCGCCGTC GACTTTTTTTTTTTTTTTTTTTTTTTTTTTT-3′.

   (b) V3 primer: 5′-NH$_2$-TATCTCGAGGGCGCGCCGGAT CCTTTTTTTTTTTTTTTTTTTTTTTTTTTT-3′.

4. RNase inhibitors: Prime RNase Inhibitor™, 30 U/μL (Eppendorf AG, Hamburg, Germany) and RNAguard RNase Inhibitor, 30 U/μL (GE Healthcare, Piscataway, NJ).

5. SuperScript® III reverse transcriptase, 200 U/μL (Invitrogen, Carlsbad, CA).

6. T4 Gene 32 Protein, 5–6 mg/mL (Roche Applied Science, Indianapolis, IN).

7. Exonuclease I, 5 U/μL, supplied with 10× Exonuclease I buffer (Takara Bio Inc., Shiga, Japan).

8. Terminal deoxynucleotidyl transferase (TdT), recombinant, 15 U/μL (Invitrogen).

9. RNase H (Invitrogen).

10. TaKaRa Ex Taq™ Hot Start Version, 5 U/μL, supplied with 10× Ex Taq buffer and dNTP mix (2.5 mM each of dATP, dCTP, dGTP, and dTTP) (Takara Bio Inc.).

11. 100 mM dATP (GE Healthcare).

12. QIAquick PCR purification kit (Qiagen).

13. QIAquick gel extraction kit (Qiagen).

14. RNAse-free microcentrifuge tubes, 0.6 and 1.5 mL.

15. NanoDrop spectrophotometer (Thermo Scientific, Waltham, MA).

2.4. Microarray Hybridization

1. Amino Allyl MessageAmp™ II aRNA Amplification Kit (Applied Biosystems/Ambion).

2. Cy3 Mono-Reactive Dye Pack (Amersham Pharmacia/GE Healthcare).

3. Cy5 Mono-Reactive Dye Pack (Amersham Pharmacia/GE Healthcare).

4. RNeasy MinElute Cleanup Kit (Qiagen).

5. Western Blocking Reagent (Roche Applied Science).

6. ArrayIt® Hybridization Cassette, Extra Deep (Telechem International, Sunnyvale, CA).

7. LifterSlips (Erie Scientific Company, Portsmouth, NH).

8. Supplies: RNase-free tips, 50-mL plastic conical centrifuge tubes, wash glasses, microscope slide holders, and slide carrier.

3. Methods

3.1. Preparation of Protoplasts (see Note 1)

The following methods have been optimized for leaves and roots of *A. thaliana*, and are adapted from refs. (8) and (11). Seeds are sterilized and grown on vertical MS⁺ sucrose plates, as previously described (24).

1. Harvest roots or aerial parts of the plants by separating these tissues using a No. 22 scalpel, and then scraping them off the surface of the agar. Collect tissues on a Kimwipe moistened with diH_2O and weigh them. Transfer to a 60×15 cm diameter plastic petri dish and add enzyme solution (solution BL for leaves and solution BR for roots); proportions of 200 mg of tissue/5 mL enzyme digestion medium are recommended.

2. Segment the tissues in the enzyme solution using a single-edged razor blade, with the goal of producing tissue pieces that are $\sim 1 \times 1$ mm (leaves) and 0.5×0.5 mm (roots). Transfer to an orbital shaker operating at 100 rpm and continue incubation for 60 min at room temperature.

3. After incubation, gently pipette the tissue fragments up and down ten times, using a disposable plastic transfer pipette with the end of the tip cut to produce a wider bore (~ 4 mm). Then filter the protoplast suspension through a 100-μm mesh cell strainer into a sterile polypropylene centrifuge tube (either 15 or 50 mL, depending on the volumes employed in step 1).

4. Pellet protoplasts by centrifugation at $100 \times g$ for 3 min. All further manipulations are done on ice.

5. Remove the supernatant using a pipette and gently resuspend the protoplasts in 20.5% (w/v) sucrose dissolved in solution

A, to a concentration of $3-10 \times 10^5$ protoplasts/mL. Transfer to a 10- or 50-mL tube as appropriate for the numbers of protoplasts. The lower (20.5% sucrose) phase should not occupy more than 50% of the volume of the tube. Gently overlay this phase with solution A, to fill the tube to about 90% of its total volume.

6. Centrifuge for 10 min at $500 \times g$.

7. Collect protoplasts from the gradient interface using a transfer pipette and dilute with solution A prior to sorting. If necessary, the protoplasts can be concentrated by centrifugation at $100 \times g$ for 3 min. Measure protoplast diameters by light microscopy using a hemocytometer.

3.2. Preparation of Plant Homogenates for Analysis of Nuclei

1. Excise plant materials (organs or tissues) and, if necessary, wash using diH_2O. Transfer to a plastic petri dish (60×15 mm). Perform the remaining procedures on ice, and preferably in a walk-in cold room. It is convenient to place the petri dish on a prechilled ceramic tile embedded in an ice filled tray.

2. Add chopping buffer (2 mL per 0.5 g of fresh weight tissue represents convenient proportions for this size of petri dish). Chop the tissues using a new razor blade for 2–3 min.

3. Filter the homogenate through a 30-μm CellTrics® disposable filter to remove tissue debris.

4. Take an aliquot (0.5 mL) of the homogenate. If to be stained using PI, add this aliquot to a labeled tube containing 2.5 μL of a 10 mg/mL solution of DNAse-free RNAse A. Incubate on ice for 10 min. Add PI to a final concentration of 50 μg/mL. If to be stained with DAPI, add the aliquot to a tube containing sufficient DAPI to give a final concentration of 20 μg/mL.

5. Incubate the stained samples on ice in darkness for 20 min prior to flow cytometric analysis.

3.3. Flow Analysis and Sorting of Protoplasts Expressing GFP

1. Switch on the flow sorter and establish conditions for sorting using 10-μm Flow-Check™ Fluorospheres and settings appropriate for the specific instrument. Based on the size estimations obtained by hemocytometry under the light microscope, select a flow tip of appropriate diameter. Particles as large as 50% of the diameter of the flow tip can be successfully sorted and at high recovery rates (31). In the interests of establishing stable sorting conditions with minimal disturbance to protoplast integrity and recovery, it is often recommended that this diameter be at least threefold larger than the mean diameter of the protoplast population. This conflicts with issues concerning the amounts of sample generated during sorting, since droplet volume evidently increases rapidly

with only slight increases in tip diameter. For the Cytomation Mo-Flo, and for *Arabidopsis* protoplasts prepared as described above, we employ a 70-μm flow tip and a sheath pressure of 40 PSI. This gives stable sorting, with a droplet delay of 20 at a piezo drive frequency of 60 kHz. These values should be considered as target ranges, since different instruments of the same type exhibit variation in responses.

2. For detection of fluorescence arising from chlorophyll autofluorescence and from GFP, employ 200 mW laser illumination at 488 nm.

3. Trigger events on side scatter and visualize the protoplasts using bivariate analysis of green fluorescence (the PMT being screened by a 530/40 bandpass barrier filter, a filter centered at 530 nm with a 40-nm half-maximal transmittance bandwith) versus red fluorescence (the PMT screened by a 630/40 barrier filter), with beam splitting at 555 nm. While most flow cytometers are routinely configured to trigger using the forward-angle light scatter (FALS) signal, we find that, for plant samples, triggering based on 90° side scatter is a better option, since FALS detection generally appears to be more noisy. This observation may be specific to the types of detector employed for FALS acquisition. One must optimize the thresholding to allow the particles of interest to be visualized while excluding as much debris as possible. Draw an amorphous sort window to include the protoplast population. Perform sorting in "Enrichment Mode," with one to two droplets being sorted; this means that one droplet is always sorted, but if the desired event is in the leading or trailing half of the droplet, one more (preceding or trailing) droplet is also sorted. Sorting in Enrichment Mode provides the best combination of sample purity and sample recovery; the inevitable presence of subcellular debris, even for protoplast samples that are gradient purified, obviates sorting in single-droplet "Purity Mode" due to the triggering of too many sort aborts. Define the position of the sort region by first analyzing the negative control, and setting a lower boundary for a positive GFP signal. Next, adjust the left and right boundaries of the sort region to exclude non-GFP-positive protoplasts; determine the effects of sort window placement by sorting a few protoplasts onto a slide and examining them under a fluorescence microscope.

4. Validate instrument sort parameters by sorting 100 protoplasts onto a microscope slide, and examining and counting these under a light microscope.

5. Sort protoplasts into 0.6- or 1.5-mL centrifuge tubes. For isolation of RNA from populations of protoplasts, we sort protoplasts into tubes prefilled with 3.5 vol. of lysis buffer

taken from the RNAqueous®-Micro Kit, assuming one final volume of sorted protoplasts. For sorting single protoplasts, prefill the tubes with 5 µL of the lysis buffer described in Subheading 3.5.2, step 1. Shake the tubes immediately to ensure that the sorted protoplast contacts the buffer.

6. Place the tubes on ice.

3.4. Sorting of Nuclei Containing Targeted GFP

1. Perform the same procedure as in Subheading 3.3, step 1.

2. For combined detection of fluorescence arising from GFP fluorescence and from DAPI-stained nuclear DNA (see Subheading 3.2, step 4), employ laser illumination at 365 nm (40 mW) and 488 nm (200 mW).

3. Trigger events on side scatter and visualize the nuclei using bivariate analysis of blue–violet fluorescence (PMT screened by a 450/65 barrier filter) versus green fluorescence (PMT screened by a 530/40 bandpass barrier filter). Position a rectangular sort window to include the desired nuclear population. Perform sorting in enrichment mode, with one or two droplets being sorted per positive event.

3.5. Amplification of Targets from Sorted Protoplasts and Nuclei

The key to success relies on rapid denaturation of the RNA samples immediately following sorting. RNA is labile, and contamination by ubiquitous RNases is easy to achieve. Gloves should always be worn, all aqueous solutions should be made using DEPC-treated diH$_2$O, and glassware should be autoclaved.

Two scenarios are described: the first involves sorting of "significant" numbers of protoplasts and nuclei, operationally defined as being 50,000–100,000; we typically recover around 100 ng of total RNA from 100,000 *Arabidopsis* nuclei. In this situation, macroscopic amounts of sorted liquids will accumulate as a function of sorting, so the issue is to dilute this with chaotropic RNA-stabilizing buffers sufficiently to ensure that transcript degradation does not occur, which would otherwise introduce variability in consequent measurements of transcript abundance. The second scenario involves sorting of individual protoplasts. In this case, manipulations of single sorted objects represent the significant issue but, ultimately, success in detection and characterization of amplified transcripts, which experimentally turns out to be a stochastic process, provides valuable insight into the issue of noise in gene expression (37).

3.5.1. Preparation of RNA Targets from Populations of Sorted Protoplasts and Nuclei

Prepare RNA using the RNAqueous®-Micro Kit, which can conveniently accommodate an input sample volume of up to 0.1 mL. This procedure is executed exactly as described in the manufacturer's manual.

3.5.2. Preparation of RNA Targets from Single Sorted Protoplasts

This protocol merges information taken from the Applied Biosystems Technical Application Note (38) and from Kurimoto *et al.* (39). The V1 and V3 primers are used for the purpose of first and second strand synthesis. Step 1 describes total RNA isolation. Steps 2–15 involve cDNA synthesis. Steps 16–20 describe the first round of cDNA amplification using PCR. Steps 21–26 provide a second round of PCR amplification. Steps 27 and 28 describe production of aRNA via in vitro synthesis from the amplified cDNA.

1. Prepare lysis buffer sufficient for 20 samples:

	μL
GeneAmp® 10× PCR buffer II	9
25 mM MgCl$_2$	5.4
5% NP40	9
0.1 M DTT	4.5
V1 primer (10 ng/μL)	1.8
2.5 mM dNTP mix	1.8
Prime RNase inhibitor (30 U/μL)	0.8
RNAguard RNase inhibitor (30 U/μL)	0.9
H$_2$O	56.8
Total volume	90

2. Add 5 μL of fresh lysis buffer to a 0.6-mL microfuge tube. Sort one protoplast into the tube, flick to mix, and centrifuge for 30 s.

3. Incubate at 70°C for 90 s. Centrifuge for 30 s and place on ice.

4. Prepare Reverse Transcription mix (sufficient for 12 samples):

	μL
Superscript® III (200 U/μL)	4
Prime RNase inhibitor (30 U/μL)	0.6
T4 Gene 32 protein (5 μg/μL)	1.4

5. Add 0.4 μL of this RT mix to each tube from step 3, mix by flicking tube, and centrifuge for 30 s.

6. Incubate at 50°C for 30 min, then inactivate the reaction by incubating for 15 min at 70°C. Centrifuge for 30 s and put on ice.

7. Prepare Exonuclease Mix (sufficient for 12 samples):

	μL
10× Exonuclease I buffer	1.2
H₂O	9.6
Exonuclease I (5 U/μL)	1.2

8. Add 1 μL of Exonuclease Mix to each tube, mix by flicking, and centrifuge for 30 s.

9. Incubate at 37°C for 30 min. Inactivate by incubation at 80°C for 25 min. Centrifuge for 30 s and put on ice.

10. Prepare poly(dA) addition mix (sufficient for 12 samples):

	μL
10× GeneAmp® PCR buffer II	7.2
25 mM MgCl₂	4.3
100 mM dATP	2.1
H₂O	52
TdT (15 U/μL)	3.6
RNase H (10 U/μL)	3.6
Total	72.8

11. Add 6 μL of poly(dA) addition mix to each tube, mix by flicking, and centrifuge in a microfuge for 30 s.

12. Incubate at 37°C for 15 min. Inactivate by incubating for 10 min at 70°C. Centrifuge in a microfuge for 30 s and put on ice.

13. Prepare Second Strand Synthesis Mixture (sufficient for ten samples):

	μL
10× Ex Taq™ buffer	76
dNTPs (2.5 mM)	76
V3 primer	15.2
H₂O	586
Ex Taq™ Polymerase	7.6

14. Add 76 μL to each tube, mix, and spin.

15. Perform one PCR cycle:

    (a) 95°C for 3 min

    (b) 50°C for 2 min

    (c) 72°C for 20 min

16. Centrifuge for 30 s and put on ice.

17. The next step involves PCR for 24 cycles for initial cDNA amplification. Prepare PCR amplification mix (sufficient for ten samples):

	µL
10× Ex Taq™ buffer	76
dNTPs (2.5 mM)	76
V1 primer	15.2
H$_2$O	586
Ex Taq™ polymerase	7.6

18. Add 76 µL to each tube, mix, and spin. Divide total volumes between two tubes, providing two tubes for each reaction (each ~80 µL).

19. Perform PCR for 24 cycles:

    (a) 95°C for 3.5 min.

    (b) 67°C for 1 min.

    (c) 72°C for 3 min (with +6 s added per cycle, for another 23 cycles, to give a total of 24 cycles).

    (d) 72°C for 10 min.

20. Purify the cDNA using the QIAquick PCR Purification Kit according to the manufacturer's instructions. Use 30 µL of elution buffer for elution, as described by the manufacturer.

21. Employ the entire sample for electrophoresis using 2% agarose gel (in 0.5× TAE buffer; the gel occupies an area of ~5 × 10 cm and the volume is 30 mL; this accommodates at least ten lanes). Excise the band at around 500–2,000 bp as defined, using a 100-bp ladder; this may only be present in a minority of the lanes.

22. Purify the cDNA from gel using the QIAquick Gel Extraction Kit, according to the manufacturer's instructions. Use 30 µL of water for elution, as described by the manufacturer. Measure cDNA concentrations in the sample using a NanoDrop spectrophotometer. Typical concentrations are around 2–6 ng/µL, with total yields being around 50 ng/ sample.

23. cDNA reamplification. This next step provides sufficient amounts of cDNA for further use in target preparation for microarrays or for NextGen sequencing. Mix for one sample:

	μL
10× Ex Taq™ buffer	5
dNTPs (2.5 mM)	5
V3 primer	1
V1 primer	1
H₂O + cDNA	37.5 (usually this involves 5 μL of cDNA, with the remainder being water)
Ex Taq™ polymerase	0.5
Total	50

24. Perform PCR for 20 cycles:
    (a) 95°C for 5.5 min.
    (b) 64°C for 1 min.
    (c) 72°C for 5 min 18 s.
    (d) 95°C for 30 s.
    (e) 67°C for 1 min.
    (f) 72°C for 5 min 24 s (+6 s for each subsequent cycle, for another six cycles).
    (g) 95°C for 30 s.
    (h) 67°C for 1 min.
    (i) 72°C 6 min.

    This cycle is repeated ten times, the total number of cycles being 18.

25. Purify the cDNA with the QIAquick PCR Purification Kit, according to the manufacturer's instructions. Use 30 μL of elution buffer for elution.

26. Employ the entire product for 2% agarose gel electrophoresis in 0.5× TAE buffer. Excise the bands that appear around 500–2,000 bp. Purify the cDNA samples using the QIAquick Gel Extraction Kit according to the manufacturer's instructions. Employ 30 μL of water for elution. Measure cDNA concentrations using a NanoDrop. Typical concentrations should be ~100 ng/μL, with totals being ~2–3 μg per sample. These are sufficient for NextGen sequencing on any of the commercially available platforms.

27. In vitro transcription of cDNA to aRNA. This step is used to provide targets for microarray hybridization. Prepare Transcription Mix; all chemicals are part of the Amino Allyl MessageAmp™ II aRNA Amplification Kit (amounts are for one sample):

	µL
cDNA	16
ATP, CTP, GTP	12
UTP	2
aaUTP	2
10× T7 buffer	4
T7 Enzyme mix	4
Total volume	40

28. Clean up aRNA according to the manufacturer's instructions.

3.6. Target Labeling and Microarray Hybridization

Coupling to Cy-3 and Cy-5, and microarray hybridization are done using protocols developed in the Galbraith laboratory.

3.6.1. Preparation of Cy3 and Cy5 Monoreactive Dye

These dyes are supplied as five aliquots; the content of each tube is sufficient for at least four labeling reactions. Dissolve the entire contents of a single tube in 22 µL of DMSO by flicking the tube several times, and leaving at room temperature for at least 30 min protected from light. Centrifuge at $1,000 \times g$ for 30 s to collect the dye at the bottom of the tube. The dye is ready for immediate use, but can be stored at –20°C for up to 1 month. Always protect the dye from light by wrapping the tubes with aluminum foil.

3.6.2. Target Labeling

1. Dissolve the dried aRNA with 5 µL of 0.2 M $NaHCO_3$ buffer by flicking the tube several times and leaving the tube at room temperature for at least 20 min.

2. Add 5 µL of the Cy3 or Cy5 solution to each tube, and mix by flicking the tube several times.

3. Spin the tubes at $1,000 \times g$ for 30 s, wrap in foil, and incubate at room temperature for 2 h.

4. Unincorporated dye is removed using an RNeasy MinElute Cleanup Kit, according the manufacturer's instructions.

3.6.3. Microarray Hybridization

The following procedure is that recommended for long oligonucleotide microarrays printed on aminosilane-coated microarray slides, and produced by our laboratory (see http://www.cals.arizona.edu/galbraith for more details on these microarrays).

1. DNA probe immobilization can be done at any time prior to hybridization. Re-hydrate the slide over a 50°C water bath for 10 s.

   (a) Hold the slide with the label side down over the water vapor.

   (b) Watch spots carefully so that they do not over-hydrate and start to merge.

2. Dry the slide by placing on a 65°C heating block for 5 s. Remove and allow to cool for 1 min.

3. Repeat steps 1 and 2 four times. (The rehydration step is important to obtain uniform spots lacking a "doughnut" effect; however, if you feel uncomfortable with the rehydration step, proceed directly to UV cross-linking).

4. Cross-link the DNA to the slide surfaces by exposing the microarrays, in batches, array-side up, to 180 mJ UV irradiation using a commercial cross-linker (we employ a Stratalinker; Stratagene, La Jolla, CA).

5. Wash the slide in 1% SDS (prepared in sterile diH_2O) for 5 min at RT on a shaker, or agitate by hand. It is convenient to employ a slide carrier for these steps.

6. Remove SDS by dipping the slides ten times into sterile diH_2O.

7. Immediately transfer the slides to 100% ethanol, dip five times, and then incubate for 3 min with shaking.

8. Spin dry slides by centrifugation. Centrifuge at no more than $200 \times g$ for 2–4 min (see Note 2).

   (a) Pack the bottom of a 50-mL plastic conical centrifuge tube with Kimwipes, occupying a packed volume of about 5 mL.

   (b) Using forceps, carefully place the slide into a tube.

   (c) Centrifuge at no more than $200 \times g$ for 2–4 min.

   (d) Repeat centrifugation if any liquid remains on the microarray surface.

9. Repeat ethanol wash if any streaks are observed on the microarray surface after step 8.

10. Store the slide in a lint-free, light-proof box at room temperature but at low humidity (use of a desiccator is recommended).

11. Prepare the following hybridization mix in a microfuge tube (see Note 3):

20× SSC	6.0 μL
Western blocking reagent	3.6 μL
2% SDS	2.4 μL
Labeled targets (volumes as from the clean-up step)	
H_2O	to 60 μL

12. Denature labeled target by incubating the tube at 65°C for 5 min.

13. Transfer the tube to ice immediately, or apply target onto the slides directly.

14. Rinse ArrayIt™ Hybridization Cassette with distilled water and dry thoroughly.

15. Make sure the flexible rubber gasket is seated evenly in the gasket channel.

16. Add 15 μL of water to the lower groove within the cassette chamber.

17. Insert the microarray (1″ × 3″ or 25 × 75 mm slide) into the cassette chamber, DNA side up.

18. Place the LifterSlip over the microarray slide (make sure the white stripe of the LifterSlip is at the lower side).

19. Apply the denatured target sample slowly to the one end of the LifterSlip and let it spread across the microarray by capillary action.

20. Quickly place the clear plastic cassette lid on top of the cassette chamber.

21. Apply downward pressure and manually tighten the four sealing screws.

22. Check all the screws to confirm a tight seal.

23. Place the cassette into a hybridization oven pre-equilibrated at 55°C (see Note 4).

24. Allow the hybridization reaction to proceed for 8–12 h.

25. After hybridization, remove the cassette, loosen the four sealing screws, and remove lid.

26. Remove the microarray slide from the cassette chamber using forceps and place the slides into the washing buffer. This is most conveniently done using a slide holder.

27. Wash the microarray slides in the following solutions for 5 min each:
    Washing is done by immersing the slides in a glass slide-staining jar containing the appropriate volume of wash buffer, followed by placing it on an orbital shaker (e.g., Belly Dancer; Stovall Life Science, Greensboro, NC) at 60 rpm. Pre-heat the first wash solution and make sure the slides are completely immersed in wash buffer.

    (a) 2× SSC, 0.5% SDS at 55°C.

    (b) 0.5× SSC at room temperature.

    (c) 0.05× SSC at room temperature.

28. After completion of the washes, spin dry the slide as described in step 8.

29. Scan slide immediately, or store in a light-tight box at room temperature under dry conditions. Immediate scanning is recommended. However, we have observed that properly stored slides (light protected, dry, RT) retain fluorescent signals for up to a month. Some reports indicate that environmental pollutants (ozone) can drastically affect fluorescence, particularly that of Cy5 (40).

30. Examine the scanned images immediately to determine the number of elements that are near zero or are saturated (for a 16-bit scanner, this represents a value of 65,400). The proportion of the elements at these extremes should be acceptably low, since information is lost in either case. It is preferable to rescan with altered gain settings on the scanner than to proceed with the analysis of images containing large proportions of zero or saturated elements. We have found that although the absolute value of the intensity values may be reduced by scanning a second or third time, the relative fluorescence distribution is preserved, so information is not lost. Scanning a second time at higher PMT/laser values can also be done to move low intensity elements higher within the dynamic range. The two intensity distributions (at low and high PMT/laser values) can then be merged to provide an increased combined dynamic range (41).

31. Save the image as a TIFF file, and implement appropriate data extraction and statistical analyses.

3.7. Target Analysis Using NextGen Sequencing

The amounts of target produced in Subheading 3.5 are sufficient for NextGen sequencing using Roche-454, Illumina-Solexa, or Life Technologies SOLiD platforms. Complete information including sequencing protocols can be found in refs. (42–44).

4. Notes

1. Protoplasts can be prepared from species other than those mentioned in this chapter, following similar procedures and principles, but the methods may require empirical optimization, whether or not described formally in the published literature. Issues include: (a) *Source of tissues.* Plants can be grown in pots as exposed plants in the greenhouse or growth chamber, or can be grown under axenic conditions on media in tissue culture vessels. If the former, the tissues will require surface sterilization prior to protoplast preparation. Tissue excision can induce systemic

wound responses, which may result in recalcitrance of the source materials to protoplast production. The change in state from vegetative to reproductive growth can also render leaves recalcitrant to digestion. (b) *Concentration and composition of the osmoticum.* Different species and tissues exhibit different cellular isotonicities, and are differentially sensitive to ionic composition. In general, low levels of Ca^{2+} (5 mM) are beneficial. Although most researchers employ polyhydric alcohols (sorbitol and mannitol) as osmotica, ionic osmotica can be compatible with protoplast viability, particularly those employing K^+ (but not Na^+) as the major cationic species. (b) *Composition and amounts of cell wall-degrading enzymes and times and temperatures of incubation that are employed for protoplast production.* This is largely a matter of trial and error, with the general aim being to employ the lowest possible levels of these enzymes, in part acknowledging their potential to act as agents signaling pathogenic attack. In general, lower levels of enzymes require longer periods of incubation. Tissue digestion and protoplast release are typically done at room temperature, although occasionally higher temperatures are recommended. In these cases, caution is warranted to avoid heat-shock responses. Vacuum infiltration is often employed, and this has no obvious deleterious effects on the tissues or protoplasts. (c) *Whether or not protoplast purification is employed.* Isopycnic step-gradient flotation is recommended to eliminate subcellular debris and broken and non-viable protoplasts, and protoplast viability can be conveniently monitored during this process using fluorochromatic dyes such as fluorescein diacetate (25).

2. If you have centrifuge adapted for microplate centrifugation, an mBox (cat. no. BX-IM-20ERIE; Erie Scientific Company) is recommended for the slide drying step, since it increases throughput and provides very uniform drying.

3. We recommend using 48 pmol of Cy5-labeled targets and 48 pmol of Cy3-labeled targets in a 60-μL hybridization volume, but this amount should be optimized empirically. We have used up to 100 pmol target with *Arabidopsis.* Increasing above this point risks Cy-dye precipitation, which leads to very high backgrounds.

4. Our combination of components in the hybridization solution (SSC buffer, SDS, and Western Blocking reagent) provides low background, high sensitivity, and highly reproducible results. Commercially available hybridization buffers containing 50% formamide can also be used, in which case hybridization should be done at 42°C instead of 55°C.

Acknowledgments

Part of the development of the methods described in this chapter was done with support from the NSF Plant Genome program.

References

1. Bell, P. R. and Helmsley, A. R. (2000) *Green Plants: Their Origin and Diversity.* Cambridge University Press, Cambridge, p. 361.

2. Harkins, K. R. and Galbraith, D. W. (1984) Flow sorting and culture of plant protoplasts. *Physiol Plant* **60**, 43–52.

3. Galbraith, D. W. (1990) Isolation and flow cytometric characterization of plant protoplasts. *Methods Cell Biol* **33**, 527–547.

4. Galbraith, D. W., Bartos, J., and Dolezel, J. (2005) Flow cytometry and cell sorting in plant biotechnology. In *Flow Cytometry in Biotechnology* (Sklar, L.A., ed.), Oxford University Press, New York, pp. 291–322.

5. Galbraith, D. W., Grebenok, R. J., Lambert, G. M., and Sheen, J. (1995) Flow cytometric analysis of transgene expression in higher plants: green fluorescent protein. *Methods Cell Biol* **50**, 3–12.

6. Sheen, J., Hwang, S., Niwa, Y., Kobayashi, H., and Galbraith, D. W. (1995) Green fluorescent protein as a new vital marker in plant cells. *Plant J* **8**, 777–784.

7. Galbraith, D. W., Herzenberg, L. A., and Anderson, M. (1999) Flow cytometric analysis of transgene expression in higher plants: green fluorescent protein. *Methods Enzymol* **320**, 296–315.

8. Birnbaum, K., Shasha, D. E., Wang, J. Y., Jung, J. W., Lambert, G. M., Galbraith, D. W., and Benfey, P. N. (2003) A gene expression map of the *Arabidopsis* root. *Science* **302**, 1956–1960.

9. Birnbaum, K., Jung, J. W., Wang, J. Y., Lambert, G. M., Hirst, J. A., Galbraith, D. W., and Benfey, P. N. (2005) Cell-type specific expression profiling in plants using fluorescent reporter lines, protoplasting, and cell sorting. *Nat Methods* **2**, 1–5.

10. Yadav, R. K., Girke, T., Pasala, S., Xie, M. T., and Reddy, V. (2009) Gene expression map of the *Arabidopsis* shoot apical meristem stem cell niche. *Proc Natl Acad Sci U S A* **106**, 4941–4946.

11. Sheen, J. (2002) A transient expression assay using *Arabidopsis* mesophyll protoplasts. http://genetics.mgh.harvard.edu/sheenweb/

12. Petersson, S. V., Johansson, A. I., Kowalczyk, M., Makoveychuk, A., Wang, J. Y., Moritz, T., Grebe, M., Benfey, P. N., Sandberg, G., and Ljung, K (2009) An auxin gradient and maximum in the *Arabidopsis* root apex shown by high-resolution cell-specific analysis of IAA distribution and synthesis. *Plant Cell* **21**, 1659–1668.

13. Galbraith, D. W., Harkins, K. R., and Jefferson, R. A. (1988) Flow cytometric characterization of the chlorophyll contents and size distributions of plant protoplasts. *Cytometry* **9**, 75–83.

14. Chalfie, M., Tu, Y., Euskirchen, G., Ward, W. W., and Prasher, D. C. (1994) Green fluorescent protein as a marker for gene expression. *Science* **263**, 802–805.

15. Snapp, E. L. (2009) Fluorescent proteins: a cell biologist's user guide. *Trends Cell Biol* **19**, 649–655.

16. Berg, R. H. and Beachy, R. N. (2008) Fluorescent protein applications in plants. *Methods Cell Biol* **85**, 153–177.

17. Galbraith, D. W. (2004) The rainbow of fluorescent proteins. *Methods Cell Biol* **75**, 153–169.

18. Nelson, B. K., Cai, X., and Nebenfuehr, A. (2007) A multicolored set of in vivo organelle markers for co-localization studies in *Arabidopsis* and other plants. *Plant J* **51**, 1126–1136.

19. Millar, A. H., Carrie, C., Pogson, B., and Whelan, J. (2009) Exploring the function-location nexus: using multiple lines of evidence in defining the subcellular location of plant proteins. *Plant Cell* **21**, 1625–1631.

20. Grebenok, R. J., Pierson, E. A., Lambert, G. M., Gong, F. -C., Afonso, C. L., Haldeman-Cahill, R., Carrington, J. C., and Galbraith, D. W. (1997) Green-fluorescent protein fusions for efficient characterization of nuclear localization signals. *Plant J* **11**, 573–586.

21. Grebenok, R. J., Lambert, G. M., and Galbraith, D. W. (1997) Characterization of the targeted nuclear accumulation of GFP within the cells of transgenic plants. *Plant J* **12**, 685–696.

22. Chytilova, E., Macas, J., Sliwinska, E., Rafelski, S., Lambert, G. M., and Galbraith, D. W. (2000) Nuclear dynamics in *Arabidopsis thaliana*. *Mol Biol Cell* **11**, 2733–2741.

23. Zhang, C. Q., Gong, F. C., Lambert, G. M., and Galbraith, D. W. (2005) Cell type-specific characterization of nuclear DNA contents within complex tissues and organs. *Plant Methods* **1**, 7, doi:10.1186/1746-4811-1-7.

24. Zhang, C. Q., Barthelson, R. A., Lambert, G. M., and Galbraith, D. W. (2008) Characterization of cell-specific gene expression through fluorescence-activated sorting of nuclei. *Plant Physiol* **147**, 30–40.

25. Harkins, K. R., Jefferson, R. A., Kavanagh, T. A., Bevan, M. W., and Galbraith, D. W. (1990) Expression of photosynthesis-related gene fusions is restricted by cell-type in transgenic plants and in transfected protoplasts. *Proc Natl Acad Sci USA* **87**, 816–820.

26. Nawy, T., Lee, J. -Y., Colinas, J., Wang, J. Y., Thongrod, S. C., Malamy, J. E., Birnbaum, K., and Benfey, P. N. (2005) Transcriptional profile of the *Arabidopsis* root quiescent center. *Plant Cell* **17**, 1908–1925.

27. Brady, S. M., Orlando, D. A., Lee, J. Y., Wang, J. Y., Koch, J., Dinneny, J. R., Mace, D., Ohler, U., and Benfey, P. N. (2007) A high-resolution root spatiotemporal map reveals dominant expression patterns. *Science* **318**, 801–806.

28. Dinneny, J. R., Long, T. A., Wang, J. Y., Jung, J. W., Mace, D., Pointer, S., Barron, C., Brady, S. M., Schiefelbein, J., and Benfey, P. N. (2008) Cell identity mediates the response of *Arabidopsis* roots to abiotic stress. *Science* **320**, 942–945.

29. Bargmann, B. O. R. and Birnbaum, K. D. (2009) Positive fluorescent selection permits precise, rapid, and in-depth overexpression analysis in plant protoplasts. *Plant Physiol* **149**, 1231–1239.

30. Galbraith, D. W. and Lucretti, S. (2000) Large particle sorting. In *Flow Cytometry and Cell Sorting*, 2nd edition (Radbruch, A., ed.), Springer-Verlag, Berlin, pp. 293–317.

31. Harkins, K. R. and Galbraith, D. W. (1987) Factors governing the flow cytometric analysis and sorting of large biological particles. *Cytometry* **8**, 60–71.

32. Galbraith, D. W., Harkins, K. R., Maddox, J. R., Ayres, N. M., Sharma, D. P., and Firoozabady, E. (1983) Rapid flow cytometric analysis of the cell cycle in intact plant tissues. *Science* **220**, 1049–1051.

33. Schena, M., Shalon, D., Davis, R. W., and Brown, P. O. (1995) Quantitative monitoring of gene expression patterns with a complementary DNA microarray. *Science* **270**, 467–470.

34. Deyholos, M. K. and Galbraith, D. W. (2001) High-density DNA microarrays for gene expression analysis. *Cytometry* **43**, 229–238.

35. Galbraith, D. W. (2006) Microarray analyses in higher plants. *OMICS* **10**, 455–473.

36. Wilhelm, B. T. and Landry, J. R. (2009) RNA-Seq-quantitative measurement of expression through massively parallel RNA-sequencing. *Methods* **48**, 249–257.

37. Larson, D. R., Singer, R. H., and Zenklusen, D. (2009) A single molecule view of gene expression. *Trends Cell Biol* **19**, 630–637.

38. Applied Biosystems Technical Application Note (2008) SOLiD™ System 2.0 Library preparation protocol for the whole transcriptome analysis of a single cell.

39. Kurimoto, K., Yabuta, Y., Ohinata, Y., and Saitou, M. (2007) Global single-cell cDNA amplification to provide a template for representative high-density oligonucleotide microarray analysis. *Nat Protoc* **2**, 739–752.

40. Fare, T. L., Coffey, E. M., Dai, H. Y., He, Y. D. D., Kessler, D. A., Kilian, K. A., Koch, J. E., LeProust, E., Marton, M. J., Meyer, M. R., Stoughton, R. B., Tokiwa, G. Y., and Wang, Y. Q. (2003) Effects of atmospheric ozone on microarray data quality. *Anal Chem* **75**, 4672–4675.

41. Skibbe, D. S., Wang, X. J., Zhao, X. F., Borsuk, L. A., Nettleton, D., and Schnable, P. S. (2006) Scanning microarrays at multiple intensities enhances discovery of differentially expressed genes. *Bioinformatics* **22**, 1863–1870.

42. http://www.454.com/.

43. http://www.illumina.com/.

44. http://www3.appliedbiosystems.com/AB_Home/applicationstechnologies/SOLiD-System-Sequencing-A/index.htm.

Chapter 21

Lentiviral Fluorescent Protein Expression Vectors for Biotinylation Proteomics

Irene Riz, Teresa S. Hawley, and Robert G. Hawley

Abstract

In vivo biotinylation tagging, based on a method in which a protein of interest is tagged with a peptide that is biotinylated in vivo by coexpression of *Escherichia coli* BirA biotin ligase, has been successfully used for the isolation of protein–protein and protein–DNA complexes in mammalian cells. We describe a modification of this methodology in which cells stably expressing the tagged gene of interest and the BirA gene can be selected by fluorescence-activated cell sorting (FACS). We recently implemented this approach to isolate and characterize proteins associated with TLX1, a homeodomain transcription factor with leukemogenic function. The modified technique utilizes two components: a lentiviral vector coexpressing the gene of interest containing a biotinylation tag on a bicistronic transcript together with a downstream yellow fluorescent protein (YFP) gene; and a second lentiviral vector encoding a fusion protein composed of bacterial BirA linked to the green fluorescent protein (GFP). This FACS-based binary in vivo biotinylation tagging system allows precise control over the levels of BirA-mediated biotinylation as well as the expression of the gene of interest, which is especially important if high-level expression negatively impacts cell growth or viability.

Key words: In vivo biotinylation tagging, *E. coli* BirA biotin ligase, Simian immunodeficiency virus-based lentiviral vectors, Fluorescent protein reporters, TLX1 homeodomain transcription factor

1. Introduction

A variety of biochemical methods exist for studying protein–protein interactions in mammalian cells (1, 2), including those based on fluorescence-activated cell sorting (FACS) (3) (see also Chapter 18). Strouboulis and colleagues developed a method in which a recombinant protein is tagged with a peptide that is biotinylated in vivo by the coexpressed *Escherichia coli* BirA biotin ligase (4). A strength of this approach is the very high affinity interaction between biotinylated substrates and streptavidin ($K_d = 10^{-15}$), which allows high stringencies to be employed

Teresa S. Hawley and Robert G. Hawley (eds.), *Flow Cytometry Protocols*, Methods in Molecular Biology, vol. 699, DOI 10.1007/978-1-61737-950-5_21, © Springer Science+Business Media, LLC 2011

during purification (5). The biotinylation tagging method described here has been successfully used for the single-step purification (6) of a number of transcription factor complexes in nuclear extracts of mammalian cells (4, 7–12).

In the initial version of the method (4), expression cassettes encoding the tagged protein of interest and BirA were subcloned into separate plasmids that also carried a selectable drug resistance gene, the neomycin phosphotransferase gene and the puromycin *N*-acetyltransferase gene, respectively. Typically, target cells would be transfected with the BirA plasmid by physical methods, such as electroporation, stable clones obtained by the selection of puromycin resistance, and then screened for BirA expression by Northern or Western blot analysis. An appropriate BirA-expressing clone would then be transfected with the vector encoding the tagged protein of interest and stable cells expressing the tagged protein biotinylated by BirA obtained by selection for cells that were resistant to the neomycin analog geneticin as well as to puromycin. A drawback of this experimental design is that the optimal concentration of drugs used for the selection depends on the cell line and needs to be determined a priori. Moreover, the selection for drug resistance of the transfected plasmids does not necessarily ensure the coexpression of BirA or the tagged gene of interest.

Hoang and colleagues reported a modification of this methodology in which the tagged gene of interest was coexpressed from a lentiviral vector on a bicistronic transcript that also contained the BirA gene (9). In addition, the BirA gene was expressed as a fusion protein with the green fluorescent protein (BirA-GFP), allowing stably transduced cells coexpressing the tagged gene of interest and BirA to be isolated by FACS. As described below, we have further modified the FACS-based strategy of Hoang and colleagues. The expression system we have developed consists of two components: a lentiviral vector coexpressing the gene of interest containing a biotinylation tag together with a downstream yellow fluorescent protein (YFP) gene via an encephalomyocarditis virus internal ribosome entry site (IRES); and a second lentiviral vector expressing GFP-BirA (12).

Our work is focused on *TLX1* (*T-cell leukemia homeobox 1*, previously known as *HOX11* or *TCL3*), an evolutionarily conserved member of the dispersed NKL (NK-Like or NK-Linked) subclass of homeobox genes (13, 14). The murine ortholog of human *TLX1* is essential for splenogenesis and required for the development of certain neurons (15, 16). Although *TLX1* is not expressed in the hematopoietic system, its inappropriate activation is a recurrent event in human T cell acute lymphoblastic leukemia (T-ALL) (17). The manner in which deregulated *TLX1* expression induces neoplastic conversion remains to be fully elucidated (18, 19). Several lines of evidence indicate that TLX1 functions as a transcriptional regulator that can either activate or repress gene expression via direct or indirect modes of action (20–28). In this regard, TLX1 has been reported to form protein–protein interactions with other transcription factors as well as with a number of transcriptional coregulators and chromatin-modifying enzymes. Among the molecules that have been identified are: CTF1, a ubiquitous transcription factor that associates with TFIIB and the basal transcription machinery (29); MEIS and PBX members of the TALE (three amino acid loop extension) superclass of homeodomain proteins (30, 31); the acetyltransferase coactivator CREB-binding protein (26); the serine/threonine phosphatases PP1 and PP2A (25, 32); and the eukaryotic initiation factor 4E (eIF4E) (33).

We recently implemented the in vivo biotinylation tagging approach to isolate and characterize the various TLX1 protein complexes in T-ALL cells (12). Our initial attempts to coexpress biotinylation-tagged TLX1 and GFP-BirA from a bicistronic lentiviral vector were unsuccessful because the cells did not tolerate high levels of TLX1. Since BirA was expressed on the same transcript, it was selected against. As a result, much higher expression levels of BirA were achieved with the empty GFP-BirA vector, making it difficult to obtain similar levels of endogenous biotinylated proteins for comparisons between the experimental and control samples. To circumvent these problems, we designed a two-component expression system using the simian immunodeficiency virus (SIV) lentiviral vector backbone pCL20cSLFR MSCV-GFP (34). In pCL20cSLFR MSCV-GFP, which was constructed from the nonpathogenic $SIV_{mac1A11}$ isolate, the GFP gene is expressed from an internal promoter derived from the long terminal repeat (LTR) of the murine stem cell virus (MSCV), which is highly active in most mammalian cell types (35). A biotinylation tagging vector expressing a COOH-terminal tagged TLX1 (TLX1[bio]) under the control of the MSCV LTR was created by inserting the TLX1 coding region in frame upstream of the coding sequences for the BirA target peptide linked to an IRES-YFP cassette, generating pCL20cSLFR MSCV-TLX1[bio]-IRES-YFP (component 1). A BirA expression vector was similarly constructed by replacing the GFP gene of pCL20cSLFR MSCV-GFP with GFP-BirA, generating pCL20cSLFR MSCV-GFP-BirA (component 2) (Fig. 1a).

In pilot studies, SupT1 cells, which are negative for *TLX1* expression but which are arrested at the same stage of T-cell differentiation as TLX1[+] T-ALL cells, were transduced with recombinant CL20cSLFR MSCV-TLX1[bio]-IRES-YFP lentiviral vector particles and/or recombinant CL20cSLFR MSCV-GFP-BirA lentiviral vector particles and stable cell lines obtained by sorting for YFP and/or GFP fluorescence (Fig. 1b). SupT1 cells expressing GFP-BirA alone served as control for the binding of any

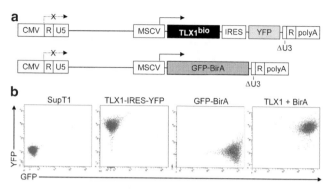

Fig. 1. FACS-based binary in vivo biotinylation tagging system. (**a**) Schematic representation of the self-inactivating SIV lentiviral vectors used to produce biotinylated TLX1. *Top*, pCL20cSLFR MSCV-TLX1^{bio}-IRES-YFP biotinylation tagging vector. *Bottom*, pCL20cSLFR MSCV-GFP-BirA vector encoding a GFP BirA fusion protein. Abbreviations: *CMV* cytomegalovirus Immediate early enhancer-promoter, *R* direct repeat, *U5* unique 5′ region of long terminal repeat, *MSCV* murine stem cell virus promoter, *TLX1^{bio}* TLX1 homeodomain transcription factor with COOH-terminal biotinylation peptide tag (the version indicated is a homeodomain deletion mutant), *IRES* internal ribosome entry site, *YFP* yellow fluorescent protein, *GFP-BirA* green fluorescent protein-BirA biotin ligase fusion protein, *ΔU3* deleted unique 3′ region of long terminal repeat, *polyA* polyadenylation signal. (**b**) Bivariate histograms depicting SupT1 cells stably expressing TLX1^{bio}-IRES-YFP and/or GFP-BirA. Cell sorting and data analysis were performed using a FACSAria instrument equipped with FACSDiva software (BD Biosciences, San Jose, CA) as previously described (44).

biotinylated endogenous proteins to the streptavidin beads. Using single-step affinity capture on streptavidin beads, followed by matrix-assisted laser desorption ionization time-of-flight (MALDI-TOF) mass spectrometry, we identified the Groucho/transducin-like Enhancer of split (Gro/TLE) family member TLE1 as an in vivo binding partner of TLX1 (12) (Fig. 2a). Gro/TLE proteins are regulated by multiple signaling cascades and serve as corepressors for different families of transcription factors (36, 37). The transcription factors that interact with Gro/TLE corepressors contain short peptide sequences related to either WRPW or to FXIXXIL (where X is any amino acid), the latter referred to as the Engrailed homology 1 (Eh1) motif, a repression domain first identified in the *Drosophila* Engrailed homeodomain protein (38). We demonstrated that TLX1 interacts with TLE1 in vitro and in vivo through a seven amino acid sequence encompassing amino acids 19–25 (FGIDQIL) that exhibits similarity to an Eh1 motif (Fig. 2b). Moreover, we found that this motif was required for optimal regulation of expression of two TLX1 target genes, *Aldh1a1* and *Fhl1* (21, 22, 27). As shown in Fig. 2a, a previously reported in vitro TLX1 interacting protein, eIF4E (33), is also efficiently precipitated by streptavidin affinity capture of in vivo TLX1^{bio} protein complexes. Thus, we believe that our adaptation

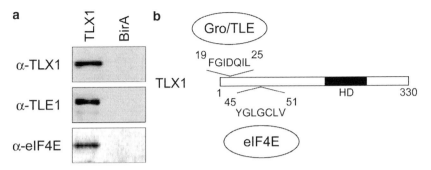

Fig. 2. Example of TLX1-interacting proteins detected in extracts of SupT1 cells coexpressing TLX1[bio] and BirA biotin ligase. (**a**) Western blot analysis for TLE1 and eIF4E after streptavidin pulldown of extracts from GFP-BirA-expressing SupT1 cells coexpressing TLX1[bio]-IRES-YFP. Antibodies used: anti-TLX1 (C-18), anti-TLE1 (M-101), and anti-eIF4E (P-2) (all from Santa Cruz Biotechnology, Inc., Santa Cruz, CA). TLE1 was initially identified as a TLX1-interacting protein by mass spectrometry (12). Mass spectrometry analysis was performed using a Kratos Axima CFR/Plus instrument equipped with Kompact software (Shimadzu Biotech, Columbia, MD) in reflectron mode. (**b**) Schematic diagram of TLX1 indicating: a seven amino acid sequence (FGIDQIL) encompassing amino acids 19–25 exhibiting similarity to an Engrailed homology 1 (Eh1) motif (FXIXXIL, where X is any amino acid) that mediates interactions involving Gro/TLE corepressors (38); a seven amino acid sequence (YGLGCLV) encompassing amino acids 45–51 representative of a consensus eukaryotic initiation factor 4E (eIF4E)-binding motif (YXXXXLΦ, where X is any amino acid and Φ is a hydrophobic residue) (33); and, the homeodomain (HD), which is deleted in the version of TLX1[bio] used in these experiments (12).

of the FACS-based in vivo biotinylation tagging system provides a powerful tool for the characterization of transcription factor and other protein–protein complexes in mammalian cells.

2. Materials

2.1. Biotinylation Tagging Vectors

1. Starting material plasmids: the SIV$_{mac1A11}$-based pCL20cSLFR MSCV-GFP lentiviral vector was provided by Arthur Nienhuis (St. Jude Children's Research Hospital, Memphis, TN) (34); coding sequences for a 23-amino acid biotinylation peptide tag (see Note 1) and the GFP-BirA fusion gene were obtained from the pTRIP/BPcter/IRES/EGFPBirA plasmid (provided by John Strouboulis, Erasmus Medical Center, Rotterdam, The Netherlands) (9).

2. SIV lentiviral vectors: pCL20cSLFR MSCV-TLX1[bio]-IRES-YFP expressing the TLX1 homeodomain transcription factor containing a COOH-terminal biotinylation peptide tag linked to a downstream YFP reporter on a bicistronic transcript (12); pCL20cSLFR MSCV-GFP-BirA expressing a fusion protein consisting of a GFP reporter and the bacterial BirA biotin ligase (GFP-BirA) (12). The pCL20cSLFR MSCV-IRES-YFP and pCL20cSLFR MSCV-GFP-BirA plasmids can be obtained from the authors upon request.

3. Restriction and modifying enzymes for PCR amplification and subcloning of genes of interest.

4. Plasmid DNA purification kits: QIAquick Gel Extraction Kit, QIAGEN Plasmid Maxi Kit (Qiagen, Valencia, CA).

2.2. Production and Titration of Vector Particles

2.2.1. Transient Transfection

1. 293T/17 (293T) human embryonic kidney cell line (CRL-11268; American Type Culture Collection [ATCC], Manassas, VA) (see Note 2).

2. 293T cell culture medium: Dulbecco's Modified Eagle's Medium (DMEM) supplemented with 4.5 g/L glucose, 2 mM L-glutamine, 50 IU/mL penicillin, 50 µg/mL streptomycin, 10% heat-inactivated fetal bovine serum (FBS). Store at 4°C and warm up to 37°C before use.

3. Phosphate-buffered saline without Ca^{2+} and Mg^{2+} (PBS): 137 mM NaCl, 2.68 mM KCl, 1.47 mM KH_2PO_4, 8.1 mM Na_2HPO_4 (adjust pH to 7.4 if necessary, using 1 N HCl or 1 N NaOH).

4. Solution of trypsin and ethylenediamine tetraacetic acid (trypsin-EDTA): 0.05% trypsin, 0.53 mM EDTA in PBS.

5. pCL20cSLFR MSCV-IRES-YFP biotinylation tagging SIV lentiviral vector plasmid DNA containing the gene of interest; pCL20cSLFR MSCV-GFP-BirA SIV lentiviral vector plasmid DNA.

6. SIV lentiviral packaging plasmid DNAs (e.g., pCAG-SIVgprre and pCAG4-RTR-SIV; provided by Arthur Nienhuis) (34).

7. Vesicular stomatitis virus G (VSV-G) glycoprotein envelope plasmid DNA (e.g., pMD.G; (39, 40)).

8. 2.5 M $CaCl_2$: Dissolve 183.7 g $CaCl_2$ dihydrate (tissue culture grade) in deionized water. Bring the volume up to 500 mL and filter–sterilize using a 0.22-µm nitrocellulose filter. Store at –20°C.

9. 2× N-(2-hydroxyethyl)piperazine-N'-(2-ethanesulfonic acid) (HEPES)-buffered saline (2× HBS): 50 mM HEPES, 280 mM NaCl, 1.5 mM Na_2HPO_4. Titrate to pH 7.05 with 5 N NaOH. Filter-sterilize using a 0.22-µm nitrocellulose filter. Store as single use aliquots at –20°C.

2.2.2. Collection and Concentration of Vector Particles

1. 0.45-µm pore-size filter units: Stericup 150-mL unit, low protein binding Durapore (PVDF) unit (Millipore Corp., Bedford, MA).

2. Ultracentrifuge, polycarbonate 70-mL ultracentrifuge bottles and caps.

2.2.3. Titration of Vector Particles

1. HT1080 human fibrosarcoma cells (CCL-121; ATCC).

2. HT1080 cell culture medium: same as 293T cell culture medium (see Subheading 2.2.1).

3. Polybrene (hexadimethrine bromide; Sigma-Aldrich Corp., St. Louis, MO) stock solution (1,000×): 8 mg/mL in sterile, deionized water. Aliquot and store at –20°C.

4. Analytical flow cytometer.

2.3. Generation of Cells Stably Expressing Biotin-Tagged Proteins

1. Replication-defective SIV lentiviral vector particles (e.g., CL20cSLFR MSCV-TLX1^{bio}-IRES-YFP and CL20cSLFR MSCV-GFP-BirA).

2. Target cells (e.g., SupT1 human T-ALL cell line, CRL-1942; ATCC).

3. Suspension cell culture medium: Iscove's modified Dulbecco's medium (IMDM) supplemented with 2 mM L-glutamine, 50 IU/mL penicillin, 50 µg/mL streptomycin, 10% heat-inactivated FBS.

4. Polybrene stock solution (see Subheading 2.2.3).

5. Fluorescence-activated cell sorter (see Note 3).

2.4. Cell Fractionation

1. Phenylmethylsulphonyl fluoride (PMSF): 0.5 M in dimethyl-sulfoxide. Store at –20°C.

2. Dithiothreitol (DTT): 0.5 M in deionized water. Store at –20°C.

3. Protease inhibitors cocktail tablets (Roche Diagnostics Corp., Indianapolis, IN).

4. PhosSTOP phosphatase inhibitor cocktail tablets (Roche Diagnostics Corp.).

5. Hypotonic wash buffer: 10 mM HEPES, pH 7.9, 1.5 mM $MgCl_2$, 10 mM KCl, 0.5 mM PMSF, 0.5 mM DTT.

6. NP40 nuclear extraction (NP40 NE) buffer: 20 mM HEPES, pH 7.9, 25% glycerol, 1.5 mM $MgCl_2$, 20 mM KCl, 0.25 M NaCl, 0.1% NP40, 5 mM EDTA, 1 mM PMSF, 0.5 mM DTT, protease inhibitors (1 tablet for 5 mL), phosphatase inhibitors (1 tablet for 5 mL).

7. Other solutions: PBS; 0.4% trypan blue in normal saline.

8. (Optional) Sonifier (e.g., Branson Sonifier 250 with microtip).

2.5. Streptavidin Affinity Precipitation and Mass Spectrometry Analysis

1. Streptavidin sepharose beads (GE Healthcare, Piscataway, NJ).

2. Chicken albumin: 20 mg/mL in deionized water. Store at –20°C.

3. HENG buffer: 20 mM HEPES, pH 7.9, 25% glycerol, 1.5 mM $MgCl_2$, 0.25 mM EDTA.

4. NP40 T buffer: 70 mM Tris–HCl, pH 8.0, 25% glycerol, 1.5 mM $MgCl_2$, 20 mM KCl, 0.25 M NaCl, 0.1% NP40, 5 mM EDTA, 1 mM PMSF, 0.5 mM DTT.

5. Tris–glycine SDS polyacrylamide gel electrophoresis (SDS-PAGE) gels, and associated running and loading buffers.

6. SDS-PAGE staining reagents (Colloidal blue staining kit; Invitrogen, Carlsbad, CA).

7. In-gel tryptic digestion reagents (Trypsin profile IGD kit; Sigma Aldrich Corp.).

8. ZipTip (0.6 μL C_{18} resin; Millipore Corp.).

9. Elution buffer: 50% acetonitrile (ACN), 0.1% trifluoroacetic acid (TFA) in Milli-Q® grade water.

10. α-cyano-4-hydroxycinnamic acid (α-cyano-CHCA; Sigma-Aldrich Corp.): 10 mg/mL in 50% ACN, 0.05% TFA.

11. Mass spectrometer.

---

3. Methods

3.1. Biotinylation Tagging Vectors

1. Subclone the gene of interest linked to a biotinylation peptide tag upstream of the IRES-YFP cassette in the pCL20cSLFR MSCV-IRES-YFP SIV lentiviral vector (see Notes 1 and 4).

2. Extract plasmid DNA using plasmid DNA purification kits.

3.2. Production and Titration of Vector Particles

293T (293 human embryonic kidney cells expressing simian virus 40 [SV40] large tumor [T] antigen) cells are highly transfectable such that transient cotransfection with the self-inactivating SIV lentiviral vectors, packaging plasmids and envelope plasmids yields high-titer, replication-defective vector particles (41). Lentiviral vectors pseudotyped with the VSV-G glycoprotein have a broad host-cell range and can be utilized to transduce all cell types (39, 40).

3.2.1. Transient Transfection

1. Propagate 293T cells in 293T cell culture medium at 37°C in a humidified atmosphere with 5% CO_2.

2. Passage the cells every 3–4 days by splitting them one fourth to one eighth. Remove the medium and rinse the cells with PBS. Remove the PBS and add enough trypsin-EDTA to cover the cells. Incubate the plate at room temperature until the cells round up and detach. Add an equal volume of cell culture medium to inactivate the trypsin. Collect the cells and centrifuge for 5 min at $375 \times g$. Resuspend the cells in fresh cell culture medium for plating.

3. Transfect 293T cells with plasmid DNA using the calcium phosphate precipitation method (42). On the day before

transfection, plate 293T cells in 7 mL of cell culture medium at a density of 5×10^6 cells per 100-mm tissue culture dish.

4. Prepare a mixture of 20 μg of plasmid DNA composed as follows: 10 μg of pCL20cSLFR MSCV-IRES-YFP biotinylation tagging SIV lentiviral vector containing the gene of interest (or pCL20cSLFR MSCV-GFP-BirA SIV lentiviral vector), 6 μg of pCAG-SIVgprre plus 2 μg of pCAG4-RTR-SIV packaging plasmids, and 2 μg of the pMD.G VSV-G glycoprotein envelope plasmid (34) (see Note 5). Bring the volume up to 400 μL with sterile, deionized water. Add 100 μL of 2.5 M $CaCl_2$ and mix. Add the DNA/$CaCl_2$ solution dropwise to 500 μL of 2× HBS in a 15-mL conical tube. Use a second pipettor and a 2-mL pipet to bubble the 2× HBS as the DNA/$CaCl_2$ solution is added. Vortex immediately for 5 s and incubate at room temperature for 20 min. Add the 1 mL of DNA/calcium phosphate mixture directly to each 100-mm dish while swirling (see Note 6). Incubate the cells at 37°C overnight (16 h). Change the medium and culture for 24–48 h (this method usually results in transfection of 50–80% of the cells).

3.2.2. Collection and Concentration of Vector Particles

Collect the culture medium containing vector particles 24–48 h after medium change. Centrifuge at $2,000 \times g$ for 10 min to remove cellular debris and filter through a 0.45 μm pore-size filter (depending on the volume, use a 150-mL filter unit or a small filter unit attached to a 5-mL syringe). Use directly for transductions or aliquot and store at –80°C (see Note 7).

Several procedures have been developed to concentrate lentiviral vector particles (43). The choice of concentration protocol depends on the envelope selected to pseudotype the particle and the quantities of particles to be produced. The stability of the VSV-G envelope protein allows the generation of high-titer lentiviral vector particles by ultracentrifugation, as described here.

1. Ultracentrifuge the vector particles at $50,000 \times g$ and 4°C for 90 min.

2. Discard supernatant. Using gentle pipetting, resuspend the pellet in ~100 μL of medium appropriate for the downstream application. To facilitate complete resuspension, vortex gently overnight at 4°C (see Note 8).

3. To remove cellular debris, centrifuge concentrated vector particles at $10,000 \times g$ and 4°C for 5 min in a microcentrifuge. Collect the supernatant containing the vector particles, aliquot and freeze at –80°C (see Note 9).

3.2.3. Titration of Vector Particles

The HT1080 human fibrosarcoma cell line can be used to determine the titer of the lentiviral vector particles.

1. Propagate HT1080 cells in HT1080 cell culture medium at 37°C in a humidified atmosphere with 5% CO_2.

2. 4–6 h prior to titrating the vector particles, plate 2.5×10^5 HT1080 cells into each well of a 6-well tissue culture dish.

3. Prepare serial dilutions of each vector preparation (e.g., 10^0, 10^{-1}, and 10^{-2} for unconcentrated vector particles; 10^{-2}, 10^{-3}, and 10^{-4} for concentrated vector particles) using cell culture medium, in a final volume of 1 mL per dilution. Add 1 μL of polybrene stock solution to each dilution. Remove the medium from the cells and add 1 mL of each dilution to each well. Incubate at 37°C for 4 h.

4. Remove the vector particles after the 4-h transduction and replace with 2 mL of fresh cell culture medium. Return to the CO_2 incubator and incubate at 37°C.

5. After 48 h, determine the relative end-point vector titer (in transducing units per mL [TU/mL]) by flow cytometric analysis. Evaluate GFP or YFP expression on an analytical flow cytometer equipped with 488 nm excitation wavelength and a 530/30-nm bandpass (BP) filter (44).

6. To determine vector titer, use the following equation:
Vector titer = number of HT1080 cells × % fluorescent protein-positive cells × dilution factor (see Note 10).

3.3. Generation of Cells Stably Expressing Biotin-Tagged Proteins

The following protocol is used to transduce suspension cell lines with the recombinant biotinylation tagging lentiviral vectors. For adherent cells, the protocol described in Subheading 3.2.3 for titration of vector particles is used.

1. Propagate cells (e.g., SupT1) in suspension cell culture medium at 37°C in a humidified atmosphere with 5% CO_2.

2. Prepare an appropriate dilution of vector particles in 1 mL of cell culture medium containing 1 μL of polybrene stock solution. Resuspend $1–2 \times 10^5$ cells in the diluted vector particles in a 14-mL conical polystyrene centrifuge tube.

3. Centrifuge the mixture of cells and vector particles at $600 \times g$ and 15°C for 2 h. Resuspend the cell pellet in 0.5 mL of fresh cell culture medium and incubate overnight. Repeat the above centrifugation-enhanced transduction once a day for 2 days.

4. Culture for 10 days. Isolate target cells stably expressing the appropriate fluorescent protein(s) on a fluorescence-activated cell sorter equipped with 488 nm excitation wavelength (see Note 11). Evaluate GFP and YFP fluorescence with a 510/20-nm BP filter and a 550/30-nm BP filter, respectively, separated by a 525-nm shortpass dichroic mirror. To determine correct settings for fluorescence compensation, use cells expressing GFP or YFP alone as single color controls (44).

3.4. Cell Fractionation

The protocol described below is a variation of the Dignam procedure for the preparation of nuclear extracts for human tissue culture cells (45) that has been modified for T-ALL cell lines (e.g., SupT1). The standard procedure is based on a starting cell number of 2×10^8 cells and can be adjusted to accommodate 2×10^9 cells by increasing the buffer volumes at steps 2 and 3 by a factor of 4. All solutions are kept on ice; prior to cell harvesting, freshly thawed PMSF and DTT, and freshly prepared protease and phosphatase inhibitors are added to the prechilled buffers. Centrifugations are performed at 4°C.

1. Harvest the cells by centrifuging at $600 \times g$ for 10 min.

2. Wash once in 10 mL of PBS and centrifuge at $600 \times g$ for 10 min.

3. Wash once in 10 mL of hypotonic wash buffer and centrifuge at $600 \times g$ for 5 min.

4. Remove supernatant immediately and discard; resuspend the pellet in 3 mL of hypotonic wash buffer.

5. Subject the cells to hypotonic shock by incubating for 30 min.

6. Check the efficiency of cell lysis under a microscope by staining an aliquot with trypan blue.

7. Centrifuge homogenate at $3,300 \times g$ for 15 min.

8. Remove supernatant (mostly composed of cytosolic and loosely-associated nuclear proteins) and adjust salt concentration by adding an equal volume of NP40 NE buffer. Freeze sample at –80°C or keep on ice until streptavidin affinity precipitation (see Subheading 3.5).

9. Resuspend the pellet in 3 mL of NP40 NE buffer.

10. Incubate for 30 min on ice to extract nuclear proteins; during the incubation period, shear DNA either by mild sonication (e.g., for 20 s using a Branson sonifier 250 set at constant duty and microtip output control limited to 3) or by passing the material through a 25-gauge needle.

11. Freeze sample at –80°C or keep on ice until streptavidin affinity precipitation (see Subheading 3.5).

3.5. Streptavidin Affinity Precipitation and Mass Spectrometry Analysis

Streptavidin affinity precipitation is performed essentially as described (6) (see Note 12).

1. Wash 120 μL of streptavidin sepharose beads (for two samples) with 1 mL of HENG buffer. Centrifuge at $10,000 \times g$ for 20 s. Resuspend the beads in 500 μL of HENG buffer containing 200 μg/mL of chicken albumin and incubate on a rotating platform for 1 h at room temperature. Centrifuge and resuspend the beads in 65 μL of HENG buffer.

2. Centrifuge the samples at $25,000 \times g$ for 30 min. To decrease nonspecific binding, transfer supernatants to new tubes even if there is no visible insoluble material.

3. Add 30 μL of streptavidin bead suspension per sample and incubate on a rotating platform for 1 h at 4°C.

4. Wash the sample twice by resuspending in 1 mL of NP40 T buffer and centrifuging at $10,000 \times g$ for 20 s.

5. Wash the sample three times by resuspending in 1 mL of NP40 T buffer, incubating on a rotating platform at room temperature for 5 min, and centrifuging at $10,000 \times g$ for 20 s.

6. Resuspend the material in 2× loading buffer and fractionate by SDS-PAGE.

7. Stain gel with colloidal blue. Excise discretely stained bands that are evident only in the presence of biotinylated protein of interest.

8. Subject each band to in-gel tryptic digestion. Desalt and concentrate resultant peptides through a ZipTip, and elute in 5 μL of elution buffer.

9. Load 1 μL of peptides and 1 μL of α-cyano-CHCA matrix onto a target plate.

10. Conduct peptide mass fingerprinting using MALDI-TOF mass spectrometry. Perform protein database searches using Mascot software (www.matrixscience.com).

11. Confirm the identity of interacting proteins revealed by mass spectrometry using Western blot analysis.

4. Notes

1. More recently, we have constructed an NH_2-terminal biotin-tagged TLX1 (bioTLX1) modeled after the bioNanog homeodomain protein and other biotin-tagged transcription factors described by Orkin and colleagues (46, 47), which is based on the 15 amino acid AviTag™ sequence (GLNDIFEAQKIEWHE) that is very efficiently biotinylated in vivo by BirA (Avidity, LLC, Aurora, CO; www.avidity.com).

2. The 293T/17 cell line is a clone of the 293T (293tsA-1609neo) human embryonic kidney cell line (48), which was selected for high transfectability and capability of producing high-titer vector stocks. The 293T cell line is a derivative of 293 cells into which the simian virus 40 (SV40) large T antigen gene was inserted. 293 cells express the adenovirus serotype 5 E1A 12S and 13S gene products, which strongly transactivate transcription from expression vectors containing

the human CMV enhancer-promoter (49). Expression of the SV40 large T antigen by 293T cells may stimulate extrachromosomal replication of plasmids containing the SV40 origin of replication during transient transfection.

3. Standard detection filters supplied with most commercial flow cytometers are unable to distinguish between GFP and YFP signals. Suppliers of custom optical filters include Omega Optical Inc. (Brattleboro, VT), Chroma Technology Corp. (Rockingham, VT), and Semrock (Rochester, NY).

4. Common recombinant DNA techniques are used for plasmid DNA preparations, restriction enzyme digestions, PCR amplifications, and subclonings. Detailed protocols for each technique can be obtained from commercial sources, as well as from various standard molecular biology manuals. Accordingly, these techniques have not been described here.

5. The pCL20cSLFR MSCV-GFP SIV-based vector system is modeled after "third generation" lentiviral vector systems that have incorporated features to enhance safety and improve the efficiency of vector particle production (50). Specifically, several viral accessory proteins have been eliminated from the pCAG-SIVgprre gag-pol packaging construct, and the LTRs have been modified to contain the human CMV enhancer–promoter and to eliminate Tat dependence during vector production (5′ LTR) and to be self-inactivating (3′ LTR) upon integration. Because both the expression of the pCL20cSLFR MSCV-GFP SIV lentiviral vector backbone as well as the *gag* and *pol* genes encoded by the pCAG-SIVgprre packaging construct are dependent on transcomplementation by the Rev protein, use of this system requires cotransfection with a rev expression construct (e.g., pCAG4-RTR-SIV).

6. Vector titer depends on the vector backbone design, the size and nature of inserted sequences as well as the efficiency of transfection. To achieve optimal transfection efficiency, use an exponentially growing culture (50–70% confluent) of 293T cells for transfection and make sure that the cells form a uniform monolayer upon plating. Transfection efficiency is also affected by the quality of the plasmid DNA used. Commercially available plasmid DNA purification kits usually yield highly purified endotoxin-free supercoiled plasmid DNA (traditionally obtained by purifying on two separate cesium chloride gradients). Sterilize the plasmid DNA by ethanol precipitation, resuspend the air-dried pellet in sterile, deionized water, and determine its concentration and quality by spectrophotometric analysis and gel electrophoresis. Pay special attention to the pH of the 2× HBS solution, which should be between 7.05 and 7.12. Upon addition of the transfection mixture to the cells, a fine precipitate should develop within a few minutes.

7. pH and temperature fluctuations as well as freeze-and-thaw frequency could have an impact on the stability of vector particles, and hence, vector titer (51). Minimize pH changes by adding HEPES at a final concentration of 10 mM to buffer the cell culture medium. Once collected, vector particles should be kept on ice at all times. For future use, they should be stored in aliquots at –80°C. Avoid repeated freezing and thawing.

8. Small (~2 to 5-mm-diameter) pellets should be visible after concentration by centrifugation. Expect a 50–75% recovery following vector concentration.

9. The vector particle preparation can be further purified if necessary by centrifuging through a sucrose cushion (52).

10. When reporter proteins are evaluated by flow cytometry, it is advantageous to wait at least 5 days before analyzing the transduced cells. This minimizes the contribution of false positive signals due to pseudotransduction, which is the direct transfer of reporter protein (adhered to vector particles or incorporated into vector particles) to the target cells; this is particularly problematic for VSV-G glycoprotein-pseudotyped vectors (53, 54). Note that it has also been shown that transgenes can be efficiently transiently expressed from unintegrated lentiviral vectors during this timeframe (10–14 days) (55).

11. When the target cells already express the protein of interest, sort YFP/GFP-positive cells that express a subendogenous level of the biotin-tagged protein so as to avoid perturbing any existing protein–protein network interactions. To further minimize competition between endogenous and biotin-tagged proteins under these circumstances, we have recently adapted the shRNA-expressing pLKO.1-puro lentiviral vector (56), by substituting the puromycin N-acetyltransferase gene with the cyan fluorescent protein (CFP) gene so that simultaneous knockdown of endogenous transcripts can be achieved by sorting CFP-positive cells.

12. While single-step purification to generate protein complexes sufficiently clean to be analyzed by mass spectrometry is a strength of the in vivo biotinylation approach (6), it might be advantageous to perform tandem-affinity purifications in some experiments. Elegant studies carried out by Orkin and colleagues in murine embryonic stem cells to elaborate the transcriptional interaction network involving nine transcription factors, including the Nanog homeoprotein, have validated and demonstrated the utility of in vivo biotinylation of tagged proteins and streptavidin affinity capture to identify downstream targets on a global scale by ChIP-on-chip (46, 47). In those studies, both single-step capture on

streptavidin beads as well as tandem purification with anti-Flag immunoprecipitation followed by capture with streptavidin were carried out. For example, candidate Nanog-interacting proteins fell into three groups: the first group included proteins present in at least two of three independent one-step purifications and tandem purification; the second group included proteins that may be part of unstable or transient complexes that dissociate during tandem purification; and the third group contained proteins that were "masked" in one-step purifications but were recovered in the tandem procedure (46). For this reason, we have also included NH_2-terminal Flag tags on our recombinant proteins (12, 23, 26), which can be used for this purpose.

Acknowledgments

We are grateful to Ali Ramezani for sharing his expertise on lentiviral vector protocols and Sara Karandish for technical assistance. This work was supported in part by National Institutes of Health grants R01HL65519 and R01HL66305, and by an Elaine H. Snyder Cancer Research Award and a King Fahd Endowed Professorship (to R.G.H.) from The George Washington University Medical Center.

References

1. Kocher, T. and Superti-Furga, G. (2007) Mass spectrometry-based functional proteomics: from molecular machines to protein networks. *Nat. Methods* **4**, 807–15.

2. Suter, B., Kittanakom, S., and Stagljar, I. (2008) Two-hybrid technologies in proteomics research. *Curr. Opin. Biotechnol.* **19**, 316–23.

3. Lievens, S., Van der Heyden, J., Vertenten, E., Plum, J., Vandekerckhove, J., and Tavernier, J. (2004) Design of a fluorescence-activated cell sorting-based Mammalian protein-protein interaction trap. *Methods Mol. Biol.* **263**, 293–310.

4. de Boer, E., Rodriguez, P., Bonte, E., Krijgsveld, J., Katsantoni, E., Heck, A. et al. (2003) Efficient biotinylation and single-step purification of tagged transcription factors in mammalian cells and transgenic mice. *Proc. Natl. Acad. Sci. U.S.A.* **100**, 7480–5.

5. Savage, M. D., Mattson, G., Desai, S., Nielander, G. W., Morgensen, S., and Conklin, E. J. (1992) Avidin-Biotin Chemistry: A Handbook. Pierce Chemical Co., Rockford, IL.

6. Rodriguez, P., Braun, H., Kolodziej, K. E., de Boer. E., Campbell, J., Bonte, E. et al. (2006) Isolation of transcription factor complexes by in vivo biotinylation tagging and direct binding to streptavidin beads. *Methods Mol. Biol.* **338**, 305–23.

7. Rodriguez, P., Bonte, E., Krijgsveld, J., Kolodziej, K. E., Guyot, B., Heck, A. J. et al. (2005) GATA-1 forms distinct activating and repressive complexes in erythroid cells. *EMBO J.* **24**, 2354–66.

8. Schuh, A. H., Tipping, A. J., Clark, A. J., Hamlett, I., Guyot, B., Iborra, F. J. et al. (2005) ETO-2 associates with SCL in erythroid cells and megakaryocytes and provides repressor functions in erythropoiesis. *Mol. Cell Biol.* **25**, 10235–50.

9. Goardon, N., Lambert, J. A., Rodriguez, P., Nissaire, P., Herblot, S., Thibault, P. et al. (2006) ETO2 coordinates cellular proliferation and differentiation during erythropoiesis. *EMBO J.* **25**, 357–66.

10. Meier, N., Krpic, S., Rodriguez, P., Strouboulis, J., Monti, M., Krijgsveld, J. et al.

(2006) Novel binding partners of Ldb1 are required for haematopoietic development. *Development* **133**, 4913–23.

11. Sanchez, C., Sanchez, I., Demmers, J. A., Rodriguez, P., Strouboulis, J., and Vidal, M. (2007) Proteomic analysis of Ring1B/Rnf2 interactors identifies a novel complex with the Fbxl10/Jmjd1B histone demethylase and the BcoR corepressor. *Mol. Cell Proteomics* **6**, 820–34.

12. Riz, I., Lee, H. J., Baxter, K. K., Behnam, R., Hawley, T. S., and Hawley, R. G. (2009) Transcriptional activation by TLX1/HOX11 involves Gro/TLE corepressors. *Biochem. Biophys. Res. Commun.* **380**, 361–5.

13. Owens, B. M. and Hawley, R. G. (2002) HOX and non-HOX homeobox genes in leukemic hematopoiesis. *Stem Cells* **20**, 364–79.

14. Holland, P. W., Booth, H. A., and Bruford, E. A. (2007) Classification and nomenclature of all human homeobox genes. *BMC Biol.* **5**, 47.

15. Roberts, C. W., Shutter, J. R., and Korsmeyer, S. J. (1994) *Hox11* controls the genesis of the spleen. *Nature* **368**, 747–9.

16. Qian, Y., Shirasawa, S., Chen, C. L., Cheng, L., and Ma, Q. (2002) Proper development of relay somatic sensory neurons and D2/D4 interneurons requires homeobox genes Rnx/Tlx-3 and Tlx-1. *Genes Dev.* **16**, 1220–33.

17. Aifantis, I., Raetz, E., and Buonamici, S. (2008) Molecular pathogenesis of T-cell leukaemia and lymphoma. *Nat. Rev. Immunol.* **8**, 380–90.

18. Hawley, R. G., Fong, A. Z. C., Lu, M., and Hawley, T. S. (1994) The HOX11 homeobox-containing gene of human leukemia immortalizes murine hematopoietic precursors. *Oncogene* **9**, 1–12.

19. Hawley, R. G., Fong, A. Z. C., Reis, M. D., Zhang, N., Lu, M., and Hawley, T. S. (1997) Transforming function of the *HOX11/TCL3* homeobox gene. *Cancer Res.* **57**, 337–45.

20. Dear, T. N., Sanchez-Garcia, I., and Rabbitts, T. H. (1993) The HOX11 gene encodes a DNA-binding nuclear transcription factor belonging to a distinct family of homeobox genes. *Proc. Natl. Acad. Sci. U.S.A.* **90**, 4431–35.

21. Masson, N., Greene, W. K., and Rabbitts, T. H. (1998) Optimal activation of an endogenous gene by HOX11 requires the NH$_2$-terminal 50 amino acids. *Mol. Cell. Biol.* **18**, 3502–8.

22. Greene, W. K., Bahn, S., Masson, N., and Rabbitts, T. H. (1998) The T-cell oncogenic protein HOX11 activates *Aldh1* expression in NIH 3 T3 cells but represses its expression

in mouse spleen development. *Mol. Cell. Biol.* **18**, 7030–7.

23. Owens, B. M., Zhu, Y. X., Suen, T. C., Wang, P. X., Greenblatt, J. F., Goss, P. E. et al. (2003) Specific homeodomain-DNA interactions are required for HOX11-mediated transformation. *Blood* **101**, 4966–74.

24. Hoffmann, K., Dixon, D. N., Greene, W. K., Ford, J., Taplin, R., and Kees, U. R. (2004) A microarray model system identifies potential new target genes of the proto-oncogene HOX11. *Genes Chromosomes Cancer* **41**, 309–20.

25. Riz, I. and Hawley, R. G. (2005) G$_1$/S transcriptional networks modulated by the *HOX11/TLX1* oncogene of T-cell acute lymphoblastic leukemia. *Oncogene* **24**, 5561–75.

26. Riz, I., Akimov, S. S., Eaker, S. S., Baxter, K. K., Lee, H. J., Marino-Ramirez, L. et al. (2007) TLX1/HOX11-induced hematopoietic differentiation blockade. *Oncogene* **26**, 4115–23.

27. Rice, K. L., Kees, U. R., and Greene, W. K. (2008) Transcriptional regulation of FHL1 by TLX1/HOX11 is dosage, cell-type and promoter context-dependent. *Biochem. Biophys. Res. Commun.* **367**, 707–13.

28. Riz, I., Hawley, T. S., Johnston, H., and Hawley, R. G. (2009) Role of *TLX1* in T-cell acute lymphoblastic leukaemia pathogenesis. *Br. J. Haematol.* **145**, 140–3.

29. Zhang, N., Shen, W., Hawley, R. G., and Lu, M. (1999) HOX11 interacts with CTF1 and mediates hematopoietic precursor cell immortalization. *Oncogene* **18**, 2273–9.

30. Allen, T. D., Zhu, Y. -X., Hawley, T. S., and Hawley, R. G. (2000) TALE homeoproteins as HOX11-interacting partners in T-cell leukemia. *Leuk. Lymphoma* **39**, 241–56.

31. Milech, N., Gottardo, N. G., Ford, J., D'Souza, D., Greene, W. K., Kees, U. R. et al. (2010) MEIS proteins as partners of the TLX1/HOX11 oncoprotein. *Leuk. Res.* **34**, 358–63

32. Kawabe, T., Muslin, A. J., and Korsmeyer, S. J. (1997) HOX11 interacts with protein phosphatases PP2A and PP1 and disrupts a G2/M cell-cycle checkpoint. *Nature* **385**, 454–8.

33. Topisirovic, I., Culjkovic, B., Cohen, N., Perez, J. M., Skrabanek, L., and Borden, K. L. (2003) The proline-rich homeodomain protein, PRH, is a tissue-specific inhibitor of eIF4E-dependent cyclin D1 mRNA transport and growth. *EMBO J.* **22**, 689–703.

34. Hanawa, H., Hematti, P., Keyvanfar, K., Metzger, M. E., Krouse, A., Donahue, R. E. et al. (2004) Efficient gene transfer into rhesus repopulating hematopoietic stem cells

using a simian immunodeficiency virus-based lentiviral vector system. *Blood* **103**, 4062–9.

35. Hawley, R. G., Lieu, F. H. L., Fong, A. Z. C., and Hawley, T. S. (1994) Versatile retroviral vectors for potential use in gene therapy. *Gene Ther.* **1**, 136–8.

36. Buscarlet, M. and Stifani, S. (2007) The 'Marx' of Groucho on development and disease. *Trends Cell Biol.* **17**, 353–61.

37. Cinnamon, E. and Paroush, Z. (2008) Context-dependent regulation of Groucho/TLE-mediated repression. *Curr. Opin. Genet. Dev.* **18**, 435–40.

38. Smith, S. T. and Jaynes, J. B. (1996) A conserved region of engrailed, shared among all en-, gsc-, Nk1-, Nk2- and msh-class homeoproteins, mediates active transcriptional repression in vivo. *Development* **122**, 3141–50.

39. Burns, J. C., Friedmann, T., Driever, W., Burrascano, M., and Yee, J. -K. (1993) Vesicular stomatitis virus G glycoprotein pseudotyped retroviral vectors: concentration to very high titer and efficient gene transfer into mammalian and nonmammalian cells. *Proc. Natl. Acad. Sci. U.S.A.* **90**, 8033–7.

40. Naldini, L., Blomer, U., Gallay, P., Ory, D., Mulligan, R., Gage, F. H. et al. (1996) In vivo gene delivery and stable transduction of nondividing cells by a lentiviral vector. *Science* **272**, 263–7.

41. Pear, W. S., Nolan, G. P., Scott, M. L., and Baltimore, D. (1993) Production of high-titer helper-free retroviruses by transient transfection. *Proc. Natl. Acad. Sci. U.S.A.* **90**, 8392–6.

42. Hawley, T. S., Sabourin, L. A., and Hawley, R. G. (1989) Comparative analysis of retroviral vector expression in mouse embryonal carcinoma cells. *Plasmid* **22**, 120–31.

43. Ramezani, A. and Hawley, R. G. (2002) Generation of HIV-1-based lentiviral vector particles. *Curr. Protoc. Mol. Biol.* **16.22**, 1–15.

44. Hawley, T. S., Herbert, D. J., Eaker, S. S., and Hawley, R. G. (2004) Multiparameter flow cytometry of fluorescent protein reporters. *Methods Mol. Biol.* **263**, 219–38.

45. Dignam, J. D., Lebovitz, R. M., and Roeder, R. G. (1983) Accurate transcription initiation by RNA polymerase II in a soluble extract from isolated mammalian nuclei. *Nucleic Acids Res.* **11**, 1475–89.

46. Wang, J., Rao, S., Chu, J., Shen, X., Levasseur, D. N., Theunissen, T. W. et al. (2006) A protein interaction network for pluripotency of embryonic stem cells. *Nature* **444**, 364–8.

47. Kim, J., Chu, J., Shen, X., Wang, J., and Orkin, S. H. (2008) An extended transcriptional network for pluripotency of embryonic stem cells. *Cell* **132**, 1049–61.

48. DuBridge, R. B., Tang, P., Hsia, H. C., Leong, P. M., Miller, J. H., and Calos, M. P. (1987) Analysis of mutation in human cells by using an Epstein-Barr virus shuttle system. *Mol. Cell. Biol.* **7**, 379–87.

49. Gorman, C. M., Gies, D., McCray, G., and Huang, M. (1989) The human cytomegalovirus major immediate early promoter can be trans-activated by adenovirus early proteins. *Virology* **171**, 377–85.

50. Ramezani, A. and Hawley, R. G. (2002) Overview of the HIV-1 lentiviral vector system. *Curr. Protoc. Mol. Biol.* **16.21**, 1–15.

51. Higashikawa, F. and Chang, L. (2001) Kinetic analyses of stability of simple and complex retroviral vectors. *Virology* **280**, 124–31.

52. Tiscornia, G., Singer, O., and Verma, I. M. (2006) Production and purification of lentiviral vectors. *Nat. Protoc.* **1**, 241–5.

53. Liu, M. L., Winther, B. L., and Kay, M. A. (1996) Pseudotransduction of hepatocytes by using concentrated pseudotyped vesicular stomatitis virus G glycoprotein (VSV-G)-Moloney murine leukemia virus-derived retrovirus vectors: comparison of VSV-G and amphotropic vectors for hepatic gene transfer. *J. Virol.* **70**, 2497–2502.

54. Gallardo, H. F., Tan, C., Ory, D., and Sadelain, M. (1997) Recombinant retroviruses pseudotyped with the vesicular stomatitis virus G glycoprotein mediate both stable gene transfer and pseudotransduction in human peripheral blood lymphocytes. *Blood* **90**, 952–7.

55. Nightingale, S. J., Hollis, R. P., Pepper, K. A., Petersen, D., Yu, X. J., Yang, C. et al. (2006) Transient gene expression by nonintegrating lentiviral vectors. *Mol. Ther.* **13**, 1121–32.

56. Moffat, J., Grueneberg, D. A., Yang, X., Kim, S. Y., Kloepfer, A. M., Hinkle, G. et al. (2006) A lentiviral RNAi library for human and mouse genes applied to an arrayed viral high-content screen. *Cell* **124**, 1283–98.

Chapter 22

Standard Practice for Cell Sorting in a BSL-3 Facility

Stephen P. Perfetto, David R. Ambrozak, Richard Nguyen, Mario Roederer, Richard A. Koup, and Kevin L. Holmes

Abstract

Over the past decade, there has been a rapid growth in the number of BSL-3 and BSL-4 laboratories in the USA and an increase in demand for infectious cell sorting in BSL-3 laboratories. In 2007, the International Society for Advancement of Cytometry (ISAC) Biosafety Committee published standards for the sorting of unfixed cells and is an important resource for biosafety procedures when performing infectious cell sorting. Following a careful risk assessment, if it is determined that a cell sorter must be located within a BSL-3 laboratory, there are a variety of factors to be considered prior to the establishment of the laboratory. This chapter outlines procedures for infectious cell sorting in a BSL-3 environment to facilitate the establishment and safe operation of a BSL-3 cell sorting laboratory. Subjects covered include containment verification, remote operation, disinfection, personal protective equipment (PPE), and instrument-specific modifications for enhanced aerosol evacuation.

Key words: Flow cytometry, High speed sorting, Biosafety, Biohazard, Safety in flow cytometry

1. Introduction

Over the past decade, there has been a rapid growth in the number of BSL-3 and BSL-4 laboratories in the USA (1). Concurrently, there has been an increase in demand for cell sorting in BSL-3 laboratories and therefore requirements to conduct these experiments safely. This chapter outlines procedures for infectious cell sorting in a BSL-3 environment, including containment verification, remote operation, disinfection, personal protective equipment (PPE), and instrument-specific modifications for enhanced aerosol evacuation. This chapter assumes that the laboratory director has performed a careful risk assessment and determined that the sample(s) (infectious agent) must be sorted in a BSL-3

Teresa S. Hawley and Robert G. Hawley (eds.), *Flow Cytometry Protocols*, Methods in Molecular Biology, vol. 699, DOI 10.1007/978-1-61737-950-5_22, © Springer Science+Business Media, LLC 2011

laboratory. General BSL-3 laboratory procedures and physical requirements are detailed elsewhere (2). The procedures presented here are considered special practices that are specific to the laboratory procedure of cell sorting and can be used as a guideline for the formulation of laboratory specific Standard Operating Procedures (SOPs). The SOPs identify hazards and specify procedures designed to minimize those hazards and are determined in part by instrument-specific design features.

In 2007, the International Society for Advancement of Cytometry (ISAC) Biosafety Committee has published *standards* for sorting unfixed cells (3). This document was driven by the following practices and procedural advances:

1. Advances in cell sorter technology made high-speed cell sorting more prevalent and changed the biohazard potential of cell sorting experiments.

2. New, easy-to-use options for the personal protection of operators have become available.

3. Instrument manufacturers responded to the need for improved operator protection and have introduced instrumentation containing novel safety features.

4. Newly designed safety modifications for cell sorters have become commercially available.

5. With the availability of compact, easy-to-operate sorters many more laboratories have incorporated cell sorting into their experimentation, but often do not have dedicated operators to perform cell sorting experiments.

6. Simpler, bead-based techniques for measuring the efficiency of aerosol containment during cell sorting have been developed.

7. The growth of research in emerging infectious disease and bioterrorism has increased the need for live infectious cell sorting.

These standards and procedures are now in-place for the flow cytometry community conducting infectious cell sorting experiments. It is important to note that the assignment of appropriate procedures and practices for samples with variable and sometimes complex levels of biohazard risk potential, such as genetically engineered cell preparations and a variety of unfixed but "pretested" samples still remain in debate. Currently, the ISAC Biosafety Committee together with a special National Institutes of Health (NIH) Safety Committee and the American Biological Safety Association (ABSA) are involved in updating these standards. A primary goal for these groups is to clarify the risk assessment procedures for the assignment of biosafety practices when sorting "questionable" samples, i.e., to match engineering controls with potential risk. Furthermore, these updates provide a foundation for modernizing the risk assessment to reflect the present knowledge and occupational safety practices as outlined in the Centers for Disease Control (CDC) and NIH sponsored document, the Biosafety in Microbiological and Biomedical Laboratories (BMBL) (2).

In 2009, the most important changes in the practice of infectious cell sorting are in the procedures for measuring containment, advances in instrument design and PPE. The common and now widely accepted procedure for measuring aerosol containment on a cell sorter is the Fluorescent Glo-Germ procedure, which has replaced the more complex bacteriophage procedure (4–8). Advances in instrument design have included the improvement of the efficiency of the aerosol evacuation system. The critical importance of respirator protection is unchanged, but for laboratories utilizing powered air purifying respiratory (PAPR) devices, there is a new model of PAPR helmet system which offers novel design features. This PAPR is approved by the National Institute for Occupational Safety and Health (NIOSH); it consists of a HEPA filter hood with face shield attached to a lightweight helmet-based fan resulting in a hose-free PAPR system. Combined with a full body Tyvek suit, this system provides full body protection and increased mobility.

2. Materials

2.1. Laboratory Design: Special Physical BSL-3 Laboratory Requirements

If the risk assessment determines that the cell sorter to be placed within a BSL-3 laboratory should be contained within a Class II Biological Safety Cabinet (BSC), then Items 1–3 are required for procedures outlined in Subheading 3.1.1.

1. Class II BSC such as the BioProtect II Class II BSC (Baker Co., Sanford, ME).

2. Table with adjustable legs (and casters), $72'' L \times 36'' D \times 36'' H$ (Glover Equipment Sales Group, LLC, Baltimore, MD).

3. Support for flow cell access door: gas spring, 20 lb force (#9417K641; McMaster Carr, Princeton, NJ).

4. Remote viewing of sample and collection tube volumes: small black-and-white CCD cameras, video capture card for the cell sorter acquisition computer.

5. Remote control software for monitoring the progress of sorting: PCAnywhere (Symantec).

6. Power backup for cell sorter and associated equipment: APC Smart UPS Model SUA2200.

2.2. Preparing a BSL-3 Cell Sorter	1. BSL-3 cell sorter, such as the FACS Aria cell sorter, equipped with an Aerosol Management System (AMS) (BD Biosciences, San Jose, CA).

2. FACS Aria modifications for improved aerosol evacuation: drill and drill bits, hand tap, female Luer thread style fitting, 0.22 μm syringe filter with Luer male fitting.

3. Service tools for the BSL-3 sorter: The SOP for most BSL-3 labs require sterilization or disinfection of articles prior to removing them from the laboratory. Many of the specialty tools required for cell sorter repair and alignment contain sensitive electronics or optics that could be damaged upon disinfection. Therefore, it is necessary to purchase these tools and place them within the BSL-3 lab for use by service engineers. It is best to consult with the service engineers for your brand of cell sorter to determine the necessary tools required; for BD FACS Aria brand cell sorters there is a single part number that encompasses all of the specialty tools (contact BD Biosciences for part number).

2.3. Measurement of Containment and Tolerances

1. Glo-Germ particles: powder/melamine copolymer resin beads (Glo Germ, Moab, UT).

2. AeroTech particle collector: A6 Single Stage Microbial Sampler (EMLab P&K, Cherry Hill, NJ, http://www.emlab.com/app/store/Store.po?event=showItem&item=141).

3. 0.45 μm vent filter with hydrophobic membrane (e.g., VacuShield™; Pall Corporation, Port Washington, NY).

4. Clean glass slides.

5. Fluorescent microscope equipped with a FITC (520–640 nm) filter (see Note 1).

2.4. Instrument Sort Procedures: Special Considerations

1. 70% ethanol (EtOH).

2. Sterile PBS with Gentimycin (Hyclone Laboratories, Logan, UT).

3. HYPE-WIPE disinfecting towel with bleach (Current Technologies, Crawfordsville, IN).

4. 10% sodium hypochlorite solution (10% bleach): 10 mL of household bleach (5.25–6.15% sodium hypochlorite) in 90 mL of water stored in a plastic opaque bottle has an expiration of 24 h. Note: Hypochlorite solutions in tap water at a pH > 8 stored at room temperature (23°C) in closed, opaque plastic containers can lose up to 40–50% of their free available chlorine level over 1 month. Thus, if a user wished to have a solution containing 500 ppm of available chlorine at day 30, he or she should prepare a solution containing 1,000 ppm of chlorine at time 0. Sodium hypochlorite solution does not decompose after 30 days when stored in a closed brown bottle (9).

2.5. Infectious Sort Procedures

1. PPE utilizing PAPR: MaxAir Disposable Medical Hood Filter System (Bio-Medical Devices Intl., Irvine, CA, http://www.maxair-systems.com/).

2. Tyvek suit.

3. Shroud for helmet.

4. Helmet with battery pack and belt.

5. Sterile 100 μm mesh filter-top tube for sample filtration (e.g., sterile BD Falcon tubes #352235; BD Biosciences).

6. Cleaning agent such as Coulter Clenz (Beckman Coulter, Inc., Fullerton, CA).

7. Warning sign for the sort facility.

3. Methods

3.1. Laboratory Design: Special Physical BSL3 Laboratory Requirements

If the risk assessment determines that the cell sorter to be placed within a BSL-3 laboratory does not require containment within a Class II BSC, skip to Subheading 3.1.2.

3.1.1. Installation of BSC and Modification of a Cell Sorter to Facilitate Operation Within a BSC

1. Install the BioProtect II Class II BSC. This model encloses the entire instrument, fluidics cart, and AMS of the FACS Aria cell sorter. The size of this hood (9′ 4″ W × 8′ 9″ H × 4′ 1″ D) necessitates careful planning before the installation of this unit.

2. Place the FACS Aria on a table with adjustable legs (and casters). Note that the table supplied with the FACS Aria I is not at the correct height to match the BioProtect II Class II BSC front grille.

3. Install a gas spring (20 lb force) on the flow cell access door of the FACS Aria. The access door to the flow cell and sort chamber is normally closed during sorting. As designed, it is hinged so that when fully opened it rests on the top of the instrument cover. However, it is not possible to easily reach the access door in this position through the sash opening of the BioProtect hood. Installation of a gas spring on the access door will permit the door to be easily opened and closed through the hood sash opening (Fig. 1).

3.1.2. Modification of a Cell Sorter to Facilitate Operation Within a BSL-3 Laboratory

Remote operation of the cell sorter is feasible using software for remote viewing of the acquisition computer. However, the levels of sample and collection tube volumes are not available from the acquisition software. Installation of CCD cameras with video capture on the acquisition computer allows remote monitoring of these parameters.

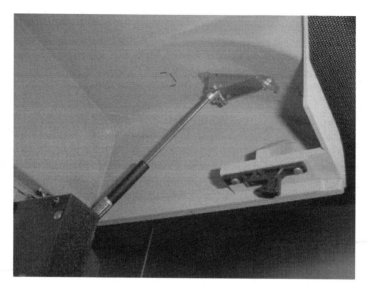

Fig. 1. Installation of a gas spring on the access door will permit the door to be easily opened and closed through the hood sash opening.

Fig. 2. Placement of cameras in front of the sample chamber and inside the collection chamber; output is sent to a video capture card in the acquisition computer.

1. Install PC remote control software packages, such as PCAnywhere on the acquisition computer. Ideally, sorting of infectious samples should be done in a location remote from the cell sorter to minimize the risk of exposure. However, full remote operation of cell sorters is not available with current models, but most can be controlled through the acquisition software. Once sorting has been initiated, sorts can be monitored remotely. Equally important is the ability of clients to view their sort sample setup without the necessity of entering the BSL-3 lab.

2. Place small black-and-white CCD cameras in front of the sample chamber and inside the collection chamber. Send output to a video capture card on the acquisition computer (Fig. 2).

3. Install an uninterruptable power supply to provide power backup for the cell sorter, acquisition computer, and AMS. It is critical when sorting infectious agents that power to the cell sorter, computer and AMS is protected from power dropouts or failures. Although many labs have automated power backup generators, switching to auxiliary systems may cause instruments and computers to shut down.

3.2. Preparing a BSL-3 Cell Sorter

3.2.1. Modification of a Cell Sorter for Improved Aerosol Evacuation

The AMS of the FACS Aria I and II is designed primarily for the evacuation of the collection chamber (area containing tube holders or 96-well plate holder). However, the sort chamber, containing the nozzle and deflection plates, is designed to contain aerosols generated during normal sorting or after a clog. With the tube holders in place, there is no direct communication between the collection chamber (and associated aerosol evacuation system) and the sort chamber. The consequence of this design is that following a clog-induced stream deviation, aerosols in the sort chamber are not efficiently evacuated, permitting aerosol escape when the sort chamber is opened. Simple modifications to the FACS Aria sort chamber and tube holder in conjunction with the existing AMS allow efficient evacuation of aerosols in the sort chamber. Procedures outlined in Subheadings 3.4 and 3.5 assume that these modifications have been performed.

1. Tube holder modification: Drill 3½″ holes in the upper portion of the tube holder (Aria I) or the Universal top component of the tube holder (Aria II) (Fig. 3).

2. Sort chamber door modification: Drill a single hole into the sort chamber door at the top using a size 3 (wire gauge size) drill bit. Thread this hole using a ¼″-28 hand tap. The tapped

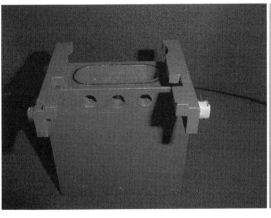

Fig. 3. Location of the holes drilled into the Aria I sort tube holder (*left panel*), and the Aria II Universal top component of the tube holder (*right panel*).

hole accommodates a female Luer thread style fitting with a 5/16″ Hex to ¼-28 UNF Thread, into which a 0.22 μm syringe filter with Luer male fitting is inserted (Fig. 4).

3.2.2. Aerosol Containment   While sorting viable infectious material (infected cells) the following guidelines must be followed for proper aerosol containment. All sort operators must be *trained and certified* by the flow core facility prior to any cell sorting operations.

1. Turn on the AMS and verify that it is functioning according to the manufacturer guidelines. Figure 5 shows the aerosol flow and the locations of the vacuum gauge and monitor. The vacuum monitor should be set to 20% and the vacuum gauge must read between 1.0 and 1.5 in. of Water Column (WC).

2. Change the HEPA filter if: (a) the vacuum monitor gauge reads 2.4 in. of WC or greater at 20% flow; (b) the red filter indicator LED is blinking; or (c) 6 weeks has passed since installation of the filter.

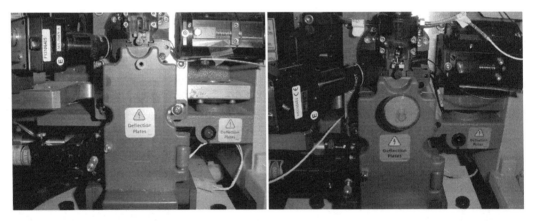

Fig. 4. Vacuum vent location on the sort door used to increase the efficiency of aerosol evacuation in the sort chamber during sorting. Notice the addition of a 0.22 μm syringe filter screwed into the vent.

Aerosol Management System

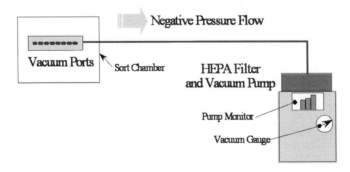

Fig. 5. Typical design of the aerosol management system showing the direction of the negative air flow and location of the HEPA filter.

3. Fill the sheath waste tank with enough sodium hypochlorite to provide a final concentration of 10% when full (1 L of household bleach to a final 10 L of waste collected). If full, empty before starting cell-sorting procedures and allow at least a 30-min contact time before disposal.

4. Verify that the sort chamber camera system is functioning normally according to the manufacturer guidelines. This camera system is used to monitor the sort stream and alerts the operator to potential increased aerosols. In this situation, the operator can correct the sort stream and reduce aerosol contamination. The FACS Aria is equipped with a "Sweet spot," which is used during all sorting operations. This device is used to monitor the normal sort stream drifts and corrects position by automatically adjusting the sort wave amplitude. If a stream blockage is detected due to cell aggregates, the "Sweet spot" will automatically shutdown the stream and close the sort drawer.

3.3. Measurement of Containment and Tolerances

The most widely accepted method of containment testing is Glo-Germ bead method (8). The AMS must be tested under simulated worse case "failure mode." In this mode, the instrument is set to 70 psi and particles are concentrated to approach speeds of approx. 20,000–50,000 particles/s. The stream is forced to glance off of waste catcher shield to create a large plume of aerosols. This is accomplished by covering the waste aspirator with a disposable rubber shield (Fig. 6) or by diverting the stream to hit the side of the waste aspirator using the sort block adjustment screws.

1. Add a clean glass slide to the AeroTech™ particle collector (Fig. 7). Place the AeroTech™ device directly on top of the sort collection chamber (Fig. 8) or in other local positions. Adjust the flow to the AeroTech™ device to 50–55 SLPM (standard liters per minute) as measured by an air flow meter attached to the laboratory vacuum source equipped with a 0.45 μm vent filter with hydrophobic membrane (such as the VacuShield™). Close the sort block door but do not install tube holders. Close the main sort collection chamber and place the AeroTech™ device directly on top of this chamber door.

2. Turn on the AMS (20%) and check for proper vacuum function (1.0–1.5 in. of WC).

3. Place Glo-Germ particles onto the sample station and adjust either the particle concentration or the flow rate to achieve a particle's rate of 20,000 particles/s. Note: When creating aerosols, which could contain Glo-Germ particles, it is recommended that the operator uses full PAPR protection (see Subheading 3.5.2).

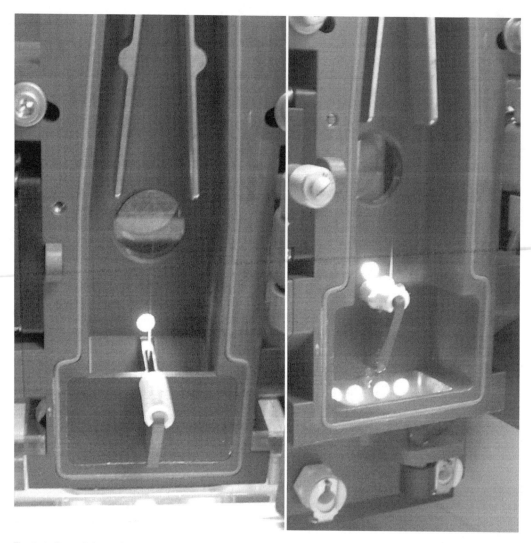

Fig. 6. *Left panel* shows the positioning of the rubber shield, relative to the waste stream (*center of picture*). When the waste collector is engaged (*right panel*), the waste stream glances off of the rubber shield producing aerosols in what is call the "failure mode" of operation.

4. Begin acquiring Glo-Germ particles and allow collection for 5 min at one or more locations. (Fig. 9, *upper panel*).

5. Turn off vacuum to the AeroTech concentrator and remove slide, mark slide as "test control" and put in a fresh slide. Continue collecting aerosols in "failure mode" with the AMS turned off for another 5 min, generating a "positive control." Stop sample acquisition and readjust waste catcher to normal position.

6. Examine glass slides for bright green fluorescence using a fluorescent microscope equipped with a FITC filter (520–640 nm) (Fig. 9, *lower panel*). In an extreme example of Glo-Germ

Fig. 7. The AeroTech collection system is separated into three sections. The *first section* is the base which holds the Petri dish and glass slide. The *middle section* is a flat medal piece with 400 finely drilled holes. The *top section* is a funnel used to direct the air flow down toward the base. Note that the base section is also where the vacuum line is attached.

Fig. 8. Location of the collection area on an Aria I, which is directly outside on the closed sorting chamber.

particles contamination, a pattern of salt circles can be found on the slide and the plate cover (Fig. 10). However, typically in the "positive control" only a few particles can be detected, and this pattern is not usually observed.

7. Scan the "test control" and "positive control" slides with the 10× objective and count all Glo-Germ particles. Record all data using the electronic report file (Fig. 11).

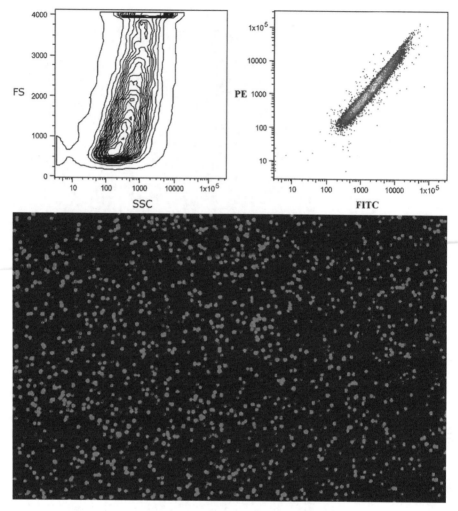

Fig. 9. The *upper panel* shows typical histograms of particle size and fluorescence of the Glo-Germ beads. Note the extreme size range of these beads. The *lower panel* illustrates the fluorescent excitation from the view of a fluorescent microscope.

8. Determine if acceptable tolerance is met. The acceptance tolerance is zero Glo-Germ particles detected after 5 min of active air sampling in front of the sort chamber door with no tube holder in place and the AMS turned on. The "positive control" slide must contain greater than 50 particles after 5 min of active air sampling with the AMS turned off and no tube holder in place.

(a) If the "test control" slide is negative for Glo-Germ particles, the operator can proceed with sort.

(b) If the "test control" slide is positive for any Glo-Germ particles, the operator must repeat the test, and if necessary, clean the AeroTech device with soap and water, followed by 70% EtOH.

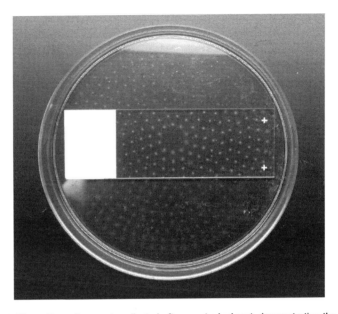

Fig. 10. The pattern of aerosols collected after an atypical sort, demonstrating the uniformity of deposit due to the holes in the middle section of the Aerotech device. The spots are dried PBS salt crystals, within each spot fluorescent Glo-Germ particles can be found.

(c) If the test fails for a second time, infectious cell sorting must be *ABORTED* until the instrument can be verified for containment.

3.4. Instrument Sort Procedures: Special Considerations

3.4.1. Instrument Startup

1. Power up cell sorter (buttons on left side of instrument), turn on computer and start DiVa program.

2. Remove probe from 70% EtOH tank and let drain. Place probe into 10 L sheath tank.

3. Select "Prime after tank refill" (select sheath tank from menu). Note: Two EtOH filters are in rotation; one in use for the day and one drying from the previous day.

4. Completely drain 70% EtOH from filter not on instrument by loosening top stopcock (A), removing bottom stopcock (B), and unscrewing bottom outlet line (C) (Fig. 12).

5. Remove sheath filter on instrument and replace with empty filter.

6. Select "Prime after tank refill" (sheath tank, repeat 1×).

7. Loosen stopcock on cart filter and wait for sheath to run out (approx. 1 min).

8. Run "Fluidics Startup," repeat 1× and turn on stream.

Quality Control for Viable Infectious Sorting

The AMS table below shows critical values measured and tolerance ranges required at the time of every viable infectious sort. Gray areas represent data information input. This form documents instrument containment and is required by the VRC laboratory before **every** infectious sample is sorted. The second table shows monthly contamination checks. This record documents assures the sorting system is free of bacterial contamination.

Infectious Cell Sorting Aerosol Containment Documentation Table (Each Sort)

Operator				
Date				
Measurement	**Particles / Slide**			
Top of Sort Chamber		Containment Vacuum		*Inches of HoH*
		AeroTech Vacuum		*L/min*
		Particle Rate		*Particles/sec*
Positive Control				

Procedure (each sort) : See BSL-3 procedure.

Critical Tolerances for the AMS:

<u>*Vacuum Tolerance*</u> = Range between 1.0 –1.5 inches of HoH . Below this tolerance, replace tubing and HEPA filter.

<u>*Particles outside*</u> = Zero tolerance, no particles on entire slide. Any positive result must be investigated, resolved and retested before proceeding with infectious sort.

<u>*Particles inside (positive control)*</u> = Greater than 100 per slide (10x objective field, result may vary with slide location).

Blood Agar Results Table (Monthly QC)

Sample Area	24 Hours	48 Hours	72 Hours
Plenum			
Stream			
Sheath Tank			

Procedure (Monthly QC): samples are taken from three locations, known to have high potential for contamination; the plenum reservoir, the sheath tank and sample stream.

Fig. 11. An example of a document form to record containment test results. In addition, the bottom section is used to record monthly checks for bacterial contamination as measured on blood agar plates.

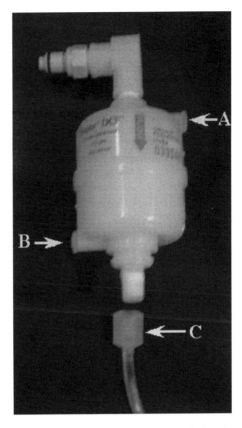

Fig. 12. A typical sheath filter stored with 70% EtOH, and the location of the vents and drain stopcocks used to remove the EtOH. To completely drain the 70% EtOH from the filter, first loosen the top stopcock (A), then remove bottom stopcock (B) and unscrew bottom outlet line (C).

3.4.2. Flow Cell Disinfection Procedure

This should be performed after each infectious sort and/or before shutdown.

1. Turn off sort stream.

2. Add a tube containing 10% bleach (with 24 h expiration) into the sample chamber.

3. Remove nozzle and place in capped tube with 10% bleach for 30 min. Wash nozzle in deionized (DI) water. Install nozzle with plugged orifice (Aria I) or closed loop nozzle (Aria II). Fill a tube with a volume of 10% bleach equal to or greater than the volume of sample that was sorted. Select from the menu – Instrument > Cleaning Modes > Clean Flow Cell. Repeat until fluid is seen in tubing exiting the flow cell. Wait 30 or more minutes with 10% bleach in flow cell. Clean Flow Cell twice with DI water. Select from the menu – Instrument > Cleaning Modes > Long Clean > Clean Bulk Injection Chamber (Aria I).

4. Replace plugged nozzle with cleaned nozzle.

5. Add PBS or sample media to the sample chamber and run for a few minutes.

6. Turn on stream and verify settings, if needed repeat Accudrop procedure to verify droplet location and proper drop delay.

7. Continue with next sort or perform shutdown.

3.4.3. Aria Shutdown Procedure

Aria I procedure:

1. Turn off sample stream.

2. Remove probe from the 10 L sheath tank and place into 10 L EtOH tank.

3. Select "Prime after tank refill" (select sheath tank from the menu). Repeat 1×.

4. Select "Fluidics Startup" (this floods the plenums with 70% EtOH). Repeat 2×.

5. Start stream and load 70% EtOH sample, run sample for 5 min.

6. With stream still on, unload sample, shut down instrument by quitting the DiVa program.

7. Place sample tube holder, nozzle cam and spring into tube containing 10% bleach for 30 min. Wash tube holder and cam in DI water and let air dry.

8. Clean all surfaces around optical bench, sort block chamber and charge plates, sort collection chamber, sample introduction area, and sample tube holder(s) with a HYPE-WIPE 10% bleach towel and/or 10% bleach from a spray bottle.

9. Turn off main power.

Aria II procedure:

1. Turn off sample stream.

2. Select "Fluidics Shutdown" in DiVa software. Remove nozzle and replace with closed loop nozzle.

3. Connect air and fluid lines to the stainless steel EtOH tank. Note: When removing the fluid line from the sheath tank, the line can be disconnected from either before or after the sheath filter. However, transferring the line and the sheath filter to the EtOH tank disinfects the filter and fluid lines with EtOH. This does not damage the filter; residual EtOH is flushed upon instrument startup.

4. Remove sheath probe from sheath tank and gasket from tank lid and spray with 70% EtOH and air dry in BSC. Autoclave sheath tank.

5. Turn off main power.

3.5. Infectious Sort Procedures (see Note 2)

3.5.1. Sorting Procedures

1. Verify that the cell sorter passes all tolerances of aerosol containment as described in Subheading 3.3, step 8. If these tolerances are not met, infectious cell sorting is not permitted.

2. For the operator, wear PPE as outlined in Subheading 3.5.2. If the operator is not protected as described, infectious cell sorting is not permitted.

3. Post a warning sign on the outside of the sort facility door (Fig. 13). Limit access to two individuals during the sort procedure.

4. Turn on the AMS and verify that it is working correctly. This device must have a flow of 1.0–1.5 in. of WC. If this tolerance is not met, infectious cell sorting is not permitted.

5. Close all barriers around the sort chamber as described in Subheading 3.3. If this is not done, infectious cell sorting is not permitted.

6. While in a BSC, filter each sample through a 100 µm mesh prior to sorting. This reduces the potential for clogging and decreases the risk of creating aerosols.

7. Remove sample from the BSC and place onto the sample station. Start sorting and monitor the sort performance using the Accudrop camera. If during the sort the stream is deflected

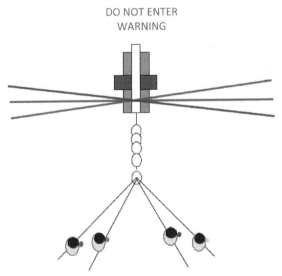

DO NOT ENTER
WARNING

Potential Infectious Aerosol Contamination

Respiratory Suit and Gloves are Mandatory

Emergency Contacts :

Fig. 13. A proper warning sign posted on the outside of the sort facility when an infectious sort is in operation.

(due in part to a clogged flow cell tip), the sorter is designed to stop automatically and block the sort tubes, at the same time the sort stream is turned off. The sort does not restart until the operator has cleared the clog. Use the following procedure to remove a clog from the cell sorter.

(a) Remove sample from the sample chamber. Increase the air evacuation rate on the AMS unit to 100% and wait at least 60 s.

(b) Turn off stream (unless turned off by the instrument in the automated shut-down mode) and then on again to see if drop delay and stream returns to normal pattern. If clog is not cleared, select the nozzle flush procedure and use a cleaning agent such as Coulter Clenz.

(c) If the nozzle cannot be cleared, open the aspirator drawer and increase the air evacuation rate on the AMS to 100%. *Wait 60 s and then close the aspirator door.* Open the sort chamber and remove the nozzle. Replace with a new nozzle or decontaminate the nozzle in 10% bleach for 30 min before placing in a sonicator (located within a Class II BSC) for 2–5 min. Check for clear hole using the microscope.

Resume sorting with new a nozzle or the same nozzle that has been cleared of the clog. Repeat Accudrop procedure to verify droplet location and proper drop delay.

8. Do not remove any sample from the sort collection chamber until the sample acquisition has been stopped. *Wait 30 s before opening the sort chamber door.* This procedure clears aerosols from the sort chamber. After this time, sorted samples can safely be removed from the sort collection chamber (see Notes 3 and 4).

3.5.2. Proper Donning of PPE

The MaxAir PAPR System is to be worn at all times within the BSL-3 lab. Figure 14. shows the proper donning of the PAPR system. The right side of this figure shows the helmet and battery design.

1. Connect battery to helmet and turn on to ensure that the helmet fan is working correctly and turn OFF. Check lights inside helmet to ensure that they are GREEN (yellow or red indicates the battery requires charging).

2. Place helmet inside face shield shroud.

3. Check to determine that the HEPA filter is located over the helmet motor.

4. Attach battery pack to belt on the inside of the respiratory suit and turn on battery.

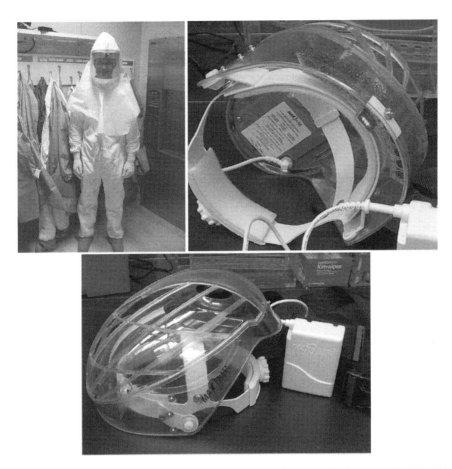

Fig. 14. The MaxAir PAPR System and the proper donning of this system. The shroud is placed over the helmet (*right side of figure*) and the battery pack is attached to the inside of the full body suit.

5. Pull on the Tyvek suit over the shoulders. Ensure that inner hood shroud is positioned under the suit and outer hood shroud is placed over the top of respiratory suit, adjusting the sleeves and booties to fit conformably.

6. Attach gloves (double gloves) and disposable sleeves and check for proper airflow. The airflow should filter from the top of the helmet and exit out of the bottom of the shroud.

3.5.3. Removal of PPE

1. Before leaving the sort facility and entering the anteroom decontaminate gloves with primary disinfectant and remove first pair of gloves.

2. Once in the anteroom, remove shroud and gown using clean gloves, place potentially contaminated Tyvek suit into an autoclave bag.

3. Remove and dispose of gloves into an autoclave bag.

4. Wash hands before leaving the anteroom.

4. Notes

1. Examination of the Glo-Germ test slides by a conventional fluorescent microscope is difficult when wearing the PAPR unit because the face shield obstructs the view of the oculars. Therefore, it is advisable to use a digital fluorescent microscope attached to the sorter acquisition computer for this procedure.

2. Individuals must meet specific BSL-3 training requirements for infectious cell sorting. The lab managers within each section provide training. After the completion of this course individuals must sign-off on all procedures. These procedures must be reviewed annually and a new signature is required to continue working in the BSL-3 facility.

3. All viable specimens (i.e., sorted cell samples) must be placed into a secondary container and wiped with a primary disinfectant, this can be followed by a spray with 70% EtOH (or any other secondary disinfectant) prior to entering the anteroom.

4. Depending on the sample risk assessment, sorted cell samples may require the inactivation of the agent before removal from the laboratory. In some cases, this may involve formaldehyde fixation or cell membrane lysis in the process of RNA or DNA isolation.

Disclaimer. The views expressed here are the opinions of the authors and are not to be considered as official or reflecting the views or policies of the Vaccine Research Center/National Institutes of Health/Department of Health and Human Services nor does mention of trade names, commercial products, or organizations imply endorsement by the US Government.

Funding. This work was supported by the Intramural Research Program of the National Institute of Allergy and Infectious Diseases, NIH; the National Cancer Institute, NIH, under contract No. HHSN261200800001E.

References

1. Rhodes, K. (2007) High-Containment Biosafety Laboratories: Preliminary Observations on the Oversight of the Proliferation of BSL-3 and BSL-4 Laboratories in the Untied States. GAO Testimony Before the Subcommittee on Oversight and Investigations, Committee on Energy and Commerce, House of Representatives 2007;GAO-08-108T.

2. Richmond, J. Y. and McKinney, R. W. (2007) CDC and NIH Biosafety in microbiological and biomedical laboratories (BMBL), 5th edition, Government Printing Office, Washington, DC.

3. Schmid, I., Lambert, C., Ambrozak, D., Marti, G. E., Moss, D. M., and Perfetto, S. (2007) Cytology ISoA. International Society for Analytical Cytology biosafety standard for sorting of unfixed cells. *Cytometry A* **71**, 414–37.

4. Schmid, I. and Dean, P. N. (1997) Introduction to the biosafety guidelines for sorting of unfixed cells. *Cytometry* **28**, 97–8.

5. Schmid, I., Kunkl, A., and Nicholson, J. K. (1999) Biosafety considerations for flow cytometric analysis of human immunodeficiency virus-infected samples. *Cytometry* **38**, 95–200.

6. Schmid, I., Merlin, S., and Perfetto, S. P. (2003) Biosafety concerns for shared flow cytometry core facilities. *Cytometry A* **56**, 113–9.

7. Schmid, I., Nicholson, J. K., Giorgi, J. V., Janossy, G., Kunkl, A., Lopez, P. A., Perfetto, S., Seamer, L. C., and Dean, P. N. (1997) Biosafety guidelines for sorting of unfixed cells. *Cytometry* **28**, 99–117.

8. Schmid, I., Roederer, M., Koup, R. A., Ambrozak, D., Perfetto, S. P., and Holmes, K. L. (2009) Biohazard Sorting, in *Essential Cytometry Methods* (Darynkiewicz, Z., Robinson, J. P., and Roederer, M., eds.), Elsevier, Oxford, UK, Chapter 8, 183–204.

9. Rutala, W. A., Weber, D. J., and the Healthcare Infection Control Practices Advisory Committee (HICPAC) (2008) CDC Guideline for Disinfection and Sterilization in Healthcare Facilities. Government Printing Office, Washington, DC.

Chapter 23

The Cytometric Future: It Ain't Necessarily Flow!

Howard M. Shapiro

Abstract

Initial approaches to cytometry for classifying and characterizing cells were based on microscopy; it was necessary to collect relatively high-resolution images of cells because only a few specific reagents usable for cell identification were available. Although flow cytometry, now the dominant cytometric technology, typically utilizes lenses similar to microscope lenses for light collection, improved, more quantitative reagents allow the necessary information to be acquired in the form of whole-cell measurements of the intensities of light transmission, scattering, and/or fluorescence.

Much of the cost and complexity of both automated microscopes and flow cytometers arises from the necessity for them to measure one cell at a time. Recent developments in digital camera technology now offer an alternative in which one or more low-magnification, low-resolution images are made of a wide field containing many cells, using inexpensive light-emitting diodes (LEDs) for illumination. Minimalist widefield imaging cytometers can provide a smaller, less complex, and substantially less expensive alternative to flow cytometry, critical in systems intended for in resource-poor areas. Minimalism is, likewise, a good philosophy in developing instrumentation and methodology for both clinical and large-scale research use; it simplifies quality assurance and compliance with regulatory requirements, as well as reduces capital outlays, material costs, and personnel training requirements. Also, importantly, it yields "greener" technology.

Key words: Minimalist cytometry, Widefield fluorescence imaging, Light-emitting diodes, Digital cameras, Flow cytometry

1. Introduction

In the late 1600s, Robert Hooke and Antoni van Leeuwenhoek both counted and measured the sizes of objects they observed; both can be said to have utilized microfluidics, observing specimens in glass capillary tubes drawn "fine as a hair." By 1900, counting chambers of defined dimensions, known as hemocytometers by virtue of the circumstances of their development, allowed an observer using a microscope to derive a reasonably

Teresa S. Hawley and Robert G. Hawley (eds.), *Flow Cytometry Protocols*, Methods in Molecular Biology, vol. 699, DOI 10.1007/978-1-61737-950-5_23, © Springer Science+Business Media, LLC 2011

accurate count of the number of cells in a unit volume of specimen, although precision was limited by counting (Poisson) statistics and dilution errors. Using eyepiece reticles, grids, etc., one could also measure the size of microscopic objects. Cytometry, as we now define it, not only includes these basic measurements, but also encompasses the use of one or more physical measurements to identify and characterize cells and other microscopic particles.

Almost all cytometers now available are designed to collect light from one cell at a time. Microspectrophotometers, the first instruments to make physical measurements of individual cells, are optical microscopes using field stops to direct light from the region of a single cell to one or more photodetectors. Smaller field stops permit measurement of smaller areas within a cell, providing higher resolution, but increasing analysis time. Comparatively slow and expensive electromechanical hardware is required for movement in the specimen plane, and also for focus control, since resolution of subcellular detail demands high numerical aperture (NA) objectives, which have low depth of focus. Using galvanometer mirror-based scanners or video cameras as photodetectors increases the rate at which cell images can be acquired, but further increases cost.

Until the 1960s, most microspectrophotometers measured absorption, rather than fluorescence, and measurements were not substantially affected even by relatively large differences in the amount of time different cells spent in an illuminating beam as long as the interval in which actual measurements were made remained constant. As fluorescence measurements became central to cytometry, it became necessary to maintain more constant total illumination intervals in order to avoid measurement errors introduced by photobleaching. This remains an advantage of flow cytometry (1, 2), in which a fluidic system mechanically much simpler than stage motion hardware is used to drive cells or particles through the measurement system in a fluid stream in single file at a constant velocity. Flow cytometry also allows whole-cell measurements to be made at a much more rapid rate than can be achieved in a microspectrophotometer.

Although flow cytometer fluidics are relatively inexpensive, they introduce two classes of problems. Clogs and bubbles may interrupt flow, and minimizing coincidences requires a low duty cycle, limiting measurement time to a few microseconds/cell if thousands of cells are to be analyzed per second. This requires fluorescence excitation sources with high optical throughput, capable of focusing substantial amounts of light through an observation region not much greater in volume than a cell, in order to generate enough fluorescence emission to produce sensitive and precise measurements. At present, lasers and compact arc lamps are most satisfactory as sources. Flow cytometry also typically requires a different photodetector, with associated optics, for each parameter measured; although photodiodes can be used to detect most light scattering and absorption or extinction signals, fluorescence detection typically requires photomultiplier tubes (PMTs) which, with their associated electronics, are substantially more expensive.

By common standards, the technology has been wildly successful. A January 2010 PubMed search on "flow cytometry" yielded over 105,000 entries. Manufacturers estimate that there are over 25,000 fluorescence flow cytometers in use worldwide. Perhaps 15% of these are cell sorters, with two to seven medium- to high-power lasers, capable of measuring fluorescence in more than 30 spectral regions, and costing at least US$200,000 and at most more than ten times that. At least 15% are benchtop (i.e., at least the size of a microwave oven) instruments designed not for analysis of cells, but for multiplexed ligand-binding assays on color-coded plastic bead substrates, and costing around $50,000. The remainder are also benchtop instruments, with one or two low-power lasers, typically measuring fluorescence in four or more spectral regions, and costing at least $35,000. Even in affluent countries, most laboratories that need and use fluorescence flow cytometry cannot afford their own instruments and utilize central resources in their institutions or nearby. Both the cost and the complexity of the instruments limit their applications in resource-poor countries and in lower budget laboratories (e.g., those devoted to microbiological analysis of drinking water and food) in more affluent ones.

Although there are alternative technologies to flow cytometry, they are not nearly as widely used and are not substantially less expensive. Scanning laser cytometers (3) make multiparameter measurements similar to those made in flow cytometers, but bring the laser beam(s) to the cells instead of bringing the cells to the beam(s). They require a separate detector for each measurement channel and use PMTs as fluorescence detectors. Cells may be measured on slides, as is done in the LSC®, iCyte™, and iCys™ instruments (CompuCyte Corporation, Westwood, MA); in a capillary of defined volume, as was done in the IMAGN™ system (BD Biosciences, San Jose, CA); or on membrane filters, as is done in the ChemScan RDI apparatus (AES Chemunex, Bruz, France). In these instruments, stage motion is used to scan in one axis of the specimen plane, typically the longer axis of the scanned area, while a galvanometer mirror moves the laser beam along the other axis. The spatial resolution of a scanning laser cytometer is determined by laser spot size and stage step size; the instruments are capable of resolving subcellular detail to varying degrees. Commercial scanning laser cytometers typically incorporate microscope components, and most have selling prices near those of flow cytometers; although cell analysis rates are typically lower by an order of magnitude, sample throughput may be comparable to that of a flow cytometer in some applications. The "Cell Tracks" instrument described by Tibbe et al. (4) uses a relatively large (5 × 14 μm) focal spot to make whole-cell measurements, and can discriminate unstained and weakly fluorescent particles somewhat better than a conventional benchtop cytometer. It incorporates optical and mechanical components developed for compact disc readers, and could probably be produced to sell for a relatively low price.

Quantitative fluorescence cytometry can be done using a fluorescence microscope with a digital camera attached; a report by Varga et al. (5) provides a good example. They used an instrument with a scanning stage and automatic focusing, and captured multiple fields through a 20×, 0.5 NA objective with epi-illumination from a mercury arc lamp. A "black" or "dark frame" image made with the camera shutter closed was used to correct for dark current error in the camera; a "white" or "flat field" image of a uniform thickness of a dye solution (6) was used for shading correction, i.e., compensation for nonuniform illumination. This reduced the coefficient of variation (CV) of measurements of uniform fluorescent calibration beads from 24.3% to 3.9%. Antifading reagents were added to both standard and specimen slides. The calculated depth of focus for the objective was 1.9 μm; however, the CV of bead fluorescence remained below 4% for images taken within 5 μm of correct focus. Data quality was not affected significantly by moderate JPEG compression of image data. These results are impressive, but, although a fluorescence microscope with computer-controlled stage motion and focus and an arc lamp is less expensive than most cytometers, its price remains at the level of tens of thousands of dollars.

In 2000, I was asked by George Janossy if I could build some relatively small, inexpensive flow cytometers for use in immunophenotyping, primarily CD4+ T-cell counting, in connection with HIV vaccine trials in Africa. George's request led me to an epiphany. For the preceding 25 years, I had been building $100,000 boxes and trying to figure out ways of making them cheaper and simpler by removing components. It suddenly occurred to me, first, that the best way to design cytometers for resource-poor areas might be to start with an empty box and put in as few components as were necessary to do the job, and, second, that, in both poor and rich countries, the best methodology for any cytometric task would be that which required the least possible amount of sample processing and the smallest possible sample. Having thus far been less successful than I hoped or anticipated in making converts, I will now provide a status report, spreading some bad news along with the good.

2. Rethinking Cytometry: Dethroning Morphology

The earliest image cytometers, for example, those used in the 1960s and 1970s for differential white blood cell counting, had to rely on morphologic information, because the dyes in the Giemsa, Wright, and other stains used on blood smears, which were essentially the only reagents then available, were not specific for any one cell type. Therefore, an instrument needed to capture a high-resolution image and then do extensive calculations on the raw input data to determine which pixels in an image represented the cell nucleus, which represented the cytoplasm, whether the cytoplasm contained granules, etc. Additional calculations yielded numerical values for parameters such as cell and nuclear size, shape, and texture, and these, in turn, were input to algorithms that, at least in theory, identified cells by type.

It was known, even 40 years ago, that the amount of white cell classification that could be done reliably even by the best human observer was limited; for example, it was clear that small monocytes were commonly misidentified as lymphocytes, and slide scanning differential counters performed no better than microscopists. As functional studies divided lymphocytes into T and B cells, and thereafter into helper and cytotoxic/suppressor T cells, etc., it became equally clear that distinguishing among subsets by examining conventionally stained smears was beyond the capacity of both observers and the slide scanning cytometers created in their image.

Modern flow cytometers, including those which have replaced slide scanning systems for differential counting, identify cells far more accurately by using more selective reagents, e.g., fluorescent dyes, fluorogenic enzyme substrates, and fluorescent-labeled monoclonal antibodies and nucleic acid probes. Although flow cytometers use high NA optics to maximize light collection, such lenses, which are essential for high-resolution microscopy, have a small depth of focus. The "images" of cells passing through a conventional flow cytometer are, therefore, blurry; however, although morphologic information is lost, light measurements are nonetheless precise enough to allow discrimination among cell types. Better quantification even allows combinations of relatively nonspecific stains, e.g., acidic and basic fluorescent dyes analogous to those in a mixture such as Giemsa's or Wright's, in the case of differential counting, to provide more accurate results than could be obtained from slide scanning (7).

From the late 1970s on, faced with the necessity of building multistation multiparameter flow cytometers as simply and cheaply as possible for our own research, my colleagues and I had pursued refinements in overall design (8–11), light sources (12–15), and hardware and software for data acquisition and analysis (16, 17).

In 1986, when we first reported work toward the "flow cytometer on a chip" (18), we were optimistic that the approach would soon yield small devices usable and affordable worldwide; more than a generation later, this has not come to pass. Although we showed in 2002 that relatively small and inexpensive flow cytometers could be used for CD4+ T-cell counting (19), and several such instruments have now reached the market (20), they remain too costly and complex to be practical for use in many of the places in which they are most needed. We, therefore, re-examined possible alternatives to both flow cytometry and the scanning and imaging techniques described above.

3. Minimalist Imaging Cytometry

An optical analytical cytometer is essentially a box into which cells are put and from which numbers are retrieved. As long as the right numbers emerge, it should not matter to the user what is in the box; however, to keep the apparatus working in an unfriendly environment with minimal maintenance, it is desirable to minimize its size, complexity (number of total and moving parts), energy consumption, and cost. This minimalist ideal is rarely achieved, or even approached, in cytometers designed by the affluent for the affluent, but the philosophy is essential in developing systems for use in resource-poor areas. It is also desirable for the user to have to do as little as possible to as few cells as possible between the time at which the cells are collected from a patient or other source and the time at which they are introduced into the apparatus, and for any reagents used to be stable over a wide range of temperature and humidity. Achieving these design goals should, not incidentally, make it possible to locate much of the capacity for manufacturing both the apparatus and reagents in or near the countries in which they will be used, with concomitant economic benefits.

Minimalism is also a good philosophy in developing instrumentation and methodology for clinical use; it simplifies quality assurance and compliance with regulatory requirements; reduces capital outlays, material costs, and personnel training requirements; and also, importantly, yields "greener" technology.

In 2006, we considered possible instrumental approaches to "personal cytometers" (21). Although that term had been applied to flow cytometers costing several tens of thousands of dollars, we explicitly restricted the definition to an apparatus that could be sold for less than $5,000. This was a critical price point for Apple, IBM, and other early personal computers; more significantly, an instrument costing more than $5,000 is considered "equipment" for grant purposes and must be specifically requested in applications and approved in the review process.

It is generally accepted that the production (labor and material) cost of an apparatus must be less than a third of the selling price; this would require that a personal cytometer be produced for less than $1,700. Our 2006 analysis (21) suggested that no apparatus in which cells were measured individually, as described above, could meet that criterion. We did, however, conclude that it would be possible to accomplish many of the simpler tasks now commonly done by flow and scanning cytometry using simple low-magnification, low-resolution imaging systems, which examined all the cells in a sample at once, and that such systems could be produced for less (and, in some cases, considerably less) than $1,700.

As noted above, the reagents and analytical methods developed as flow cytometry evolved have made it possible to accomplish even highly sophisticated procedures for cell identification and characterization without recourse to morphologic information. It was recognized in the early 1990s that whole-cell fluorescence and luminescence measurements with sensitivity approaching those made in flow cytometers could be made with low-magnification, low-resolution imaging systems using video and digital cameras (22–25); although this yielded an extremely simple instrument design, the cameras, lenses, and light sources used at that time cost thousands of dollars each, which would have made production devices no cheaper than flow cytometers.

The Fluorescence Array Detector (FAD) described by Wittrup et al. in 1994 (24) used camera lenses to form a 1:1 image of a 1-cm^2 area of a microscope slide on a cooled 512×512-pixel charge-coupled device (CCD) detector with 20 μm^2 pixels. Because each pixel collected light from an area larger than the area of a typical cell, no morphologic information was available. Low-intensity (1 mW/cm^2) illumination of the field was obtained from the expanded beam of a 488-nm air-cooled argon ion laser. Rather than using an epi-illuminated configuration in which the fluorescence collection lens also transmitted excitation light, these investigators illuminated the surface obliquely at Brewster's angle, minimizing light scattering. Although a software shading correction was used in an attempt to compensate for the uneven illumination, this was only partially successful; the CV of fluorescence measurements of highly uniform 6-μm polystyrene beads was 12.9%. Measurement sensitivity, however, was impressive; when low-fluorescence quartz slides were used, noise due to dark current and stray light was equivalent to the fluorescence of only a few hundred fluorescein molecules in solution.

Although light-emitting diodes (LEDs) had been described earlier as fluorescence excitation sources for both fluorescence microscopes and flow cytometers (26), it was not until 2003–2004 that LEDs bright enough to provide excitation for relatively low-intensity fluorescence measurements became available (27).

By this time, a relatively inexpensive, low-resolution fluorescence imaging cytometer designed for cell counting had been introduced as a commercial product. The Nucleo-Counter (ChemoMetec A/S, Allerød, Denmark), which appeared in 2002, is a small benchtop instrument using an array of eight green LEDs to excite fluorescence of propidium iodide in cell samples introduced into the system in plastic carriers. The carrier design allows cells in the sample to mix with the dye before entering a windowed counting area approximately 1 cm² in area, where they are imaged onto a CCD camera at low magnification and are counted with simple image analysis software. The counting area contains a volume of diluted sample corresponding to 1 μL of the original specimen. Total cell counts are obtained by adding a lysing agent to the sample before its introduction into the chamber; if the lysing agent is omitted from another sample, a count of "nonviable," i.e., membrane-damaged cells, is obtained, and subtraction of this value from the total cell count yields a "viable" cell count. The Nucleo-Counter was originally priced at approximately $8,000 but now it costs considerably more; sample carriers are a few dollars each.

We have established (21) that an apparatus using LED excitation, relatively inexpensive achromats or low-power microscope lenses for light collection, and one or more CCD or complementary metal oxide semiconductor (CMOS) camera chips as a detector could detect low-level fluorescence signals in the range that would be expected from cells stained with fluorescent antibodies, as well as the substantially stronger signals associated with cells stained with nucleic acid stains, fluorogenic substrates, and fluorescent redox or membrane potential probes. The basic design of the apparatus and the details of an illuminator design using a highly efficient but very inexpensive (under $3) lens (Fraen Srl, Settimo, Italy) to collect light from the LED are presented in Fig. 1.

Interference filters now readily available for fluorescence microscopy and cytometry are sufficiently well blocked outside the

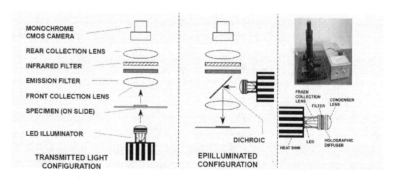

Fig. 1. Schematic of a simple imaging cytometer. The *left* and *middle panels* illustrate transmitted light and epi-illuminated configurations; the *right panel* shows a picture of an epi-illuminated prototype, *above*, and the details of an illuminator module, *below*.

passband to allow the transmitted light instrument configuration shown on the left of Fig. 1 to be used without penalty for brightly fluorescent specimens on media with good light transmission characteristics, e.g. slides, hemocytometers, and multiwell plates with flat, optically clear bottoms. The use of separate optics for illumination and fluorescence collection is advantageous when the light source is an LED, as this enables more light to be delivered to a large area than can be done using conventional epi-illumination (27). For examination of specimens collected on filters or in filter-bottom plates, which would impede transmission of excitation from an illuminator placed beneath the specimen, we use the epi-illuminated configuration, shown in the center of Fig. 1, with a dichroic, providing further background reduction, advantageous for measurement of weaker fluorescence from labeled antibodies or gene probes. The oblique illumination originally used by Wittrup et al. not only reduces background but also provides very uneven intensity across the field; it is, however, desirable for generating large angle scatter signals. The large field of view and depth of focus obtained with any of the illumination configurations eliminate the need for any stage motion or focus adjustment components, making the image cytometer considerably smaller and more robust than a standard microscope.

The prototypes illustrated in Fig. 1 incorporate a single LED and a single camera chip; we have used this simple design for single-parameter fluorescence measurements in a project on rapid (<48 h) diagnosis and drug susceptibility testing of tuberculosis and for other work on the detection of responses of bacteria and parasites to drugs (28). Our CD4+ T-cell counter designs, shown schematically in Fig. 2, could use a single LED and two camera

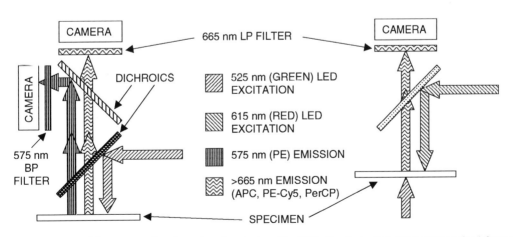

Fig. 2. Schematics of fluorescence imaging cytometers designed for CD4+ T-cell counting; the system on the *left* uses CD4 and CD3 antibodies labeled with phycoerythrin (PE) and a PE-Cy5 tandem conjugate; the one on the *right* uses antibodies labeled with allophycocyanin (APC) and PE-Cy5, or, with a blue LED substituted for the green one, APC and peridinin chlorophyll protein (PerCP).

chips, or two LEDs and a single camera chip, the latter configuration eliminating the problem of registering images made with two chips. None of the prototypes has moving parts; this makes them more robust for use in unfriendly environments. Similar, although mechanically and optically more complex, dedicated fluorescence imaging-based CD4+ T-cell counting apparatus, developed by commercial organizations (LabNow, Austin, TX; Inverness Biomedical, Waltham, MA) which have acknowledged our influence in publications and/or patents, will shortly be introduced to the market. Similar multibeam designs should also be usable for the diagnosis of malaria based on measurements of parasite nucleic acid content and base composition and pigment production, and for general laboratory use for two- to four-color fluorescence measurements of cells and beads, making capabilities of some lower-end flow cytometric systems accessible to more laboratories in more places than can now afford and maintain these more complex and more expensive instruments.

4. Prospects for Low- (and High-) Cost Flow Cytometers; Conclusions

Both LED light sources and digital camera chip detectors for minimalist fluorescence imaging cytometers are extremely inexpensive (typically tens of dollars). The optical filters and dichroics used in these instruments, which cost no more than those used in flow cytometers, are typically the most expensive single components, as data acquisition and processing can be accomplished using low-power chips such as are now used in mobile phones and media players instead of personal computers, and these microcontrollers also typically only cost tens of dollars.

An LED is well suited for illuminating a relatively large field of view, but even the highest throughput LEDs can only deliver fractions of a milliwatt of power to areas the size of a cell, making even relatively low-power lasers superior sources for flow cytometry. Although laser diodes and some solid-state laser modules (e.g., the green frequency-doubled YAG devices used in laser pointers) are relatively inexpensive (tens to hundreds of dollars), once the stabilized power supplies and illumination optics necessary to use them in flow cytometers have been added, the total cost of each laser illuminator typically rises to hundreds or thousands of dollars. So-called silicon photomultipliers, actually avalanche diode arrays, some available for less than $200, can replace PMTs as detectors for some low-level fluorescence measurements, but their temperature sensitivity, higher noise, and lower dynamic range limit their applicability. A single red-sensitive PMT detector and associated housing and power supply cost at least $700, making it difficult to produce a multiparameter fluorescence flow cytometer that could be sold for under $5,000, even if simple microcontrollers (29) can be used for data acquisition. The recent introduction of a $500 projector incorporating blue, green, and red lasers, each capable of emitting tens of milliwatts (Microvision, Redmond, WA), suggests that laser sources suitable for flow cytometry could be made substantially less expensive than they now are, but, although production methods for small PMTs are expected to improve in coming years, it is not clear if this will drastically lower prices. Work on smaller "micro" flow cytometers continues nonetheless, and the field is surveyed in a new book edited by Ligler and Kim (30).

High-end, multibeam, multiparameter flow cytometer/sorters continue to be in high demand, their prices notwithstanding, but the fact that more than half of the time available on many of these instruments is spent sorting cells bearing fluorescent proteins, utilizing only one or two fluorescence parameters, has attracted some manufacturers' attention. I would expect to see at least one simpler sorter become available for well under $100,000 within the next few years.

The dominance of the high-end flow cytometers, at least in analytical applications not requiring isolation of viable cells, is now being threatened by the introduction of an apparatus (the CyTOF Mass Cytometer; DVS Sciences, Toronto, ON) in which mass spectroscopy is used to detect several times as many labeled ligands as can now be detected by fluorescence flow cytometry (31).

For point-of-care and field research applications in poor countries and richer ones, however, I'm still betting on low-resolution imaging. The 20-teens should, all in all, prove to be an interesting decade for cytometry.

References

1. Gucker, F. T., Jr., O'Konski, C. T., Pickard, H. B., and Pitts, J. N., Jr. (1947) A photo-electronic counter for colloidal particles. *J. Am. Chem. Soc.* **69**, 2422–2431.

2. Shapiro, H. M. (2003) *Practical Flow Cytometry*, 4th edn. John Wiley & Sons, Hoboken, NJ, pp. 1–681. A printable, searchable copy in pdf format can be downloaded free of charge from http://coulterflow.com/bciflow/research01.php.

3. Shapiro, H. M. (2004) Scanning laser cytometry, in *Current Protocols in Cytometry* (Robinson, J. P., ed.), John Wiley & Sons, Hoboken, NJ, Unit 2.10., pp. 2.10.1–2.10.12.

4. Tibbe, A. G. J., de Grooth, B. G., Greve, J., Dolan, G. J., Rao, C., and Terstappen, L. W. M. M. (2002) Cell analysis system based on compact disk technology. *Cytometry* **47**, 173–182.

5. Varga, V. S., Bocsi, J., Sipos, F., Csendes, G., Tulassay, Z., and Molnár, B. (2004) Scanning fluorescent microscopy is an alternative for quantitative fluorescent cell analysis. *Cytometry* **60A**, 53–62.

6. Model, M. A. and Burkhardt, J. K. (2001) A standard for calibration and shading correction of a fluorescence microscope. *Cytometry* **44**, 309–316.

7. Shapiro, H. M. (1977) Fluorescent dyes for differential counts by flow cytometry: does histochemistry tell us much more than cell geometry? *J. Histochem. Cytochem.* **25**, 976–989.

8. Shapiro, H. M. (1983) *Building and Using Flow Cytometers: The Cytomutt Breeder's and Trainer's Manual*. Howard M. Shapiro, M.D., P.C., West Newton, MA.

9. Shapiro, H. M., Feinstein, D. M., Kirsch, A. S., and Christenson, L. (1983) Multistation

multiparameter flow cytometry: some influences of instrumental factors on system performance. *Cytometry* **4**, 11–19.

10. Shapiro, H. M. (1988) *Practical Flow Cytometry*, 2nd edn. Alan R. Liss, Inc., New York, NY, pp. 211–265.

11. Olson, R. J., Frankel, S. L., Chisholm, S. W., and Shapiro, H. M. (1983) An inexpensive flow cytometer for the analysis of fluorescence signals in phytoplankton: analysis of chlorophyll and DNA distributions. *J. Exp. Mar. Biol. Ecol.* **68**, 129–144.

12. Shapiro, H. M., Glazer, A. N., Christenson, L., Williams, J. M., and Strom, T. B. (1983) Immunofluorescence measurement in a flow cytometer using low-power helium-neon laser excitation. *Cytometry* **4**, 276–279.

13. Shapiro, H. M. (1986) The little laser that could: applications of low power lasers in clinical flow cytometry. *Ann. N. Y. Acad. Sci.* **468**, 18–27.

14. Shapiro, H. M. and Stephens, S. (1986) Flow cytometry of DNA content using oxazine 750 or related laser dyes with 633 nm excitation. *Cytometry* **7**, 107–110.

15. Shapiro, H. M. and Perlmutter, N. G. (2001) Violet laser diodes as light sources for cytometry. *Cytometry* **44**, 133–136.

16. Margolick, J. B., Scott, E. R., Chadwick, K. R., Shapiro, H. M., Hetzel, A. D., Smith, S. J., and Vogt, R. F. (1992) Comparison of lymphocyte immunophenotypes obtained simultaneously from two different data acquisition and analysis systems on the same flow cytometer. *Cytometry* **13**, 198–203.

17. Shapiro, H. M., Perlmutter, N. G., and Stein, P. G. (1998) A flow cytometer designed for fluorescence calibration. *Cytometry* **33**, 280–287.

18. Shapiro, H. M. and Hercher, M. (1986) Flow cytometers using optical waveguides in place of lenses for specimen illumination and light collection. *Cytometry* **7**, 221–223.

19. Janossy, G., Jani, I. V., Kahan, M., Barnett, D., Mandy, F., and Shapiro, H. M. (2002) Precise CD4 T-cell counting using red diode laser excitation: for richer, for poorer. *Cytometry (Clin. Cytom.)* **50**, 78–85.

20. Janossy, G. and Shapiro, H. (2008) Overview: simplified cytometry for routine monitoring of infectious diseases. *Cytometry B Clin. Cytom.* **74 Suppl 1**, S6–S10.

21. Shapiro, H. M. and Perlmutter, N. G. (2006) Personal cytometers – slow flow or no flow? *Cytometry* **69A**, 620–630.

22. Masuko, M., Hosoi, S., and Hayakawa, T. (1991) A novel method for detection and counting of single bacteria in a wide field using an ultra-high-sensitivity TV camera without a microscope. *FEMS Microbiol. Lett.* **65**, 287–290.

23. Masuko, M., Hosoi, S., and Hayakawa, T. (1991) Rapid detection and counting of single bacteria in a wide field using a photon-counting TV camera. *FEMS Microbiol. Lett.* **67**, 231–238.

24. Wittrup, K. D., Westerman, R. J., and Desai, R. (1994) Fluorescence array detector for large-field quantitative fluorescence cytometry. *Cytometry* **16**, 206–213.

25. Yasui, T. and Yoda, K. (1997) Imaging of Lactobacillus brevis single cells and microcolonies without a microscope by an ultrasensitive chemiluminescent enzyme immunoassay with a photon-counting 17elevisión camera. *Appl. Environ. Microbiol.* **63**, 4528–4533.

26. Olson, R. J., Chekalyuk, A. M., and Sosik, H. M. (1996) Phytoplankton photosynthetic characteristics from fluorescence induction assays of individual cells. *Limnol. Oceanogr.* **41**, 1253–1263.

27. Mazzini, G., Ferrari, C., Baraldo, N., Mazzini, M., and Angelini, M. (2005) Improvements in fluorescence microscopy allowed by high power light emitting diodes, in *Current Issues on Multidisciplinary Microscopy Research and Education*, Vol 2 (Méndez-Vilas, A. and Labajos-Broncano, L., eds.), FORMATEX, Badajoz, Spain, pp. 181–188. (http://www.formatex.org/micro2003/papers/181-188.pdf).

28. Shapiro, H. M. and Perlmutter, N. G. (2008) Killer applications: toward affordable rapid cell-based diagnostics for malaria and tuberculosis. *Cytometry B Clin. Cytom.* **74 Suppl 1**, S152–S164.

29. Naivar, M. A., Wilder, M. E., Habbersett, R. C., Woods, T. A., Sebba, D. S., Nolan, J. P., and Graves, S. W. (2009) Development of small and inexpensive digital data acquisition systems using a microcontroller-based approach. *Cytometry A* **75**, 979–989.

30. Ligler, F. S. and Kim, J. S. (eds.) (2010) *The Microflow Cytometer*. Pan Stanford Publishing Pte. Ltd., Singapore. ISBN-13 978-981-4267-41-0, ISBN-10 981-4267-41-4.

31. Tanner, S. D., Bandura, D. R., Ornatsky, O., Baranov, V. I., Nitz, M., and Winnik, M. A. (2008) Flow cytometer with mass spectrometer detection for massively multiplexed single-cell biomarker assay. *Pure Appl. Chem.* **80**, 2627–2641.

INDEX

A

Apoptosis

caspase activation.............................204–208, 212–214, 218, 220–222, 225

cell permeability........................204, 205, 208, 209, 216, 219, 224, 228

fluorochrome combination 205, 206, 208–211

instruments...209–211

phosphatidylserine flipping......................204, 338, 339, 345–347

simultaneous assays.. 204

B

Biosafety level-3 (BSL-3) cell sorting

infectious sort procedures 453, 465

laboratory design

aerosol containment....................................456–457

biological safety cabinet...................................... 451

cell sorter 453, 455

personal protective equipment

gloves..413, 418, 467

helmet with shroud............................453, 466, 467

powered air purifying respirator........................ 451

Tyvek suit 453, 467

tolerances and measurement of containment

Glo-Germ particles 452

particle collector 452

C

Cell cycle analysis

antibodies reactive with cell cycle regulated proteins...233, 234

antibody validation 191

cell cycle regulated proteins: cyclin A2, cyclin B1, phospho-S10-histone H3 ....230, 233, 237, 238, 241, 244

DNA-binding dye (DAPI)......................233–235, 237, 240, 241, 253, 258, 259

mitosis 120, 233, 242

Cytometry

flow cytometry

data analysis

gating.............................19, 104, 314, 331, 348, 351

histogram...................15, 18, 19, 62, 106, 126, 127, 186, 328, 376, 379–381, 434, 460

parameters,

plots 19, 20, 32, 75, 102, 106, 138, 144, 174, 191, 198, 207, 212, 214, 215, 284, 348, 351, 379

fluidics 7–10, 454, 461, 464, 472

historical perspective... 2

lasers 5, 12, 13, 204, 222, 472

listmode file 32, 36–42, 44–47, 302

minimalist imaging cytometry

camera chip detector....................................... 480

charge coupled device (CCD)..................451, 453, 454, 477, 478

fluorescence measurement 472, 477, 479, 480

interference filters 478

light-emitting diode (LED)............................. 477

microscope lenses.. 478

phtotodetectors

photodiode...11, 473

photomultiplier tube (PMT) 13, 87, 473

signals

amplification..15, 16, 18

analog-to-digital convertor (ADC) 15, 18

fluorescent light 8, 12, 13

scattered light 3, 5, 12, 26

sorting

fluorescence-activated cell sorter (FACS)............ 21

Mack Fulwyler.. 21

F

Fluorescence resonance energy transfer (FRET)

choice of fluorophores

acceptor molecule 372, 374, 389

donor molecule 372, 373

custom made evaluation software 376

Teresa S. Hawley and Robert G. Hawley (eds.), *Flow Cytometry Protocols*, Methods in Molecular Biology, vol. 699, DOI 10.1007/978-1-61737-950-5, © Springer Science+Business Media, LLC 2011

Fluorescence resonance energy transfer (FRET) (*Continued*)

data analysis

based on donor and acceptor

fluorescence...380–382

based on donor quenching.................373, 374, 376,

380, 388

with cell-by-cell autofluorescence

correction...383–384

labeling

extracellular epitopes... 378

intracellular epitopes... 378

protein-protein interactions...................................... 371

Fluorescent protein-assisted identification of rare cells

labeling

of melanoma tumor tissues 399, 401

of rare skin stem cells...................................394–395

reporters

green fluorescent protein (GFP)..............394, 409,

432, 434

luciferase/GFP (Luc/GFP) 395

sorting GFP+ cells.. 400

Fluorescent proteins

non-toxic DsRed derivatives: DsRed-Express2,

E2-Orange, E2-Red/Green,

E2-Crimson...........................355–357, 359, 363

screening

for aggregation... 359

for brightness... 361

for cytotoxicity...363–365

whole cell labeling

cells..365–368

lentiviral vectors...367–368

Fluorescent proteins in biotinylation proteomics

biotinylation tagging vector particles

concentration 436, 439

production 436–440

titration...................................436–437, 439–440

cells stably expressing biotin-tagged proteins

FACS purification 394, 395, 431, 432

fractionation 437, 441

MALDI-TOF mass spectrometry.............437–438,

441–442

streptavidin affinity precipitation...............437–438,

441–442

in vivo biotinylation tagging 433–435, 444

lentiviral fluorescent protein expression vectors

MSCV-GFP-BirA 434, 435

MSCV-TLX1^bio-IRES-YFP.................................... 435

protein-protein interactions.............. 431, 433, 435, 444

H

High-dimensional data analysis

examples

B cell progression... 39–41

CD4 T-cell antigen-dependent

development 39–44

erythroid progression..................................... 47–48

myeloid progression 44–45

GemStone analysis software 209

probability state modeling 33–39

I

Immature myeloerythroid progenitors

anti-mouse antibodies: CD41, CD105,

CD150, cKit, FcgRII/III, ScaI 279, 291

fluorescence minus one (FMO).......................260, 280,

283, 285–287, 289, 291

internal reference populations.........................285, 286

lineage depletion.....................................282, 283, 290

mouse bone marrow........................278, 279, 282–283

Immune cell function

cell tracking dyes

membrane dyes: CellVue® Claret,

PKH26, PKH67.. 123

protein dye: CFSE.................................124, 125

cytotoxicity

lymphokine-activated killer

(LAK) cells 131, 133, 139

percent cytotoxicity............................134, 158, 159

target cells.......................... 131, 133, 134, 137, 138

proliferation

percent suppression.. 147

proliferation fraction........... 127, 131, 144–147, 149

proliferative index 127, 131, 144–147, 149

T_eff cells...139–148

T_reg cells...139–148

Intracellular cytokine staining

antibody panel design 167–168

fluorochrome brightness 167

procedural variables for different functional markers

fixation and permeabilization 176

secretion inhibitors: brefeldin A,

monensin 166, 170, 171, 175, 176

stimulation conditions 183

M

Malignancies

hematolymphoid neoplasia

antibody panel design 298–301

immunophenotypic characteristics..................... 307

internal controls.. 302

specimens.. 296

primary immunodeficiency diseases

assays

CD154 upregulation...................321–322, 329–333

immunophenotyping 320, 322

oxidative burst....................................321, 327–328

quality control....................................326–327, 333

underlying genetic abnormality
    functional abnormality.......................................319
    marker abnormality...319
    subset abnormality....................................318, 319
Microparticles
    flow cytometry
        conventional flow cytometer...............347–349, 352
        imaging flow cytometer 349–352
    generation of microparticles
        from purified cells......................................343–345
        from whole blood..343–345
    labeling of microparticles....................................345–347
Multiplexed analyses
    bead-based analysis of analytes
        analyte affinity 93–94
        analyte quantification.......................................92–93
        analyte titer... 91
        bead conjugation.. 89
        capture molecule 89–90, 94, 95
    combinatorial antibody profiling
        229 directly conjugated antibodies to human cell
            surface proteins 98
        categorization of expression levels 110–111
        high throughput.......................................97–99, 102
        R analysis software... 103

P

Phospho flow
    Cytobank analysis software...................... 182, 186, 189,
        196, 198, 199
    fixation and permeabilization 165, 166,
        176, 180
    fold change 183, 186, 189,
        196, 197
    phospho-specific antibodies: p-p38, pStat1, pStat3,
        pStat6 .... 184, 186, 189, 195–197
    sequential staining 180, 183,
        191–195, 201
    surface marker antibodies 180, 181, 185, 195

Q

Quantitative multi-color flow cytometry
    antibodies bound per cell (ABC)........ 55, 56, 58, 59, 62
    equivalent number of reference fluorophores
        (ERF)..54–57, 61
    fluorescent calibration microspheres
        CS & T microspheres 58, 60, 62, 63
        quantum dot-labeled microbeads..............72, 77–78
        quantum™ FITC MESF beads 67, 69,
        70, 79, 82
        Ultra Rainbow microspheres.............56–58, 62, 123
    molecules of equivalent soluble fluorophores
        (MESF) 54, 67–70, 79, 82, 88, 93, 95

R

Rare event analysis
    Anti-human antibodies: CD3, CD14, CD31,
        CD33, CD34, CD44, CD45, CD73, CD90,
        CD105, CD117, CD133, CD146,
        glycophorin A, pan cytokeratin.............253–255
    data analysis
        classifier parameters....................................263, 269
        elimination of sources of interference 263–269
        outcome parameters....................................263, 269
    DNA-binding dye (DAPI)..............................272, 279
    number of events to acquire..................................... 302
    tissue sources: adipose pericytes, bone marrow
        mesenchymal cells, pleural fluid............253–256

T

Transcriptional profiling of plants expressing fluorescent
        proteins
    microarray hybridization...........................415, 423–426
    RNA extraction and amplification....................415, 423
    sorting
        nuclei containing targeted GFP......................... 418
        protoplasts expressing GFP 416–418
    target labeling: Cy3, Cy5, Leader 423–426